Interactions between Electromagnetic Fields and Cells

NATO ASI Series
Advanced Science Institutes Series

A series presenting the results of activities sponsored by the NATO Science Committee, which aims at the dissemination of advanced scientific and technological knowledge, with a view to strengthening links between scientific communities.

The series is published by an international board of publishers in conjunction with the NATO Scientific Affairs Division

A Life Sciences **B Physics**	Plenum Publishing Corporation New York and London
C Mathematical and Physical Sciences	D. Reidel Publishing Company Dordrecht, Boston, and Lancaster
D Behavioral and Social Sciences **E Engineering and Materials Sciences**	Martinus Nijhoff Publishers The Hague, Boston, and Lancaster
F Computer and Systems Sciences **G Ecological Sciences**	Springer-Verlag Berlin, Heidelberg, New York, and Tokyo

Recent Volumes in this Series

Volume 91—Recent Advances in Biological Membrane Studies: Structure and Biogenesis, Oxidation and Energetics
edited by Lester Packer

Volume 92—Evolutionary Relationships among Rodents: A Multidisciplinary Analysis
edited by W. Patrick Luckett and Jean-Louis Hartenberger

Volume 93—Biology of Invertebrate and Lower Vertebrate Collagens
edited by A. Bairati and R. Garrone

Volume 94—Cell Transformation
edited by J. Celis and A. Graessmann

Volume 95—Drugs Affecting Leukotrienes and Other Eicosanoid Pathwavs
edited by B. Samuelsson, F. Berti, G. C. Folco, and G. P. Velo

Volume 96—Epidemiology and Quantitation of Environmental Risk in Humans from Radiation and Other Agents
edited by A. Castellani

Volume 97—Interactions between Electromagnetic Fields and Cells
edited by A. Chiabrera, C. Nicolini, and H. P. Schwan

Series A: Life Sciences

Interactions between Electromagnetic Fields and Cells

Edited by

A. Chiabrera

and

C. Nicolini

University of Genoa
Genoa, Italy

and

H. P. Schwan

University of Pennsylvania
Philadelphia, Pennsylvania

Plenum Press
New York and London
Published in cooperation with NATO Scientific Affairs Division

Proceedings of a NATO Advanced Research Workshop entitled
Interactions between Electromagnetic Fields and Cells,
held September 17-28, 1984,
at the Ettore Majorana Center for Scientific Culture in Erice, Sicily, Italy

Library of Congress Cataloging in Publication Data

Main entry under title:

Interactions between electromagnetic fields and cells.

(NATO ASI series. Series A, Life Sciences; v. 97)
"Proceedings of a NATO advanced research workshop entitled Interactions between electromagnetic fields and cells, held September 17-28, 1984, at the Ettore Majorana Center for Scientific Culture in Erice, Sicily, Italy"—T.p. verso.
Includes bibliographies and index.
1. Cell physiology—Congresses. 2. Electromagnetism—Physiological effect—Congresses. I. Chiabrera, A. II. Nicolini, Claudio A. III. Schwan, Herman P. IV. North Atlantic Treaty Organization. Scientific Affairs Division. V. Series.
QH656.I57 1985 574.87'6041 85-16788
ISBN 0-306-42083-X

A Division of Plenum Publishing Corporation
233 Spring Street, New York, N.Y. 10013

Printed in the United States of America

PREFACE

From 17 to 28 September, 1984, an international meeting devoted to "Interactions Between Electromagnetic Fields and Cells" was held at the E. Majorana Centre for Scientific Culture in Erice (Italy), under the sponsorship of the International School of Pure and Applied Biostructure (ISPAB).

During the initial five working days, advanced and intensive lectures were offered and discussed, as a NATO Advanced Research Workshop (ARW). The ARW was the first truly interdisciplinary forum in this research area, where principles, experimental methods, results and criticisms, independently developed in contiguous fields, were shared and discussed by the participants. Interactions mechanisms and biological effects of electromagnetic fields, from the molecular to the cellular level, were assessed and recommendations for new research directions and experiments were formulated.

After one intermission day, a satellite meeting took place as the 5th ISPAB Course, which offered complementary lectures during four additional working days. These lectures provided a specific tutorial background concerning some key areas. Special attention was given to the discussion of recent research results presented by participants and to measurement techniques and instrumentation. Finally, the main findings, criticisms and recommendations of the NATO ARW and of the 5th ISPAB Course were summarized, so that their effectiveness was amplified, immediately spreading these results among a wide number of scientists.

The scientific and technical results discussed during the meeting constitute the contents of this book. Before summarizing them, we wish to acknowledge the financial sponsorship of the following agencies and companies whose contributions, besides the main NATO support, allowed the organization of both the NATO ARW and the 5th ISPAB Course, the publication of this book and its widespread distribution:

- ANSALDO elettronica biomedicale (Italy)
- Biophysical and Electronic Engineering Department at the University of Genoa (Italy)
- DIGITAL EQUIPMENT S.p.A. (Italy)
- GENERAL ELECTRIC COMPANY (U.K.)

- IBM Italia S.p.A. (Italy)
- ITALCAD Selenia-Autotrol (Italy)
- Italian Ministry of Education
- Italian Ministry for Scientific and Technological Research
- Italian National Research Council
- RAYTHEON COMPANY (U.S.A.)
- Office of Naval Research (U.S.A.)
- Sicilian Regional Government

Moreover, we are greatly indebted to Professor E. Grant (King's College, University of London, U.K.) and to Professor Shiro Takashima (University of Pennsylvania, Philadelphia, U.S.A.) for their contributions to the organization of the meeting.

As the area of the interactions between electromagnetic fields and cells is at the interface of biological and physical sciences, only few studies have so far satisfied the up to date rigorous standards of both disciplines. The meeting offered a multidisciplinary professional evaluation of the biological effects of non-ionizing electromagnetic exposure reported in this book. The participants were scattered in different areas, ranging from Biophysics to Bioelectrochemistry, Bioelectromagnetics, Bioengineering, Biology, Biochemistry, Physiology, and Medicine. Accordingly, a scientific interdisciplinary consensus on the contents of this book was established.

The biological targets of electromagnetic exposure, chiefly at the membrane and nuclear level, are reviewed from the biophysical standpoint, also paying attention to very sensitive membrane systems of some species which can detect voltage drops across the membrane, in the submicrovolt range and over narrow frequency bandwidths.

Measurements of the equivalent complex dielectric constant appear to be useful in order to characterize the electromagnetic linear properties of biological media, to evaluate the average field inside a tissue or a cell culture and to identify mechanism of interaction at cellular and macromolecular levels. Electrical and electrochemical modelling of electromagnetically exposed cells demonstrates strong frequency-dependence and the necessity to consider the field problems at the microscopic level. These contributions complement those included in a previous book, devoted to "Modelling and Analysis in Biomedicine" (C. Nicolini Ed., World Scientific Publishing Co., Singapore, 1984), improving the physical modelling of biological media from the electromagnetic point of view, with possible fringe benefits for dosimetry, microwave thermography, hyperthermia, and nuclear magnetic resonance.

The survey of the mathematical properties of nonlinear systems provides interesting examples not only of ordered structures resulting from the propagation of coherent electrical waves in living systems, but also of irregular responses; as a consequence, various cell processes under electromagnetic stimulation could fit into the class of chaotic systems.

Classical kinetic rate theory suggests that significant interactions can be anticipated and have been observed by applying to proteins and like-sized molecules short pulses of electric field at the kilovolt-per-centimeter level, and at the volt-per-centimeter level in the case of cells.

A satisfactory theory for the dielectro-electrophoretic forces and torques involved is finally established, pointing out the need for a critical review of the interpretations of the experimental data so far available, and the need for new experiments as well.

Field effects on the interaction between ligands and their binding sites can occur in the millivolt-per-centimeter range, and human lymphocytes are good candidates for their detection. The prospective importance of dc and ac magnetic fields per se attracted the attention of some authors, who consider the effects of magnetic forces on the microscopic movements of charged particles.

Polyelectrolyte theory and related experiments offer examples of abrupt conformational transitions of macromolecules , as those already observed in the chromatin of mammalian cells during cell cycle progression, which can be triggered by small changes in ion concentrations, enzymatic protein modifications, or by small exogenous electric fields.Polyelectrolyte dynamics is of paramount importance for future progress in this area.

The onset of nonlinear effects of possible biological significance across excitable membranes, driven by sinusoidal signals, is considered, together with their influence on cell excitability. Computer simulation of nonlinear ionic fluxes on sinusoidally driven Hodgkin-Huxley membranes shows a significant ac-to-dc conversion.

Field force effects, and the biological effectiveness of pulsed or sinusoidal low-amplitude electromagnetic fields, are demonstrated on a number of occasions, pointing out the necessity for further systematic experiments in different laboratories and the potential of electrical cell manipulation for various purposes.

The reported results on electric field-induced perturbation of Ca^{++} fluxes prove the need to consider the contribution of dc and ac magnetic fields, which should henceforth be taken into greater account in the design of related experiments.

Other important applications concern the possible use of electromagnetic exposure for therapeutic purposes. There is a paucity of reliable data and a lack of satisfactory theories in the case of anticipated applications for tissue growth and control of cell proliferation. This, as well as the strong impact on the public opinion, explain our conservative and critical attitude toward such applications. Electromagnetic therapy of bone fractures has only been included as the first consolidated application. The book provides the right framework for stimulating a better understanding of the mechanisms involved.

There are many applications which can benefit from this book. The debate which has accompanied the recent proposal of a new Safety Standard for exposure to radio frequencies and microwaves by the American National Standards Institute proves the need for a new "data base". This should include more biomedical experience and search for explicative mechanisms. Some believe that the so-called "thermal effects" of electromagnetic exposure on humans are not solely responsible for observed bioeffects, even though they are the best established ones. Most of the biological effects of electromagnetic exposure that are not directly caused by a temperature increase have not yet been accepted, without further considerations, as reliable data on which safety standards should be based.

Finally the attention paid to field effects at the subcellular and cellular levels can induce beneficial effects in other areas, such as molecular electronics and biotechnologies. The studies of field effects at the molecular level can contribute to bring out new ideas for the design of molecular electronic devices, while electrophoretic and dielectrophoretic effects are of great interest for cell fusion and cell sorting technologies.

A spreading of knowledge among the readers can be expected which should help in establishing, in their countries, the needed common scientific basis for studying biological effects of electromagnetic exposure.

January, 1985

Alessandro Chiabrera

Claudio Nicolini

Herman P. Schwan

CONTENTS

MATHEMATICAL ASPECTS OF NONLINEAR INTERACTIONS

INTERACTION MECHANISMS OF ELECTROMAGNETIC FIELDS

CELL MEMBRANES UNDER EM PERTURBATION

FIELD-INDUCED FORCE EFFECTS

CELL RESPONSES TO LOW LEVEL EM EXPOSURE

HISTORICAL REVIEW, STATE OF THE ART, OPEN PROBLEMS

H. P. Schawn

Department of Bioengineering, University of Pennsylvania
Biomedical Engineering and Science Institute, Drexel
University, Philadelphia, PA 19104

INTRODUCTION

Historical accounts are based on personal experience and literature study. They are therefore limited to the extent that such experience and study is incomplete. I had the good fortune to participate over a lifetime in the development of biophysical and biomedical engineering sciences, including the nonionizing radiation field.

The perceptions of the scientific method, if there is such a thing, and the value of science to mankind vary greatly, even among scientists. And such variance is at the very root of present day discussions about safety standards, guide numbers and recommended health policies. Standards may be based on all sorts of considerations, including scientific, rational, educated guess, "properly" selected or total relevant literature, political and "practical" considerations. To fully assess the interplay of all these considerations is not easy.

The total effort in the nonionizing radiation field during the past decades may be conveniently divided into three parts: The biomedical approach, the bioengineering approach and the biophysical approach. Biomedical research reports clinical results and exposures of animals or cells and evaluates whatever clinical end point. The bioengineering research has been eminently productive during the past three decades. It relates external field strength values or radiation fluxes to absorbed energies, studies on a macroscopic scale the mode of propagation of EM-signals into the body and provides insight as to how EM-energy is absorbed. It is based on the knowledge of the electrical characteristics of body tissues.

The biophysical research is concerned with microscopic, cellular and macromolecular responses to EM-fields, i.e. attempts to determine the mechanism responsible for the biological response. All three approaches are needed of course and complement each other.

In order to better understand the relative contributions of these three approaches, let us consider the sequence of events which takes place when man is exposed to RF- or MW-fields. We list first the "primary" event of absorption of EM-energy by the tissues. It is analyzed by the contributions of the bioengineers. As a consequence of this absorption of energy all sorts of events at the cellular and biopolymer level take place, events which require analysis by the biophysicist, the physiologist and eventually the biochemist. For example, heat may be generated which initiates vascular responses and appropriate reactions through the complex thermoregulatory system of man. If local temperature increase is significant, the hypothalamus may take corrective action by adjusting peripheral blood flow and sweat rate. Changes in temperature may also affect the permeability of cell membranes and thereby ion transport. We know now a good deal how to model elevated temperature responses; however, little is known about other direct field interaction at the cellular and biopolymer level except at fairly high field strength levels. Finally, there is the clinical observation.

Much has been accomplished by the bioengineers with regard to the primary event. A good deal is known about the biophysical events and some about the thermoregulatory response thanks to recent advances by the physiologists interested in thermoregulation. However more needs to be done and little is known about the biochemical events triggered eventually. Thus, the need to close the gap in the sequence of events from the primary interaction process to the clinical observation remains to this day, even though much progress has been made.

I believe it is appropriate to briefly outline the development of interdisciplinary sciences in the area of biophysics and bioengineering. I hope that it shall become clear that it was largely based on the belief that the scientific method of the physicist and his tools, instrumental as well as analytical, have much to contribute to the life sciences and ultimately permit the rationalization of experimental data. The original concept, that life processes are far too difficult to permit such rationalization, has been virtually defeated by the exciting advances of modern biophysics and molecular biology, and the modeling approach is now considered appropriate in most biophysical areas.

Early Developments in Biophysics and Bioengineering

Biomedical engineering, biophysics and medical physics emerged almost simultaneously. Early beginnings date about 40 to 50 years back and took place at a variety of different places in different countries. At the time no real differences were apparent between the three disciplines. No formal training was available and motivations to enter these interdisciplinary fields varied from case to case. I came in contact with these fields almost by accident. Originally studying physics and some engineering, I was forced to interrupt my studies for financial reasons. I worked for some time as an engineer with electronic companies until I met by accident a former university teacher who spread the word about my plight. Eventually I was offered a technician's job with permission to continue my studies at the Oswalt Institute for Physics in Medicine, established 1921 in Frankfurt, and today the Max Planck Institut für Biophysik. I entered the Institute with some hesitation. Biology appeared to me extremely complex and it seemed almost impossible to apply the scientific approach of the physicist to such "messy" systems.

The Institute rapidly developed into a prototype of the institutions to develop later. It was made possible by a donation by the Oswalt family in Frankfurt, which has retained a continued interest in the institution to this day with members serving on the Board of Trustees of the Institute. Its founder was Dessauer, one of the early pioneers in the X-ray field, with a particular interest in the biological effects of ionizing radiations. He created interest for the fields of bioeffects of radiation and medical instrumentation. The Institute became affiliated with the University in Frankfurt, and a doctoral program leading to a Ph.D. in biophysics was in existence by the late 1930s. The total staff numbered about 20 people, including university lecturers, research fellows, assistants and technicians. Aside from research about the biological effects of X-rays, efforts focused on electrical properties of tissues and cells and applications to medicine such as ultrashortwave diathermy, electrical accidents and electrocardiography.

The diathermy related work was largely concerned with dosimetry and the deposition of electromagnetic energy in tissues, i.e. the distribution of SAR (specific absorption rate). It is well summarized in Rajewsky's book on ultrashortwaves, which also includes related earlier developments in electrophysiology and work at other institutions of the time. Soon after I entered the Institute in 1937, Rajewsky gave me some twenty reprints of Cole and Fricke. Fricke was probably the first during the late 1920s to rigorously analyze the dielectric properties of cell suspensions in the RF-range, and is recognized as a pioneer in this field as well as radioiology. I became well acquainted both with Fricke and Cole shortly after coming to the USA and profited much from their counsel and help. Cole had analyzed a number of cellular systems from an electrical point of view, then concentrated on the giant squid axon and eventually extended with Curtis measurements into the nonlinear range. This work led eventually to the Hodgkin-Huxley equations which contributed so much to our knowledge of the nonlinear electrical characteristics of excitable biological membranes.

Fricke's early work had been conducted at a laboratory established by Otto Glaser at the Cleveland Clinic. He then moved to the Cold Spring Harbor laboratories of the Carnegie Foundation for the rest of his life. His work stopped for some time during World War II. In the early 1950s I was able to support him from ONR-funds available to me, and he published a number of highly significant papers. Cole had spent many summers at Cold Spring Harbor and his laboratories first at Columbia and then Chicago and Bethesda became known worldwide. No doubt there existed other interdisciplinary laboratories. But the total number was small, with only few universities participating.

In most of these early institutions, engineers joined with physicists and medical doctors. They were attracted by the possibility to apply the analytical power of the physical sciences and its instrumentation to problems in biology and medicine.

The early efforts indicated above combined interests in the effects of ionizing radiation and related therapeutic and diagnostic technology with many other topics. These included nonionizing therapeutic and diagnostic techniques and more basically oriented work of a biophysical nature, including the study of the electrical and acoustic properties of cells and biopolymers and electrophysiological research. There was a growing interest since d'Arsonval in the healing potential of RF-fields. Esau and Schliephake introduced ultrashortwave diathermy, which rapidly became popular before the second world war.

The medically oriented experimental work before the second world war led almost immediately to a split between two different schools of thought. One had it that whatever therapeutic or other effect was observed was caused by a noticeable temperature increase and that, therefore, the results observed were due to heat and not caused by electrical fields per se. The other school of thought believed that direct field interactions termed nonthermal or athermal are important. This controversy led to some hundred publications of variable quality. The debate was never settled and these papers are largely forgotten. Proponents of the athermal school in Europe included Liebesny, who published a book on this topic, Schliephake, the pioneer of ultrashortwave therapy, Krasny-Ergen, who published theoretical papers on pondermotoric cellular force effects, and N. N. Malov in Russia, who propagated the idea of point heating effects at the cellular or even molecular level. A review of these efforts as well as then available dielectric tissue data are contained in Rajewsky's monograph "Ultrashortwave, Vol. 1 of Results of Biophysical Research", 1938. As World War II started related efforts came to an abrupt end.

Technology developed during the second world was almost immediately applied for health purposes and ultrasonic as well as microwave diathermy were added to ultrashortwave diathermy about 1950.

Efforts to optimize these techniques were undertaken, first at the University of Pennsylvania and the Mayo Clinic, later also at the University of Washington and eventually at the University of Utah. Since the 1970s a rapidly growing number of other institutions entered the field, all now primarily concerned with the potential health hazards resulting from exposure to nonionizing radiation. But many are also interested in the advancement of therapeutic techniques and local hyperthermia which provide, together with X-rays, synergistic treatment of cancer.

Interest in the electrical and conducting properties of tissues and their response to electricity developed already during the second part of the past century. They are of a more biophysical nature and discussed next.

This field can be conveniently divided between work concerned with cell suspensions and tissues. The DC and low frequency properties of erythrocytes were extensively investigated and it was soon recognized that Maxwell's equation for the conductivity of a suspension of poorly conducting particles in a highly conducting medium could be used in good approximation.

Thus cells were recognized to have a very low electrical conductivity at low frequencies. Hoeber was the first to recognize that at very high frequencies erythrocytes conduct very well and postulated that they are surrounded by membranes as suggested by Bernstein. These thin membranes were believed to be capacitatively bridged by the high frequency current, thus enabling the interior to participate in the conduction process.

The early tissue work led almost immediately to the recognition that nervous and contractible tissue response to electrical currents is far more complex than that of erythrocytes. When voltage steps were applied, the response was not entirely instantaneous suggesting that the conductivity was readjusting with time. A "polarization" property was thought to be indicative of capacitative elements and believed to be similar to the one observed at the interface between current injecting electrodes and electrolytes. It was also noted that excited tissues had different conductivity values than unexcited ones. Anisotropy of the conductivity of muscular tissues was also noted and the ratio of longitudinal to transversal values reported to be nearly ten. This fits very well with modern data. Hysteresis effects were noted, but "dead" tissue yielded much simpler response characteristics. The hypothesis was stated that tissue "interfaces" were responsible for many of the observed results. At the same time Tesla noted that HF-currents do not excite tissue and are much less dangerous than low frequency ones. He thought that HF-currents do not penetrate into the body. Later work revealed that the unique sensitivity to ELF-currents is caused by several effects. They are related to the β-dispersion mechanism, the redistribution of membrane currents which takes place as the membrane time constant is approached while varying the frequency and the unique nonlinear properties of many cellular membranes. Much of the earlier work was initiated by the well-known physiologist Du Bois-Reymond and his students. In particular the work of Herrmann stands out by its quality and the many remarkable results obtained. These efforts have been summarized by Rajewsky, by Cole and others.

The developments between the two world wars are indicated next. Gildemeister investigated excitability as a function of frequency, using the newly acquired ability to produce high frequency currents of sinusoidal form with the vacuum tube. Gildemeister noted that excitability decreases rapidly above some KHz and that it is virtually impossible to excite tissue as one approaches a frequency of 100 KHz. Equivalent electrical circuits were suggested to mimic the electrical behavior of tissues, including "polarization" elements of constant phase angle as typical for the electrode polarization impedance. Sapegno was to our knowledge the first to note that the anisotropy of tissue's conductance is strongly frequency dependent. It diminishes as the frequency advances through the β-dispersion range. A preferential conduction of electricity in tissues along the direction of the fibers is suggested by the conductivity anisotropy existing at ELF frequency. But it should vanish with increasing frequency. Rushton noted that excitation with fields parallel to the elongated cells requires lower field strength values than excitation across the cell. The factor is given by the ratio of cell length constant and its diameter, i.e. can be easily more than 10. Here the length constant is defined by cable theory. It is the distance required to diminish a small signal applied at one point to the cylindrical cell by a factor of e. Preferential conduction of ELF induced currents in the fiber direction and the Rushton rule suggest lower excitability thresholds than one might anticipate for isotropic tissues.

Extensive tissue conductivity data throughout the RF-range were collected by Rajewsky and his colleagues up to 100 MHz. Additional data including dielectric constants were provided by Osswald at 25, 50 and 100 MHz. All these data are internally consistent. However, the techniques used prevented recognition of anisotropy and, therefore, are values averaging over all directions. Fricke, Cole and

Curtis applied rigorous theory and advanced instrumentation to investigate both conductivity and dielectric constant of cell suspensions and some tissues. They first formulated the theory of the β-effect. In the case of tissues the existence of a constant phase angle element associated with the cell membrane was postulated, extending the earlier proposals about tissue "polarization" indicated above. Cole and Curtis then turned to detailed studies of the giant squid axon membrane, observing that excitation changes the membrane conductance but not capacitance if determined at 100 KHz. This work was to pave the way for the extensive effort undertaken after the second world war to delineate the nature of the processes responsible for the response of nerve cell membranes to electrical signals.

During the period between the two world wars interest in dielectric properties was by no means restricted to biological media. A large effort on dielectric properties was undertaken resulting in substantial advances in the understanding of such properties. Debye developed his concept of polar molecules and received the Nobel Prize for this singular achievement. During these years dielectric theory was perfected. In the late 1930s the concept of polar molecules was applied to biological molecules. First Arrhenius and shortly thereafter Oncley studied protein properties. Particularly Oncley and his colleagues measured properties with precision over most of the RF-range. They then applied the Debye theory, modifying it by incorporating the rotational diffusion constants calculated before by the well known French physicist Perrin for ellipoids. The results explained the dispersive behavior of proteins. In summary then, two important theories of dielectric dispersive behavior had been successfully applied to biological materials by the beginning of the 1940s: The theory of inhomogenous media, i.e. the Maxwell Wagner effect, accounted for cellular behavior in the RF-range and appropriately considered the existence of biological membranes. The Debye theory of polar molecules, in modified form, explained protein data at RF-frequencies.

Biophysical and Bioengineering Contributions After the Second World War

After the second world war biophysicists and biologists concentrated increasingly on membrane biophysics, investigating how membranes are composed and how they respond to electrical fields at ELF-frequencies. Hodgkins and Huxley formulated a complex set of equations which predicts the response of membranes to electrical signals. Their model of the membrane includes the existence of local "channels" for N_a and K ions at separate locations. This model is still accepted even though its physical basis remains uncertain. The Hodgkins-Huxley model is not restricted to small signals since it is highly nonlinear. It reduces to a linear model for small disturbances of the resisting potential, i.e. if potentials applied to the membrane are of the order of some mV or less. This means that ELF-fields which induce less than 1 mV across the membrane could not trigger any response of the nervous system. Much experimental work has been carried out to test the HH-equations. This work was very much aided by the development of the voltage clamp technique by Cole and his colleagues. This technique has recently been refined to permit the application of voltage steps to very small areas of the membrane.

The tissue and cell suspension work was extended to cover the total range from very low to microwave frequencies. Schwan and his co-workers observed additional dispersion effects at ELF- and MW-frequencies and developed a comprehensive characterization of the dielectric behavior over the total frequency range. At low frequencies they applied ultrasensitive technology to measure the dielectric constant and conductivity and observed first the existence of the α-dispersion. A variety of new relaxation mechanism were discovered including the relaxation of counter ions near charged membrane surfaces and the relaxation of protein bound water in the UHF range. Abundant tissue data were obtained providing sufficient data for our purposes. This work has been very useful in determining SAR distributions at ELF- and MW-frequencies and in developing coupling considerations at the microscopic level.

Particular attention was accorded the problem of how electrical fields may directly interact with cells. Fields can be shown to exert all sorts of forces, termed pondermotoric forces. They may demonstrate themselves as rearrangements of cells, such as blood cells forming pearl chains. Or they may cause deformation, cell destruction, fusion, rotation or changes in cell shape, protoplasmic streaming, etc. Electrically mediated cell fusion is now rapidly becoming the favorite technique of achieving cell fusion, so important in biotechnology. Our work was both experimental and theoretical and we investigated minimum field strength criteria needed to achieve the quoted effects. These nonthermal effects require fields usually somewhat larger than those required to meet safety standards. Pohl investigated dielectrophoresis over a lifetime and Zimmermann and others rejuvenated the field with their work on field effects on membranes and cell fusion.

The historical development of relevant biophysics clearly indicates several lines of inquiry. The collection of tissue data over the entire frequency spectrum and the explanation of the observed relaxation effects have been useful in dosimetry and in biophysical considerations to what extent various tissue compartments at the microscopic level may be affected by electrical stimuli. The other efforts almost entirely concentrated on biological membranes, their structure and function and response to electrical stimuli. This work so far would indicate that "subtle" responses to ELF-field, corresponding to membrane potentials below 1 mV, are not anticipated.

Initial bioengineering contributions originated largely from our laboratory. First, we determined how deep microwaves penetrate into tissues. It was recognized that the depth of penetration decreases rapidly as the frequency increases and is less than 1 cm at a frequency of 2450 MHz. Next, reflection effects at tissue interfaces such as skin-subcutaneous fat-skin-air, subcutaneous fat-muscle were investigated. These reflections are also strongly dependent on frequency. Third, SAR values in skin, subcutaneous fat and muscle were evaluated and also noted to be strongly dependent on frequency and other parameters. As a consequence of these studies we recognized that the 2450 MHz frequency is not well suited for microwave diathermy and that 915 MHz would be more suitable. At this lower frequency greater penetration and better reproducibility could be realized.

All above mentioned contributions address the case of partial body irradiation as is the case when a diathermy generator is directed to a limited small part of the total body surface. However, during the late 1950s potential hazards caused by accidental exposure became of increasing interest. We therefore investigated the absorption of EM waves by man both mathematically and experimentally using the Mie scattering theory and models of man exposed in a large unechoic chamber. We noted that man's relative absorption cross section can be greater than one, i.e. man can absorb more energy than illuminates him. This phenomena, termed today "the resonance effect", occurs when man's height is about half a wavelength. We also noted that above the resonance frequency the absorption cross section approaches a value of about 0.5 and that it decreases rapidly when the frequency decreases below the resonance range.

We next addressed the possibility that hotspots may be generated under appropriate conditions in man, i.e. focal concentrations of absorbed energy which might cause strong local temperature increases even though average temperatures remain low. We found that these hotspots occur either if body segments such as the head resonate with the field, or if circumstances permit radiation to be focused towards a focal spot due to the curvature of the body surface.

Finally, we investigated theoretically what temperature increases result in such hotspots due to the absorption of EM-energy. We first used computer techniques to solve the bioheat equation and finally a modern multi-compartmental model devised by thermoregulatory physiologists to state how man's thermoregulatory system responds to a local hotspot. We recognized that for incidents energy fluxes of about 10 mW/cm^2 temperature changes of the order of 0.5°C or more are possible, but only in the frequency range of about 0.5 to 2 GHz.

As a consequence of these and those of other laboratories' research efforts, it was recognized that man's exposure to incident EM-exposure may be divided into four frequency ranges:

1. The below resonance range. Here man absorbs little radiation. Most is reflected. But the absorbed energy increases rapidly with frequency (<30 MHz).

2. The resonance range. In this range the absorbed energy reaches its peak and exposure may most readily result in fairly large amounts of total absorbed energy (30-300 MHz).

3. The hotspot range. It ranges from about 0.3 to 3 GHz. In this range local spotlike temperature increases are most likely.

4. The surface heating range. As the frequency increases beyond 3 GHz, most energy is absorbed in the skin and microwave induced temperature elevations are comparative to those induced by sunlight of the same intensity.

The recognition of these different ranges has now been incorporated into revised radiation standards. They are frequency sensitive with a minimal permissible flux of only 1 mW/cm^2 in the resonance range from 30 MHz to 300 MHz and replace in this range the earlier 10 mW/cm^2 standard originally proposed by us.

The bioengineering and biophysical accomplishments listed above began originally in our laboratory. As time advanced other laboratories made substantial contributions which complemented, supported, and extended our work. In the area of dielectric properties the laboratories of Grant and the Stuchleys made significant contributions, coming to almost identical conclusions. In the bioengineering area the laboratories of Guy in Seattle and Gandhi and Durney in Salt Lake City have made major contributions, particularly during the past decade. Many others contributed to the bioengineering effort and this activity is now well established.

Biomedical Contributions and Present Status

Biomedical contributions to the field were already made in the early 1930s and gave rise to much speculation. Clearly established were the thermal effects of ultrashortwave therapy, then in vogue. But a small but significant number of researchers held that direct field interactions, so-called nonthermal or athermal effects, had also to be considered. During the late 1950s first reports from Russia strongly supported such claims. But much of the earlier work was thought to be ill-controlled with questionable statistics and extensive studies in this country, particularly at Rochester under

Michaelson, did not support nonthermal claims. During the 1960s and 1970s more material has been collected and the controversy about the significance of nonthermal effects continues. Thousands of articles have by now been published which report biomedical responses to EM-fields. I am not able to summarize all this work but shall concentrate on a few results which have sparked particular interest:

1. The auditory RF-phenomena. Man can perceive pulsed MW-fields as a hearing sensation. The effect was originally thought to be caused by a direct field interaction with the central nervous system. Subsequent work by Foster and Finch and Guy's group strongly indicated that the effect is a classical macroscopic thermoacoustic effect. It is the microwave equivalent of the photoacoustic effect first noted in 1881 by Alexander Graham Bell, who discovered that a periodically interrupted beam of light shining on a solid produces acoustic signals.

2. A large number of reports both from Russia and the USA leaves but little doubt that intensities in the range of 1 to 10 mW/cm^2 can evoke all sorts of biological responses. This intensity range is only weakly thermogenic and the classification of the observed effects as thermal or nonthermal is often difficult. Many subtle and not so subtle effects may be expected with small test animals. Hence reported responses are not necessarily contradicting a human tolerance near 10 mW/cm^2. It must also be stated that much of the earlier work in this field was of poor quality. Dosimetry was not well developed, statistical evaluation unconvincing and controls often missing.

3. A considerable number of papers have been published stating certain biological responses to be dependent on modulation frequency. They include effects on the EEG power spectrum, reaction time and C_a-efflux. These responses are particularly pronounced at certain modulation frequency and power "windows". Surprisingly these effects appear to be much less dependent on the carrier frequency since they have been observed at ELF, RF and MW frequencies. Peak alterations typically occur at ELF or modulation frequencies somewhere between 10 and 100 Hz. In situ field strength levels are highly frequency dependent ranging astonishingly from values near 10^{-7} V/cm to about 1 V/cm as the carrier frequency increases from ELF to MW values. In contrast, the ambient field strength values in air are much more comparable, suggesting that the external field is more important than the in situ value.

4. In 1973 Russian experiments at their Academy of Sciences Institute of Biophysics indicated sharply frequency sensitive influences on cell growth at mm-wavelength frequencies. Experiments of this sort were conducted at the Max Planck Institut für Festkörperforschung and at Neuherberg by Grundler and Keilmann. Extraordinary frequency sensitivities of the order of 1 part in 10,000 on bacterial growth were noted. These effects curiously complement those reported under item 3. The former appear not much to depend on the carrier frequency while the latter are highly sensitive to it. This experimental work was theoretically predicted by Froehlich. Attempts to confirm it are presently in progress.

5. It also was pointed out that all sorts of life species can sense electric and magnetic fields with astonishing sensitivity. Sharks and rays can detect field as small as 0.01 μV/cm; birds can detect the magnetic field of the earth and use this ability for orientation purposes. Even some bacteria have recently been discovered which move in the direction of the earth magnetic field.

6. A variety of cellular and tissue responses to weak fields, particularly at low frequencies, have been reported. They include by now well established effects on the speed of bone repair, on cellular growth and limb regeneration in more primitive species, and a variety of bioeffects suspected to be caused by magnetic fields. I shall not delve into a review of these fields since they are somewhat more removed from my primary interests. But they certainly will be discussed at this meeting.

7. Forces which are generated by electromagnetic fields have been demonstrated to be significant at the cellular level as already indicated above. These field induced force effects have been used to develop novel and effective techniques to fuse cells, a technique of great importance in biotechnology. They may result in a variety of other responses of biological interest and their threshold is now believed to possibly extend below the thermal level.

Some of the observations reported above cannot be readily explained by existing classical biophysical concepts about membranes and biopolymers and their sensitivity to electrical fields. The modulation frequency sensitive results have yet to be clearly demonstrated to be caused by events at the membrane and biopolymer level and the mm-frequency selective results await further independent confirmation. An inherent problem with most of these studies is that experimental observations in the absence of any valid interpretation tend to be subjected to inquiry or to be forgotten eventually. Only if they can be properly explained can they be of use in predicting effects under different circumstances. Indeed, this work has not been applied yet to health hazard standard considerations.

Hyperthermia and Diathermy

The comparative merits of available diathermy techniques have been frequently outlined by us. MW-diathermy unfortunately has not yet advanced to the state that it is possible to predictably cause a temperature rise at a given location. A major problem has been the unfortunate frequency choice of 2450 MHz. A lower frequency of 915 MHz would certainly be much more suitable. Ultrasonic diathermy on the other hand provides better penetration and deep heating. But it is more cumbersome to apply since careful contact between transducer and body surface must be maintained constantly.

Renewed interest in diathermy as a tool for deep heating or hyperthermia, the presently preferred term, has developed in recent years. This interest was strongly motivated by the observation of synergistic effects of combinations of X-rays and heat on cancer tissues. But the quantitative problems of precise temperature control are considerable. A temperature elevation in the abnormal tissue mass to about 42°C must be precisely maintained. This cannot be done without local temperature determinations.

Present day analytical techniques can now be applied to predict what modality is best used to achieve the desired local temperature increase. Some favor ultrasound in view of its great ability to be directed towards the target and its better penetration. But microwaves operating in the hotspot frequency range at about 1 GHz or 915 MHz appear also to have great potential. The analytical techniques to precisely predict specific absorption rates (SAR) for a given modality have been developed in recent years and the translation of these SARs into temperature elevations is possible using computers to evaluate the bioheat equation and to apply modern compartmental models of man's thermoregulatory system.

Early Interest in Hazards

Already during the second world war the U.S. Navy had considered potential health problems. The then available sources of microwave radiation as used by planes and ships were rather weak and considered, therefore, not dangerous. The development of more powerful oscillators advanced rapidly after the war and the Navy decided to hold a meeting on the topic of hazards in 1953. To my knowledge, the question of a standard was raised probably for the first time. Some of the attendees included K. S. Cole, who had contributed much to bioelectricity, J. Hardy, who was then a recognized leader in the field of human thermoregulation, D. Goldman, who already had done much in the field of bioelectricity and was to contribute with Ely to hazard studies and the first ANSI standard, and myself. There was at first a tendency to declare a safe limit at the 100 mW/cm^2 level. In a memorandum which I submitted after the meeting I suggested a tenfold lower level of 10 mW/cm^2. My inputs included several considerations, all mutually supportive. For example, our data indicated that absorbed power could be as high as illumination at certain frequencies. I was also aware that in microwave diathermy total applied power in partial body irradiation could approach 100 watts and felt that thermal tolerance was probably least, if all the energy was concentrated. Yet there was no reported damage from microwave diathermy exposure. A third input came from some early estimates of temperature increase, published by Cook in England. And finally we knew that tolerable exposure to skin absorbed sunlight or the deep heat created by heavy work probably would bracket the cases of microwave exposure, depending on frequency. Altogether a load comparable to the base metabolic rate appeared quite safe, providing a probable margin of safety of 2 or 3. It has been indicated that I was unaware of the possibility of nonthermal dangerous effects and that, therefore, our rationale was of limited usefulness. This was not so. As indicated above, I was probably more aware than any of my younger colleagues of all the work before 1940 and the thermal-athermal controversy. But I simply could not see how any rationale could be developed from these inconsistent and sometimes anecdotal data at nonthermal energy levels.

It has also been claimed that we naively or purposefully directed our research interests to purely thermally related topics. But since the earlier work by Krasny-Ergen and others on pearl chain formation we had been interested in these direct field interaction phenomena and we addressed a major portion of our work effort from the late fifties to the middle sixties to this area. Many papers were published, with support from ONR and NIH. This work did not immediately provide clues as to how fields, well below the thermal threshold range, might affect biosystems. This type of work has now again become of great interest as scientists better learn how to manipulate cells by AC field induced pondermotoric forces for all sorts of useful purposes.

I have to regress in time for a moment. At the end of World War II Germany was restricted for some time to use electronic equipment. Hence, Siemens, a leader in medical technology and ultrashortwave equipment, had for a brief time to abandon these fields. It instead successfully introduced ultrasonics into medicine and produced its first ultrasonic diathermy machines, based on the recognition that the absorption coefficient for ultrasonic waves in tissues is suitable for therapeutic and diagnostic purposes. At the same time Raytheon, having developed a rugged 2450 MHz Magnetron, introduced microwave diathermy. This motivated clinical interest in the potential of the technique and sustained clinical evaluations and research at the Mayo Clinic and other places.

The Department of Physical Medicine at the University of Pennsylvania hired me to investigate biophysical principles and the Mayo Clinic got Lehmann from Frankfurt. Lehmann eventually became chairman of Physical Medicine in Seattle and attracted Bill Guy to this field. The Mayo team was headed by Krusen at the time, a senior figure in the field of physical medicine and rehabilitation and of great influence. One of his colleagues at Mayo was Julia Herrick, one of the founders of the IEEE Engineering in Medicine and Biology Society and a prime mover to get biomedical engineering going. With two colleagues from von Hippel's laboratory at MIT she presented in the early 1950s the first dielectric tissue data in the 1 to 8.5 GHz range. These data were never published by her but made available to me and included in my reviews of such data and eventually confirmed in our laboratory recently. It was also Julia Herrick who first took the initiative to organize at the Mayo Clinic a meeting dedicated to microwave hazard and related topics. This was in 1955 and I presented our opinion on the 10 mW/cm^2 which the Navy adopted at about the same time. At the same meeting some of the earliest observations on suspected nonthermal bioeffects in the USA were reported. The IEEE Transactions Eng. Med. Biol. issue dedicated to the meeting still makes interesting reading. From then on, interest in the hazards area increased.

I mentioned to Julia the prewar interest in pearl chain formation and she subsequently published a paper on erythrocytes forming pearl chains when exposed to microwaves or shortwaves, thereby confirming earlier work by Liebesny. It created great interest and was no doubt in part responsible for rekindling interest in nonthermal effects. Again, I state this to reemphasize our and others' awareness and continued interest in nonthermal mechanisms. Some not acquainted with these earlier events believe that the work at the time was only interested in thermal effects. Nothing could be further from the truth.

Soviet Contributions

It has been stated by some that we in this country neglected to recognize Soviet and Eastern European efforts in the field. My first awareness of Dr. Gordon, then the leader of the field in Russia, goes back to the late 1950s. I received a paper of hers published in an East German journal on "Hygiene problems of work with cm-wave generators" in 1957. I also made frequent contact with Russian investigators including Gordon at the 1959, 1960, 1961 and 1964 International Conferences on Medical Electronics in Paris, London, New York, and Paris again. At the 4th International Conference in

New York in 1961, Dr. Gordon participated upon my invitation. I was program chairman of the large meeting and four of the 36 sessions were dedicated to microwaves. I personally chaired the first of these sessions which was totally dedicated to athermal effects, including papers by Bach, Heller, Saito-Sher-Schwan, Gordon et al., Obrosov et al., Barber and Frey. Allan Frey gave a paper about his early work and probably was aware of our solicitation of two Russian papers in this session and others in other sessions. At the 1961 meeting I talked with Dr. Gordon who had together with her coauthors Kitsovskaia and Tolgskaia presented a lecture entitled, "Biological effects of microwaves of small intensity". I pointed out that not much was known yet about her work in the USA or England and motivated her to submit for publication in those countries.

In short, I hope that I have deflated three myths:

1. that there was no interest in athermal effects prior and during Tri-Service period of Col. Knauff;

2. that there was no work going on in this field;

3. that there was no interest in and awareness of the Russian work and, worse, that there was almost a conspiracy to neglect such work.

Awareness of the Russian work increased rapidly from then on. Already the early 1957 article by Lifshitz on "The Role of the Nervous System in Reactions to UHF Electromagnetic Fields" published in the Russian Biophysics Journal was available. A translation of a detailed manuscript by Letavet and Gordon from February 1960 (The Biological Action of Ultra High Frequencies) and a detailed evaluation were available in English by 1962. From there on the connection never ceased all through to 1971 when Tolgskaia and Gordon published yet another book on "Pathological Effects of Radio Waves".

Tri-Service and ANSI

Support for research in the field of bioeffects of nonionizing radiations had been very limited during the 1950s. It was almost entirely due to the interest of the Navy through its Office of Naval Research. I believe that they supported us and two to four other programs. But in the late 1950s the Air Force became interested, largely due to the efforts of Col. Knauff and under his leadership the Tri-Service program started in the late 1950s. But even this program was small by comparison with present day efforts, in spite of the valiant efforts of Col. Knauff to increase support. Col. Knauff has been unjustifiably singled out as wanting to "prove" the thermal hypothesis and to neglect the Eastern literature. I well remember presenting papers at his frequent meetings on direct nonthermal effects and ventured the possibility of field induced nonthermal effects during mitosis. I also remember that he intended a large literature review while he served as chairman of subcommittee IV of the ANSI C 95 committee. Unfortunately he was denied the funds needed to conduct this research and evaluation effort.

The ANSI C 95 committee came about during the late 1950s, largely as a result of the Navy's and Glen Heimer's efforts. It was sponsored by both the Navy and the IEEE and I accepted to be its first chairman until about 1965 when the first health standard had been finally adopted. There is no need to review this activity. Dr. Stenneck has researched this development in detail as published in a lead article in Science some years ago. I largely agree with his account and points of disagreement are not of sufficient importance to warrant consideration here. But I should make some general comments which characterize the then existing climate both at Knauff's and the ANSI meetings. The Tri-Service effort was largely dedicated to about half a dozen efforts at various universities con-entrating on different frequency values, one of the largest being the Rochester program under Sol Michaelson. These efforts did not disclose any significant effects at clearly nonthermal intensity levels. One exception was the results by Heller and colleagues, who presented their results at various occasions. But dosimetry was not well developed and the results of Heller's group were debated. As far as the Russian work is concerned: most of this work was carried out at intensity levels between about 1 and 10 mW/cm^2. In this range subtle and tolerable thermal effects may as well occur as nonthermal effects and the health implications of this work were difficult to evaluate. The American work in contrast permitted much closer scrutiny as it was reported in much greater detail. The standard which evolved was conceived with due consideration of all materials known to us. It had become clear that 100 mW/cm^2 originally suggested at the 1953 ONR meeting was close to dangerously high. There was no conclusive evidence that 10 mW/cm^2 was dangerous. We also felt that we should not set down figures different from decade numbers such as 1 or 10 or 100, thereby indicating our limitations. We also should not address the problem of frequency dependence even though the early work on cross-sections was strongly indicative of things to come. We also felt we should not set standards numbers but rather guide numbers separating safe from unsafe and leaving it up to whomever to use these numbers with due adjustments depending on circumstances. We took great pains to formulate these points.

The revised ANSI standard which became effective 1982 appears to be a logical extension from the previous one. It incorporates the resonance phenomena which has been investigated with great care during the past decade. Nevertheless it has been the subject of heated debate and apparently faced as many difficulties as the previous one.

In 1973 WHO organized an international meeting under the able chairmanship of Dr. Cerski. I had the opportunity to spend more time than ever before with Dr. Gordon and we drafted a memo on how to assess the field and what ought to be done. This memo was approved by the assembly and published later, however in significantly changed form. One of our recommendations had been that more biophysical research is needed, aimed at clarifying mechanism of interaction. Without understanding mechanism of interactions extrapolation of experimental data is difficult. This is as true today as it was then.

Clearly, we have learned a great deal about nonionizing radiation, We have now a rather refined insight into the thermal aspects and we can fairly well evaluate what levels of radiation may establish an undue thermal hazard. In addition, basic biophysical insight is highly advanced. It does not suggest subtle direct interactions of nonionizing fields with biomolecules and cells so far. But, a variety of medical and biological observations have not yet been explained by this extended body of microscopic and submicroscopic biophysics as indicated above.

CELL MEMBRANE STRUCTURE AND FUNCTION

Harvey M. Fishman

Department of Physiology and Biophysics
University of Texas Medical Branch
Galveston, Texas 77550

INTRODUCTION

The existence of a hypothetical structure called the "plasma membrane," which encompasses cells and separates the intracellular from the extracellular space, was proposed as early as the mid-nineteenth century. The membrane premise was based on observations of cellular uptake of pigments and other substances and from demonstrations of the effects of osmotic pressure changes on cell volume (Nägeli and Cramer, 1855). The first physical concepts concerning the function of cell membranes were presented by Bernstein (1902) in his membrane hypothesis. He suggested that (1) cells have an electrolytic interior surrounded by a thin membrane, which is relatively impermeable to ions, (2) the cell maintains a high internal potassium ion concentration and consequently an electrical potential difference exists across a resting membrane and (3) electrical activity results from an increase in permeability of the membrane to potassium ions thereby reducing the resting membrane potential to zero. Since Bernstein's hypothesis, the cell membrane (Fig. 1) has undergone intense scrutiny because of its recognized function as "the barrier between life and death." A wide variety of physiological processes depend upon membranes. Membrane transport of ions underlie conduction, excitation and junctional transmission in nerve and muscle. The movement of calcium ions across the membranes of the sarcoplasmic reticulum in a muscle fiber controls contraction. Production of energy by oxidative phosphorylation of adenosine diphosphate to adenosine triphosphate occurs in mitochondrial membranes. Phototransduction of light to an electrical response in photoreceptor cells involves interaction of membrane-bound, light-sensitive molecules which alter membrane ion transport. Epithelial transport, which depends upon membrane

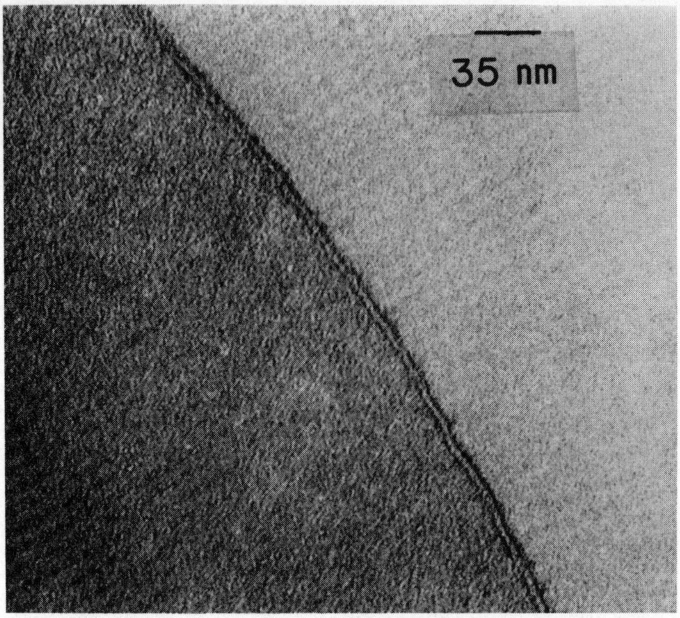

Fig. 1. Micrograph of permanganate-fixed section of a red blood cell showing the "unit-membrane" structure bounding the cell. (From J.D. Robertson in: Hendler, 1971). (By permission of Physiol. Rev.).

function, is responsible for the uptake of nutrients, the formation of sweat and urine and plays an important role in the maintenance of fluid volume and electrolyte balance. In short, membranes are the structures that perform the tasks necessary to sustain living organisms.

CHEMICAL COMPOSITION

The three major types of biochemical constituents of plasma membranes are: protein, lipid and carbohydrate. Table 1 shows the lipid composition of representative animal and bacterial membranes. Cholesterol (Ch), phosphatidylethanolamine (PE) and phosphatidylcholine (PC) account for most of the lipid in animal cell membranes whereas the lipid in bacterial membranes is almost entirely PE. The structure of some of the most common phospholipid molecules is shown in Fig. 2. In modeling membrane structure the phospholipid molecule as represented in Fig. 3 (a PC molecule with one oleic acid chain and one palmitic acid chain) is usually written as the skeleton diagram in the inset. To what extent the differences in lipid composition cause differences in membrane properties is not known.

Because soluble proteins are known to accomplish many important functions in the fluid media of cells, membrane-bound proteins are assumed to be the main constituents of biological membranes which determine function. This assumption, which will be discussed later, has been verified for the acetylcholine-receptor at the neuromuscular junction. Difficulties in the dissociation of membrane-bound

Table 1. Lipid Composition of Animal and Bacterial Membranes. Values given are approximate percentages of total lipid.

Lipid \ Source	Myelin	Erythrocyte	Mitochondria	Microsome	Microsome	*Azotobacter agilis*	*Escherichia coli*	*Agrobacterium tumefaciens*	*Bacillus megaterium*
Cholesterol	25	25	5	6	*	0	0	0	0
Phosphatidylethanolamine	14	20	28	17	18	100	100	90	45
Phosphatidylserine	7	11	0	0	9	0	0	0	0
Phosphatidylcholine	11	23	48	64	48	0	0	10	0
Phosphatidylinositol	0	2	8	11	6	0	0	0	0
Phosphatidylglycerol	0	0	1	2	0	0	0	0	45
Cardiolipin	0	0	11	0	2	0	0	0	0
Lysyl phosphatidylglycerol	0	0	0	0	0	0	0	0	10
Sphingomyelin	6	18	0	0	9	0	0	0	0
Cerebroside	21	0	0	0	0	0	0	0	0
Cerebroside sulfate	4	0	0	0	0	0	0	0	0
Ceramide	1	0	0	0	0	0	0	0	0
Unknown or other	12	2	0	0	0	0	0	0	0

* Not analyzed.

proteins from membranes has hindered their analysis until very recently. The procedures employed in dissociating proteins from membranes has led to the definition of two broad classes of proteins. "Peripheral" proteins are released by mild treatment such as increasing the ionic strength of the solution or addition of a chelating agent. The remaining proteins, called "integral" proteins, are extracted only by use of detergents or solvents that disrupt membrane lipids or lipid-protein interactions.

Table 2 lists the lipid/protein ratio for a variety of biological membranes. From the table an obvious functional correlation can be made for the membranes that have extremes in lipid-protein ratio. That is, myelin, which serves primarily as an electrical insulating wrap around the internodes of myelinated nerve fibers, has a high ratio (3.5:1) whereas the ratio for inner mitochondrial membrane, which is the site of oxidative phosphorylation is low (0.3:1). Most of the other membranes listed in Table 2 have more protein than lipid, which is consistent with the notion that membrane functions are determined by specific proteins that are bound to the membrane.

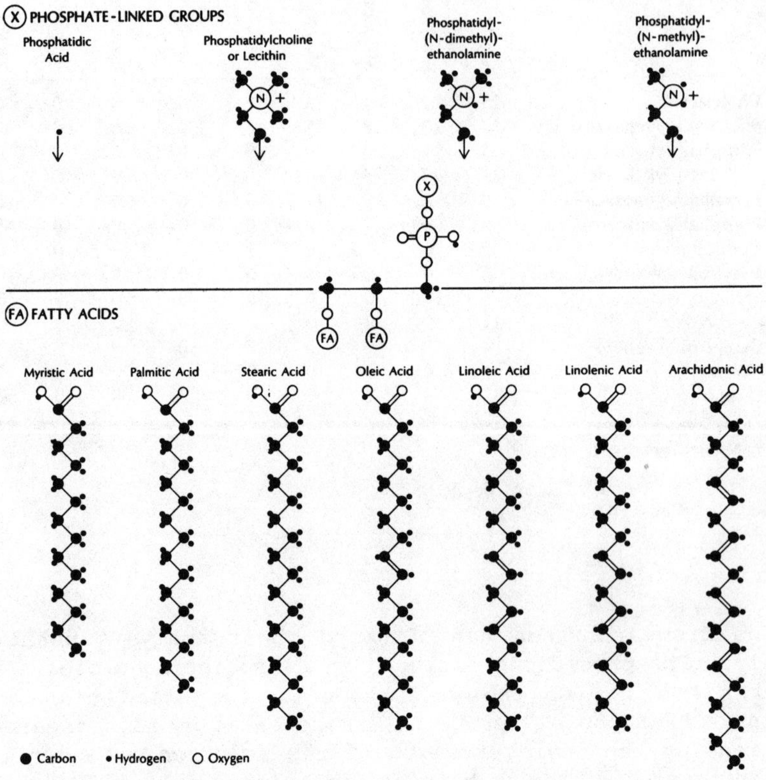

Fig. 2. Variety in phospholipid molecules arises through variations in fatty acids attached to two glyceryl carbons and variations in the groups attached to the phosphate. (From Chapman, 1975). (By permission of HP Publishing).

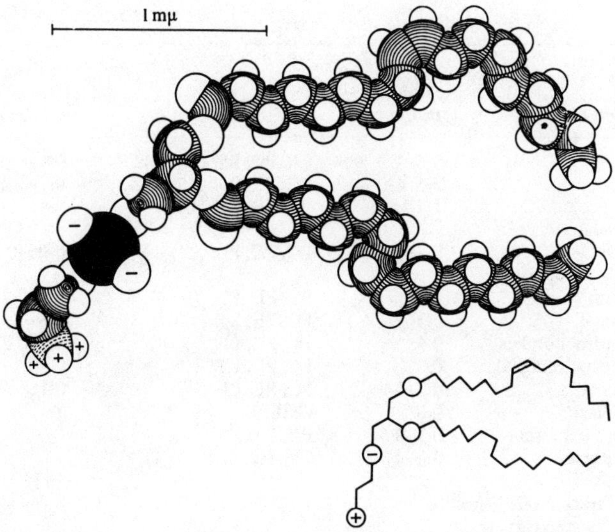

Fig. 3. A molecular model of a phosphatidylcholine (lecithin) molecule with one oleic and one palmitic acid hydrocarbon chain and the "stick" model diagrammatic representation. (From Chapman, 1975). (By permission of HP Publishing).

Carbohydrates are the third major class of constituents of biological membranes. Carbohydrates are found joined to peptides to form glycoproteins and also form glycolipids by O-glycosidic bonds to lipids. In both cases, monosaccharides are believed to form branched oligosaccharide chains but these structures are not well characterized. The main recognized function of glycoproteins and glycolipids is in the immunological properties of membranes and in the structural "cement" between cells in a tissue.

DETERMINATION OF MEMBRANE THICKNESS AND ELECTRICAL PARAMETERS

The first estimates of the thickness of biological membranes came from alternating-current impedance measurements on a suspension of red blood cells by Fricke (1923). At high frequency (100kHz) stimulation the cell membrane behaved as a dielectric with a specific capacitance of $0.8\mu F/cm^2$. Assuming a dielectric constant for the membrane similar to that of oil ($\kappa=3$), Fricke estimated the membrane thickness to be about 30A. Subsequent measurements of membrane capacitance by Cole (1972) and others on many different cell types have confirmed a specific capacitance of about $1\mu F/cm^2$ for most plasma membranes. In addition, optical measurements of the reflectance of lipid bilayers has yielded dielectric constants of about 2.2 making the value of 3 reasonable for the membrane as a whole. Thus a thickness of 30-50A is generally accepted.

Table 2. Composition of membranes.

Membrane	mg lipid/mg protein	Major lipids	Phospholipid/ cholesterol ratio mole/mole
Plasma	0·5–1·0	Ch, PC, PS, PE, Sph.	0·4–1·0
Myelin	3·5–4·0	PC, PE, Sph, Ch.	0·7–1·2
Golgi	1·2	PC, PE, Sph, Ch.	0·45–0·5
Rough endoplasmic reticulum	0·2–0·5	PC, PE, PI.	0·06–0·1
Smooth endoplasmic reticulum	1·2	PC, PE, PI.	0·1–0·2
Lysosomal	0·3	PC, PE, Sph, Ch.	0·5
Outer mitochondrial	0·4	PC, PE, PI.	0·1–0·2
Inner mitochondrial	0·3	PC, PE, DPG.	0·06
Nuclear	0·2–0·6	PC, PE, PI	0·1
Chloroplast	0·6	GalL, PG, Chl, PC.	
Gram+ve bacteria	0·3–0·5	PE, DPG.	
Mycoplasma	0·3	Ch, TG, PG.	

PL : Phosphatidyl choline
PS : Phosphatidyl serine
PE : Phosphatidyl ethanolamine
PI : Phosphatidyl inositol
Sph : Sphingomyelin
Ch : Cholesterol
DPG : Diphosphatidyl glycerol
GalL : Galactolipid
PG : Phosphatidyl glycerol
Chl : Chlorophyll
TG : Triglyceride

Impedance measurements on tissue slices and on small cells also showed that cell membranes have a substantial conductance, which appeared to have major effects on the impedance at frequencies below 1kHz. Cole and Hodgkin (1939) measured a DC resistance of 1000 $\Omega\cdot cm^2$ for squid axons between two long electrolyte-filled electrodes surrounding an axon at its extremities as a function of distance between the electrodes. However, in contrast to the capacitance, it is now known that the conductance of various cell membranes can differ by orders of magnitude, depending on the type of cell and the its function. For example, the specific conductance of an eel electroplaque membrane is 1 $\Omega\cdot cm^2$ whereas that of a chromaffin cell is 100,000 $\Omega\cdot cm^2$.

SOME NERVE MEMBRANE FUNCTIONAL CONSIDERATIONS

The first indication that electrical activity in nerve was related to changes in membrane conductance (ion permeability) came from transverse impedance measurements of the squid axon during propagated action potentials (Cole and Curtis, 1939). From the near constancy of membrane capacitance and the unlikelihood of uniform membrane conduction without a change in capacitance, they suggested that only a small fraction of the total membrane area may be involved in conduction. This was the first suggestion that conduction occurs in discrete portions of the membrane and is confined to specific structures which have come to be known as "ion channels."

Development of the "voltage clamp" technique (see Fishman, 1982) allowed direct observations of evoked ion conductance changes in a membrane. Hodgkin and Huxley (1952) then provided a quantitative description of ion conductances in the squid axon membrane. The concept of ion channels regulating membrane permeability to ions evolved directly from the success of the Hodgkin and Huxley formulation. The membrane current was separated into independent Na and K currents, each of which could be "blocked" by subsequently discovered specific agents, with the implication that separate pathways for conduction of Na and K ions existed in the membrane. Furthermore, voltage clamp data of conduction processes in squid axons and many other preparations became most widely interpreted in terms of the phenomenological ion channel shown in Fig. 4. Electric-field initiated alterations in the membrane that allow conduction of specific ions is called "gating." After ions gain access to the channel, the ease with which ions move through the channel is the conductance. The driving force for a specific ion channel is the gradient of electrochemical potential of the ion across the membrane. Each of these processes is assumed distinct and independent of the others.

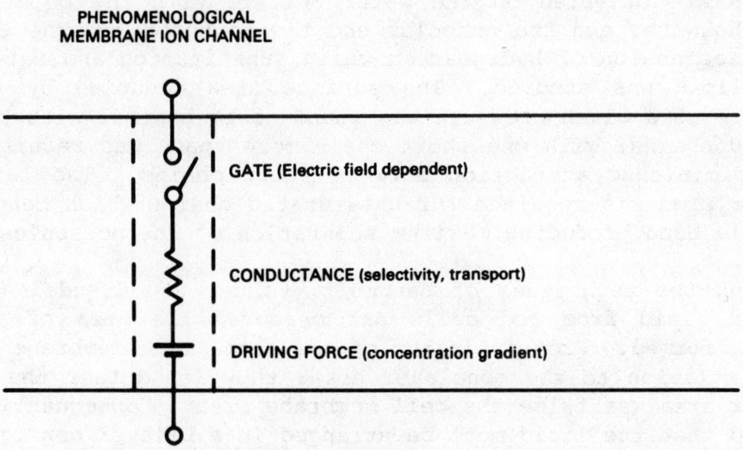

Fig. 4. Division of a membrane ion channel into three elements (gate, conductance, and driving force) representing processes that are thought to be essential to an understanding of conduction.

Implicit in the notion of gating is that specific, charged structures in the membrane respond to electric fields of sufficient strength by initiating events that allow membrane channels to conduct ions. Consequently, the displacement (capacitive) current, arising from external changes of membrane potential, includes (1) ion redistribution at the membrane surfaces (Fig. 5a), (2) electron redistribution in membrane molecules (Fig. 5b), and (3) reorientation of membrane bound charges (Fig. 5c) or dipoles (Fig. 5d). The measurement of a nonlinear portion of membrane displacement current preceding and associated with conduction in ion channels is assumed to be due to (3) above and is called "gating current" (Armstrong and Bezanilla, 1974). Furthermore, recordings of elementary current fluctuations of single ionic channels in various membranes (Hamill et al., 1981) show random transitions between discrete conducting and non-conducting states (See, for example, Fig. 11). Thus "single channel currents" indicate that membrane macroscopic ion conductances are controlled by an increase or decrease of the number of conducting channels rather than by a change in channel conductance.

THE BILAYER MODEL OF BIOLOGICAL MEMBRANES

In addition to the measurements mentioned above, several other observations influenced models of membrane structure. Permeability studies showed cell membranes to be preferentially permeable to lipid soluble substances (e.g., Collander, 1949).

I. Langmuir developed a method for studying lipid monolayers on a water surface. He and his colleagues showed that the lipid molecules are oriented on the water surface with the polar end toward the water and the nonpolar end toward the air. The effects on surface tension of hydrocarbon chain substitution and saturation of the lipid was studied. The surface area occupied by the monolayers at a given pressure was found to be minimal with two long saturated chains; with one short chain more space was required due to the diminished attraction between short chains. The largest monolayer area was required for unsaturated chains which bend about the double bond producing maximum separation of the molecules.

Using the techniques of Langmuir, Gorter and Grendel (1925) extracted lipid from red cells and measured the area of the monolayer formed. From estimates of the total cell membrane surface area in relation to the monolayer area, they found that the lipid monolayer area was twice the cell membrane area. Consequently, they concluded that the lipid must be arranged in a bilayer configuration. More recent estimates (Bar, Deamer, and Corwell, 1966) showed that the membrane lipid was incompletely extracted but compensatory underestimate of the total surface area still yielded a 2:1 ratio.

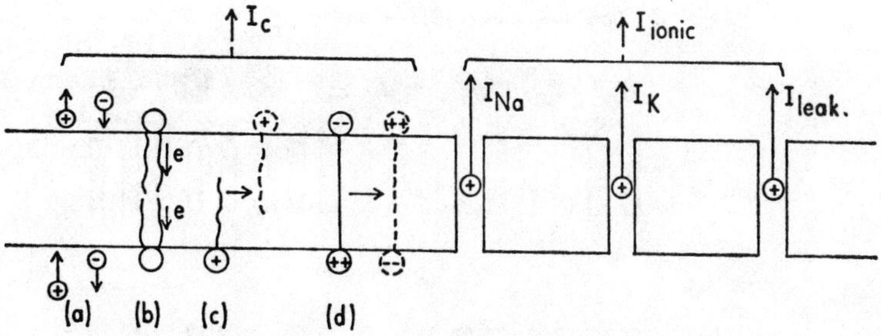

Fig. 5. In addition to ionic current in channels, the capacitive (displacement) current associated with the dielectric portion of the membrane arises from (a) redistribution of ions at surfaces, (b) redistribution of electrons on membrane molecules, (c) reorientation of membrane-bound charges or, (d) reorientation of membrane dipoles. Charge movement of type (c) and (d) that precedes and is associated with channel conduction is called "gating current." (From Armstrong and Bezanilla, 1974). (By permission of J. Gen. Physiol.)

Cole (1932) compressed marine eggs and computed a surface tension of about 0.1 dyne/cm which was two orders below the tension expected from a bimolecular leaflet or even a single oil-water interface. This observation led Danielli and Davson (1935) to postulate a lipid bilayer model (Fig. 6) with a layer of protein adsorbed to both the external and internal membrane surface to reduce the tension to the measured level. This basic model remained viable over a 40 year period without significant modification. Thompson (1964) then showed that addition of phospholipids to saturated hydrocarbon reduces the surface tension in water of the resulting complex to about 1 dyne/cm. Thus phospholipids are good surfactants without addition of proteins.

MODIFICATION OF THE BILAYER MODEL

By the late sixties, evidence of functional proteins embedded in membranes was accumulating. For example, in exposed freeze-fractured surfaces of most functional membranes (Fig. 7), a large number of particles (Branton, 1969), which were subsequently shown to be proteins, appeared to be embedded in a smooth matrix. In addition, at about the same time, peptide molecules such as gramicidin A, when incorporated into biological (Pressman, 1965) and artificial bilayer membranes (Hladky and Haydon, 1970), were shown to increase cation permeability in these membranes. These and many other facts led to modification of the bilayer model to allow for globular proteins interspersed in the lipid matrix. This

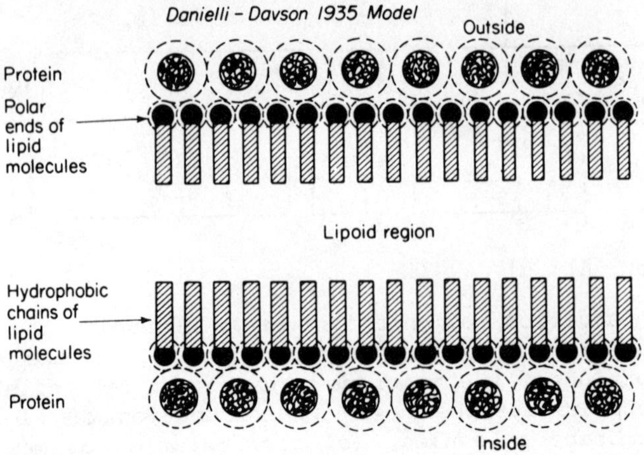

Fig. 6. The Danielli-Davson lipid-bilayer model of cell membrane structure. This model was viable for nearly 40 years. (From Danielli and Davson, 1935). (By permission of J. Cell Physiol.).

modification became known as the lipid globular-protein mosaic model (Singer and Nicolson, 1972). In this model (Fig. 8) the globular molecules of integral protein are structurally asymmetric with a highly polar end oriented toward the aqueous phase and a nonpolar end embedded in the hydrophobic interior of the membrane. An integral protein of appropriate size and structure or an aggregate of integral proteins may traverse the entire membrane and make contact with the aqueous phase on both sides of the membrane.

The most convincing support for the validity of the mosaic model came from progress in the isolation and structural determination of several functionally relevant important membrane proteins. First, a structural model of the cation-conducting gramicidin A channel formed in artificial bilayers was deduced (Urry, 1971). The transbilayer channel consisted of a dimer which is formed by the head-to-head association of two monomer beta-helices. The central pore along the helical axis is 4A in diameter and is lined with the peptide carboxyl groups, whereas the hydrophobic residues lie on the exterior of the helix. The dimer model was shown to be consistent with the kinetics of gramicidin A channel formation derived from the relaxation of the electrical current after a step change in voltage across the bilayer (Bamberg and Läuger, 1973).

Next, in biological membranes, the structure of several membrane proteins became clearer. A three-dimensional model of bacteriorhodopsin, a protein that uses the energy from light to transport hydrogen ions across the membrane of the salt-loving

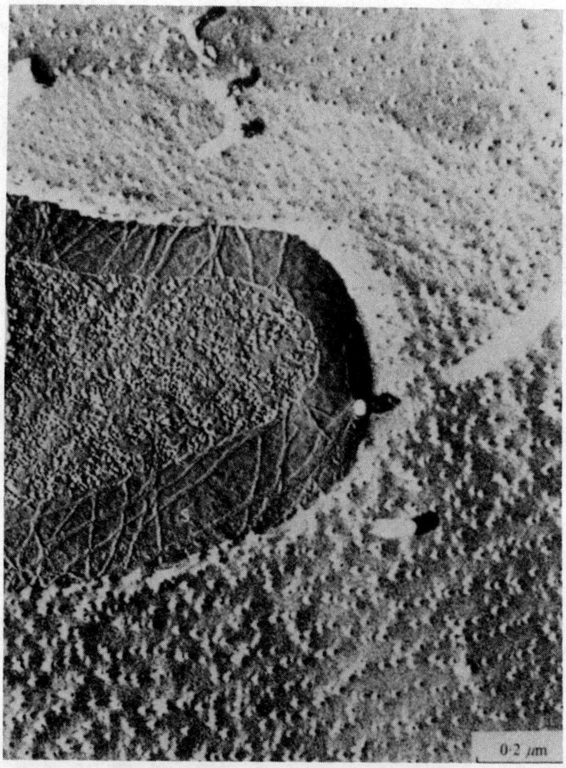

Fig. 7. View of the surfaces of a fragment of a human red blood cell ghost after freeze-fracture and etching. The observation of particles in the rough face in this and many other replicas of membranes, prepared by the same technique (Branton, 1969), preceded the notion of functional globular-protein structures embedded in a lipid matrix. (From Robinson, 1975). (By permission of Clarendon Press.)

bacterium Halobacterium halobium, was determined by Henderson and Unwin (1975) to a resolution of 7A by new techniques that permitted imaging in three dimensions. The protein consisted of 248 amino acids in a monomeric polypeptide chain, which is linked to one light-absorbing retinal group. Seven segments of the chain are coiled into a regular alpha-helix conformation which spans the membrane (Fig. 9). Using the same techniques for three dimensional image reconstruction, the structure of gap junction channels between cells were examined (Unwin and Zampighi, 1980). Each hemichannel (connexon) is composed of a hexamer of identical subunits arranged somewhat like the shutter of a camera (Fig. 10). In the presence of Ca, which is known to cause closure of the channels, a constriction of the channel was observed.

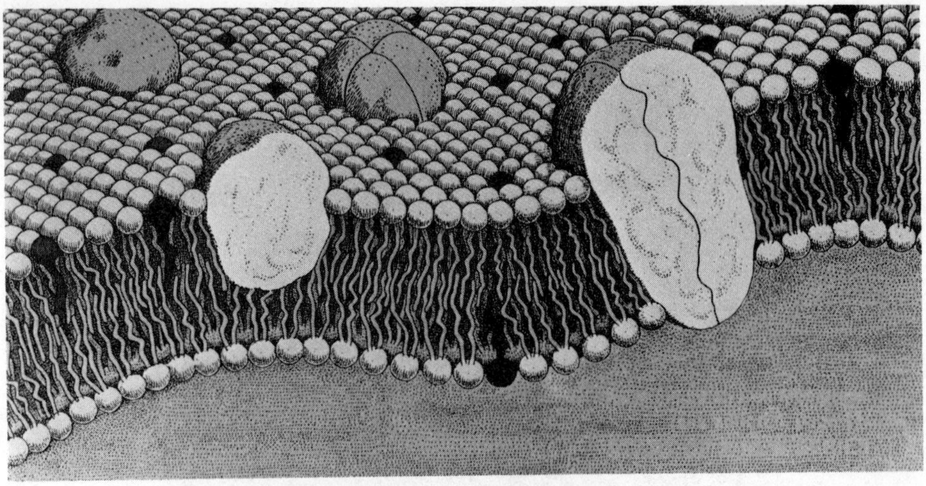

Fig. 8. The lipid globular-protein mosaic model of membrane architecture (Singer and Nicolson, 1972). Globular proteins are amphipathic with their hydrophilic ends protruding into the aqueous phase and their hydrophobic ends embedded in the bilayer of lipid (gray) and cholesterol (black). (By permission of HP Publishing).

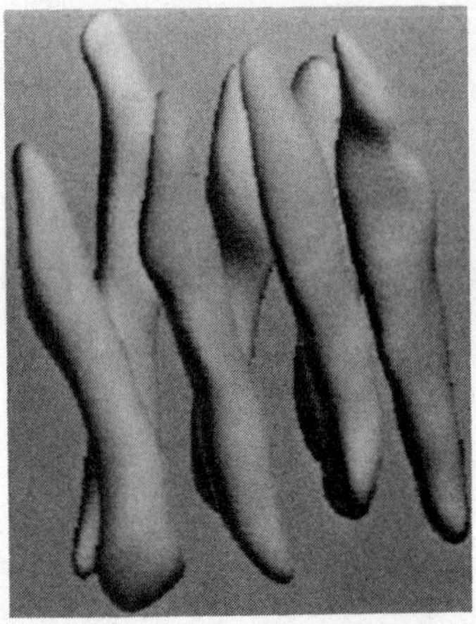

Fig. 9. The structure of bacteriorhodopsin, a protein that uses energy from light to transport hydrogen ions across the membrane of a bacterium. Seven alpha-helical segments of a single polypeptide chain that is linked to one light-absorbing retinal group, which is not visible at a resolution of 7A. (From Unwin and Henderson, 1984). (By permission of Scientif. Amer.).

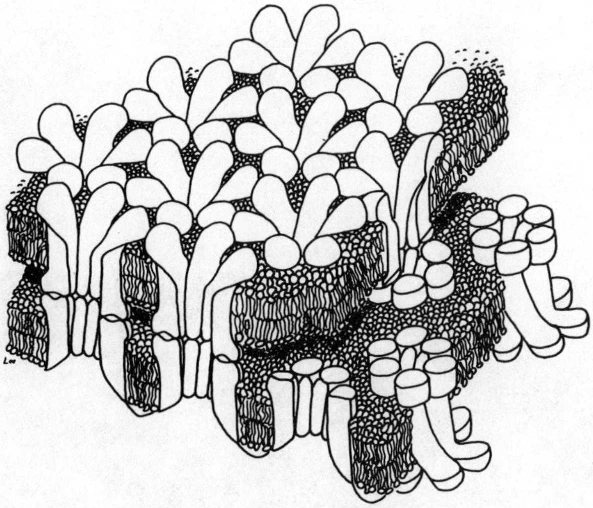

Fig. 10. Drawing of the structure of gap junctions isolated from mouse liver as deduced by x-ray diffraction. (Makowski et al., 1984). (By permission of Biophys. J.).

The best characterized structure of any membrane protein is the acetylcholine (AcCh) receptor from postsynaptic membranes at neuromuscular junctions. Electron micrographs such as the one in Fig. 11 show structures much like the petals of a flower extending into the extracellular space about 50A above the bilayer (Kistler et al., 1982). Structural analysis led to a three-dimensional model (Fig. 12) in which a monomeric receptor consisting of 5 subunits (designated α, β, α, γ, and δ) are arranged around a central ion channel which traverses the entire 110A length of the molecule (Kistler et al., 1982). Radioactively labeled α-bungaratoxin, which is known to bind to the AcCh receptor and blocks conduction was used to identify the αsubunit as the location of the agonist binding site. Recently the primary structure of the AcCh receptor molecule was determined (Noda et al., 1983ab).

Furthermore, purified AcCh receptor molecules have been successfully reconstituted in planar artificial lipid bilayers. Records of the spontaneous fluctuations of current in the presence of agonist show (Fig. 13) a binary waveform with random transitions between two conducting states that is similar to the one observed from AcCh-activated channels in native membrane (Labarca et al., 1984). Thus, the reconstituted functional properties have confirmed that the functional AcCh receptor and ion channel, whose primary structure is now known, was successfully identified and removed.

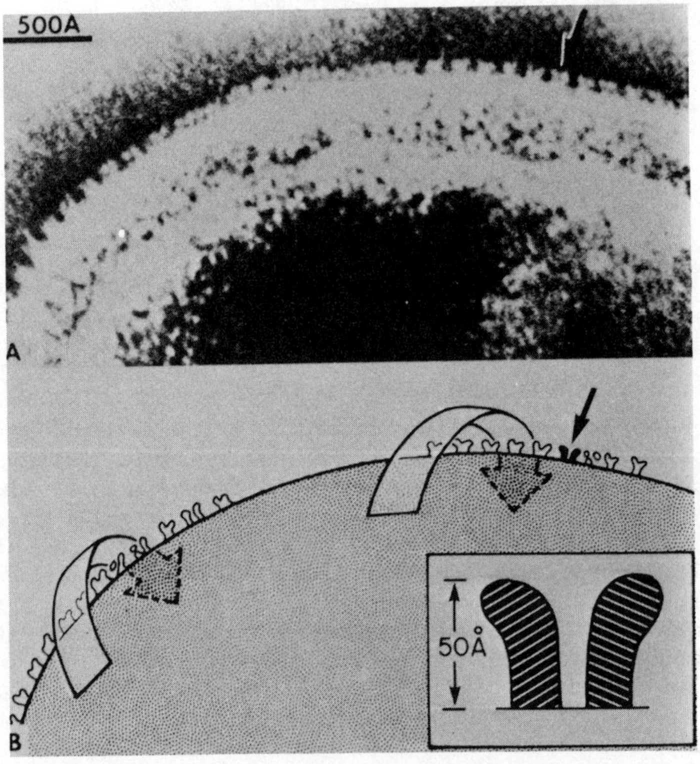

Fig. 11. (A) Synaptic edge of AcCh receptor membrane vesicle stained with uranyl acetate. (B) Drawing of the observed structures in (A). (From Kistler et al., 1984). (By permission of Biophys. J.).

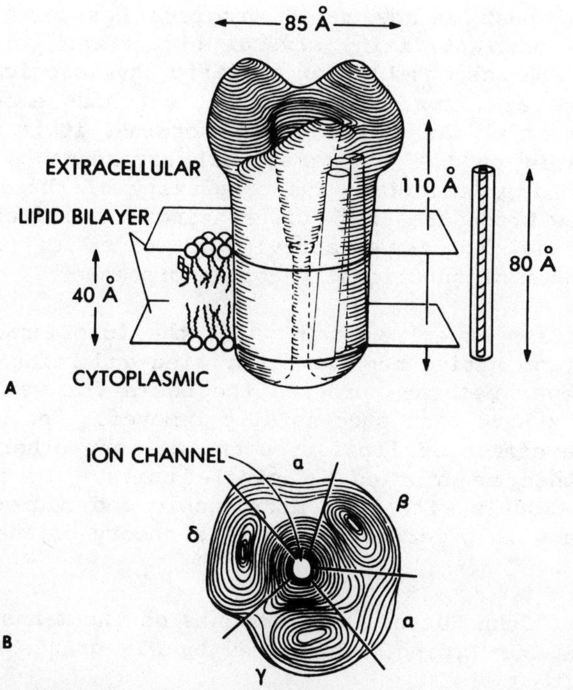

Fig. 12. (A) Three dimensional model of the AcCh-receptor (AcChR) molecule in a lipid bilayer (Kistler et al., 1984). The structure was deduced from densities in computer filtered images of AcChR tubular lattices, from side views and from x-ray diffraction. (B) Synaptic side view of AcChR molecule with assignment of 5 subunits labeled α, β, α, γ, and δ. The primary structure of the AcChR molecule has been determined (Noda et al., 1983ab). (By permission of Biophys. J.).

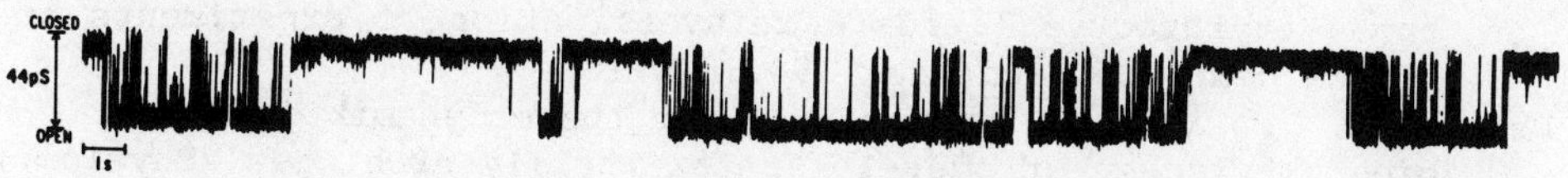

Fig. 13. Current fluctuations in a single AcCh-activated channel which was extracted from electric organ of *Torpedo californica*, purified and reconstituted in a planar lipid bilayer. The single-channel current was activated with 0.1 μM suberyldicholine and recorded at constant applied voltage of -100 mV. The waveform is similar to that recorded from AcCh-activated channels in native membrane. (From Labarca et al., 1984). (By permission of J. Neurosci.).

CONCLUSIONS

During the past decade rapid progress has been made in identifying and characterizing several structures in biological membranes that are responsible for specific physiological functions. These structures are, for the most part, proteins embedded in a lipid bilayer matrix. As with soluble proteins, it is presumed that function is based on the conformational arrangements of these proteins in the lipid matrix. The complexity of these structures will require new procedures for "visualizing" the structural changes that occur during functional activity in order to provide a structural account of functional membrane processes.

Reconstitution experiments in which the functional structures are extracted from native membrane, purified and reinserted into an artificial bilayer membrane provide the basis for verifying that functional units have been successfully removed. In the artificial environment the effect of lipid type and of many other factors on function can then be studied in single units. In addition, comparisons of models with both microscopic and macroscopic data will be necessary in order to arrive at a theory of charge transfer across membranes.

I thank Dr. John Russell for comments on the manuscript and Ms. Sarah Adams for typing. Supported by NIH grants NS11764 and NS13778.

REFERENCES

Armstrong, C.M., Bezanilla, F., 1974, Charge movement associated with the opening and closing of the activation gates of the Na channel, J. Gen. Physiol., 63:533-552.

Bamberg, E., and Läuger, P., 1973, Channel formation kinetics of Gramicidin A in lipid bilayer membranes., J. Membr. Biol., 11:177-194.

Bar, R.S., Deamer, D.W., and Corwell, D.G., 1966, Surface area of human erythrocyte lipids: reinvestigation of experiments on plasma membrane, Science, 153:1010-1012.

Bernstein, J., 1902, Untersuchungen zur thermodynamik der bioelektrischen strome, 1 (Erster Theil), Arch. ges. Physiol., 92:521-562.

Chapman, D., 1975, Lipid dynamics in cell membranes, in: "Cell Membranes - Biochemistry, Cell Biology, and Pathology," G. Weissmann and R. Claiborne, eds., HP Publishing, NY.

Cole, K.S., 1932, Surface forces of the Arbacia egg, J. Cell. Comp. Physiol., 1:1-9.

Cole, K.S., 1972, "Membranes, Ions and Impulses," University of California Press, Berkeley.

Cole, K.S., and Curtis, H.J., 1939, Electric impedance of the squid giant axon during activity, J. Gen. Physiol., 22:649-670.

Cole, K.S., and Hodgkin, A.L., 1939, Membrane and protoplasm resistance in the squid giant axon, J. Gen. Physiol., 22:671-687.

Collander, R., 1949, The permeability of plant protoplasts to small molecules. Physiol. Plantarum, 2:300-311.

Danielli, J.F., and Davson, H., 1935, A contribution to the theory of permeability of thin films, J. Cell. Comp. Physiol., 5:495-508.

Fishman, H.M., 1982, Current and voltage clamp techniques, in: "Techniques in Cellular Physiology, Part II," P.F. Baker, ed., P119:1-42, Elsevier-North Holland, Ireland.

Fricke, H., 1923, The electric capacity of cell suspensions, Phys. Rev., 21:708-709.

Gorter, E., and Grendel, F., 1925, On bimolecular layers of lipoids on the chromocytes of the blood, J. Exp. Med., 41:439-443.

Hamill, O.P., Marty, A., Neher, E., Sakmann, B., and Sigworth, F.J., 1981, Improved patch-clamp techniques for high-resolution current recording from cells and cell-free membrane patches, Pflugers Arch., 391:85-100.

Henderson, R., and Unwin, P.N.T., 1975, Three-dimensional model of purple membrane obtained by electron microscopy, Nature, 257:28-32.

Hendler, R.W., 1971, Biological membrane ultrastructure, Physiol. Rev., 51:66-97.

Hladky, S.B., and Haydon, D.A., 1970, Discreteness of conductance change in bimolecular lipid membranes in the presence of certain antibiotics, Nature, 225:451-453.

Hodgkin, A.L., and Huxley, A.F., 1952, A quantitative description of membrane current and its application to conduction and excitation in nerve, J. Physiol., 117:500-544.

Kistler, J., Stroud, R.M., Klymkowsky, M.W., Lalancette, R.A., and Fairclough, R.H., 1982, Structure and function of an acetylcholine receptor, Biophys. J., 37:371-383.

Labarca, P., Lindstrom, J., and Montal, M., 1984, The acetylcholine receptor channel from Torpedo californica has two open states, J. Neurosci., 4:502-507.

Makowski, L., Caspar, D.L.D., Philips, W.C., Baker, T.S., and Goodenough, D.A., 1984, Gap junction structures. VI. Variation and conservation in connexon conformation and packing, Biophys. J., 45:208-218.

Nägeli, K., and Cramer, K., 1855, "Pflanzenphysiologische Untersuchungen Schultess", Zürich.

Noda, M., Takahashi, H., Tanabe, T., Toyosato, M., Kikyotani, S., Furutani, Y., Hirose, T., Takashima, H., Inayama, S., Miyata, T., and Numa, S., 1983a, Structural homology of Torpedo californica acetylcholine receptor subunits, Nature, 302:528-532.

Noda, M., Takahashi, H., Tanabe, T., Toyosato, M., Kikyotani, S., Hirose, T., Asai, M., Takashima, H., Inayama, S., Miyata, T., and Numa, S., 1983b, Primary structures of β- and δ-subunit precursors of *Torpedo californica* acetylcholine receptor deduced from cDNA sequences, *Nature* 301:251-255.

Pressman, B.C., 1965, Induced active transport of ions in mitochondria, *Proc. Nat. Acad. Sci. (USA)*, 53:1076-1083.

Robinson, G.B., 1975, Principles of membrane structure, *in*: "Biological Membranes," D.S. Parsons, ed., 33-54, Clarendon Press, Oxford.

Singer, S.J., and Nicolson, G.L., 1972, The fluid mosaic model of the structure of cell membranes, *Science*, 175:720-731.

Thompson, T.E., 1964, The properties of bimolecular phospholipid membranes, *in*: "Cellular Membranes in Development," M. Locke, ed., 83-96, Academic Press, N.Y.

Unwin, N., and Henderson, R., 1984, The structure of proteins in biological membranes, *Scientif. Am.*, 250(2):78-94.

Unwin, P.N.T., and Zampighi, G., 1980, Structure of the junction between communicating cells, *Nature*, 283:545-549.

Urry, D.W., 1971, The Gramicidin A transmembrane channel: A proposed $\Pi_{(L,D)}$ helix, *Proc. Nat. Acad. Sci. (USA)*, 68:672-676.

CELL NUCLEUS AND EM FIELDS

Claudio Nicolini

Chair of Biophysics
School of Medicine
University of Genoa, Genoa, Italy

This chapter is aimed to present a broad overview of the interaction between electromagnetic waves and living matter, particularly stressing their role in the acquisition of the most recent understanding of the structure and of the molecular mechanisms controlling the function of the single most important cell compartment - i.e. the cell nucleus. Many aspects of the interaction between EM fields and mammalian cells can be of interest, in the range $0\text{-}10^{11}$ Hertz. Within such broad range we may have to discuss the entire physics and the entire life sciences, so we will confine ourselves to a short overview of few pertinent and correlated topics, which apply to the characterization of the cell nucleus and its constituents, namely DNA.

Few details are initially given on the various order of chromatin-DNA organization within the intact nucleus, from the primary (base sequence), secondary (A-, B- and Z- forms) up to the tertiary (nucleosome subunits) and higher order structures, underlying for each level the critical role of the physical state of water, ions and enzymatic protein modifications in determining the overall DNA structure and the cell function (also through the mediation of interlinked changes in the cell cytoskeleton and of the nuclear membrane).

The complexity of the structural and molecular alterations, which are correlated with neoplastic transformation and normal cell proliferation, are then exemplified by the study conducted on isolated nuclei with independent non-invasive biophysical probes - all using EM fields (NMR spectroscopy and imaging, complex impedence measurements and differential scattering of circularly polarized light).

Finally and comfortingly the abrupt nature of both the experimentally observed nuclear transitions - both in functional and structural (phase-like transitions) terms - and of their association with localized or global changes in the electromagnetic environment are shown to be explained from first physical principles within the EM-based polyelectrolyte theory and mean field theory.

NUCLEAR AND GENE STRUCTURE

One of the biggest steps forward was the discovery that a piece of chemical substance (called DNA) was selectively attacked and cut by certain enzymes, called "restriction enzyme", which could recognize in the DNA very specific sites; these extracted regions were then recognized as functional units, known from the Mendellian time as genes.

We don't know yet exactly what is a gene at molecular level in terms of 3-D structure. There are only few hypotheses still to be proved. Surely, we can identify pieces of fragments which can be associated with specific messenger RNA molecules, which then make specific proteins; we can therefore identify functional units and associate them with specific fragments of given base pairs sequence. The very first structure, or primary structure, is then the base pairs sequence.

The secondary DNA structure is polymorphic and is given either by the B-form (predominant), Z-form or A-form, all discovered by neutron and X-ray diffraction of crystals of very short fragments down to 4 nucleotides long. These localized secondary structures are possibly the sites where restriction enzymes actually may cut and identify genes: we may indeed obtain a map of the genome by using restriction enzymes.

The B-form DNA, which is dominant in the mammalian cells, is a right-handed helix with the base pairs perpendicular to the helix axis. This enormous giant molecule, two-three meters long and able to replicate itself with the precision of the angstrom, has a negative electric charge due to the phosphate groups and can be today described at molecular level in detail. This permits the use of theory, such as polyelectrolyte theory, to make a molecular model, which can make predictions on the DNA interaction with ions and with proteins. I want to stress that this structure, the B-form DNA, is strongly maintained and critically dependent on ions and on water. Are indeed the amount of bound water and the amount and charge of ions to critically determine transitions at the level of very short DNA fragments (4 or 5 base pairs) from the B- to a A- form or a Z-form. Water, ignored for many years, has been recently shown to be correlated even with changes in the DNA higher order structure by the introduction of EM-based powerful techniques, i.e. NMR and complex impedence measurements.

While most (i.e. A- and B-) DNA secondary structures are right-handed, Z DNA is left-handed, and this along with the different geometrical parameters of the helix may create constraints and conditions for the creation of negative superhelicity - which appears to be one of the mechanisms controlling gene expression.

But what happens in the cell when the DNA interacts with the nuclear proteins, being either histones (positively charged) or non-histones (negatively charged or neutral)? At the first level this interaction causes the formation of the so-called 'nucleosome' subunits, where the DNA folds around the octamer which in turn is caused by self-aggregation of histone proteins (with the exception of H1); these nucleosomes, interconnected by free strands of DNA, give rise to the tertiary structure, or nucleofilament.

Then, upon interaction with H1 histones and with non-histones, the DNA folds itself in a supercoil or two order superhelix, which corresponds to the quaternary structure[1,2].

Rather than have one nucleofilament, you may well have more than one nucleofilament going at the same time around the superhelix axis[3]; what is important however is that this 300 A° fiber, corresponding to the quaternary structure, has regions which do open up under physiological conditions and regions which never open up, and which appear adjacent to the nuclear envelope. This creates a condition for another fifth order folding, possibly drapery-like, of the 300 A° fibers, which appear to lay down either from the nuclear envelope near the pores or from the nucleoli[3,4].

We have to emphasize that chromatin-DNA structure critically depends on the function of the cell, but the general architecture described above applies for all cell types, with only the fraction of total DNA in any given structure being different among the different cell types.

All these transitions between the various chromatin-DNA structures in the cell take place in seconds: furthermore, the entire replication of this giant DNA is a regular process, which starts in two directions, but appears to begin in fixed sites - possibly where DNA is originally in the condensed state and then locally starts to open up, successively going from the 300 A° to the 100 A° nucleofilament to the free DNA.

Other structures, the pores at the nuclear membrane, were a mystery for many years and were thought to be like channels needed for the communication between the nucleus and the cytoplasm. Instead, recent data show that these pores are filled with a lot of DNA, proteins and RNA[5].

Nuclear pores have as many clusters as there are metaphase chromosomes and interphase chromatin islands (about 20 in the rat hepatocytes). The number of these clusters is invariant with respect to the different state of a given diploid cell, whenever either goes through the cell cycle or through neoplastic transformation[6]. The mitosis last very little, about half an hour, while the cell cycle has quite a long and variable duration, but during all this period the number of chromatin-DNA islands remain constant changing only the degree of DNA supercoiling.

Analyzing the chromatin fiber, regions of variable length (which never open up) appear to alternate to regions which do open up and have constant length of about 17.000 base pairs. Each of these 100.000 units (called fibrosomes) in mammalian cells appears fixed at the extremes and to have a different amount of negative superhelixes per unit DNA, which in turn could create the thermodynamic conditions for local gene expression. It is suggested that gene expression takes place whenever chromatin undergoes a localized transition from the highly packed structure to the unfolded structure (nucleofilament), followed by a successive unfolding to free DNA with the octamer histones coming out.

For many years the number of genes was considered equivalent to the number of messenger RNA molecules (about 1.000.000 from the average protein size). Instead we know now that the number of genes is about 100.000 both from genetic engineering and classical genetics, which would be compatible with an average size of genes of about 17.000 bp, as stikingly independently suggested by the "fibrosome" structural model[10]. Of course the linkage between what appears from the structural studies and what is seen from the functional studies has yet to clarify the high order structure of a gene. An electron microscopy picture of a ribosomal gene while transcribing was taken by Miller many years ago, indicating a typical Christmas tree (transcribing) structure of fixed length alternating with a highly packed (non-transcribing) DNA structure of variable length. The size of this transcribing unit is in mammalian cell similar to the size of fibrosome unit which at present is only a strong but circumstantial evidence in favor of the equation fibrosome (structural unit) = gene (functional unit).

The important implication of the fibrosome model is that each structural unit is independent and may be affected specifically by the local concentration of ions and bound versus free water, by local enzymatic modification of proteins (like H1, H3 and H4 histones and non-histones) and generally by the local electrostatic environment; there appears indeed to be significant difference of concentration of various substances from region to region of the nucleus which creates the electromagnetic conditions needed for the change in chromatin-DNA structure and thereby for the expression or repression of each individual unit.

NUCLEAR MECHANISMS CONTROLLING CELL FUNCTION

One common expectation is that genetic engineering may solve the problem of cancer, by finding the genes which control specific functions and specific neoplastic transformations.

Numerous oncogenes or "cancer-causing" genes have been identified today, but for each of them there are homologues in the normal counterparts which are expressed at different rates during normal cell cycle progression. The rate by which these "oncogenes" is expressed doesn't only depend indeed on neoplastic transformation, but depends also on cell proliferation. It appears then that the expression of these genes (and any other gene) is really varying with cell cycle progression, likewise the structure of chromatin. In addition to gene and chromatin-DNA several things dramatically vary along the cycle, and some of these have been associated (from enzymes to membranes) with the control of the transition from a normal to a transformed state. Instead most of these changes appear mere reflections of different proliferation rates between the normal and the transformed counterparts.

There appears to be, however, a common denominator in the jungle of changes, namely the changes in the degree of negative superhelicity of DNA high order structure. All the variations in ions, either monovalent, divalent or trivalent, in bound versus free water, in enzymatic modifications of proteins, induce changes in the electrostatic interaction with DNA which could ultimately cause the observed changes in the number of negative superhelixes. Each fibrosome may then act like a spring, whereby in a normal resting state each cell fibrosome has a different degree of negative superhelicity; there could be certain fibrosomes which have zero superhelicity and they may be transcribing, and others (the large majority) which have some degree of superhelicity and which are not transcribing. It has been experimentally observed that over the one hundred thousands of genes only a very small fraction appears active at any given time, while most of the others are inactive: when you induce global or local molecular alteration you affect all these units at a different degree, depending on their initial state. You may then either induce differentiation or transformation, by repressing certain units and expressing other units. In terms of the numbers of superhelixes, we may have in each fibrosome a limit value like in a spring, which accumulates a great amount of energy so that when we go above this limit the energy can be released all at once, i.e. during local gene expression. Far from being proven, this model - where the normal cell can be alternated by global or local variations in the cell constituents which in turn affect, through changes in the EM environment, the negative superhelicity of each chromatin structural units - appears to agree with experimental observations. We have obviously to remember that gene expression can be regulated in principle not only at transcriptional level, but also at the level of processing and translation (not discussed here).

The cell nucleus has a way to communicate with the cytoplasm through the nuclear pores and a way to communicate with the cell membrane and the outside through the cytoskeleton (microtubules and macrofilaments). In the latter case we have something which appears unique to cell differentiation and transformation, whereby the microtubules are highly organized/oriented in normal cells and disrupted in transformed cells.

What is unique to a normal cell is indeed the perfect correlation between the changes occurring at the level of the cell and the changes occurring at the level of the nucleus, whereby the cytoskeleton acts as the structure controlling cell geometry and cell growth. Both normal and transformed cell change structurally and morphologically along the cycle; but what is unique is that while the morphometric and structural changes at the level of the nucleus and of the cell occur simultaneously (in a perfect synchrony) in the normal cells, they appear highly uncorrelated in the transformed cancer cells[7].

COMPLEXITY OF STRUCTURAL ALTERATIONS AND EM-BASED APPROACHES

An example of the complexity of structural alterations during the life of a cell is given by the opposite variations induced at cellular and nuclear level in rat liver, which can be understood only by a variety of independent biophysical probes all undestructive and EM-based (Circular Intensity Differential Scattering, NMR spectrometry, complex impedence measurements and cytometry). This is apparent by monitoring the cellular and nuclear physical state of water in correlation with changes in chromatin structure.

By proton NMR spectrometry at 42 megahertz of rat hepatocytes, the resting cell population appears to have only one relaxation time T1, while in the proliferating cells we recognized two types of populations - a population still with the same T1 as resting cells and a population with much larger T1.

This is comforting since after partial hepatectomy only a small fraction (about 40%) responds to the stimulation, while the rest remain quiescent.

'much larger T1, however, means less bound water per hepatocyte, that, if confirmed for the corresponding isolated nuclei, means a more supercoiled chromatin in contradiction with findings on the nuclei from the same G1 hepatocytes, indicating with respect to G0 nuclei an increase in bound water (by complex impedance measurements) and a decrease in chromatin supercoil (by CIDS) and measurements and a decrease in chromatic supercoil by (CIDS) and condensation (by cytometry.)

We then looked to the isolated nuclei from the same cells and we found in G1 still two populations with two different T1, but while one was 830 milliseconds (corresponding to the G0 nuclei,) the other was significantly decreased to less than 600 milliseconds; a decrease in T1 means an increase in bound water per unit nuclei, as expected from the parallel studies on the same rat liver nuclei

EPILOGUE

Phase like transitions in nuclear volume and in chromatic-DNA structure - occurring either naturally in a mammalian cell in the control of its function or artificially upon induced modifications in pH, temperature, electric field and ionic strength - can be explained within the framework of a theory based on first physical principles, namely the mean field theory[8].

Similarly theoretical explanations can be found for changes on the high order structure of DNA occurring at the two levels of ion concentration, either monovalent or divalent, and of chromosomal proteins (particularly the H1 histones.) There are two different alterations acting on the same macromolecule, and another theory, polyelectrolyte theory, has indeed been used to find an explanation, by building a molecular model in terms of the free energy derived from both electrostatic and mixed contributions. As for the mean field theory the results are encouraging[9], since the polyelectrolyte theory can explain by itself how and why the nucleofilament DNA become a supercoil in presence of both proper ionic strength and of the H1 molecule, and how and why the degree of supercoil changes as a function of ions and enzymatic protein modifications.

A comprehensive and tutorial treatment, from experimental and theoretical standpoints, of these and many other related topics at the crossing of life and physical sciences can be found elsewhere[10,11] - with all due references to the enormous literature existing in this interdisciplinary field..

Acknowledgements

This work has been supported by grants awarded by the "Oncology" finalized project of the Italian 'National Research Council (C.N.R.) and by the Italian Ministry of Public Education (M.P.I.).

REFERENCES

1. J. T. Finch and A. Klug, PNAS-USA, 73:1897 (1976).
2. C. Nicolini and F. Kendall, Phys.Chem.Phys., 9:265 (1977).
3. C. Nicolini et al., J.Cell Science, 62:103 (1983).

4. A. Belmont, et al., J.Cell Science, 65:123 (1984).
5. P. N. Unwin and R. A. Milligan, J.Cell Biology, 93:63 (1982).
6. C. Nicolini et al., J.Cell Science, 67 (1984).
7. C. Nicolini and F. Beltrame, J.Cell Biol.Intern.Rep.,6:63 (1982).
8. T. Sun et al., J.Chem.Phys., 73:5971 (1980).
9. A. Belmont and C. Nicolini, J.Theor.Biol., 90:169 (1981).
10. C. Nicolini, ed., "Nobel Symposium on New Frontiers at the Crossing of Life and Physical Sciences", Plenum Publishing Corp., New York (1985).
11. C. Nicolini, "Biophysics and Cancer", Plenum Publishing Corp., New York (1985).

ELECTROMAGNETIC ORIENTATION : A RELATIVISTIC APPROACH

Ad. J. Kalmijn

Scripps Institution of Oceanography
University of California, San Diego
La Jolla, CA 92093

Through interaction with the earth's magnetic field, tidal and wind-driven ocean currents induce electric fields, which may inform sharks, skates, and rays about their drift with the water. Furthermore, through their swimming movements, the animals induce electric fields, which may provide the physical basis of their magnetic compass sense. Both the ocean-current fields and the fields of the animals' own movements result from the motion of a conductor, either the water or the fish, through the earth's magnetic field.

To detect the motional-electric fields, the elasmobranch fishes presumably use their highly sensitive electroreceptors, the ampullae of Lorenzini. According to physical law, however, the animals can measure only the electric fields existing in the frames of reference in which the sense organs are at rest. But, if the fish are not moving in the rest frames of their own sense organs, what are the electric fields from which they could infer their drift and magnetic compass headings relative to the earth's frame of reference?

This question assumes that it would be physically feasible for fishes, when drifting or swimming with constant velocity, to sense their motion electrically by the fields they detect in their own frames of reference. However, would not such an idea violate the principle of relativity as originally proposed by Galileo[1] for mechanical phenomena and extended by Einstein[2] to include electromagnetic phenomena as well? Sharks, skates, and rays may have been the first living creatures to resolve these fundamental problems.[3]

The theory of electromagnetic orientation has been introduced previously.[3-6] The present article addresses some of the subtle issues that the author has struggled with for a long time. The reader will be encouraged to view the world through the animals' senses, rather than remain an outside observer. The difficulties will not so much be in the physics itself as in its application to fishes moving about in large volumes of seawater, where neither the animals nor the water are perfect insulators or perfect conductors.

ELECTRIC AND MAGNETIC FIELD TRANSFORMATIONS

In special relativity, electric and magnetic fields are measured in different inertial frames of reference, and the results are related to one another mathematically by transformation formulas. In particular, if in the earth's frame of reference $\mathbf{E}_{e\perp}$ and $\mathbf{B}_{e\perp}$ are the electric and magnetic field components perpendicular to the velocity $\mathbf{v}$ of a moving animal, then the corresponding components in the animal's frame of reference are[7]

$$\mathbf{E}_{a\perp} = \gamma(\mathbf{E}_{e\perp} + \mathbf{v} \times \mathbf{B}_{e\perp}) \quad \text{and} \tag{1}$$

$$\mathbf{B}_{a\perp} = \gamma(\mathbf{B}_{e\perp} - \frac{1}{c^2}\mathbf{v} \times \mathbf{E}_{e\perp}), \tag{2}$$

where c is the speed of light and γ denotes $[1 - (^v/_c)^2]^{-1/2}$. The components $\mathbf{E}_{e\parallel}$ and $\mathbf{B}_{e\parallel}$ parallel to $\mathbf{v}$ remain unaltered under transformation from the earth's to the animal's frame of reference, or

$$\mathbf{E}_{a\parallel} = \mathbf{E}_{e\parallel} \quad \text{and} \tag{3}$$

$$\mathbf{B}_{a\parallel} = \mathbf{B}_{e\parallel}. \tag{4}$$

At speeds much lower than the speed of light, expressions (1) and (2) simplify considerably, as γ tends to unity and $^1/_{c^2}\, \mathbf{v} \times \mathbf{E}_{e\perp}$ becomes negligibly small compared to $\mathbf{B}_{e\perp}$. Thus, the perpendicular and parallel field components together yield to good approximation

$$\mathbf{E}_a \simeq \mathbf{E}_e + \mathbf{v} \times \mathbf{B}_e \quad \text{and} \tag{5}$$

$$\mathbf{B}_a \simeq \mathbf{B}_e, \tag{6}$$

in which $\mathbf{v} \times \mathbf{B}_e \equiv \mathbf{v} \times \mathbf{B}_{e\perp}$. If, in the earth's frame, the electric field $\mathbf{E}_e$ is zero, equation (5) reduces further to

$$\mathbf{E}_a \simeq \mathbf{v} \times \mathbf{B}_e. \tag{7}$$

Hence, an animal moving in the earth's frame of reference with a velocity $\mathbf{v}$, in the presence of a magnetic field $\mathbf{B}_e$, is in its own frame of reference subject to an electric field $\mathbf{E}_a$, which aproximately equals the electric field $\mathbf{E}_e$ in the earth's frame plus the $\mathbf{v} \times \mathbf{B}_e$ induced by the animal's motion (Fig. 1). However, the magnetic field $\mathbf{B}_a$ in the frame of the moving animal, is practically the same as the magnetic field $\mathbf{B}_e$ existing in the frame of the earth.

Conversely, the electric field $\mathbf{E}_e$ and the magnetic field $\mathbf{B}_e$ in the earth's frame of reference can be expressed in terms of the electric and magnetic fields in the animal's frame of reference by

$$\mathbf{E_e} \simeq \mathbf{E_a} - \mathbf{v} \times \mathbf{B_a} \quad \text{and} \tag{8}$$

$$\mathbf{B_e} \simeq \mathbf{B_a}, \tag{9}$$

where the minus sign indicates that the earth's frame of reference moves with a velocity $-\mathbf{v}$ with respect to the animal's frame.

As at low velocities $\mathbf{B_e}$ approximately equals $\mathbf{B_a}$, the subscripts will be omitted. Nevertheless, in the earth's frame of reference, the magnetic field enters the description of the electric field that a swimming fish is subjected to, whereas in the animal's frame, it does not, as is indicated by equation (7). That is to say, the $\mathbf{B}$ in the vector product $\mathbf{v} \times \mathbf{B}$ refers to the magnetic field in the frame of reference with respect to which the motion takes place.

Approximate equations (5) through (7) express the classical results of Faraday's 1832 research[8] on motional electricity, and led Einstein[9] in 1905 '.. more or less directly to the special theory of relativity ..' For elasmobranch fishes, $\mathbf{v} \times \mathbf{B}$ indeed is '.. nothing else but an electric field ..' $\mathbf{E_a}$. The origin of this purely electric field will be discussed relativistically under the heading of 'electric versus magnetic field detection.'

ELECTRIC FIELDS OF OCEAN CURRENTS

In the reference frame of the earth, a tidal or wind-driven ocean current, flowing with a horizontal velocity $\mathbf{v}$, induces, by interaction with the vertical component of the earth's magnetic field

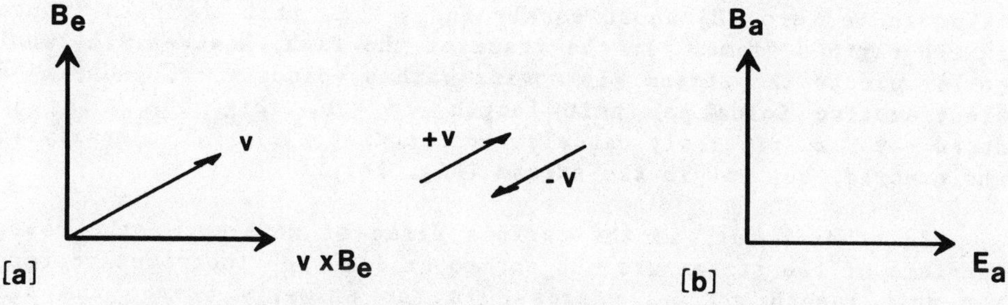

Fig. 1. Electric and magnetic field vectors at position of fish: (a) in earth's frame of reference, in which fish moves with uniform velocity $\mathbf{v}$, and (b) in animal's frame of reference, in which fish is at rest, assuming $\mathbf{E_e} = \mathbf{0}$. The velocity $+\mathbf{v}$ refers to the motion of the animal's frame and equals the fish's velocity $\mathbf{v}$ as observed in the earth's frame. The velocity $-\mathbf{v}$ refers to the motion of the earth's frame as observed in the animal's frame. Cf. equations (6) and (7).

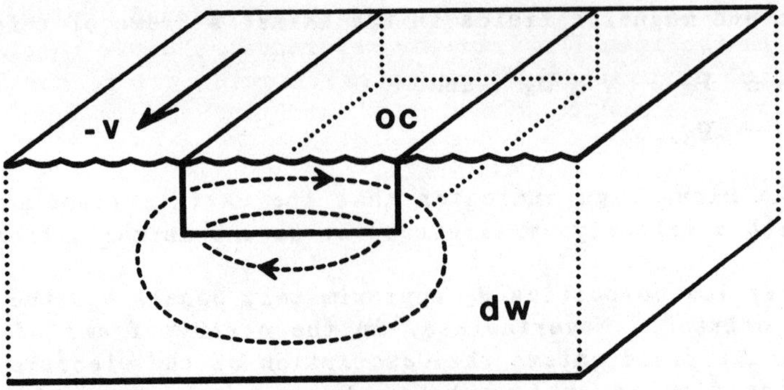

Fig. 2. In reference frame of drifting animal, ocean current (oc) is at rest, while, outside the stream, the whole world is moving with velocity $-\mathbf{v}$. Indicated is the induced electric current giving rise to the ohmic $\rho\mathbf{J}$ field that the drifting fish detects. The induced current flows across the stream and completes its electrical circuit through deeper water layers (dw) and bottom sediments. For southern hemisphere.

$\mathbf{B}_V$, a motional electromotive force per unit length $\mathbf{v} \times \mathbf{B}_V$ in the moving water and in the animals drifting with it.[10] The results of interaction with the horizontal component of the earth's magnetic field will be discussed later in this section.

In the frame of an animal drifting with the flow of water, the ocean current is at rest (Fig. 2), while, according to equation (7), the world is invaded by an electric field $\mathbf{E}_a$ (Fig. 3a). If $\mathbf{E}_e$ is taken to be zero, $\mathbf{E}_a$ about equals the $\mathbf{v} \times \mathbf{B}_V$ that the fish induces in the earth's frame. In the frame of the fish, however, the whole world, except the stream, is moving with a velocity $-\mathbf{v}$, inducing an electromotive force per unit length $-\mathbf{v} \times \mathbf{B}_V$ (Fig. 3b). The induced $-\mathbf{v} \times \mathbf{B}_V$ precisely cancels the electric field $\mathbf{E}_a$ in nearly the whole world, but not in the stream (Fig. 3c).

In other words, in the earth's frame of reference, the charge carriers of the stream are subject to an induced electromotive force per unit length $\mathbf{v} \times \mathbf{B}_V$, whereas, in the animal's frame, the same charge carriers are subject to a purely electric field $\mathbf{E}_a$. In the earth's frame, the force on the charge carriers is attributed to the local flow of water. In the animal's frame, however, in which the water is at rest, the force has its cause outside the stream.

If, in the earth's frame, the velocity $\mathbf{v}$ and the magnetic field $\mathbf{B}_V$ in the vicinity of the drifting animal are independent of time and uniform, then the electric field $\mathbf{E}_a$, in the animal's frame, is locally uniform as well. Thus, integrated around any measuring loop at rest in the animal's frame of reference, whether completely within the fish or partially through the flowing water, the $\mathbf{E_a}$ field as such fails to produce measurable potential differences across the extremely thin electroreceptive membranes.

However, the $\mathbf{E_a}$ field of the stream causes an electric current of density $\mathbf{J}$ to flow in the water and through the fish (Fig. 3d). This current gives rise to accumulations of electrical charges along the boundaries of the stream, where the $\mathbf{E_a} - \mathbf{v} \times \mathbf{B_v}$ field is discontinuous (Fig. 3c). The electric field $\mathbf{E_\pm}$ of those charges (Fig. 3e) provides the driving force for the electric current to complete its circuit through deeper water layers and bottom sediments, and, at the same time, counteracts the $\mathbf{E_a}$ field within the stream, thereby yielding a continuous current density $\mathbf{J}$ around the circuit.

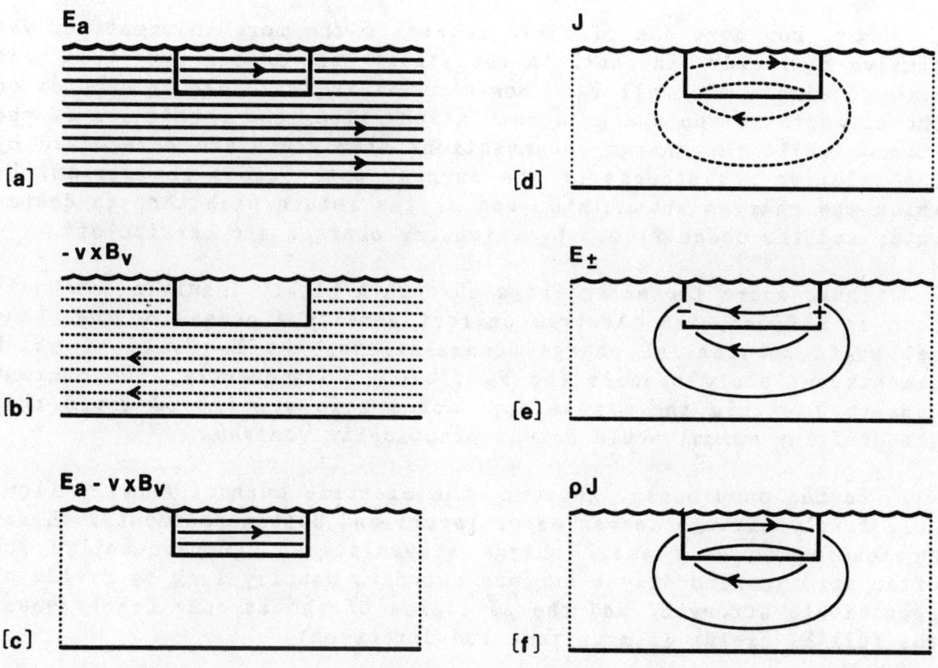

Fig. 3. Motional electric field of ocean current observed in reference frame of drifting animal. (a) $\mathbf{E_a}$ field imposed upon fish, assuming $\mathbf{E_e} = \mathbf{0}$, (b) $- \mathbf{v} \times \mathbf{B_v}$ field induced by apparent motion of outside world, (c) vector sum of imposed $\mathbf{E_a}$ field and induced $- \mathbf{v} \times \mathbf{B_v}$ field, (d) current density field $\mathbf{J}$, (e) $\mathbf{E_\pm}$ field of charge accumulations, and (f) ohmic $\rho\mathbf{J}$ field, equal to vector sum of imposed $\mathbf{E_a}$ field, induced $- \mathbf{v} \times \mathbf{B_v}$ field, and $\mathbf{E_\pm}$ field of charge accumulations.

Along its path, the electric current develops an ohmic field $\rho\mathbf{J}$, where ρ denotes the resistivity of the medium (Fig. 3f). The $\rho\mathbf{J}$ is the force needed to overcome the viscous drag of the charge carriers. The current density $\mathbf{J}$ adjusts itself so that the viscous drag just balances the total electrical force. Thus, outside the stream, $\rho\mathbf{J}$ equals the $\mathbf{E}_\pm$ field of the charge accumulations, driving the external current density $\mathbf{J}$, whereas in the stream, $\rho\mathbf{J}$ equals the sum of the $\mathbf{E}_a$ field, which drives the current density $\mathbf{J}$, and the $\mathbf{E}_\pm$ field of the charge accumulations, which tends to diminish $\mathbf{J}$.

Because the local $\rho\mathbf{J}$ field depends on the resistivities of the water and the fish, it certainly is not uniform, even when $\mathbf{E}_a$ is. Therefore, integrated around any of the measuring circuits at rest in the animal's frame of reference, the $\rho\mathbf{J}$ can lead to measurable potential differences across the receptor epithelia, if the fish's sensory system is sufficiently resistive relative to the seawater medium. It is the strategically distorted $\rho\mathbf{J}$ field of the stream, not the nearly uniform $\mathbf{E}_a$ field, that the drifting animal detects.

But, how does the $\rho\mathbf{J}$ field relate to the more informative, yet elusive $\mathbf{E}_a$ field? In fact, in the stream, the $\rho\mathbf{J}$ can vary from near zero to almost the full $\mathbf{E}_a$. How much $\rho\mathbf{J}$ is less than $\mathbf{E}_a$ depends on the strength of the charge accumulations along the boundaries of the stream, while the charge accumulations themselves are determined by the relative resistances of the current path across the stream, by which the charges accumulate, and of the return path through deeper water and the ocean floor, by which the charges are carried off.

Thus, where the water flows through a highly insulated channel, such as a fjord, the electric current density $\mathbf{J}$ causes on the channel walls substantial charge accumulations, the $\mathbf{E}_\pm$ field of which almost completely cancels the $\mathbf{E}_a$ field. Consequently, the current density $\mathbf{J}$ within the stream approaches zero and the $\rho\mathbf{J}$ field that the drifting animal would detect practically vanishes.

In the open ocean, however, the electric current usually flows more freely through deeper water layers and bottom sediments, thereby leading to only small charge accumulations. Consequently, the often strong, wind-driven surface currents usually lack $\mathbf{E}_\pm$ fields of appreciable strength, and the $\rho\mathbf{J}$ fields of the streams nearly equal the full $\mathbf{E}_a$ fields in magnitude and direction.

Under transformation from the animal's to the earth's frame of reference, $\rho\mathbf{J}$ remains virtually the same for the low speeds considered. That is, whether the situation is analyzed in the animal's or the earth's frame of reference, the drifting fish always detects just the $\rho\mathbf{J}$ field. Fortunately, $\rho\mathbf{J}$ attains its greatest strength in the open ocean, where orientational cues are needed most.

Interacting with the horizontal component of the earth's magnetic field $\mathbf{B}_h$, ocean currents also give rise to vertically directed $\mathbf{v} \times \mathbf{B}_h$ fields in the earth's frame, and corresponding $\mathbf{E}_a$ fields in the animal's frame. The ensuing $\rho\mathbf{J}$ fields remain weak, however, as the air-water interface prevents sizeable current densities from developing. Thus, the $\rho\mathbf{J}$ fields of wind-driven ocean currents result mainly from interaction with the vertical component of the earth's magnetic field and are predominantly horizontal in direction.

INTERPRETATION OF FIELDS INDUCED BY OCEAN CURRENTS

The question remains whether the drifting animal, by sensing the ohmic $\rho\mathbf{J}$ field in its own frame of reference, really can establish the magnitude and direction of its velocity $\mathbf{v}$ with respect to the earth's frame of reference. According to the principle of relativity, the animal cannot,[11] at least not without further information about just that part of the world relative to which it wishes to determine the velocity of its drift.

In the open ocean, the fish may tentatively assume that the $\rho\mathbf{J}$ field is exclusively due to the drift of the water, or $\mathbf{E}_e = \mathbf{0}$, and that deeper water layers offer suitable, low-ohmic return paths, so that little charge accumulation occurs and $\rho\mathbf{J}$ nearly attains the full strength of $\mathbf{E}_a$. If correct, the animal could infer the existence of another frame of reference, that of the deeper water, in which the $\mathbf{v} \times \mathbf{B}_v$ it induces about equals the $\rho\mathbf{J}$ it detects.

The drifting fish might furthermore assume that deeper water layers, forming the main part of the return path, are at rest in the earth's frame of reference. If correct again, the animal could, to good approximation, take the reference frame in which the $\mathbf{v} \times \mathbf{B}_v$ it induces equals the $\rho\mathbf{J}$ it detects as that of the earth, and identify the velocity $\mathbf{v}$, derived from the $\rho\mathbf{J}$ estimate of $\mathbf{v} \times \mathbf{B}_v$, with the strength and direction of its drift with respect to the bottom. An electric method to determine $\mathbf{B}_v$ will be introduced in next section.

By thus inferring its velocity $\mathbf{v}$ relative to the earth's frame of reference, the fish does not violate the principle of relativity, since the interpretation of the ohmic $\rho\mathbf{J}$ field in terms of its drift depends entirely on the assumptions that the fish has to make about the ambient electric-field conditions in the earth's frame of reference, the relative resistances of the stream and the return path, and the motion of deeper water layers with respect to ground.

If the resistance of the return path is not negligible compared to the resistance across the stream[10], then the field of the charge accumulations partially cancels the original $\mathbf{E}_a$ field, leaving a lesser $\rho\mathbf{J}$ for the fish to detect. Thus, to estimate the velocity of its drift correctly, the animal must know the degree of field cancelation due to the finite resistance of the return path.

If, in the earth's frame of reference, the ambient electric field $\mathbf{E}_e$ differs from zero due to the presence of extraneous fields of various nature,[3] the fish would, without correction, make a corresponding error in its inferred drift velocity $\mathbf{v}$. Moreover, if deeper water layers are not at rest, the inferred drift would not refer to ground, but rather to the vertically averaged, resistivity-weighted velocity of the water and bottom sediments.[12]

In deep-ocean waters, the $\rho \mathbf{J}$ fields of fast, wind-driven surface currents may provide the drifting animal with excellent estimates of its ground velocity, if indeed the ambient electric field $\mathbf{E}_e$ in the earth's frame of reference and the field $\mathbf{E}_{\pm}$ due to the charge accumulations contribute little to $\rho \mathbf{J}$. When the deeper water layers are moving as well, although with lesser velocity, the animal's drift would refer to them rather than to ground.

In coastal waters, the conditions are usually less ideal. Yet, the animal can make equally good use of the $\rho \mathbf{J}$ field it detects, if it occasionally checks the actual relation between $\rho \mathbf{J}$ and its drift velocity $\mathbf{v}$ by diving to sufficient depth or, preferably, to the bottom, as sharks and other fishes often seem to do. Even when the magnitude of the inferred drift velocity $\mathbf{v}$ is not very accurate, its general direction may still present invaluable orientational cues.

In protected waters, the $\rho \mathbf{J}$ fields of nearby ocean currents may be detected even in the absence of any local water flow. By diving to the bottom, the fish could readily identify the $\rho \mathbf{J}$ with the ambient electric field $\mathbf{E}_e$ and treat it as such.[6] Thus, the motional-electric fields of ocean currents offer potentially important orientational cues in those waters as well.

In short, the subtle difficulties of detecting and interpreting the oceans' motional electric fields must not be taken too lightly. Even though the animals receive detectable signals from their drift with the water, they still need, in order to interpret those signals, further knowledge of the physical conditions about which they wish to gather more specific information.

Nevertheless, despite these complexities, the motional-electric fields of ocean currents offer reliable orientational cues for animals that are familiar, either by exploring or by experience, with their physical environment. Actually, sharks, skates, and rays have exhibited a cunning ability to orient with regard to electric fields simulating those they normally encounter in the oceanic environment, as will be detailed in a later section.

Thus, sharks, skates, and rays may have correctly applied the electromagnetic theory long before it was formulated scientifically. Naturally, the fishes need not have any knowledge of the physical principles on which they base their behavior, despite the author's objectionable manner of speaking, which often seems to impart great wisdom to the animals that he respects so highly.

ELECTRIC FIELD OF SWIMMING FISH

When, in the reference frame of the earth, a fish swims with a horizontal velocity **v** relative to the water, while the water is at rest, the animal induces, by interaction with the horizontal component of the earth's magnetic field $\mathbf{B}_h$, a motional electromotive force per unit length $\mathbf{v} \times \mathbf{B}_h$, parallel to its dorsoventral axis.[3] Interaction with the vertical component of the earth's magnetic field $\mathbf{B}_v$ leads to similar physical results, although of different biological import, as will be explained later in this section.

Even though swimming relative to the water, the fish is at rest in its own frame of reference (Fig. 4), in which, according to equation (7), it is exposed to an electric field $\mathbf{E}_a$ (Fig. 5a), where $\mathbf{E}_a$ about equals the $\mathbf{v} \times \mathbf{B}_h$ that the animal induces in the frame of the earth, assuming $\mathbf{E}_e$ to be zero. As observed by the fish, however, the whole world is moving with a velocity $-\mathbf{v}$, inducing an electromotive force per unit length $- \mathbf{v} \times \mathbf{B}_h$ (Fig. 5b), which precisely cancels the $\mathbf{E}_a$ field everywhere, except in the fish (Fig. 5c).

Thus, in the earth's frame, the charge carriers of the fish are subject to an electromotive force per unit length $\mathbf{v} \times \mathbf{B}_h$, due to the motion of the fish. In the animal's frame, however, the same charge

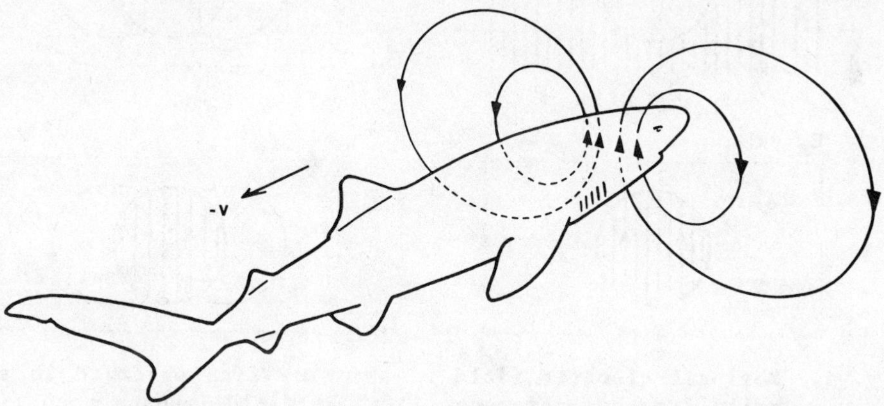

Fig. 4. In reference frame of swimming fish, the animal itself is at rest, while the whole world appears to move with velocity $-\mathbf{v}$. The electric current, induced by interaction with the horizontal component of the earth's magnetic field $\mathbf{B}_h$, generates the $\rho\mathbf{J}$ field that the swimming fish detects. The current flows across the body of the animal to complete its circuit through the seawater medium. Fish heading east.

carriers are subject to a purely electric field $\mathbf{E_a}$, imposed upon the fish. Although of outside origin, the $\mathbf{E_a}$ field is under control of the swimming fish. Moreover, the medium relative to which the motion takes place is directly accessible to the animal.

If the fish's velocity $\mathbf{v}$ and the magnetic field $\mathbf{B_h}$ are independent of time and locally uniform, the $\mathbf{E_a}$ field in the vicinity of the animal is uniform as well and, integrated around any measuring loop at rest in the reference frame of the animal, as such does not yield measurable potential differences across the thin electroreceptive membranes. Consequently, the $\mathbf{E_a}$ field itself escapes detection by the swimming fish. Note that, to measure the $\mathbf{E_a}$ field directly,

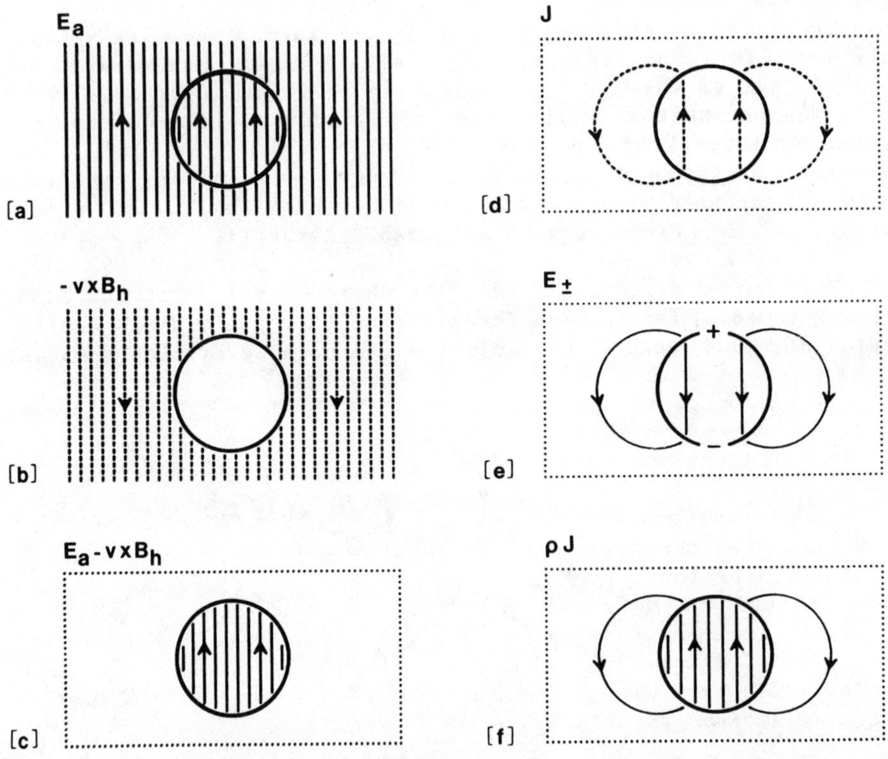

Fig. 5. Motional electric field of swimming fish observed in animal's frame of reference. (a) $\mathbf{E_a}$ field imposed upon fish, if in earth's frame $\mathbf{E_e}$ and velocity of water are zero, (b) $-\mathbf{v} \times \mathbf{B_v}$ field induced by apparent motion of water, (c) sum of imposed $\mathbf{E_a}$ field and induced $-\mathbf{v} \times \mathbf{B_v}$ field, (d) current density field $\mathbf{J}$, (e) $\mathbf{E_{\pm}}$ field of charge accumulations, and (f) ohmic $\rho\mathbf{J}$ field, equal to sum of imposed $\mathbf{E_a}$ field, induced $-\mathbf{v} \times \mathbf{B_v}$ field, and $\mathbf{E_{\pm}}$ field of charge accumulations.

the entire path of integration, even if partially outside the fish, must be at rest in the frame of the animal.

However, the $\mathbf{E_a}$ field causes an electric current of density $\mathbf{J}$ to flow within the fish (Fig. 5d), giving rise to electrical charge accumulations at the animal's ventral and dorsal surfaces, where the $\mathbf{E_a} - \mathbf{v} \times \mathbf{B_h}$ field is discontinuous (Fig. 3c). The electric field $\mathbf{E_\pm}$ of these charges provides the driving force for the electric current to complete its loop through the seawater medium (Fig. 5e). Since the resistance of the return path is much lower than the resistance across the fish, only weak charge accumulations are required to produce a continuous current density $\mathbf{J}$ around the circuit.

The current density $\mathbf{J}$ develops an ohmic $\rho\mathbf{J}$ field in the animal and the seawater medium (Fig. 5f). In the fish, the $\rho\mathbf{J}$ field equals the sum of the $\mathbf{E_a}$ and $\mathbf{E_\pm}$ fields, whereas in the water it equals just the $\mathbf{E_\pm}$ field. Unlike $\mathbf{E_a}$, the $\rho\mathbf{J}$ field depends on the resistivity distribution of the seawater medium and the tissues of the fish, and is highly non-uniform. Therefore, integrated around above-mentioned measuring loops, the $\rho\mathbf{J}$ can, and usually will, lead to detectable potential differences, given that the sense organs are sufficiently resistive. Thus, the stimulus derives again from the $\rho\mathbf{J}$ field.

Averaged over the swimming fish, the $\rho\mathbf{J}$ nearly equals $\mathbf{E_a}$, for the $\mathbf{E_\pm}$ field of the charge accumulations, opposing $\mathbf{E_a}$, remains weak at all times. This favorable situation is due to the relatively high resistance across the body of the fish and the low resistance of the return path through the seawater medium.[6] In other words, the efficiency of the proposed electromagnetic orientation mechanism crucially depends on the short-circuiting action of the seawater medium. For marine elasmobranchs, it is close to one hundred percent.

Upon transformation from the animal's to the earth's frame of reference, the change in $\rho\mathbf{J}$ field is negligible. While in the frame of the fish $\rho\mathbf{J}$ tends to $\mathbf{E_a}$, in the earth's frame it tends to $\mathbf{v} \times \mathbf{B_h}$, where $\mathbf{E_a}$ and $\mathbf{v} \times \mathbf{B_h}$ refer to the same field observed in different reference frames. In previous papers, the situation was analyzed in the frame of the earth, with identical results, of course. Biologically, however, the real challenge is to explain electromagnetic orientation from the fish's point of view.

By interaction with the vertical component of the earth's magnetic field $\mathbf{B_v}$, the fish also induces a $\mathbf{v} \times \mathbf{B_v}$ parallel to its left-right, or transverse body axis. The corresponding $\mathbf{E_a}$ field is again nearly matched by the average $\rho\mathbf{J}$ field that the animal detects, as hardly any charges accumulate, in this instance, on the lateral surfaces of the fish. Thus, the $\rho\mathbf{J}$ fields resulting from interaction with the horizontal and vertical components of the earth's magnetic field are physically very similar, although different in direction.

INTERPRETATION OF FIELDS INDUCED BY SWIMMING FISH

Could the swimming fish, by sensing the strength and direction of the $\rho\mathbf{J}$ field in its own frame of reference, really determine its magnetic compass heading in the earth's frame of reference? The animal might, according to the principle of relativity, but only if it has further information about the world relative to which it is moving. Such information, however, is readily available in this case, as the swimming fish is not only in direct contact with that world, but also controls its velocity with respect to it.

If the $\rho\mathbf{J}$ field that the animal detects results solely from its own swimming motion, and if the animal takes into account that the seawater medium offers an excellent return path, so that virtually no charges accumulate and the $\rho\mathbf{J}$ closely approaches the strength of $\mathbf{E}_a$, then the fish may conclude that the $\rho\mathbf{J}$ it detects nearly equals the $\mathbf{v} \times \mathbf{B}$ it induces in the earth's frame of reference, considering, for the moment, $\mathbf{E}_e$ to be zero and the water to be at rest.

Consequently, by identifying the vertical $\rho\mathbf{J}$ in its own frame of reference with the $\mathbf{v} \times \mathbf{B}_h$ in the frame of the earth, the animal may, knowing its swimming velocity $\mathbf{v}$, determine the component of the horizontal magnetic field perpendicular to the direction of motion. Furthermore, by its swerving mode of swimming, the fish may gather from those components the strength and direction of the resultant horizontal magnetic field and thereby its compass heading.[6]

In short, given the speed of swimming and the low resistivity of the seawater medium, the fish need not infer, but may actually determine the strength and direction of the ambient magnetic field. The normal swerving mode of swimming suffices to remove any directional ambiguity. Except for the direction of gravity, no further information is required to comply with the principle of relativity.

To verify whether the $\rho\mathbf{J}$ field is solely due to its swimming motion, the fish could, for example, stop momentarily or seek the compass direction in which the field is weakest, that is, by heading due north or due south. If the remaining field is non-zero, then it must be corrected for. Although the fish may not, in fact, follow these particular procedures, they show the feasibility of separating the cues pertaining to the compass heading and to the drift.

The electric fields most relevant to the magnetic compass sense derive from the horizontal component of the earth's magnetic field and are dorsoventrally directed, whereas the fields due to the animal's drift derive mainly from the vertical component of the earth's magnetic field and are usually horizontal in direction. Moreover, the stimuli that the fish receives from its own locomotion and from its drift are entirely different functions of its mode of swimming.

The electrical stimuli that the fish receives from interaction with the vertical component of the earth's magnetic field are at all times transversely, or left-right directed, whereas those from ocean currents change in direction with respect to the body axes when the fish turns. That marine stingrays can really distinguish between the two types of motional-electric fields has been demonstrated in behavioral tests, as will be reported in a later section.

As a word of caution, it must be noted that only the instantaneous, uniform, translational velocity of the fish is considered in this article. Certainly, while swerving the fish changes the direction of its velocity from moment to moment, thus probing, as it were, its physical environment. However, the electric fields induced by these turning movements as such do not seem to constitute physically adequate stimuli for the electroreceptors of elasmobranch fishes.[6] The consequences of relative motion between different body parts during swimming remain to be evaluated.

By sensing the horizontal as well as the vertical component of the earth's magnetic field, the fish may also determine the magnetic inclination and, thereby, the local magnetic latitude.[3] The animal also needs the strength of the vertical magnetic component in order to interpret the electric fields of its drift with the water, as was pointed out before. Thus, the electromagnetic orientation mechanism is not only highly versatile, but also largely self-sufficient.

ELECTRIC VERSUS MAGNETIC DETECTION

If elasmobranch fishes use the ampullae of Lorenzini to sense their magnetic compass heading electrically, should this be referred to as magnetic or electric detection? Certainly, the magnetic field in the animal's frame of reference does not differ noticeably from the magnetic field in the earth's frame at the fish's speed of swimming. Nonetheless, the proposed orientation theory does not require the electrical charges of the receptor epithelia, on which the stimulus acts, to move in the reference frame of the animal. And, without moving, electrical charges are not affected by the mere presence of a static magnetic field. Thus, from the point of view of the recipient fish, electric detection is the proper term.

If the earth's magnetic field as such is not detected, where then are the sources of the purely electric fields that the animals orient to? It is learned from the special theory of relativity that electric currents, having zero charge density in the earth's frame of reference, appear negatively charged to an observer moving in the direction of the current, and positively charged to an observer moving in the opposite direction.[13] Consequently, a current loop appears negative in the one half and positive in the other to the moving observer,[7] thereby creating an electric dipole field (Fig. 6

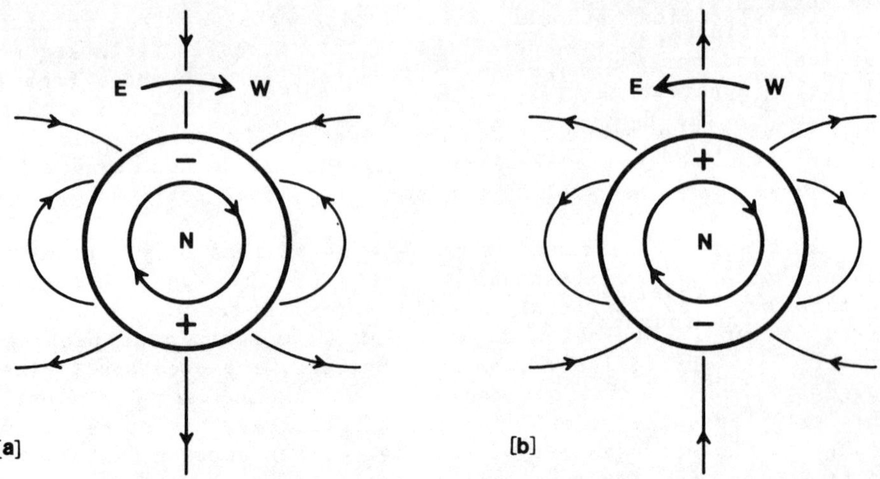

Fig. 6. Earth's sphere viewed from a position over the north pole. To a fish drifting or swimming along the equator (a) in westerly and (b) in easterly direction, the current loops generating the earth's magnetic field appear electrically charged, producing part of the vertical electric field that the animal may infer from the $\rho\mathbf{J}$ field it detects. Other apparent charges not discussed here (see text) decrease the field strength at the position of the fish by a factor two. Note that the fields depend on the direction of motion.

and 7). The apparent charges on the parts of the current loop that the observer approaches or recedes from perpendicularly contribute to the field as well, but will not be detailed here.[15]

According to geophysical theory,[14] the earth's magnetic field results from electric currents, which can be thought to flow in concentric circles from east to west around the N-S axis of the globe. Consequently, for a fish swimming or drifting along the equator in westerly direction, the part of the current loop on the near side of the earth appears negatively charged and the part of the loop on the far side of the earth appears positively charged, producing a global electric dipole field $\mathbf{E}_a$ pointing vertically downwards (Fig. 6a). For a fish heading east, the electric field $\mathbf{E}_a$ points vertically upwards at the same equatorial position (Fig. 6b).

When a fish swims or drifts in polar waters, where feasible, the current loops creating the earth's magnetic field appear negatively charged on the left and positively charged on the right relative to the direction of motion in the northern hemisphere, and positively charged on the left and negatively charged on the right in the southern hemisphere, thereby producing electric $\mathbf{E}_a$ fields that are horizontally directed at the position of the fish (Fig. 7a, b). At mid-latitudes, the animal receives electric $\mathbf{E_a}$ fields having both vertical and horizontal components. The $\mathbf{E_a}$ fields of the relativistically apparent charges, including those not discussed here, are the same as the $\mathbf{E_a}$ fields derived from the transformation formulas and again correspond to the $\mathbf{v} \times \mathbf{B}$ fields in the earth's frame.

If, in special relativity, magnetic fields may be considered to express the relativistic aspects of electric fields, what then could be meant with the term magnetic detection? The relativistic electrical effects described by the aid of magnetic fields occur only when both the source charges and the test charges are moving in the rest frame of the observer. In the frame of the swimming fish, the source charges, those producing the earth's magnetic field, are moving, but the test charges, those of the electroreceptors, need not in the proposed theory. The equivalent[15] of circulating, or rather spinning test charges occurs, however, in the magnetite crystals of various sediment bacteria, causing these organisms to orient to the magnetic field present in their own frames of reference.[16] Thus, for the bacteria, magnetic detection would be the proper term.

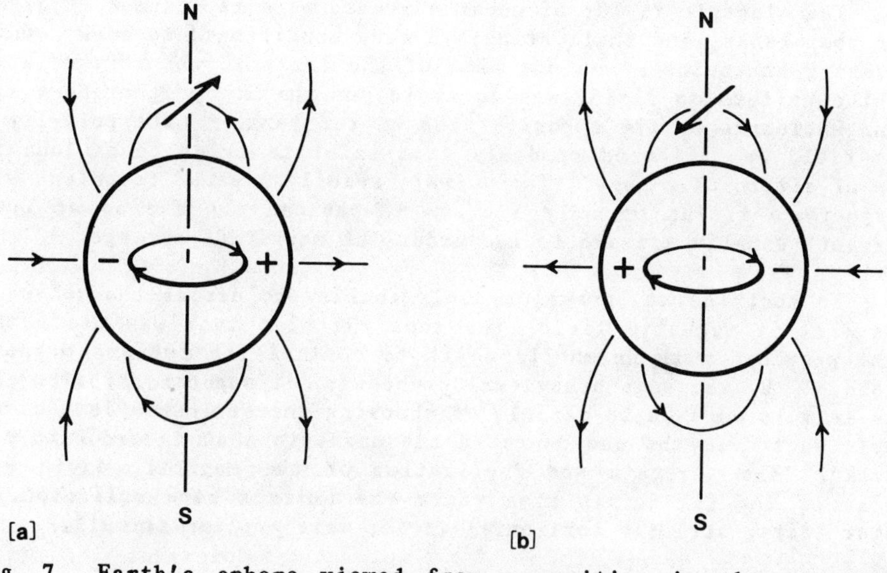

Fig. 7. Earth's sphere viewed from a position in the equatorial plane. When a fish swims over the north pole in any direction, as in (a) or (b), the loop currents generating the earth's magnetic field appear electrically charged, producing part of the horizontal electric field that the fish may infer from the $\rho\mathbf{J}$ field it detects. The apparent charges not discussed increase the field strength at the position of the fish by a factor two. Note that the fields differ in direction in the northern and southern hemispheres.

IMPLEMENTATION OF ORIENTATION MECHANISMS

Elasmobranch fishes are extremely sensitive to electric fields, showing meaningful behavioral responses at voltage gradients as low as 5 $nV.cm^{-1}$ in the dc to 8 Hz frequency range.[17,18] The electric sense enables the fishes to detect the bioelectric fields of prey, the motional electric fields of ocean currents, and the fields that the fishes themselves induce while swimming. These fields are not only potentially of great biological importance, but also well within the operating range of the elasmobranchs' electric sense. Other, less-well studied fields remain to be evaluated.[3]

Bioelectric fields of prey fish and simulations thereof have been presented to sharks, skates, and rays both in the laboratory[19] and at sea.[18] When motivated by odor, these largely nocturnal predators strike at the electric-field source from distances up to half a meter, even when receiving dc voltage gradients of 5 $nV.cm^{-1}$ or less. As a rule, the attacks are remarkable fast, very accurate, and, once initiated, executed without further course correction.[20] Night-time observations at sea were made on bottom-dwelling dogfish and pelagic blue sharks, off Cape Cod, Massachusetts.

The electric fields of ocean currents were reproduced in circular test tanks, and small stingrays were conditioned to seek food by entering an enclosure on one side of the tank, to the left relative to the uniform dc field, and to avoid punishment by ignoring a similar enclosure on the opposite side of the tank.[18] The polarity of the field was selected randomly from trial to trial to exclude the use of alternative cues. The animals readily learned to orient with respect to fields of only 5 $nV.cm^{-1}$, whereas the fields of ocean currents usually are one to two orders of magnitude stronger.[3]

To test the stingrays on their ability to direct themselves to the earth's magnetic field, the same circular tank was used, this time provided with Helmholtz coils to control the ambient magnetic field.[21] By the same behavioral procedure, it was demonstrated that the animals were quite capable of choosing the enclosure in the magnetic east over the enclosure in the magnetic west in order to earn reward. The strength and inclination of the magnetic field were those of the California area where the animals were collected. In later tests, strictly horizontal fields were applied as well.[6]

Although the latter experiments proved that the stingrays are capable of detecting the direction and polarity of the earth's magnetic field, as expected theoretically, they were not designed to reveal the physical nature of the orientation mechanism. However, that the animals are familiar with the fields they induce themselves follows indirectly from the electric-field experiments, in which the stingrays learned to orient to imposed electric fields up to twenty times weaker than the fields induced by their own locomotion.[18] Research to verify the electromagnetic mechanism is in progress.

The elasmobranchs' behavioral responses to electric fields are mediated by the ampullae of Lorenzini,[22] each of which consists of a small pore in the skin, giving access to an often long, jelly-filled canal, ending in a blind sensory swelling, the ampulla proper. The canal wall has high resistance, while the gelatinous core is a good conductor.[23,24] The ampullae proper are clustered in a few discrete capsules, allowing the fishes to detect the fields of interest differentially and to suppress undesired, common-mode fields.[3] The sense organs are high-ohmic input devices, while the skin and body tissues are, on the average, about a hundred times more resistive than the seawater medium.[6] In short, the ampullae of Lorenzini seem well-suited for the proposed orientational functions.

The receptor cells in the wall of the ampulla proper are lodged in between supporting cells, together with them forming a simple, single-layered sensory epithelium. The apical membrane of the re-

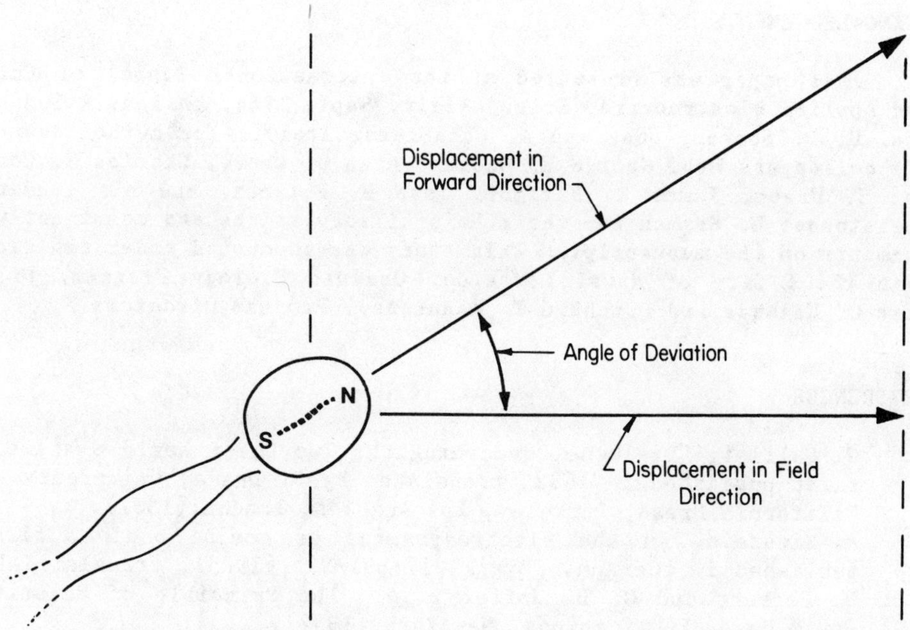

Fig. 8. When separated from the substrate, magnetic bacterium is forced to swim along earth's magnetic field lines, although it departs randomly from strict alignment due to thermal agitation (angle of deviation). Indicated are the string of magnetite crystals and the polarity of the northern-hemisphere bacteria tested on their time of travel over a measured distance (between dashed lines, not to scale). Average cell diameter 1.75 μm. From Kalmijn, 1982, ref. 5.

ceptor cell detects the electrical stimulus and passes the excitation on to the basal membrane, from where, by chemical transmission, the signal crosses the synapse to act on the afferent nerve fiber.[25] At a 5-nV.cm^{-1} threshold field in the water, the apical membrane of the receptor cell responds to a potential difference of, at most, 50 nV, which is of the same order as the thermal noise of the detection circuit.[6] This small patch of membrane thereby shows the highest electrical sensitivity presently known in the animal kingdom.

By comparison, the magnetite-based orientation mechanism of magnetic bacteria is extremely simple, although amazingly effective. Rather than detecting the magnetic field and acting accordingly, the bacteria evidently are forced to swim along the earth's magnetic field lines when separated from the sediment.[16] The passive nature of the orientation mechanism has been scrutinized by analyzing the migration rates of single, living cells as a function of the ambient magnetic field strength (Fig. 8).[5] The role of magnetite in the orientation of higher organisms remains shrouded in uncertainty.

ACKNOWLEDGEMENTS

This paper was presented at the International School of Pure and Applied Biostructure, Erice, Italy, Sept. 1984, on invitation of Drs. H. P. Schwan, USA, and A. Chiabrera, Italy. The author thanks his colleagues Drs. George E. Backus, Alan D. Chave, Charles S. Cox, Per S. Enger, James R. Enright, Jean H. Filloux, and his student Christopher L. Seaman for their helpful suggestions and constructive comments on the manuscript. This study was conducted under contract with the Office of Naval Research, Oceanic Biology Program, Drs. Eric O. Hartwig and Bernhard J. Zahuranec, Program Directors.

REFERENCES

1. G. Galilei, "Dialogue Concerning the Two Chief World Systems," first published in 1632, translated by S. Drake, University of California Press, Berkeley, Los Angeles, London (1967).
2. A. Einstein, On the electrodynamics of moving bodies, first published in the Ann. Physik, 17:891-921 (1905), translated by W. Perrett and G. B. Jeffery, in: "The Principle of Relativity," Dover Publications, New York (1952).
3. A. J. Kalmijn, The detection of electric fields from inanimate and animate sources other than electric organs, in: "Handbook of Sensory Physiology," Vol. III/3, A. Fessard, ed., pp. 147-200, Springer Verlag, Berlin, Heidelberg, New York (1974).
4. A. J. Kalmijn, Electric and magnetic sensory world of sharks, skates, and rays, in: "Sensory Biology of Sharks, Skates, and Rays," E. S. Hodgson and R. F. Mathewson, eds, pp. 507-528, Government Printing Office, Washinton DC (1978).

5. A. J. Kalmijn, Biophysics of geomagnetic field detection, IEEE Trans. Magnetics, 17:1113-1124 (1981).
6. A. J. Kalmijn, Theory of electromagnetic orientation: a further analysis, in: "Comparative Physiology of Sensory Systems," L. Bolis, R. D. Keynes, and S. H. P. Maddrell, eds., pp. 525-560, Cambridge University Press, Cambridge (1984). This and the present paper were together announced in ref. 5 under the title "A relativistic approach to animal orientation."
7. W. K. H. Panofsky and M. Phillips, "Classical Electricity and Magnetism," Addison-Wesley, Reading, Massachusetts (1962).
8. M. Faraday, Experimental researches in electricity, 1st and 2nd series, Phil. Trans. Roy. Soc. Lond., 122(1):125-194 (1832).
9. R. S. Shankland, Michelson-Morley experiment (for quote of Einstein's statement), Am. J. Phys., 32:16-35 (1964).
10. M. S. Longuet-Higgins, M. E. Stern, and H. Stommel, The electrical field induced by ocean currents and waves, with applications to the method of towed electrodes, Papers Phys. Oceanog. Meteorol., 13:1-37 (1954).
11. J. A. Stratton, "Electromagnetic Theory," McGraw-Hill, New York and London (1941).
12. T. B. Sanford, Motionally induced electric and magnetic fields in the sea, J. Geophys. Res., 76:3476-3492 (1971).
13. R. Resnick, "Introduction to Special Relativity," Wiley, New York, London, Sydney (1968), and texts listed herewith.
14. F. D. Stacey, "Physics of the Earth," Wiley, New York, Sydney, Toronto (1969)
15. E. M. Purcell, "Electricity and Magnetism," Berkeley Physics Course, Vol. 2, McGraw-Hill, New York and London (1965).
16. R. P. Blakemore, Magnetotactic bacteria, Science, 190:377-379 (1975).
17. A. J. Kalmijn, Electro-perception in sharks and rays, Nature, 212:1232-1233 (1966).
18. A. J. Kalmijn, Electric and magnetic field detection in elasmobranch fishes, Science, 218:916-918 (1982).
19. A. J. Kalmijn, The electric sense of sharks and rays, J. Exp. Biol., 55:371-383 (1971).
20. A. J. Kalmijn and M. B. Weinger, An electrical simulator of moving prey for the study of feeding strategies in sharks, skates, and rays, Ann. Biomed. Eng., 9:363-367 (1981).
21. A. J. Kalmijn, Experimental evidence of geomagnetic orientation in elasmobranch fishes, in: "Animal Migration, Navigation, and Homing," K. Schmidt-Koenig and W. T. Keeton, eds., pp. 347-353, Springer Verlag, Berlin, Heidelberg, New York (1978).
22. S. Dijkgraaf and A. J. Kalmijn, Untersuchungen ueber die Funktion der Lorenzinischen Ampullen an Haifischen, Z. Vergl. Physiol., 47:438-456 (1963).
23. R. W. Murray, The response of the ampullae of Lorenzini of elasmobranchs to electrical stimulation, J. Exp. Biol., 39:119-128 (1962).

24. B. Waltman, Electrical properties and fine structure of the ampullary canals of Lorenzini, Acta Physiol. Scand., 66, Suppl. 264, pp. 1-60 (1966).
25. M. V. L. Bennett and W. T. Clusin, Physiology of the ampulla of Lorenzini, the electroreceptor of elasmobranchs, in: "Sensory Biology of Sharks, Skates, and Rays," E. S. Hodgson and R. W. Mathewson, eds., pp. 483-505, Government Printing Office, Washington DC (1978).

DIELECTRIC PROPERTIES OF WATER IN BIOLOGICAL SOLUTIONS

E.H. Grant, V.E.R. McLean*, N.R.V. Nightingale+
and C. Gabriel
Physics Department, King's College
London, U.K.,WC2R 2LS

INTRODUCTION

Interest in the dielectric properties of water in biological solutions dates back to the classic work of Oncley in the 1930's which is reported in Cohn and Edsall (1943) and elsewhere. In these studies the relative permittivity and conductivity of various protein solutions were measured at radiofrequencies and interpreted in terms of the molecular dipole moment and the shape and hydration of the protein molecule. In the 1960's these kinds of measurement were extended in to the microwave region by Schwan (1965), Grant (1965) and others and since then the dielectric method of measuring hydration has been improved continuously as more sophisticated experimental techniques and computer methods have become available. The strengths and limitations of the dielectric method have been summarised fairly recently by Grant (1982) and the purpose of the present paper is to describe progress made since then and to give some indications in respect of the future of the techniques.

ASSUMPTIONS UNDERLYING THE DIELECTRIC METHOD OF MEASURING HYDRATION IN BIOLOGICAL MATERIAL

The dielectric behaviour of pure water can be characterised by a single Debye dispersion with a dielectric increment (Δ) of around 75 and a relaxation frequency of 25GHz at 37°C (Grant et al.1978). The mechanism underlying the process of dielectric relaxation in water is the breaking of a hydrogen bond, which

* Now at Lloyds Register of Shipping,Fenchurch St.London EC3
+ Now at Mercury Communications Ltd,Long Acre, London WC2

corresponds to an activation enthalpy of about 16 kJ/mole. Therefore since the value of the relaxation frequency depends upon the potential barrier that has to be surmounted, which in turn is directly related to the strength of the intermolecular forces, it follows that changes in the local environment of a water molecule will affect its rotational correlation time. In other words a given molecule will have a different dielectric relaxation time if the adjacent molecules are different.

In an aqueous protein solution some of the water molecules immediately adjacent to the macromolecules will be under the influence of strong intermolecular forces and will therefore have a longer rotational correlation time (lower relaxation frequency) than those molecules in the bulk of the liquid. These latter water molecules should, from a dielectric standpoint, behave similarly to those in pure water. It would, however, be too simple a picture to regard a protein solution as necessarily containing only two types of water molecule with just two correlation times. The very tightly bound water will clearly not be able to rotate at 25GHz; this may be termed 'irrotationally' bound at this frequency. In addition there will be water molecules which are influenced by the protein molecules but which may possess a degree of rotational freedom higher then that associated with the irrotationally bound water. Such water will have a relaxation frequency transitional between that of irrotationally bound water and pure water. Moreover there could, in theory, be several different types of irrotationally bound water and non-irrotationally bound water, which would further complicate the situation by causing overlapping dispersion regions. This means that a dielectric relaxation region observed for any aqueous component in a biological solution will relate to the average behaviour of a group of similarly bonded water molecules which are present in the system. For example if the solution is dilute then it would be reasonable to expect that a substantial fraction of the water present will behave dielectrically as pure water and will give rise to one dispersion region. As for the remaining water fraction it may be possible to resolve it into more than one component; however in other cases the complexity of the biological material and the lack of resolution of the measuring equipment and in the techniques of analysis may mean that all of the non-bulk water has to be treated as one entity.

Another problem is the presence of other mechanisms of polarisation in addition to that due to the aqueous component. For example in the frequency range 10MHz-1GHz, where dielectric measurements have suggested the presence of bound water relaxation (Pennock and Schwan 1969, Essex et al.1977), there may also occur, in principle, other dispersion processes such as side chain rotation or proton fluctuation. There may also be contributions from the tail of the dispersions due to rotation of the macromolecules themselves or, with whole biological tissue, from membrane polarisation. The problems of interpreting unambiguously dielectric data on their own have been well explained recently by Foster et al. (1984) who showed how much more rigorous an interpretation is normally possible if dielectric data are combined with those obtained using other techniques such as thermal conductivity studies and NMR. In view of these considerations it is important to evaluate dispassionately what can be achieved by performing dielectric measurements in isolation on biological solutions or biological tissue.

Taking the case of biological solutions (the principal concern of this paper) it is obvious that the simpler the system then the easier it will be to provide an unambiguous interpretation of the dielectric data. For example with an aqueous solution of protein molecules of globular configuration whose shape and size are known from X-ray studies there is sufficient information available to allow certain important characteristics of the protein dispersion to be fixed. By making measurements over a wide range of concentration, frequency and temperature and by varying the pH and ionic strength, any existing polarisation mechanisms other than those due to dipolar rotation of water and protein molecules can be allowed for or eliminated. Considering the specific case of the rotation of water molecules it is reasonable in the light of present knowledge to conclude that any dispersion observed above about 1GHz is due largely, if not entirely, to these. Therefore accurate measurements performed at frequencies in the vicinity of 1GHz and above can show how the behaviour of the water molecules in a biological solution has been modified by the presence of the macromolecules.

Assuming that accurate measurements of permittivity and conductivity are available over a wide frequency range, and given that interfering mechanisms of polarisation have been eliminated or corrected for, it remains to interpret the data in terms of structure. Here again it is greatly advantageous to know the size and shape of the macromolecules present in the solution. If this knowledge is available then dielectric mixture relationships based on Maxwell's equations can be used to calculate the bound water by assuming an appropriate model for the hydrated macromolecule. Otherwise the quantity of water of hydration may have to be estimated by comparison of the dielectric increment of the dispersion, or dispersions, due to the bound water with that due to the bulk water. In the rest of this paper we report recent measurements which, despite being carried out at only one concentration, pH and ionic strength provide new information on the dielectric behaviour of water in myoglobin solution.

DIELECTRIC BEHAVIOUR OF WATER
IN MYOGLOBIN SOLUTION

Results and Analysis

High purity whale myoglobin obtained from Miles Laboratories (PTY) Limited was dissolved in deionised water. The solution, to which no further purification was necessary, was prepared at a concentration of 170mg/ml and controlled at a pH of 6.9. The dielectric measurements were carried out over a frequency range 100kHz-18GHz and at temperatures of 9,15, 25, 35 and 45°C. At each temperature and frequency point determinations were made on at least two independent samples. Between 100kHz-100Mhz measurements of relative permittivity (ε') were made at 21 frequencies using bridge methods that have been described previously (Essex et al.1975). Over the remainder of the frequency range coaxial line techniques were employed (Nightingale et al.1981, Szwarnowski 1982) and measurements of ε' and ε'' were made at 44 different frequencies. Hence at each temperature there are 109 data points over the total frequency range 100kHz-18GHz and the error associated both with ε' and ε'' is typically around 1%.

Although the dispersion characteristics of the myoglobin molecule are well known from previous work (South and Grant 1972) it was decided in the first instance to fit the present dielectric data to the simplest general model, without any preconceived ideas. The first model to be tried was a sum of two Cole-Cole dispersions $F(\Delta_1, f_1, \alpha_1)$ and $F(\Delta_2, f_2, \alpha_2)$ where f is the relaxation frequency, α is the Cole-Cole distribution parameter and the number of the suffix increase as the frequency increases. The fit was satisfactory and the values of the fitted parameters are shown in Table 1. The important conclusion to be drawn here is that in the absence of any other independent evidence the fit shown in Table 1 would be a perfectly acceptable interpretation of the dielectric data. However in the case of a myoglobin solution there are good grounds for stating that the double Cole-Cole representation is incorrect. It is known from X-ray studies that the myoglobin molecule has the shape of an oblate spheroid of axial ratio 2:1 which should exhibit dielectric relaxation behaviour indistinguishable from that of a sphere. Therefore it must conform to a Debye dispersion with its associated single relaxation time. Accordingly a three dispersion fit of the form Debye, Debye, Cole-Cole in order of ascending frequency was tried and a good fit was obtained with a root mean square error comparable to that found for the double Cole-Cole representation, but the parameters appropriate to the low frequency dispersion are altered significantly. For example at 25°C the relaxation frequency and dielectric increment which were 2.9MHz and 27.6 (Table 1) are now 2.2MHz and 18.4

Table 1. Parameters for the double Cole-Cole Fit

Temperature (°C)	Δ_1	f_1(MHz)	α_1
9	31.2±1.5	1.7±0.2	0.28±0.04
15	27.1±0.7	2.3±0.2	0.23±0.03
25	27.6±1.0	2.9±0.2	0.20±0.04
35	28.6±1.5	3.6±0.5	0.34±0.03
45	28.0±1.3	4.2±0.5	0.37±0.03
	Δ_2	f_2(GHz)	α_2
9	67.4±0.3	11.3±0.1	0.06±0.01
15	66.4±0.2	13.5±0.1	0.06±0.01
25	63.1±0.2	17.7±0.2	0.06±0.01
35	60.2±0.2	23.3±0.2	0.05±0.01
45	57.6±0.2	29.2±0.3	0.07±0.01

respectively. These are in excellent agreement with the values obtained previously by South and Grant (1972) from measurements of permittivity and conductivity taken on one bridge apparatus and over a wide range of concentrations and temperatures at frequencies confined to the dispersion of the myoglobin molecules. The effect of this fitting procedure at higher frequencies is to introduce an extra dispersion with (at 25°C) an increment of around 6-7 and with a relaxation frequency of about 20MHz. This is then followed at the top end of the frequency range by the Cole-Cole dispersion, with dielectric parameters equal to within experimental error with those in the lower part of Table 1. Therefore the addition of the extra dispersion which was necessary to obtain the correct parameters for the myoglobin molecule has had no effect on the dispersion centred around 20GHz, which is due to the aqueous component.

Examination of this dispersion (lower part of Table 1) reveals small differences at each temperature between these dielectric parameters and those of pure water. The values of relaxation frequency are lower than those of pure water and the value of the Cole-Cole parameter is significantly greater than zero. For example at 25°C the relaxation frequency of pure water is 19.2GHz compared with 17.7GHz in the Table. With a myoglobin solution of concentration about 17% by weight it would be reasonable to expect on chemical grounds that a large proportion of the water molecules would be unaffected by the myoglobin.

More will be said about this in the next Section. Assuming the presence of an aqueous component having the same dielectric properties as those of pure water the dispersion illustrated in the lower part of Table 1 can be alternatively represented by two dispersions, one of which has its dispersion parameters clamped at the known values for pure water. The result of this exercise is shown in Table 2, which identifies the presence of an extra dispersion at around 3-4GHz. This is of course equivalent to fitting the data over the whole frequency range 100kHz-18GHz to four Debye dispersions and, again, the root mean square deviation of this fit did not differ significantly from that found in the previous fits. The situation for the myoglobin solutions may therefore be summarised as follows. At each temperature, more than a hundred data points were available in the frequency range 100kHz-18GHz and these could be represented adequately, in theory, by a model composed of the sum of two Cole-Cole dispersion. In practice, however, it is known from the shape of the myoglobin molecule that it must exhibit single relaxation time behaviour. It is also reasonable to assume that for a solution of mass concentration as low as 17% a substantial fraction of the water should behave dielectrically as pure water, which exhibits Debye behaviour with values of relaxation time which are known accurately. Including all this known information the best fit becomes four Debye dispersions. At 25°C the values of the dielectric parameters in order of ascending frequency are: dielectric increment;- 18.4, 6-7, 5-6 and 57.6; relaxation frequency:- 2.2MHz, 20MHz, 3-4GHz and 19.2GHz. These figures will now be interpreted in terms of the various states of water present in the system.

Dielectric behaviour of water in myoglobin solution

In a previous study it has been shown (South & Grant 1972) that the measured value of the relaxation time of the myoglobin molecule coupled with the knowledge that the molecule has the shape of an oblate spheroid of axial ratio 2:1, requires the presence of a layer of water of hydration width 0.35nm. This corresponds to 1-2 layers of water molecules surrounding the protein and is equivalent to a hydration factor of about 0.6g/g. Thus when the myoglobin molecule rotates with its characteristic frequency of 2MHz it carries round with it a mass of water more than half that of the myoglobin itself. The fraction of the total water volume occupied by this water of hydration can be estimated as follows. The volume of the anhydrous myoglobin molecule is known from X-ray crystallographic studies to be $15nm^3$. Assuming for simplicity that the shape of the molecule can be approximated to a sphere this would lead to a radius of 1.53nm. Adding 0.35nm for the width of the bound water shell the volume occupied by the water of hydration can be calculated and compared with the total water content for a solution of given

Table 2. Which identifies the presence of an extra water dispersion at around 3-4GHz

Temperature (°C)	Bulk water Δ_B	f_B (GHz) (values clamped)
9	62.0±0.7	12.1
15	60.5±0.9	14.5
25	57.6±0.9	19.2
35	57.0±0.3	24.3
45	55.0±0.3	29.6
	Water influenced by macromolecules	
	Δ_H	f_H (GHz)
9	5.0±0.6	2.9±0.6
15	5.6±0.8	3.7±0.7
25	5.4±0.8	4.0±0.7
35	3.4±0.3	4.1±0.7
45	2.9±0.3	4.1±0.7

concentration. For the solutions considered in the present work (170mg/ml) it turns out that the water of hydration occupies about 13% of the total water present. This means that nearly 90% of the water is more than 1-2 molecular diameters away from any macromolecule, which is ample justification for assuming that a large fraction of the water present behaves dielectrically as pure water.

Turning now to the high frequency dispersion (Table 2) the following can be noted. The first observation is that the sum of the dielectric increments of the two dispersions at any temperature is equal to the increment of the single dispersion indicated in the lower half of Table 1. Hence this description represents the average behaviour of the given water component whereas the interpretation in Table 2 provides more information by dividing the same quantity of water into two categories. The smaller dispersion corresponds to water under the influence of the macromolecules i.e. it is at least part of the total water of hydration. Assuming a value of $\Delta_H = 5.4$ the quantity of bound water can be calculated using the Maxwell-Fricke mixture formula and this gives a value of hydration of about 0.3g/g with an associated error of around 10%. Comparison of this with the value obtained above (0.6g/g) from the myoglobin dispersion shows that about one half of the bound water can be accounted for by the small dispersion centred around 4GHz. The rest is relaxing at lower frequencies and it would account for the data well if this remaining fraction of the bound water were responsible for the dispersion centred around 20MHz. The dispersion amplitude (5-6) is similar in magnitude and therefore would be produced by the same quantity of water. Moreover the ratio of the sum of the increments of the bound water to the increment of the total water content is 14-18%. Making the justifiable assumption that the dielectric increment is proportional to the quantity of substance present this figure is, to within the limits of assumed model and experimental error, in good agreement with the value of 13% obtained independently from consideration of the myoglobin dispersion parameters.

Comparison of the present work with previous dielectric studies on myoglobin and other proteins provokes the following comments. First, the dielectric parameters of the dispersion of the myoglobin molecules (the β dispersion) have been confirmed. Second, the water of hydration has been shown to distribute itself between two resolvable dispersions rather than being represented as one broad dispersion, as was assumed for myoglobin in the previous studies of Grant et al. (1974). The idea of having two resolvable dispersions at frequencies between the β dispersion and the bulk water dispersion is not in itself new. It was proposed by Pennock and Schwan (1969) for haemoglobin and by Essex et al. (1977) for bovine serum albumin. However these previous studies suggested that side chain rotation or proton fluctuation might play a major role in the dispersion occurring at the lower frequency whereas the present work would account for most, if not all, of this dispersion in terms of bound water relaxation. More detailed information on the characteristics of dispersion could be obtained by extending the present measurements to include variation of concentration, pH and ionic strength. The third point is that the relaxation frequency of water of hydration can be as high as 4GHz which is only a factor of five below that of free water. Although a frequency of 4GHz is 5-10 times greater than had been proposed in previous work on protein solutions, relaxation frequencies as high as this have been observed by Epstein et al.(1983) for water associated with microemulsions. Similarly high values of bound water relaxation frequency have been observed by us in ocular tissues in work involving the comparison of their dielectric properties in the normal, supercooled and frozen states (Gabriel 1982, Gabriel and Grant 1985).

The general conclusion to the present work is that dielectric studies can be employed to obtain unambiguous information on the state of water present in an aqueous solution of globular protein molecules or other molecules of known structure, provided that good frequency coverage and a temperature variation are adopted. More specific information on the nature of the binding may be obtained if other parameters such as concentration, pH ionic strength and viscosity are varied. With whole tissue the situation is much more complicated, mainly because of the presence of the membranes and in consequence the large β dispersion which dominates the dielectric behaviour up to frequencies of several tens of MHz. In addition the complexity of the structure of tissue makes it difficult to set up an adequate molecular or structural model which is accessible to mathematical analysis. These problems may be alleviated by freezing the tissue which removes the β dispersion and allows the aqueous component to be studied in isolation.

ACKNOWLEDGEMENTS

Thanks are due to the United States Office of Naval Research for supporting NRVN (Contract N00014-77-G-0075) and to the United States Army for financing VMcC (DAMD 17-81-C-1083: USAMRDC).

We also acknowledge the support to CG from the Institute of Bioelectrical Research, Edinburgh,U.K.

REFERENCES

Cohn,E.J. and Edsall,J.T., 1943, "Proteins, Amino-Acids and Peptides", Reinhold,New York.

Epstein,B.R., Foster,K.R., and Mackay,R.A., 1983,Microwave dielectric properties of ionic and nonionic microemulsions, J.colloid,interf.sci.,95; 218-227.

Essex,C.G., South,G.P., Sheppard,R.J. and Grant,E.H.,1975, A bridge technique for measuring the permittivity of a biological solution between 1 and 100MHz, J.Phys. E.(Sci.Instrum)., 385-389

Essex,C.G., Symonds,M.S., Sheppard,R.J., Grant,E.H., Lamote,R., Soetewey, F., Rosseneu,M.Y. and Peeters, H., 1977, Five component dielectric dispersion in bovine serum albumin solution., Phys.Med.Biol., 22: 1160-1167.

Foster,K.R., Cheever,E., Leonard,J.B. and Blum,F.D., 1984, Transport properties of polymer solutions - a comparitive approach, Biophys.J., 45: 975-984.

Gabriel,C., 1982, Dielectric study of ocular tissue using time domain spectroscopy, Ph.D.Thesis, University of London.

Gabriel and Grant 1985. To be published.

Grant,E.H.,1965, The structure of water neighbouring proteins, peptides and amino acids as deduced from dielectric measurements, Ann.N.Y.Acad.Sci.125:418-427.

Grant,E.H., 1982, The dielectric method of investigating bound water in biological material: an appraisal of the technique, Bioelectromagnetics, 3: 17-24.

Grant,E.H.,Mitton,B.G.R.,South,G.P. and Sheppard,R.J.1974. An investigation by dielectric methods of hydration in myoglobin solution, Biochem.J.139, 375-380.

Grant,E.H., Sheppard,R.J. and South,G.P., 1978, Dielectric behaviour of biological molecules in solution, Oxford University Press.

Nightingale,N.R.V.,Szwarnowski,S., Sheppard,R.J. and Grant, E.H., 1981, A coaxial line cell for measuring the permittivity of medium to high loss liquids in the frequency range 2-15GHz, J.Phys.E (Sci. Instrum) 14: 156-160.

Pennock,B.E. and Schwan,H.P., 1969, Further observations on the electrical proteins of haemoglobin bound water, J.Phys.Chem. 73: 2600-2610.

Schwan,H.P., 1965, Electrical properties of bound water, Ann.N.Y.Acad.Sci. 125: 344-354.

South,G.P. and Grant,E.H., 1972, Dielectric dispersion and dipole moment of myoglobin in water, Proc.R.Soc.A. 328: 371-387.

Szwarnowski,S., 1982, A transmission line cell for measuring the permittivity of liquids over the frequency range 90MHz-2GHz, J.Phys.E (Sci.Instrum) 15: 1068-1072.

DIELECTRIC PROPERTIES OF CELLS AND TISSUES

H. P. Schwan

Max Planck Institut fur Biophysik, Frankfurt a.M.
Germany, and Department of Bioengineering/D3*
University of Pennsylvania, Philadelphia, PA 19104

INTRODUCTION AND PRINCIPLES

In this lecture we shall summarize the electrical properties of biological cells and tissues over the total investigated frequency range. We also discuss mechanisms responsible for observed frequency dependencies and indicate the most probable sites for EM-field interactions.

Figure 1 presents the display of the frequency dependent dielectric constant ε and conductivity K of muscle over the total frequency range from a few Hz to GHz[1]. The dielectric constant ε is presented, since it characterizes the ability of the tissue to store electricity compared to free space and the conductivity is simply a measure of the ability to transport charges with the field.

---

* Mailing Address

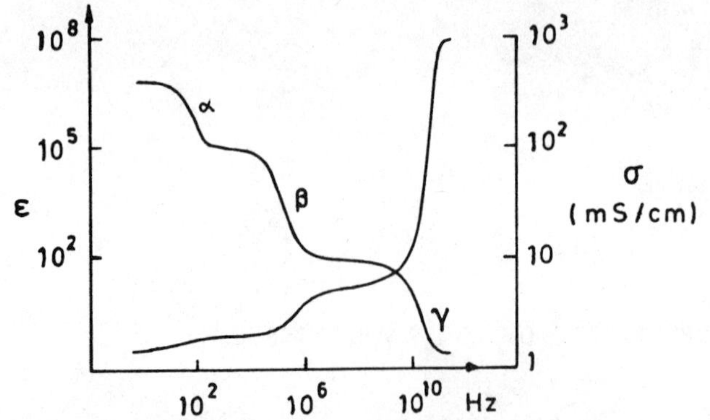

Fig. 1. Dielectric constant ε and conductivity σ of muscle vs. frequency.

These two properties solely characterize the electrical characteristics of matter. The magnetic properties of tissue are not considered as they are virtually identical with those of free space. The curves display two unique features: a) the dielectric constant reaches enormous values of more than a million as the frequency decreases below 100 Hz, b) the data change in three distinct steps each of them of a clearly "dispersive" nature. Figure 2 indicates some of the mechanism responsible for the observed behavior[1]. The γ-effect is accounted for by a corresponding change in the abundant tissue water properties. The β-change is caused by the polarization

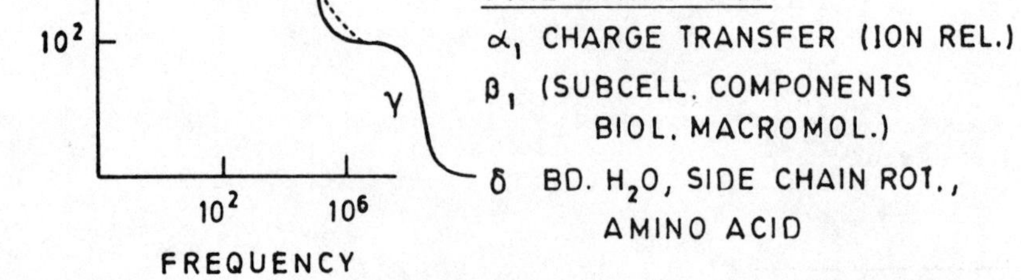

Fig. 2. Summary of various relaxation mechanisms.

of cellular membranes with the cytoplasmic and extracellular media serving as access paths for the charging currents. It is a "Maxwell-Wagner" relaxation effect. The α-effect may have several causes, including the charging of intracellular structures connecting with the outer membrane (tubular system?), counter ion relaxation effects and frequency dependencies of the membrane itself as may be anticipated from a linearization of the Hodgkin-Huxley equations. In addition various "fine" structure contributions have been identified which broaden the tail of the β-dispersion. They include relaxation of protein bound water, polarization of cellular organelles such as nucleus and mitochondria and dielectric relaxation of the abundant tissue proteins. The relative contributions of these effects have been studied in some detail recently[2].

Most of the studies indicated above have been carried out with excised tissues and the question has been raised to what extent results obtained with "dead" tissues characterize live data. Care has to be taken that excised tissues be investigated promptly. This is demonstrated in figures 3 and 4 which display low frequency data for muscle tissue as function of frequency at various times after excision from the live animal. It appears that the data are not strongly affected for some hours and days are required before

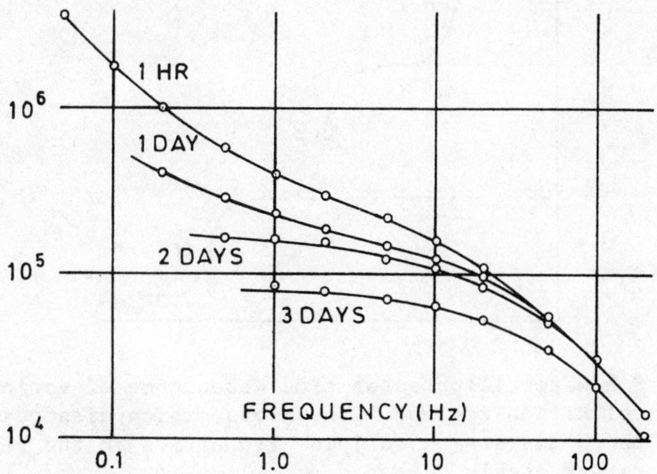

Fig. 3. Dielectric constant vs. frequency. Parameter time.

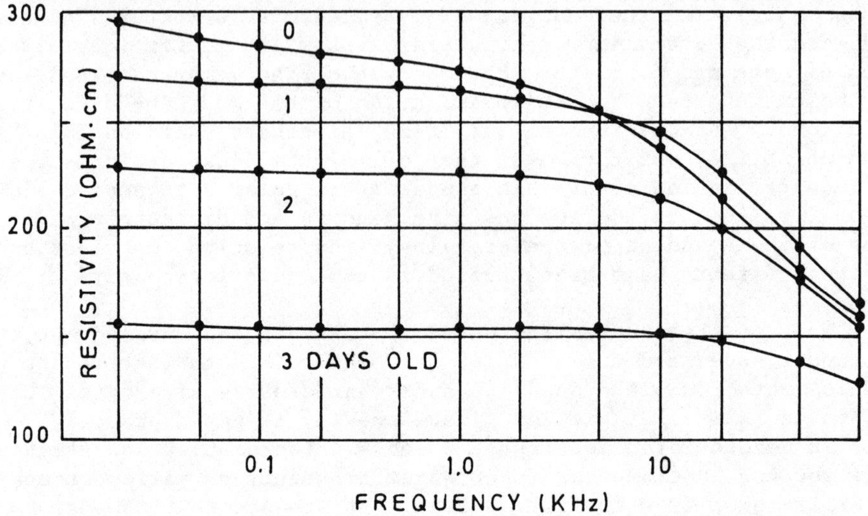

Fig. 4. Resistivity vs. frequency. Parameter time.

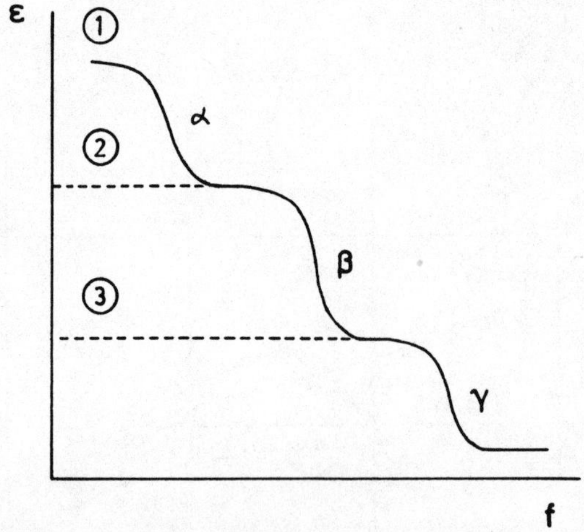

Fig. 5. Schematic illustrates time dependence of various relaxation regions. The α-dispersion disappears first, the β-dispersion follows. Properties in the γ-range are rather stable.

complete breakdown of membrane structures accounts for disappearance of the low ELF-and RF-dispersion effects. There is also a degree of redistribution of the blood pool which takes place as circulation ceases, causing only small changes in the electrical data for most tissues. Comparison of live data obtained by insertion of electrodes into live animals with data after death supports the conclusion that data are not immediately affected after death. Comparison from live data with excised data at very high frequencies are also now available and confirm that such data are insensitive to the live state. Perusal of all pertinent literature indicates that the α-effect may disappear in a few hours, the β-dispersion in two or three days and the γ-dispersion to remain stable for longer periods of time. This is schematically indicated in figure 5.

A few words are next in order about the nature of the observed dispersive effects. In figure 6 we indicate how matter usually responds if a voltage step is applied. A sudden surge charge is often followed by a more gradual increase towards a steady state value. In the simplest case this transient response is of an

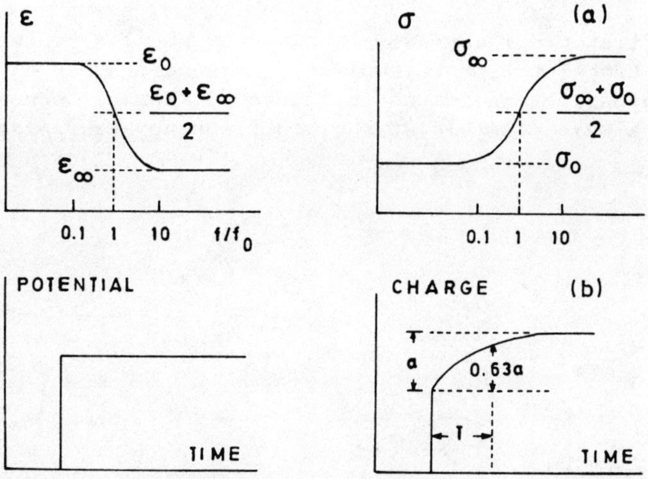

Fig. 6. Dielectric behavior in frequency and time-domain for a single relaxation process.

exponential nature. In this case the corresponding behavior in the frequency domain, i.e. as function of frequency, is readily calculated and results in the dispersive behavior indicated in the lower part of the figure and expressed by the following equations

$$\varepsilon = \varepsilon_\infty + \frac{\varepsilon_o - \varepsilon_\infty}{1 + (\omega T)^2} \qquad (1)$$

$$K = K_o + (K_\infty - K_o) \frac{(\omega T)^2}{1 + (\omega T)^2} \tag{2}$$

$$K_\infty - K_o = (\varepsilon_o - \varepsilon_\infty)\, \varepsilon_v / T \tag{3}$$

Here the ε denotes dielectric constants relative to free space, ε_v the dielectric constant of vacuum, the subscripts o and ∞ indicate low- and high-frequency limits, ω is the angular frequency $2\pi f$ and T is the same time constant which characterizes the approach to the steady state in the time domain.

It should be noted that the changes in dielectric constant and conductivity are interrelated by the time constant T. It can indeed be shown that in general the frequency dependence of the dielectric constant determines that of the conductivity, i.e. both are casually related. The mathematical formulation of this relationship has been given by Kramers-Kronig.

β-DISPERSION

As a first order approximation we may model a biological cell as a conducting interior (cytoplasm) surrounded by an external shell (membrane) as indicated in figure 7. Maxwell demonstrated that it is always possible at a given frequency to replace this

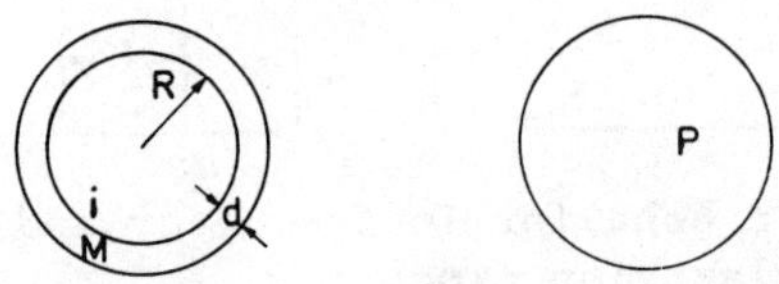

$$\frac{K_p - K_m}{K_p + 2K_m} = \left(\frac{R}{R+D}\right)^3 \frac{K_i - K_m}{K_i + 2K_m}$$

Fig. 7. Membrane surrounded particle and equivalent homogeneous particle. Maxwell's equation relates the properties of the two particles. The subscripts indicate membrane, interior and equivalent homogeneous particle.

configuration by a homogeneous sphere such that the external field remains unchanged. He also derived the bulk conductivity of a suspension of spherical particles in a conducting medium. Combining these two approaches it is then possible to derive an equation for the case of a suspension of spherical biological cells in a conducting medium. Maxwell's approach can be readily extended to the case of alternating fields where all three phases (cytoplasm, membrane, medium) have complex properties, i.e. dielectric constants and conductivities. This has been done using assumptions typical for cells such as a membrane thickness small compared to cell radius and a predominantly conducting medium. A set of dispersion equations is readily derived which characterizes the β-dispersion. An equivalent circuit of physical significance which describes the same dispersive behavior is shown in figure 8. It consists of a parallel combination of a series and a parallel R-C combination.

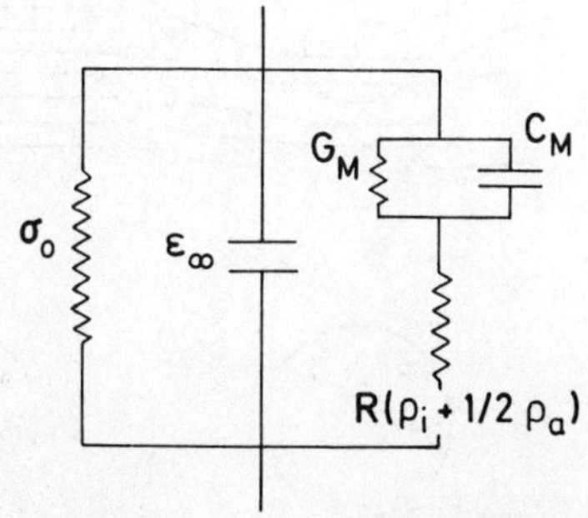

Fig. 8. Electrical circuit modelling spherical cells in suspension. The influence of the membrane capacitance G_m is usually very small and can be neglected.

The parallel branch represents the dielectric constant and conductivity of the path conducting electricity bypassing the cell, i.e. the extracellular current path and the series branch consists of a capacity representing the membrane capacity and a resistance representing the cytoplasm. This branch determines the part of the total current entering the cell. At low frequencies the membrane capacity blocks this current from entering the cell, but at high frequencies this capacitance is shortcircuited and the cytoplasmic resistance fully participates in the conduction of electricity through the cell suspension. The validity of the model responsible for the β-dispersion has been extensively tested and has not been successfully challenged to the knowledge of this author.

Figure 9 once more summarizes the principle of the β-dispersion. At low frequencies the current is blocked from entering the cell and the membranes become charged, resulting in a high induced dipole moment for the entire cell, while at high frequencies the total interior participates in the conduction mechanism and the dipole moment of the cell is very low. The β-dispersion range is the frequency range where the induced moment changes from high to low values. We recognize already here that field interactions with the interior are more likely at high than low frequencies, since the membrane effectively shields the interior

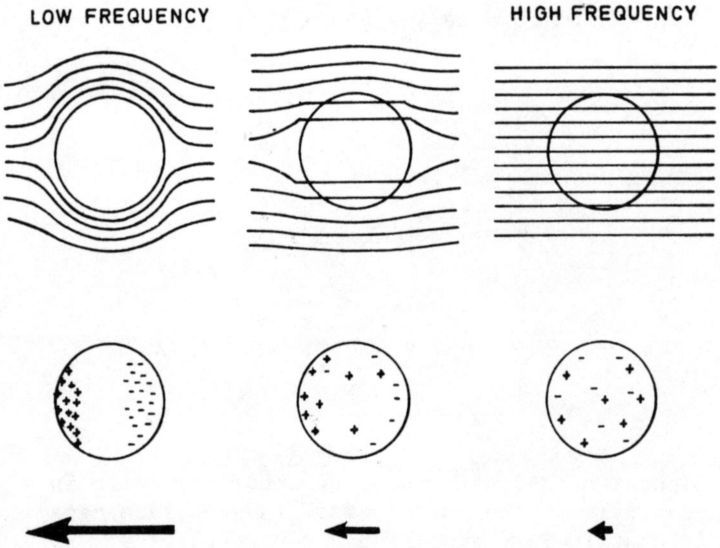

Fig. 9. Schematic indicating current flow at various frequencies and magnitude of induced dipole.

compartments from field exposure at low frequencies. Since the characteristic frequencies of the β-range vary between 0.1 and 10 MHz depending on cellular size, the total ELF-range must be considered "low" in frequency.

α-DISPERSION

We shall now examine the implications of the β-dispersion model for the α-dispersion. The limit equations for the β-dispersion for low frequencies can be written in the form given in figure 10. Here the limit values of the dielectric constant ε and conductivity K are indicated by subscripts o and ∞ for frequencies low and high compared to the characteristic frequency of the β-dispersion $f_c = 1/2\ \pi T$, with T the time constant of the β-effect.

R is the cell radius, assumed to be spherical and C_m and G_m are membrane capacitance and conductance per cm^2 area as indicated in the figure. K_a is the conductance of the external fluid and p the volume fraction occupied by the cells. Clearly, if the dielectric properties of the cell suspension change at frequencies well below the β-dispersion range, then some of the parameters in the equations must change correspondingly. The medium conductivity K_a, particle radius R and volume fraction p do not change with frequency. Hence, the α-dispersion indicates a corresponding change of the membrane parameters C_m and G_m. These "apparent" membrane property changes may have several reasons as will be indicated below.

$$\varepsilon_0 = \varepsilon_\infty + \frac{9}{4} pRC_m$$

$$\sigma_0 = \sigma_a \frac{1-p}{1+p/2} + \frac{9}{4} pRG_m$$

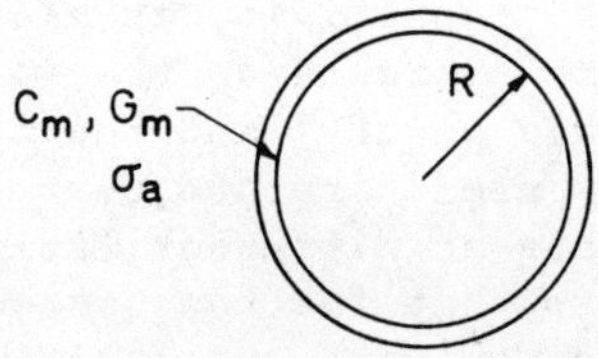

Fig. 10. Low frequency dielectric constant and conductivity for a suspension of spherical cells. R radius, p cell volume fraction, σ_a extracellular conductivity, C_m and G_m membrane capacitance and conductance per cm^2 membrane surface.

Figure 11 displays in detail the α-dispersion of muscle tissue and the beginning of the β-dispersion effect. The strong change in dielectric constant and the comparatively weak one in conductivity are interrelated as demanded by the Kramer-Kronig relationships. The equations in figure 10 would indicate that the "apparent" membrane capacitance changes from the usual value of about 1 $\mu F/cm^2$ to a value near 30 $\mu F/cm^2$ as the frequency decreases and that the corresponding membrane conductance change is about 20 mS/cm^2. These are rather large changes for the membrane per se and alternate suggestions for the effect will be stated below since they provide insight how currents penetrate through cellular structures.

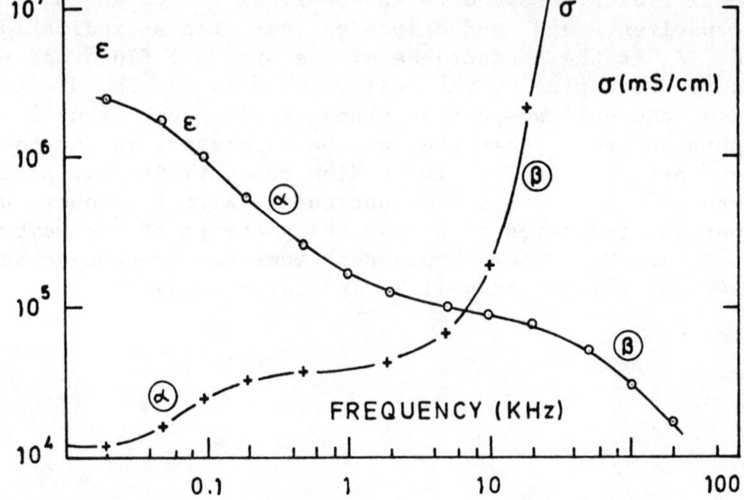

Fig. 11. α-dispersion of ε and σ for muscle tissue.

Figure 12 indicates how dielectric data depend on the applied field strength. The resistivity is presented as function of applied voltage and shown to decrease as the voltage exceeds about 1 Volt. The cell constant was about 1 and the corresponding field strength and current density are therefore near 1 V/cm and 1 mA/cm^2 respectively. Measurements at different ELF-frequencies support these results. It appears that for the quoted field strength and current

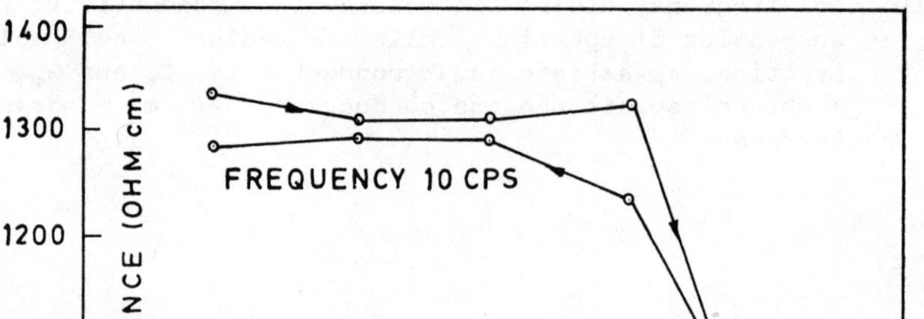

Fig. 12. Field strength dependence of specific resistance of muscle at ELF frequencies.

density levels tissue data are affected. We shall see that corresponding membrane potentials are a significant fraction of the resting potential and that, therefore, nonlinear properties at the quoted levels are not unexpected.

A brief summary of available ELF data for dielectric constant and loss angle is given in Table 1. The dielectric data were obtained in situ on dogs, using small electrodes inserted into the tissue of interest. The dielectric data are in reasonable agreement with excised tissue data, indicative of the strong α-dispersion effect of the dielectric constant. Loss tangent data are presented in terms of products RωC with R resistance and C capacitance of tissue sample. This product is equal to the ratio of capacitive to conductive current. The ratios are negligibly small at 0.1 KHz and below, but approach values near 20% at 10 KHz. Clearly, tissues, in spite of their high dielectric constants, behave almost like conductors with minimal dielectric losses at ELF-frequencies.

Table 1. Dielectric constant ε and loss tangent $\tan\delta = R\omega C$ of live tissues

	ε/1000			RωC		
	0.1 KC	1 KC	10 KC	0.1 KC	1 KC	10 KC
Lung	600	120	30	0.02	0.06	0.13
Liver	1000	160	50	0.04	0.05	0.17
Muscle	800	130	50	0.04	0.06	0.21
Heart Muscle	800	300	100	0.04	0.15	0.32
Fatty Tissue	150	50	20	0.01	0.03	0.15

MECHANISM FOR α-DISPERSION

Various mechanism have been indicated as responsible for the α-effect[3] and have pronounced effects on the interaction of ELF-fields with various cellular compartments associated with the membrane. They will now be discussed.

Counter Ion Relaxation

The dielectric properties of colloidal systems display at frequencies in the ELF range pronounced dispersion effects in many respects quite similar to those displayed by cell suspension. Figure 13 shows the dispersive behavior of the dielectric constant of two polysterene suspensions with different particle diameters.

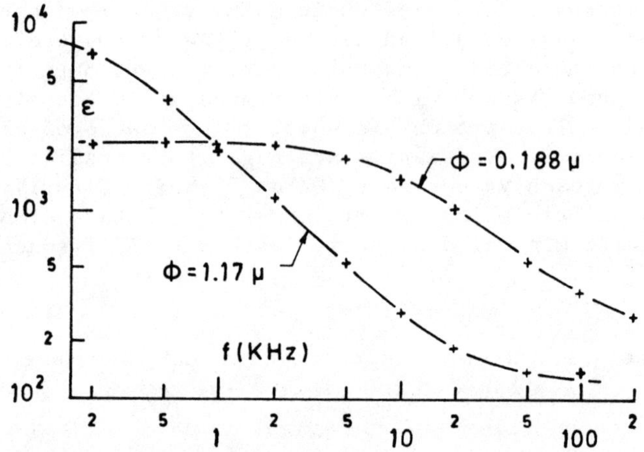

Fig. 13. Dielectric constant vs. frequency for two polysterene particle suspensions.

Figure 14 demonstrates that the relaxation frequency for suspensions of particles of different size varies inversely proportional

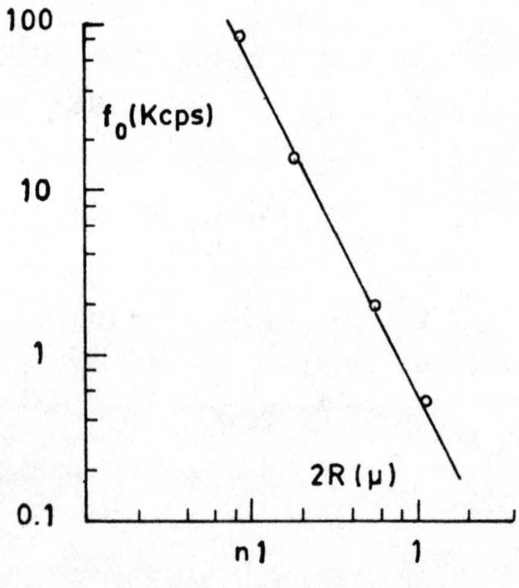

Fig. 14. Characteristic frequency $f_o = 1/2\pi T$ vs. particle size for polysterene particle suspensions.

to the square of the particle size. The data can be explained by assuming a frequency dependent surface admittance due to counter ions with

$$Y_s = G_s + j\omega C_s \tag{4}$$

with

$$Y_s = Y_\infty + \frac{Y_o - Y_\infty}{1 + (j\omega T)^\alpha} \tag{5}$$

undergoing dispersive behavior. This quantity reduces for low frequencies to the usual surface conductance so well known in electrophoretic theory. The subscripts o and ∞ indicate limit values at low and high frequencies, T is the average of a distribution of time constants and equal to $1/2\pi f_c$ with f_c the characteristic frequency and α is the Cole-Cole parameter which characterizes the spread of relaxation times about the average value T[4]. Typically G_s-values range from 10^{-10} to 10^{-9} MHo and are of the same order of magnitude as known from electrophoretic studies.

Surface conductance effects are typical for charged particles. They usually have isolating properties or poorly conducting ones as cells. Particles are kept individually in suspension by appropriate surface treatment, providing fixed charges and consequent repellant properties. Such particles and cells, which are also well known to possess many fixed charges at their surface, surround themselves with counter ions of opposite sign to achieve overall electrical neutrality. These counter ions can move tangentionally along the surface of the particle or cell. They cannot penetrate due to the isolating properties of the particle. Such a tangential movement may be induced either by a DC field or low frequency alternating fields. As a consequence a polarization of the particle is achieved, i.e an induced dipole moment of often substantial magnitude is created. The effectiveness of this polarization mechanism is highly frequency dependent causing the dispersive behavior of the electrical bulk properties noted above. The theory of the counter ion relaxation effect has been developed by Schwarz[5], using reasonable assumptions valid for colloidal particles. But they are not necessarily valid for biological cells with their intrinsic high dielectric constants at the low end of the α-dispersion effect. Figure 15 indicates schematically the counter ion polarization effect by the asymetric distribution of counter ions under the influence of an applied field and equations for low frequency dielectric constant and time constant. The above mentioned proportionality of T with R^2 is stated. e is the elementary charge, σ its surface density, K and T Boltzmann constant and absolute temperature and p the particle volume fraction.

The counter ion relaxation effect is of importance in assessing ELF effects on cells. Presumably a part of the potential applied to the membrane will be consumed by the counter ion environment. But no detailed theory has as yet addressed itself to this problem.

$$\varepsilon_0 - \varepsilon_\infty = \frac{9}{4}\,\frac{P}{(1+\frac{P}{2})^2}\,\frac{e^2 R\sigma}{KT}$$

$$T = \frac{R^2}{2uKT}$$

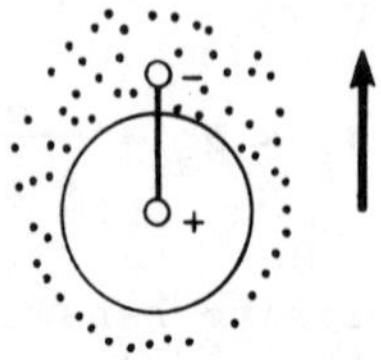

Fig. 15. Principle of counter ion relaxation process, its characteristic frequency and magnitude.

For low frequencies the bulk conductivity of a suspension of nonconducting charged particles or cells can be written in the form

$$K_o = K_a \frac{1 - P}{1 + P/2} + \frac{9}{2}\,\frac{P}{(1 + P/2)^2}\,\frac{G_s}{R} \tag{6}$$

with K_a the conductivity of the medium and the other symbols defined before. If we combine this expression with the one derived previously for a cell suspension with the cells surrounded by a conducting membrane quoted in figure 10 we obtain

$$K_o = K_a \frac{1 - P}{1 + P/2} + \frac{9}{2}\,\frac{P}{(1 + P/2)^2}\,\frac{G_s}{R} + \frac{9}{4}\,\frac{P}{(1 + P/2)^2}\,RG_M \tag{7}$$

provided that the terms containing G_s and G_m, i.e. surface conductance and membrane conductance, do not contribute very much to the first terms. This is usually true for cells. The two mechanisms due to G_s and G_m are indicated schematically in the lower part of figure 16. On the left the tangential current due to the surface conductance term is indicated and on the right the term indicating the small current entering the cell at low frequencies through the membrane. The two contributions are equal if the equation in the center of the figure is fulfilled. For typical cellular parameters this equation frequently applies, i.e. surface conductance and membrane conductance contribute about equal to the total conductance.

This creates a dilemma: Since the terms frequently are of comparable magnitude it is not possible from bulk measurements to determine their relative contribution.

We conclude: It is difficult to establish how the total ELF-potential applied to the greater membrane complex including its counter ion atmosphere will partition itself to membrane proper and ionic environment. This will complicate attempts to relate biological effects to either factor.

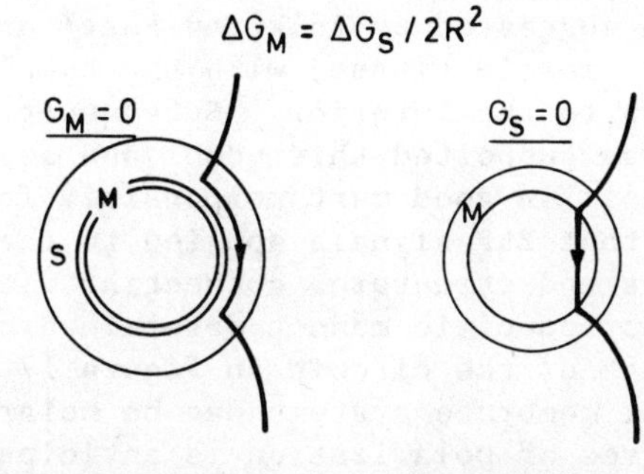

Fig. 16. Surface conductance G_s and membrane conductance G_m.

Above discussions assume a fixed charge distribution on the surface of the cell. However, the details of the fixed charge contribution are more complex. Attached to the membrane surface are glycoproteins which extend 100 Å or more with branches which contain many fixed charges. The total charge density is of the order of one elementary charge per 10 Å squared or more. A degree of flexibility of these charges can be assumed and the counter ion distribution is no doubt more complex than considered in the Schwarz model. However, the discussions above should still be useful in the absence of a more refined model at least as far as the general conclusions are concerned. Adey[6] has suggested that ELF fields might be able to significantly interact with the counter ions. No quantitative arguments have been advanced about minimal field levels required to interact as suggested.

Connecting Membrane Systems

Figure 17 indicates a folded membrane system inside a cell which connects with the outer membrane. An equivalent electrical circuit is presented below. The arrow indicates how current might enter into the subsystem, C_m and G_m are membrane capacitance and conductance per unit area of the outer membrane, R_R indicates the resistive entrance to the reticulum structure and C_R its distributed membrane capacitance. Clearly, at very low frequencies the total observed capacitance is determined by the sum C_m+C_R, i.e. the total membrane area available, and at frequencies higher than $1/2\pi\ R_R C_R$ only the outer membrane capacitance C_m counts since the impedance of the inner system (folded capacitance) is in series with the larger resistance R_R. Its contribution to C_m is correspondingly attenuated if referred to the input of the total circuit. This mechanism has been suggested by Falk and Fatt[7] as responsible for the α-dispersion of muscle tissue, with the tubular system being the site for access to the interior. Subsequent work, not reviewed in this context, has supported this model and suggests that this mechanism is at least in good part responsible for the ELF-changes. We thus recognize that ELF signals applied to the membrane reach internal organelles and structures connecting with the outer membrane in a frequency specific manner determined by the time constant of the series branch of the circuit in figure 17. Clearly, some connecting internal membrane systems can be polarized by the ELF-field, but the degree of polarization is anticipated to be a strong function of frequency for the ELF-range.

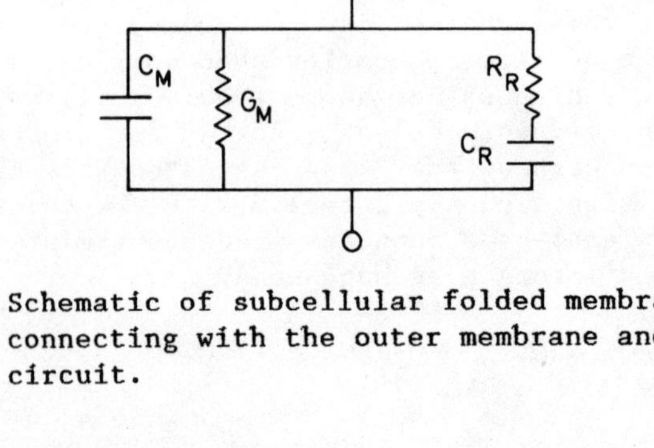

Fig. 17. Schematic of subcellular folded membrane structure connecting with the outer membrane and equivalent circuit.

Membrane Changes

As indicated by the equations in figure 10, membrane property changes at ELF-frequencies will correspondingly affect the dielectric properties of tissues and cell suspensions. Such changes can be readily anticipated from a linearization of the Huxley-Hodgkin equations if it is assumed that membranes of interest here display features similar to those of the squid axon membrane investigated by the English investigators. They occur below 100 Hz and may increase or decrease the capacitance and conductance depending on the interplay of the membrane channels for the abundant ion species.

An additional membrane dispersion has been observed by Takashima and Schwan[8]. It occurs near 2.5 KHz and its nature has not been determined as yet.

MICROWAVE DISPERSION

Dielectric studies of cell suspensions and tissues have been extended well into the microwave frequency range. Data up to 8500 MHz have been summarized by Schwan already in 1957[1]. More recently dielectric studies have been extended up to 18 GHz on muscle tissue[9]. It immediately became apparent that the dielectric behavior of tissues and cell suspensions in the microwave frequency range is largely determined by the strongly dispersive behavior of water. Indeed, from the analysis of the RF-properties summarized above, it follows that cell membranes impose no hindrance on the flow of electricity through the cell interior for frequencies above 100 to 300 MHz. Consequently it is apparent that as far as dielectric properties are concerned, tissues of high water content and cells simply behave as water containing some 20% proteins. This predictive model served well to describe numerically the dielectric properties observed at frequencies from about 100 MHz to many GHz. This approach tacitly assumes that the dielectric properties of tissue water are identical with those of normal water. Furthermore the availability of dielectric data for the cellular protein solutions is assumed. These two items are summarized now. They will be discussed in greater detail by Grant in these Proceedings.

Water and Tissue Water

The dielectric properties of pure water have been well established from DC up to microwave frequencies, approaching the infrared[10]. For all practical purposes they are characterized by a single relaxation process centered near 20 GHz at room temperature. Static and infinite frequency permittivity values are, at room temperature, close to 78 and 5, respectively. Hence the microwave conductivity increase predicted by Eq. (3) is close to 0.8 mho/cm above 20 GHz, much larger than typical low-frequency conductivities of biological fluids which are about 0.01 mho/cm. The dielectric properties of water are independent of field strength up to fields of the order 100 kV/cm.

The dielectric properties of electrolytes are almost identical to those of water with the addition of a σ_o term in Eq. (2) due to the ionic conductance of the dissolved ion species. The static dielectric permittivity of electrolytes of usual physiological strength (0.15 N) is about two units lower than that of pure water, a negligible change.

Hoeber[11] had already obtained a value for the conductance of the red cell interior. The work of Fricke[12] and that of Cole[13] provided additional data with no indication that tissue water might be differently structured from normal water. Earlier data frequently had to rely on extrapolations to higher frequencies where membrane properties no longer affect dielectric properties. But such extrapolations are difficult since tissue properties in the RF range include secondary dispersion phenomena caused by mitochondria, cell nuclei, and biopolymers. A direct and more precise determination of internal conductance values requires frequencies near 100 MHz, high enough to short-circuit and thereby eliminate membrane contributions and yet low enough to exclude conductance contributions from the rotation of water molecules. We measured the dielectric properties of various tissues and cell suspensions over the frequency range from 100 to 1000 MHz[14]. The results may be described as follows:

1. The dielectric constant of tissues of high water content and blood are identical to that of normal water if allowance is made for the space occupied by tissue proteins and hemoglobin.

2. The conductance values are approximately comparable to those calculated from the concentrations of ionic species. But there appears to be a systematic deviation with experimental values smaller than calculated ones. A detailed study was conducted with erythrocytes and ghosts containing varying amounts of Hb[15]. In the ghosts, agreement between theory and experiment was achieved at Hb concentrations less than half normal. At high concentration levels internal friction with proteins reduces the mobility of ions noticeably. This effect, if adequately considered, entirely removes the discrepancy between theory and experiment. No comparable analysis has yet been conducted for tissues and other cell suspensions.

3. Dielectric data for tissues and blood from 1000 to 8500 MHz[16] strongly indicated that the relaxation frequency for tissue water and normal water are identical even though the range of observation did not extend to the relaxation frequency of normal water.

Thus it appears that dielectric constants near 100 MHz, ion mobilities obtained from the conductance data, and tissue water relaxation compare well with properties characteristic of normal water if we correct for the small volume taken by protein bound water. However, this point of view has been challenged. Ling and his colleagues[17] stated that tissue water is differently structured from normal water, a view shared by a small but active group of biophysicists. Drost-Hansen also believes that the identification of tissue water with normal water is naive and has developed his own concepts of structured water[18]. Hovey et al.[19] attempted to determine the internal conductance of aplysia cells and derived values which simply could not be understood unless one assumed vastly different properties of cellular and normal water. However, in a subsequent paper by Foster et al.[20] this earlier work was corrected and the original internal conductance values were replaced by values supportive of our views stated above. Then there appeared a series of papers by Masszi et al.[21,22] stating that the relaxation frequency of tissue water is appreciably shifted from that of normal water. Their work was carried out at a frequency of 2.45 GHz and based on an evaluation of conductance data. At this frequency the Debye contribution to the low frequency conductivity is already appreciable and sensitive to the value of the relaxation frequency. But in a more recent detailed analysis of available tissue data we could not find a confirmation of this conclusion[9].

Techniques have recently become available in our laboratory which extend the range of observation to 18 GHz, close to the relaxation frequency of normal water at room temperature and twice that of a 9°C water. We decided to check the previously unpublished data by Herrick[16], which had been made available to us, to extend tissue data to 18 GHz, and to review the question of tissue water with the precision microwave techniques now available using a microwave network analyzer. The dielectric properties of barnacle muscle were investigated first, concentrating on the relaxation frequency of tissue water[9]. Two approaches were chosen. The first one is based on an evaluation of the Cole-Cole circle in the complex dielectric phase. The static dielectric constant obtained is as one might predict using Maxwell's mixture equation to account for the protein volume. The peak of the circle gives the relaxation frequency and is identical to that of free water. The second approach evaluates conductance data alone. The conductance is plotted versus a normalized conductance axis. The result is a straight line as may be seen from the relaxation Eq. (2) for the conductance and its slope is given by the relaxation frequency and thus provides f_c. At high frequencies values were obtained for the slope providing a relaxation frequency identical with that of normal water. But at lower frequencies the slope changes. This analysis demonstrates that this change is caused by a subsidiary protein UHF dispersion. For a more detailed discussion, see[9].

We conclude therefore that Masszi's results, obtained at much lower frequencies, were in error since he did not appropriately correct for the UHF-dispersion effect unrelated to tissue water and probably caused by proteins[23,24,25].

More recently, Jenin and Schwan[26] have pointed out how to determine the time constant of cell water from lower frequencies data. Again, a time constant was noted which compared well with that of normal water. At the same time the dielectric properties of the hemoglobin solution inside the red cell were noted to be identical with those of a solution of hemoglobin in normal water of equal concentration.

We therefore conclude that cellular water is, at least from a dielectric point of view, identical with normal water.

Protein Solutions

The dielectric properties of proteins and nucleic acids have been extensively reviewed[27,28]. Protein solutions exhibit three major dispersion ranges. One occurs at RF's and is believed to arise from molecular rotation in the applied electric field. Typical characteristic frequencies range from about 1 to 10 MHz, depending on the protein size. Dipole moments are of the order of 200-500 Debyes and low-frequency increments of dielectric permittivity vary between 1 and 10 units/g protein/100 ml of solution. The high-frequency dielectric permittivity of this dispersion is lower than that of water because of the low dielectric permittivity of the protein leading to a high-frequency decrement of the order of 1 unit/g protein/100 ml. This RF dispersion is quite noticeable in pure protein solutions, but in tissues and cell suspensions it only contributes slightly to the large β-dispersion found in these materials.

At microwave frequencies the dielectric properties of tissues are dominated by the water relaxation centered near 20 GHz. The magnitude of this water dispersion in tissues is typically diminished by some 20 dielectric units due to the proteins which displace a corresponding volume of water.

Between these two readily noticeable dispersions is a small one termed the δ-dispersion by Grant. It was first noted for hemoglobin[1,23] and then carefully examined for hemoglobin[24] and albumin[25]. This dispersion is characterized by a fairly broad spectrum of characteristic frequencies extending from some hundred to some thousand MHz. Its magnitude is considerably smaller than that of the other two dispersions, and it is thought to be caused by a corresponding dispersion of the protein bound water and/or partial rotation of polar subgroups. Its contribution to the dielectric properties of cell suspensions and tissues is therefore small.

SOME CONCLUSIONS

Above summarized principles which determine the frequency dependent dielectric properties of cells have consequences of particular interest in this context. They can be stated as follows:

1. At low frequencies the cell interior is effectively shielded by the cell membrane and the membrane receives the total potential applied across the cell. Hence external fields are multiplied by a factor of about R/d (R cell radius, d membrane thickness) across the membrane. These facts appear to rule out any direct interaction of external low frequency fields with the cell interior and make the membrane responsive to small field values of the order of 1 V/cm.

2. At microwave frequencies the membranes are shortcircuited and no longer profit from the large amplification factor R/d available at frequencies below the β-range. Hence membrane interactions become less likely, but the cell interior is now fully exposed to the field.

3. Each component of the cell displays a frequency sensitivity resulting from the "coupling" considerations which can be based on above presented analysis and which have been presented elsewhere[29]. The overall frequency dependence of the tissue response to a frequency independent field becomes extremely complex as a result of these specific sensitivities[29].

4. Significant displacement of counter ion atmospheres has been estimated to occur at V/cm[30]. This might well have some biological significance. But further work is needed to substantiate this estimate.

REFERENCES

1. H. P. Schwan, Electrical properties of tissue and cell suspensions, in: "Advances in Biological and Medical Physics," Vol. 5, J. H. Lawrence and C. A. Tobias, eds., Academic Press, New York (1957).
2. R. D. Stoy, K. R. Foster, and H. P. Schwan, Dielectric properties of mammalian tissues from 0.1 to 100 MHz: A summary of recent data, Phys. Med. Biol. 27:501-513 (1982).

3. H. P. Schwan, Electrical properties of cells: Principles, some recent results and some unresolved problems, in: "The Biophysical Approach to Excitable Systems," W. S. Adelman and D. Goldman, eds., Plenum Press (1981).
4. K. S. Cole and R. H. Cole, Dispersion and absorption in dielectrics. I. Alternating current characteristics, J. Chem. Physics 9:341-351 (1941).
5. G. Schwarz, A theory of the low-frequency dielectric dispersion of colloidal particles in electrolyte solution, J. Physical Chemistry 66:2636 (1962).
6. W. R. Adey, Tissue interactions with nonionizing electromagnetic fields, Physiological Reviews Vol. 61, No. 2 (1981).
7. G. Falk and P. Fatt, Linear electrical properties of striated muscle fibers observed with intracellular electrodes, Proc. Royal Soc. B160:69 (1964).
8. S. Takashima and H. P. Schwan, Passive electrical properties of squid axon membrane, J. Membr. Biol. 17:51-68 (1974).
9. K. R. Foster, J. L. Schepps and H. P. Schwan, Microwave dielectric relaxation in muscle, Biophys. J. 29:271 (1980).
10. M. N. Afsar and J. B. Hasted, J. Opt. Soc. Amer. 67:902 (1977).
11. R. Höber, Eine methode die electrische leitfähigkeit im innern von zellen zu messen, Arch. ges. Physiol. 148:189 (1912).
12. H. Fricke, Phys. Rev. 21:708 (1923).
13. K. S. Cole, "Membranes, Ions and Impulses," University of California Press, Berkeley (1972).
14. H. P. Schwan and K. Li, Capacity and conductivity of body tissues at ultra-high frequencies, Proc. of I.R.E. 41:1735 (1953).
15. H. Pauly and H. P. Schwan, Dielectric properties and ion mobility in erythrocytes, Biophys. J. 6:621 (1966).
16. J. F. Herrick, D. G. Jelatis, and G. M. Lee, Personal communication as cited in Ref. 1.
17. G. N. Ling, C. Miller, and M. M. Ochsenfeld, The physical state of solutes and water in living cells according to the association induction hypothesis, Ann. New York Acad. Sci. 204, C. F. Hazlewood, ed. (1973).
18. W. Drost-Hansen, Water and biological interfaces: Structural and functional aspects, Phys. Chem. Liq. 7:243 (1977).
19. M. M. Hovey, A. F. Bak, and D. O. Carpenter, Low internal conductivity of aplysia neuron somata, Science 176:1329 (1972).
20. K. R. Foster, J. M. Bidinger, and D. O. Carpenter, The electrical resistivity of cytoplasm, Biophys. J. 16:991 (1976)
21. G. Masszi, A. Szuarto, and P. Grof, Investigations on the ion- and water-binding of muscle by microwave measurements, Acta Biochim. Biophys. Acad. Sci. Hung. 11:129 (1976).

22. G. Masszi, Dielectric relaxation and water structure in gelatin solutions, Acta Biochim. Biophys. Acad. Sci. Hung. 7:349 (1972).
23. H. P. Schwan, Electrical properties of bound water, Ann. New York Acad. Sci. 125:344 (1965).
24. B. Pennock and H. P. Schwan, Further observations on the electrical properties of hemoglobin bound water, J. Phys. Chem. 73:2600 (1969).
25. E. H. Grant, The structure of water, neighbouring proteins, peptides and amino acids as deduced from dielectric measurements, Ann. New York Acad. Sci. 125:418 (1965).
26. P. C. Jenin and H. P. Schwan, Some observations on the dielectric properties of hemoglobin's suspending medium inside human erythrocytes, Biophys. J. 30:285 (1980).
27. S. Takashima, Dielectric properties of proteins. 1. Dielectric relaxation, in: "Physical Principles and Techniques of Protein Chemistry," J. S. Leach, ed., Academic Press, New York (1969).
28. S. Takashima and A. Minikate, Dielectric behavior of biological macromolecules, in: "Digest of Dielectric Literature," National Research Council, Washington, D.C. 37:602 (1975).
29. H. P. Schwan, Biophysical principles of the interaction of ELF-fields with living matter. II. Coupling considerations and forces, in press, Plenum Press (1984).
30. H. P. Schwan, Interactions of ELF-fields with excitable tissues, in press, Plenum Press (1984).

ELECTRICAL BEHAVIOR OF HUMAN SERA BETWEEN 50 - 1000 MHz

Aldo Casaleggio[1], Claudio Martini[1], Mauro Morando[1], Sandro Ridella[1], Guiseppe Sandro Mela[2], Livio Spiga[2] and Ettore Intra[3]

[1]Istituto per i circuiti Elettronici, C.N.R., Genova
[2]I.S.M.I. Cattedra di Clinica Medica R, Universita' di Genova, Genova
[3]Ospedale Galliera, Genova, Italy

INTRODUCTION

The interaction between electromagnetic fields and biological systems may be used for both modifying the behavior of the living matter and for measuring properties at the protein or at the cell level. The dielectric constant measurement may give information about surface behavior, membrane properties and protein characteristics[1]. Using the radio-frequency range it is well known that it is possible to obtain information about the behavior of the protein surface[2,3]. This has an interest for biochemical and biophysical research in order to understand protein structure and chemistry. Work has been done to correlate complex dielectric constant measurements with the pathological conditions of the individuals from whom the sample has been obtained[4,5,6]. In this paper we describe:

1. a procedure especially developed to obtain the complex dielectric constant of small biological samples of human sera in the frequency range 50 - 1000 MHz in real time;
2. a model which fits the obtained data in order to characterize the surface properties of the protein in the suspension and the water-ions' behavior near the macromolecules;
3. the correlation between the properties of the water bound to the proteins and the pathological conditions of the subjects. This research may be important since it can be used with other measurement techniques (NMR, Calorimetry, Viscosimetry, etc.) for modelling biological tissues by fitting together the results.

COMPLEX DIELECTRIC CONSTANT MEASUREMENT

The first thing is to define what is the complex dielectric constant. It is well known that in a material we can have moving charges and electrical dipoles which can rotate: from the theoretical point of view, the best matter is to define as conductivity all the properties which originate from the first and as dielectric constant those from the second. With electrical measurements, it is impossible to separate these two phenomenon at a single frequency; then, in this field of research, the measured electrical quantity, defined as complex dielectric constant, may be written as:

$$\varepsilon = \varepsilon' - j\varepsilon''$$

In another field of research a similar quantity, the complex conductivity has been defined as:

$$\sigma = \sigma' + j\sigma''$$

ε', ε'', $\sigma'$and σ'' are real, positive, frequency dependent values and the relation between ε and σ is $\sigma + j2\pi f\varepsilon_v \varepsilon$ where f is the frequency and ε_v,= 8.85×10^{-12} F/m. It is easy to obtain the relation between the real and the imaginary parts:

$$\sigma' = 2\pi f\varepsilon_v\varepsilon''; \quad \sigma'' = 2\pi f\varepsilon_v\varepsilon'$$

By using frequency sweeping it is possible to fit models having information on charge and dipole separately. The main problems in measuring complex dielectric constants of human biological materials in radio and microwave frequency range are:

1. the values of ε' are generally high (ε' = 50-100) and may enhance the efforts of errors in the geometrical dimensions of the sample holder;
2. in broad band measurements, the dissipation factor $\varepsilon''/\varepsilon'$ may have large variations;
3. the material to be tested is usually available in small quantities;
4. in general the material is highly sensitive to thermal variations;
5. the measurements have to be taken very quickly in order to characterize a large number of samples to make significant statistical analysis of the results;
6. moreover, quick measurements avoid denaturation problems due to heating.

A schematic longitudinal view of the sample holder used is shown in Figure 1. The holder is essentially a cylinder with circular cross section, 7×10^{-3} m in inner diameter, divided in two parts. Materials to be measured are placed in the upper part, which is actually a coaxial line (length 4×10^{-3} m, diameter 1.3×10^{-3} m) terminated in a circular waveguide. A teflon dish prevents leakage into the lower part, which is a coaxial line with a characteristic - impedance of 50Ω, terminated at section A-A' by a standard APC 7 connector. At section A-A' the structure is connected to the measuring instrument which, in this case, is a Hewlett Packard 4191A RF Impedance Analyzer controlled by an HP85 Computer using IEEE 488 Bus. Logarithmic swept measurements were taken at 39 points between 50-1000 MHz. Repeated testing showed that it was sufficient to fill the holder with a liquid column a few millimeters higher than the top of the inner conductor: more material did not show any influence on the complex reflection coefficient at section A-A'. This result was expected according to previous measurements[6] and from theory[7]. Then the quantity of material which is needed to fill the holder is less than 5×10^{-7} m^3 (0.5 ml.) Following Bussey[7], the holder of Figure 1 can be modelled with a chain of two lines terminated in an open circuit. The first line (characteristic impedance R_1, electrical length L_1, both in vacuum) represents the segment from A-A' to B-B'. The second line represents the structure from B-B' (characteristic impedance R_2, electrical length L_2, both in vacuum): it is filled with the material to be tested. The length L_2 takes into account the length of the inner conductor plus about 1×10^{-3} m which represents the length of the terminating circular waveguide[7]. Repeated measurements of NaCl solution in distilled water (normality 0.0, 0.077, 0.098, 0.154, 0.196 Eq/l) were made: these values were stored in a CDC Cyber 170/720 computer. Using expressions of the complex dielectric constant of NaCl solutions, obtained by modifying Stogryn expressions[8] (Appendix 1), which are functions of NaCl normality, frequency and temperature (which is measured, for every sample, with thermistor using a digital ohmmeter) and using the optimization procedure due to Hook-Jeeves[9], the unknown parameters R_1, R_2, L_1, L_2 were obtained: $R_1 = 47.93\Omega$, $R_2 = 102.9\Omega$, $L_1 = 9.063 \times 10^{-3}$ m, $L_2 = 5.266 \times 10^{-3}$ m. These values compare favorably with the geometrical dimensions of the holder. The procedure was the following:

1. four values for R_1, R_2, L_1, L_2, were considered;
2. the values of the complex dielectric constant of the measured NaCl solutions were obtained from the complex reflection coefficients solving the complex transcendental equations obtained from the two line chain model previously mentioned;
3. these values were compared with those obtained from modified Stogryn (Appendix 1) expressions and the magnitude of the differences were summed;
4. the last quantity was minimized by varying the values of R_1, R_2, L_1, L_2.

The difference between the complex dielectric constant, obtained from the measured complex reflection coefficients using the two line chain with the optimized values of R_1, R_2, L_1, L_2, and the modified

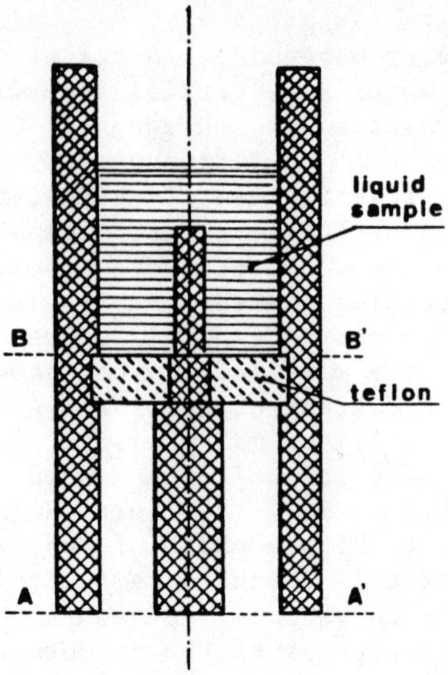

Fig. 1. A schematic lateral view of the sample holder.

Stogryn (Appendix 1) value was 1%, which was in the range of the accuracy of the HP4191A. The procedure actually used is different, since to solve the complex transcendental equations to obtain the complex dielectric constant from the complex reflection coefficient- is a heavy computation task for a personal computer, if the result has to be obtained in real time (less than one minute), in order to let the operator check the results immediately. We used the procedure previously described[6]:

1. during the calibration task three known NaCl solution in distilled water (0.0, 0.077, 0.154 Eq/l) were measured and the complex reflection coefficients were stored in the HP85 Computer (ρ_x, ρ_y, ρ_z);
2. the complex dielectric constants of the three NaCl solutions were computed from modified Stogrym (Appendix 1)[8] (ε_x, ε_y, ε_z);
3. the complex reflection coefficient of the unknown material was measured and stored in the HP85 computer (ρ);
4. the value of the complex dielectric constant of the material under test (ε) was computed from the cross ratio expression:

$$\frac{(\varepsilon - \varepsilon_x)(\varepsilon_y - \varepsilon_z)}{(\varepsilon - \varepsilon_y)(\varepsilon_x - \varepsilon_z)} = \frac{(\rho - \rho_x)(\rho_y - \rho_z)}{(\rho - \rho_y)(\rho_x - \rho_z)} \tag{1}$$

This procedure is very quick: the time required for step 3 is about 40 seconds and the time required for step 4 is about 30 seconds. In order to show that using the previously described procedure, yields the same results that are obtained from the two line model, a simulation on the HP85 Computer was made using the following procedure:

1. a typical sample was considered: its complex dielectric constant was known ($\bar{\varepsilon}$);
2. using the two line model, the complex reflection coefficient at section A-A' was calculated when the testing material was that of a previous point (ρ);
3. in the same way, the complex reflection coefficients were obtained (ρ_x, ρ_y, ρ_z) when the holder was filled with the three NaCl solutions of the calibration task (0.0, 0.077, 0.154, Eq/l), whose complex dielectric constants were known from modified Stogryn (Appendix 1)[8](ε_x, ε_y, ε_z);
4. using Eq.(1) ε was compared for each frequency with $\bar{\varepsilon}$: the magnitude of the difference ($\varepsilon - \bar{\varepsilon}$) was always lower than 0.001.

The goodness of this result depends on the choice of the calibration solutions, with regard to the samples which have to be measured. In fact the simulation just described was put into an optimization procedure to find the best values of the three NaCl solution normalities for a typical human serum.

THE PROTEIN SURFACE MODEL

The problem of obtaining information of biological interest from the complex dielectric constant measurements of protein suspension has been discussed by many Authors[2,3]. A well accepted model is the following (Figure 2)[10]:

1. The protein is modelled as a sphere with $\varepsilon_p = 2\text{-}j0$; the volume fraction of the proteins in the whole suspension is v_p[2,3]:
2. A shell of immobilized water ($\varepsilon_{iw} = 5\text{-}j0$) surrounds the protein and the volume fraction in the whole suspension is v_{iw}[13];
3. The complex dielectric constant of the bound water (whose volume fraction in the suspension is v_{bw}) is given by:

$$\varepsilon_{bw} = \varepsilon'_\infty + \frac{\varepsilon_o - \varepsilon_\infty}{1 + jf/f_\delta} + \frac{\sigma_\delta}{j2\pi f\varepsilon_v} \qquad (2)$$

where

$$\varepsilon'_\infty = \varepsilon_\infty + \varepsilon_\Delta \qquad (3)$$

$$f_\delta = \beta_1 \cdot f_o \qquad (4)$$

$$\sigma_\delta = \beta_2 \cdot \sigma_o \qquad (5)$$

where the values of ε_∞ (high frequency dielectric constant), ε_o (static dielectric constant), f_o (relaxation frequency) and σ_o (d.c. conductivity) are obtained from modified Stogryn (Appendix 1), as functions of temperature and equivalent NaCl normality; ε_Δ, β_1 and β_2 are the parameters characterizing the protein surface, and they will be obtained from the measurement of the complex dielectric constant of the protein suspension.

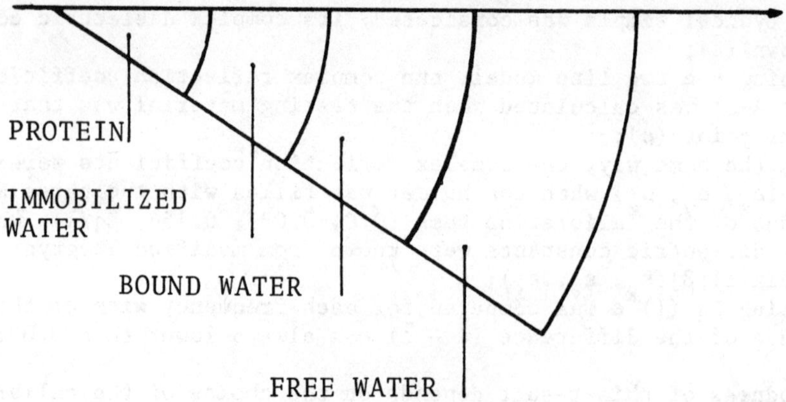

Fig. 2. The protein surface model.

4. The expression of ε_{fw} for free water is the same as modified Stogryn (Appendix 1); its volume fraction in the suspension is v_{fw}.

Some comments may be stressed about previous definitions:

1. Obviously,

$$v_p + v_{iw} + v_{bw} + v_{fw} = 1 \tag{6}$$

2. The NaCl equivalent normality means the NaCl normality which behaves, as far as the complex dielectric constant is concerned, like the solution under test, which contains many different types of ions: the true reason why modified Stogryn (Appendix 1) expressions are used is that they are the only available analytical expressions which give ε as a function of frequency, temperatures and normality;
3. The expressions of the complex dielectric constant of the bound water, Eq.(2),(3),(4),(5), fit experimental results very well. In order to justify these expressions, the following considerations, deduced from literature[2,3], may be done:

a) the relaxation frequency of the bound water is smaller than that of free water due to the increased water rotational viscosity near protein surface[11];
b) from fitting our experimental data we found that the best approach was to take the bound water conductivity equal to the free water conductivity ($\beta_2 = 1$).
c) the high frequency dielectric constant of the bound water is greater than in the free water, and the difference has been estimated $\varepsilon_\Delta = \varepsilon_o - \varepsilon_\infty \overset{N}{\cong} 75$ by fitting our data as well as from literature[2,3].

We have used the mixture expression proposed by Maxwell-Wagner Fricke[12] in order to obtain the complex dielectric constant of the suspension:

(7)

where ε, ε_1, ε_2 are the complex dielectric constants of the suspension, suspending and suspended media, respectively; v_2 is the volume fraction of the suspended medium and x is a form factor which is equal to 2 for the sphere. Thus starting from the protein and the immobilized water using Eq. (7) one can obtain the equivalent complex dielectric constant of the mixture. This result is the value of ε_2 into Eq. (7) for the mixture with the next shell and so on. In this way the complex dielectric constant of the protein suspension can be obtained and it can be compared with the measured value. The parameter of biological interest can be fitted by this model but the optimization is time expensive and the results cannot be obtained in real time on a personal computer like HP85. It is easy to show (Appendix 2) that Eq. (7) is equivalent to

(8)

if $|\varepsilon_1|\cdot(v_2+x)>>|\varepsilon_2|\cdot(1-v_2)$, which is always true for the three mixtures which have to be done. Then the complex dielectric constant of the suspension is a linear combination of ε_p, ε_{iw}, ε_{bw} and ε_{fw}. The complex dielectric constant of the suspension obtained from ε_p, ε_{iw}, ε_{bw} and ε_{fw} using Eq. (8) (Appendix 3) is the ratio between two polynomials of third degree of the complex frequency, whose coefficients are functions of the 4 parameters of biological interest[13]:

1. the quantity of immobilized water,
2. the quantity of bound water,
3. the bound water relaxation frequency,
4. the NaCl equivalent normality in the suspension.

The volume fraction of the protein in the suspension was obtained from biochemical measurements and the volume fraction of free water may be obtained from Eq. (6). Then the coefficients of the polynomials may be fitted easily from the experimental data[14] and from these one may obtain the 4 unknown parameters. On an HP85 this procedure needs 35 seconds and it allows the operator to obtain, a few seconds after measurements, the values of the parameters which characterize a serum. In Figure 3 and Figure 4 a typical result for ε' and ε'' of a human serum from normal subject is presented. A parameter is evaluating the goodness of the fitting has been defined as

$$C = (\Sigma_{freq}((\Delta\varepsilon')^2 + (\Delta\varepsilon'')^2)/39)^{\frac{1}{2}}$$

where $\Delta\varepsilon'$ and $\Delta\varepsilon''$ are, at each measured frequency, the differences between experimental and theoretical values. The range of the values of C parameter is 0.1-0.2 for all the measured human sera: this is also the range of the difference $|\Delta\varepsilon'|$ and $|\Delta\varepsilon''|$ for a serum as a function of frequency. For the data of Figure 3 and Figure 4, the measured temperature is 23.04°C and the biochemical protein volume is v_p = .054; the optimized values (C = .16) are v_{iw} = .006, v_{bw} = .029, f_δ^p = 244 MHz, and N = .130 Eq/l.

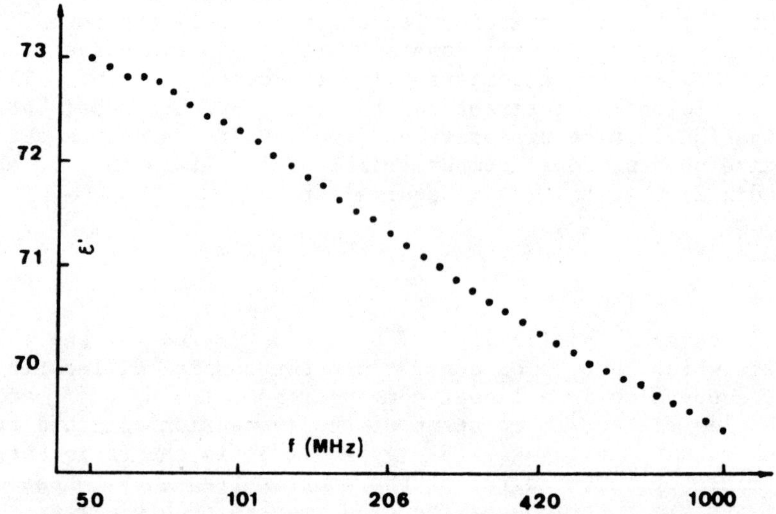

Fig. 3. The real part of the complex dielectric constant of a human serum.

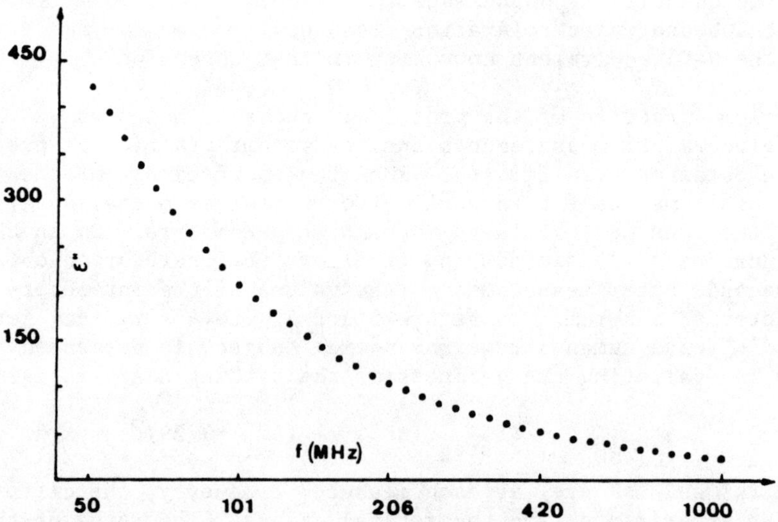

Fig. 4. The imaginary part of the complex dielectric constant of a human serum.

SUBJECTS, BIOLOGICAL MATERIALS AND METHODS

Subjects

About 1200 normal young blood donors voluntarily agreed to participate in this clinical trial. For each subject blood was sampled before breakfast, after at least twenty minutes rest. Excluded from this normal group, were the cases who evidenced any illness six months before and/or after the blood was drawn, as well as any abnormalities in routine clinical and laboratory examinations. After screening, we selected 697 normal subjects. 61 patients with untreated, histologically confirmed cancer of the bowel without any sign of methastases, 8 patients with untreated chronic active hepatitis, and 5 individuals with untreated cirrhosis of the liver and ascitis, volunteered to participate in this project. Also in these cases, blood was obtained before breakfast, at resting.

Biochemical Measurements

All sera were prepared by drawing blood from the elbow vein into a clean plastic syringe. The blood was transferred to a glass tube and allowed to clot at 22°C for six hours. The tubes were centrifuged for 15 minutes at 3000 rpm. Serum was removed with a Pasteur pipet and immediately divided into several 2 ml aliquots in tightly sealed plastic vials. Aliquots were frozen slowly at -4°C for 24 hours, and then stored at -20°C. Sera were used within six months of acquisition and thawed only once. Sera were stored until biochemical and dielectric spectra determination, which was made at the same day for each group of samples. Specimens were allowed to melt slowly at room temperature, reaching 24°C, the temperature at which the tests were performed. On one aliquot, total serum protein determinations were done by the Biuret-Lowry method. Protein electrophoresis pattern were accomplished on a Technicon strip electrophoresis apparatus. Stained stripes were read for densitometer to determine the concentration of serum protein fractions. From the weight/weight proteinemia, P_{cw}, and the standard electrophoresis one can calculate the weight concentration of each serum protein. From these values, from the average molecular weight and from the average partial specific volume of each serum protein (Table Geigy), the molar, P_{cm}, and the volume, P_{cv}, concentration of total serum protein in the samples can be calculated, according to the following formulas:

$$P_{cm} = P_{cw}*(alb/69+\alpha_1G/150+\alpha_2G/560+\beta G/695+\gamma G/156)/100$$

$$P_{cv} = P_{cw}*(.733*alb+.767*\alpha_1G+.714*\alpha_2G+.838*\beta G+.739*\gamma G)/100$$

Obviously, P_{cw}, P_{cv}, and P_{cm} cannot be considered equivalent values, being related to a mixture of several different proteins. The NaCl equivalent normality of the samples was assumed to be N_c = Na + K, being Na (in mEq/l) and K (in mEq/l) the total concentration of sodium and potassium in the whole serum, as obtained with flame determination.

Dielectric Measurements

Calibration of automatic impedance analyzer was always tested using as reference standard the commercially available Emagel (Behringer), a very stable 10^{-3} molar solution of 35000 molecular weight gelatine in saline. Calibration was accepted only when Emagel dielectric measurements did not differ significantly from a set of these, previously measured with high precision degree[15].

RESULTS

Measurement Accuracy

Standard reference solution Emagel was used to study measurement accuracy. A set of 158 dielectric determinations of Emagel were performed at standard temperatures (24±.5°C) under different calibrations. The optimized parameters v_{iw}, v_{bw}, N, and f_δ showed a present standard deviation lower than 1%. Lower and upper values of each parameter never spread up to 2% of the mean values.

Model Accuracy

Model accuracy can be tested in several ways, for example, verifying that results are physically consistent and like those previously reported in literature. However, the more powerful tool to state the accuracy of the present model to describe the water and the protein interaction in solution, is to keep in mind that from more than 1300 dielectric determinations of human sera, we never found negative values of v_{iw}, v_{bw}, N and f_δ. Moreover, the optimized NaCl equivalent normality was always found to be lower than the chemically assessed ones, N_c.

Freezing-time Effect on Dielectric Measurements

To verify the effect of the freezing time on the dielectric measurements, ten normal sera and ten samples of Emagel were divided into 50 aliquots and then frozen. Dielectric measurements of these frozen aliquots were made every month for four years. For each serum and Emagel sample, optimized parameters of these different freezing-time aliquots have never showed any statistical difference nor a trend behavior due to time. Therefore, one can deduce that human sera and Emagel can safely be stored for at least four years at -20°C.

Results Presentation

The total amount of protein in sera varies, although within narrow limits, from subject to subject: therefore, the optimized values of immobilized and bound water must be normalized for the protein concentration. It is customary among protein chemists to give the protein hydration, H, in terms of weight/weight ratio between the coupled water and the protein. Nevertheless, we chose the molar/ molar ratio for the following reasons. The frequency distribution of the protein bound water does not significantly differs from the gaussian distribution (χ^2 = 15.4, f.d. = 11, $p(\chi^2)$ = .17). The frequency distribution of the proteinemia significantly differ from the normal distribution when the protein amount is expressed as weight p_{cw}, (χ^2 = 28, f.d. = 10, $p(\chi^2)$ = .002), and when it is expressed as volume, P_{cv}, (χ^2 = 19.7, f.d. 10, $p(\chi^2)$ = .03). On the contrary, the frequency distribution of P_{cm} does not evidence any significant difference when compared with the guassian distribution (χ^2 = 6.5, f.d. = 11, $p(\chi^2)$ = .83). Of course, the molar/ molar ratio R between the bound water and the protein concentration is normally distributed too (χ^2 = 11.8, f.d. = 10, $p(\chi^2)$ = .3). The parameter R was defined as follows:

$$R = w_{bw}/P_{cm}; \text{ where } w_{bw} = v_{bw}/18$$

The weight/weight ratio H distribution does not fit the normal distribution. The normal distribution of the R parameter is of major interest, because this is a fundamental condition to perform statistical analysis in order to assess difference, if any, between R of two or more groups of subjects, i.e. normal and pathological individuals.

Effect of Serum Protein Mixture Composition on the Bound Water

Human serum may be regarded as a mixture of several kinds of protein in saline. It seems therefore, useful to verify the effect on the optimized parameters of large variations in the protein mixture ratios. Unfortunately, normal sera usually evidence a rather constant composition and cannot therefore, be used for this purpose. We choose sera from patients with chronic active hepatitis and cirrhosis of the liver with ascitis because these diseases are not malignant illnesses, becuase they usually show an albumin/gamma-globulin ratio significantly lower than that of normal, and becuase clear serum proteins alteration are not reported for both these pathological conditions. As already reported[16,17], both chromic active hepatitis and cirrhosis patients evidenced an albumin/gamm-globulin ratio (1.49 ± .40 and 1.12 ± .39, respectively) significantly lower than the normal one, 3.15 ± .64. They showed proteinemia (773 ± 188 and 599 ± 30) and R ratio (2131 ± 688 and 2681 ± 321) not significantly different from the values of normal control groups. This result suggests that the R parameter is uninfluenced by albumin/gamma-globulin variations, at least in these illnesses and in these protein concentration ranges.

Correlation Between Bound-water and Total Serum Protein

It seems reasonable to assume that the more the protein concentration, the more the protein bound-water will be. In normal subjects, proteinemia varies from .570 to .889 mMol/l, with mean of .707 ± .052. Both p_{cm} and w_{bw} are normally distributed, thus a linear regression can be calculated. Two models of linear relationships may be used: the first is characterized by two parameters, the slope and the intercept, the second is characterized only by the slope, passing the straight line through the origin of the coordinates. The first model suggests that the water may be bound not only to both serum proteins and other soluted molecules, the second model states on the contrary that the bound-water is proportional to the protein concentration. The interpolation of w_{bw} as function of the p_{cm} with the linear regression with the intercept gives a good correlation coefficient, being r=.435 (t(r)=12.7), and the elevation does not differ significantly from the zero (t=.436). The interpolation with the linear model without the intercept gives a correlation coefficient r=.997 (t(r)=354.9), with slope of 2191±6. These findings strongly suggest that the water molecules which behave as bound from a dielectrical point of view are mainly coupled to the serum proteins, and that w_{bw} is directly proportional to p_{cm}. Such a result cannot be obtained if total protein weight or volume concentration amounts are considered.

The Hydration of Each Kind of Serum Proteins

Bound-water and total serum proteins were found to be proportionally correlated. Because the total proteins are the sum of several types of proteins, it follows that bound-water may also be proportional to each of these kinds of serum proteins. The hydration of the protein mixture components may be assessed using the multiple regression analysis without the constant factor. Such a relationship was found to be very significant, being F=42249 (p(F)≅0). The relevant results are reported in Table 1. As one can see, 1409 ± 38 molecules of water are bound to one molecule of albumin, with a weight/weight ratio of about .37, and 3111 ± 200 to a molecule of gamma-globulin, with a weight/weight ratio near to .36. These results agree very well with those already reported in the literature, although they were obtained with different experimental techniques.

Difference Between Normal and Cancer Serum Proteins

In normal subjects the R parameter showed an average value of 2206 ± 165, whereas in cancer subjects the R mean was 2564 ± 342: these two mean values strongly differ between themselves ($t=14.4$, $p(t)<10^{-7}$). Therefore, one can state that cancer serum proteins are more hydrated than normal serum proteins. Calculating with the multiple regression analysis the number of molecules of water bound to each serum protein type in cancer serum, one cannot find any significant differences, apart from the gamma-globulin, which can bound 3283 ± 695 molecules of water, behaving thus as more hydrated than the normal gamma-globulins ($t=4.7$, $p(t)<10^{-6}$).

DISCUSSION

The biological importance of interactions of suspending medium and proteins in solution is discussed in[10]. The role of suspending these interactions during cancer conditions has been previously reported[18]. Present data suggest that among cancer serum proteins, mainly gamma-globulins are more hydrated than normal proteins, thus giving further evidence of an immunological disorder during malignancies. These abnormalities might be characteristic features of some diagnostic value. Blood serum protein from patients with untreated solid cancer[19] and plasma cell myeloma[20] were shown to be more hydrated than normal ones. At the present state of the art, such a higher-than-normal hydration was found exclusively in the serum protein of patients with malignancies, whereas the serum protein hydration is found to be normal in the individuals without malignancies, and it does not depend on the total protein composition of sera[16,17]. Moreover, the relaxation frequency of the bound water of serum proteins from cancer and plasma cell myeloma is found to be lower than normal, thus suggesting that these molecules may rotate at a rate lower than normal serum proteins[13,16,17].

Table 1. The Hydration of Serum Proteins.

Protein	R	s.d.	w/w ratio
Albumin	1409	38	.37
α_1-globulins	906	886	.11
α_2-globulins	18784	1802	.60
β-globulins	20636	1976	.53
γ-globulins	3111	200	.36

CONCLUSION

The properties of water bound to proteins are becoming an important tool for biochemical research and for clinical purposes. Biochemical measurements of protein bound water are very troublesome. Many biophysical methods are suitable to study protein bound water, e.g. calorimetry, viscosimetry and, overall, NMR. Dielectric techniques compare favorably with them, being in many cases quicker, cheaper and more reliable.

APPENDIX 1: MODIFIED STOGRYN EXPRESSION

The expression of the complex dielectric constant of NaCl solutions was modified in order to reduce computer time during optimization computation. The temperature range of validity was reduced to the range from 20 to 30 degrees centigrade and the normality range was reduced to the range from 0 to 0.2 Eq/l. The expression for the d.c. conductivity was obtained by fitting the experimental data obtained from[21] in the range above specified: it is much more accurate than the corresponding Stogryn expression since this is valid in the range 0-3 Eq/l of normality.

where $\varepsilon_\infty=4.9$

$$\varepsilon_o(T,N)=\varepsilon_o(T,0)\cdot a(N)$$

$$\varepsilon_o(T,)=78.35-(T-25)\cdot .3431$$

$$a(N)=1-.25\cdot N$$

$$f_o(T)=(5.118\cdot 10^{-11}-1.323\cdot 10^{-12}\cdot (T-25))^{-1}$$

$$\sigma_o(T,N)=\sigma_{NaCl}(25,N)\cdot b(25-T)$$

$$\sigma_{NaCl}(25,N)=N\cdot (12.645-6.12\cdot (N)^{\frac{1}{2}})$$

$$b(25-T)=1-.0208\ (25-T)$$

f being the frequency (in Hz), N the normality (in Eq/l) and T the temperature (in °C).

APPENDIX 2: APPROXIMATED MIXTURE EXPRESSION

It is well known from literature[22,23] that the mixture expression (7) is equivalent to the circuit of Figure 5 where:

$$Y_o = A\cdot\varepsilon_1 \quad ; \quad A = \frac{\chi\cdot(1-v_2)}{\chi+v_2}$$

$$Y_1 = B\cdot\varepsilon_1 \quad ; \quad B = \frac{v_2\cdot(1+\chi)^2}{(1-v_2)\cdot(\chi+v_2)}$$

$$Y_2 = C\cdot\varepsilon_2 \quad ; \quad C = \frac{v_2\cdot(1-\chi)^2}{(\chi+v_2)^2}$$

The series of Y_1 with Y_2, when $B\cdot\varepsilon_1 >> C\cdot\varepsilon_2$, is reduced to Y_2 alone since Y_1 is equivalent to a short circuit. Then, making the parallel of Y_o with Y_2 one obtains (8).

APPENDIX 3: COMPLEX DIELECTRIC CONSTANT OF THE SUSPENSION

Using the mixture expression (8) and the given expression of ε_p, ε_{iw}, ε_{bw} (2) and ε_{fw} (Appendix 1), the final expression of the complex dielectric constant of the suspension is given by:

$$\varepsilon = \left(\varepsilon_\infty + \frac{\varepsilon_o-\varepsilon_\infty}{1+jf/f_o} + \frac{\sigma_o}{j2\pi f\varepsilon_v}\right)\cdot\frac{\chi\cdot(1-v_p-v_{iw}-v_{bw})}{(\chi+v_p+v_{iw}+v_{bw})} +$$

$$+ \frac{(1+\chi)^2\cdot(v_p+v_{iw}+v_{bw})}{(v_p+v_{iw}+v_{bw}+\chi)^2}\cdot\left(\left(\varepsilon'_\infty + \frac{\varepsilon_o-\varepsilon_\infty}{1+jf/f_\delta} + \frac{\sigma_\delta}{j2\pi f\varepsilon_v}\right)\cdot\right.$$

$$\left[\frac{\chi\cdot\left(1-\dfrac{v_{iw}+v_p}{v_{iw}+v_p+v_{bw}}\right)}{\left(\chi+\dfrac{v_{iw}+v_p}{v_{iw}+v_p+v_{bw}}\right)} + \frac{(1+\chi)^2\cdot\left(\dfrac{v_{iw}+v_p}{v_{iw}+v_p+v_{bw}}\right)}{\left(\dfrac{v_{iw}+v_p}{v_{iw}+v_p+v_{bw}}+\chi\right)^2}\right]\cdot\quad(5-j\phi)\cdot$$

$$\left.\left(\frac{\chi\cdot\left(1-\dfrac{v_p}{v_{iw}+v_p}\right)}{\left(\chi+\dfrac{v_p}{v_{iw}+v_p}\right)} + (2+j\phi)\cdot\frac{(1+\chi)^2\cdot\left(\dfrac{v_p}{v_{iw}+v_p}\right)}{\left(\dfrac{v_p}{v_{iw}+v_p}+\chi\right)^2}\right)\right)$$

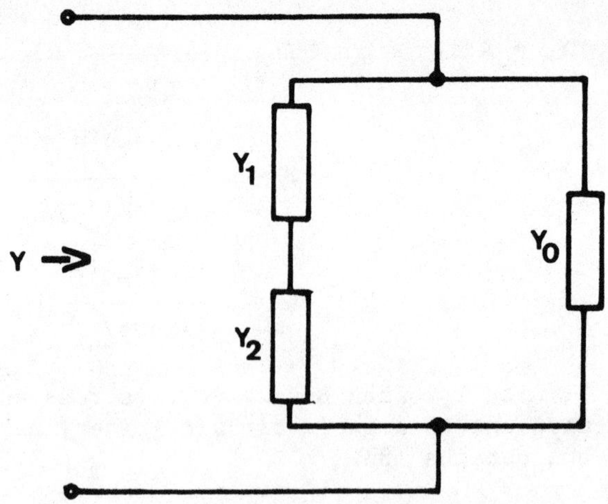

Fig. 5. Mixture expression equivalent circuit.

REFERENCES

1. H. P. Schwan, Electric properties of tissue and cell suspensions, in: "Advances in Biological and Medical Physics", C.A. Tobias, ed., Academic Press In., New York (1957).
2. E. H. Grant, S. E. Keefe and S. Takashima, The dielectric behavior of acqueous solutions of bovine serum albumin from radiowave frequencies, J.Phys.Chem., 72-13:4373 (1968).
3. B. E. Pennok and H. P. Schwan, Further observations on the dielectric properties of hemoglobin-bound water, J.Phys.Chem., 73-8: (1969).
4. C. G. Essex, E. H. Grant, R. J. Sheppard, G. P. South, M. S. Symonds, G. L. Mills and J. Slack, Dielectric properties of normal and abnormal lipoproteins in acqueous solutions, Ann.NY Acad.Sci., 303:142 (1977).
5. V. A. Zore, O. D. Kimel'fel'd, A. I. Shul'mina and P. V. Afanas'ev, Change in the dielectric properties and resistance to heating of human serum proteins in normal conditions and in certain diseases, Biofiz., 3:469 (1973).
6. B. Bianco, G. P. Drago, M. Marchesi, C. Martini, G. S. Mela and S. Ridella, Measurements of complex dielectric constant of human sera and erythrocytes, IEEE Trans.Instr.Measur., IM-28:290 (1979).
7. H. E. Bussey, Dielectric measurements in a shielded open circuit coaxial line, IEEE Trans.Instr.Measur., IM-29-2:120 (1980).
8. A. Stogryn, Equations for calculating the dielectric constant of saline water, IEEE Trans.Microwave Theory Tech., MTT-19:733 (1971).

9. G. Barabino and M. Marchesi, "An Optimization Algorythm Improving Hook and Jeeves' Pattern Method", Optimization Days, Montreal (1979).
10. R. Cook and I. D. Kunz, The properties of water in biological systems, Ann.Rev.Biophys.Bioeng., 3:95 (1974).
11. H. Elmgren, The mobility profile around polymer segments as calculated from microviscosity data on polymer solutions and gels, J.Polymer Science, 18:339 (1980).
12. M. Parodi, S. Serpico and S. Ridella, Non interacting elipsoidal bodies suspensions, Alta Frequenza, LI-5:287 (1982).
13. S. Ridella, E. Intra, G. S. Mela and L. Spiga, Measurements of complex dielectric constant to detect differences of hydration and ions bound to serum proteins in normal and pathological individuals, Proc.Vth ICEBI Tokio., 261 (1981).
14. E. C. Levy, Complex-curve fitting, IRE Trans.Automatic Control, 37 (1959).
15. M. Marchesi, Emagel as an accuracy standard for human sera dielectric measurements, in: "Rapporto CNR-ICE Genoa 6/1981", CNR-ICE, Genoa (1981).
16. G. S. Mela, B. Bianco, M. Calamai et al., Hydration and protein bound-water dielectric relaxation of blood serum proteins of human subjects with untreated chronic active hepatitis, IRCS Med.Sci.Bioch., 8:227 (1980).
17. L. Spiga, G. S. Mela and S. Ridella, Dielectric direct measurements of the hydration state of blood serum proteins: differences between normal and pathological subjects with cirrhosis of the liver and ascites, IRCS Med.Sci.Bioch., 9:949 (1981).
18. P. T. Beal, B. B. Asch, D. Medina and C. Hazlewood, Distinction of normal, preneoplastic and neoplastic mouse mammary cells and tissues by nuclear magnetic resonance techniques, in: "The Transformed Cells", Academic Press Inc., New York (1981).
19. G. S. Mela, B. Bianco, G. P. Drago et al., Differences in the hydration of blood serum proteins between normal and untreated cancer human subjects, IRCS Med.Sci.Bioch., 7:537 (1979).
20. G. S. Mela, M. Calamai, B. Bianco et al., Amount and physical state of the hydration water bound to human serum total proteins: differences between normal ad plasma cell myeloma subjects, extending to the "non-secretory" condition, IRCS Med.Sci.Bioch., 8:194 (1980).
21. D. E. Gray, "American Institute of Physics Handbook", McGraw-Hill, New York (1972).
22. H. Fricke, The complex conductivity of a suspension of stratified particles of spherical or cylindrical form, J.Phys.Chem., 59:168 (1955).
23. G. P. Drago, M. Marchesi, S. Ridella and M. Parodi, Electrical modelling of cells in: "Modelling and Analysis in Biomedicine", C. Nicolini, ed., World Scientific Publishing Co. PTE Ltd., Singapore (1984).

9. [illegible] and M. [illegible], [illegible] "[illegible] Hodgkin [illegible], [illegible]" [illegible] Research (1979).

10. [illegible] and [illegible] H. [illegible], The [illegible] *Military [illegible]* (1983).

11. [illegible], [illegible] [illegible] [illegible] *[illegible]* (1980).

12. [illegible], [illegible] and [illegible] (1980) [illegible] *Progress [illegible]*.

13. [illegible], M. [illegible], [illegible] [illegible] of [illegible] of [illegible] [illegible] *[illegible]*, 247 (201).

14. [illegible] *[illegible] Research* [illegible] (1979).

15. [illegible] of [illegible] standard for human [illegible] [illegible], (1981), [illegible] (1981).

16. [illegible] M. [illegible] [illegible] protein [illegible] of blood serum proteins in man [illegible] *IEEE Trans. Bio. Eng.* 27 (1980).

17. [illegible] [illegible] and S. [illegible] [illegible] [illegible] [illegible] between [illegible] [illegible] of the liver and [illegible] *[illegible]* (1982).

18. [illegible] [illegible] and [illegible] Blood [illegible] of [illegible] and [illegible] [illegible] by nuclear magnetic resonance [illegible] (1983).

19. [illegible] [illegible] et al., [illegible] [illegible] of blood serum proteins between normal and [illegible] human subjects, *IEEE [illegible]* (1983).

20. [illegible] et al., [illegible] [illegible] of the [illegible] [illegible] [illegible] between normal ad [illegible] subjects, [illegible] of the "non-secretory" condition, *IEEE [illegible]* (1983).

21. [illegible] "[illegible] [illegible] of [illegible] [illegible]" [illegible] New York (1978).

22. [illegible] [illegible] of [illegible] [illegible] *[illegible]* (1975).

23. [illegible] M. [illegible] [illegible] and [illegible] [illegible] of [illegible] [illegible] [illegible] [illegible] (1979).

A SIX-OCTAVE EXPERIMENTAL SYSTEM FOR IN-VIVO AND IN-VITRO DIELECTRIC MEASUREMENTS

Maria A. Stuchly* and Stanislaw S. Stuchly**

*Radiation Protection Bureau, Health & Welfare Canada
Room 233, EHC, Tunney's Pasture, Ottawa, Ontario K1A 0L2
**Dept. of Electrical Engineering, University of Ottawa
Ottawa, Ontario, K1N 6N5

INTRODUCTION

Electrical properties of biological substances and systems have been of great interest ever since the discovery of the bioelectric effect in early 1800's. Very generally, the properties can be divided into two groups, the active properties which describe the ability to generate electrical potentials and fields, and the passive properties, which characterize the responses to externally applied electrical stimuli.

There are numerous and diversified reasons for interest in the passive electrical properties of biological materials, subsystems and systems. On the fundamental research level, the electrical properties provide an insight into the structure and function of molecules, cells and subcellular systems, as well as organized systems such as tissues. Dielectric properties are essential in elucidating interaction mechanisms of electromagnetic fields with biological systems. On the applied level, knowledge of electrical properties is imperative in order to assess the rate of the energy absorption (the specific absorption rate - SAR) and its spatial distribution in biological bodies exposed to electromagnetic fields. The specific absorption rate is used in quantifying the effects of electromagnetic fields on various species including man. The SAR is considered sufficient, in most cases, to quantify the effects of the deposition of thermal energy in the biological system resulting from exposure to electromagnetic fields. In those cases where the effect depends on the rate of temperature change, the temporal changes of SAR have to be determined. The SAR and other essential exposure parameters can also be used in defining exposure conditions for nonthermal biological effects, e.g. calcium efflux, or stimulation of bone growth.

Dielectric properties of biological substances including tissues have been investigated extensively since the 1950's, with the paramount contribution by H.P. Schwan and his colleagues, as well as important studies by others [1].

Current rapid progress in instrumentation and computers has facilitated development of broadband, accurate, automatic and easy to use systems for dielectric measurements. We describe here a computer-controlled system for measurements of passive electrical properties of tissues in vivo and in vitro, and of cell suspensions. The system has been developed and used by our research group at the University of Ottawa. It operates in a frequency range of six-octaves from 20 kHz to 18 GHz. Only two sensors are needed for determination of properties of typical high-water-content tissues in this frequency range.

PRINCIPLE OF OPERATION

The measurement system consists of a sensor and a computer-controlled network analyzer. The sensor translates the permittivity of a test sample into the input reflection coefficient. The reflection coefficient is then measured by a network analyzer.

Various coaxial line arrangements can be used as sensors. For instance, for cell suspensions or tissues in vitro a small shunt capacitor, as shown in Figure 1, can be used [2]. For tissues in vivo, an open-ended coaxial line shown in Figure 2 constitutes a suitable sensor [3]. The equivalent circuit of the sensor usually consists of one or two elements (Figures 1 and 2), a capacitor $C(\hat{\varepsilon})$ ($C(\hat{\varepsilon})=C_0\hat{\varepsilon}$, where C_0 is the capacitance when the line is in air), and a capacitor C_f which accounts for the fringing field [1]. This equivalent circuit is valid at frequencies where the dimensions of the line are small compared with the wavelength.

The input reflection coefficient at the plane of the discontinuity can be calculated from

$$\hat{\Gamma} = \Gamma e^{j\phi} = \frac{1-j\omega Z_0[C(\hat{\varepsilon})+C_f]}{1+j\omega Z_0[C(\hat{\varepsilon})+C_f]} . \qquad (1)$$

Solving for the permittivity we get

$$\hat{\varepsilon} = \frac{1-\hat{\Gamma}}{j\omega Z_0 C_0(1+\hat{\Gamma})} - \frac{C_f}{C_0} . \qquad (2)$$

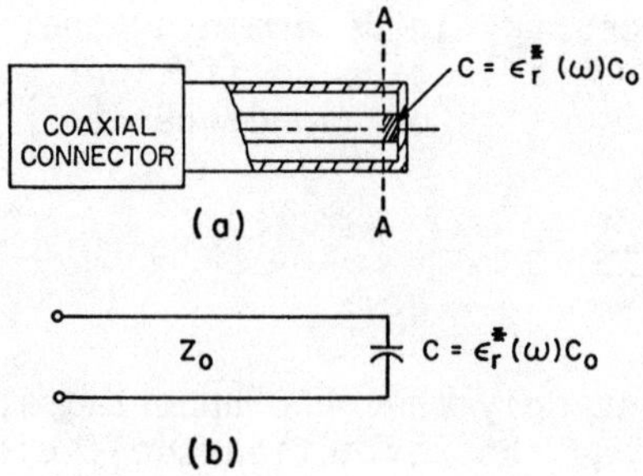

Fig. 1. Shunt capacitor coaxial-line sensor (a) and its equivalent circuit (b).

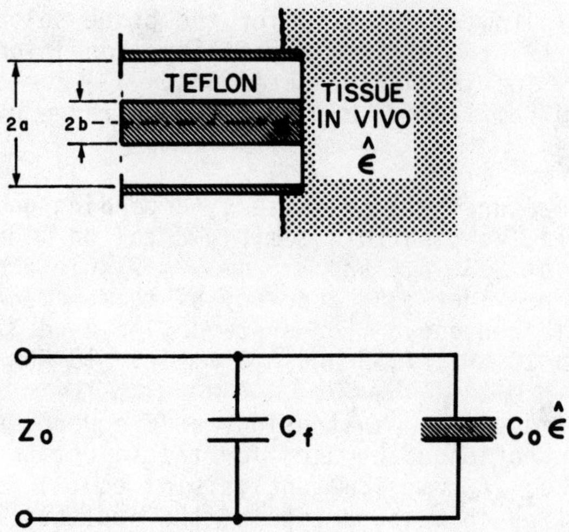

Fig. 2. Open-ended coaxial-line sensor for in-vivo measurements (a) and its equivalent circuit (b).

The reflection coefficient of the sensor with the test sample can, therefore, be used to determine the permittivity, once the capacitances C_0 and C_f are known.

Sensors

The greatest accuracy in determining the permittivity for a given accuracy of the reflection coefficient measurement for materials having $\varepsilon'' < \varepsilon'$ is obtained when [4]

$$C_0 = \frac{1}{\omega Z_0 \sqrt{\varepsilon'^2 + \varepsilon''^2}} \tag{3}$$

This expression holds only when the uncertainities in the magnitude and phase of the reflection coefficient are approximately the same, i.e., $\Delta\phi \simeq \Delta\Gamma/\Gamma$. In other cases this is still a good compromise as the optimum C_0 is different for ε' and ε'', these values being smaller and larger or vice versa than the value given by [3].

The total capacitance in air, $C_T = C_0 + C_f$, is a function of the line dimensions, the permittivity of the dielectric filling of the line, and the frequency. These capacitances can be calculated analytically [5] and also determined experimentally [6].

The following criteria for the probe selection should be applied: (1) the optimum capacitance condition; (2) size, especially for in vivo measurements; (3) compatibility with liquid and semiliquid samples; and (4) range of operating frequencies.

For measurements of tissues, depending on the range of frequencies, we used 50-Ω semirigid teflon lines with external diameters of 8.3, 6.4 and 3.6 mm. A single probe, when properly selected, provides good accuracy of measurements over two to three decades of frequency. For instance, we used 3.6 mm line from 100 MHz to 10 GHz [7,8] and 8.3 mm from 10 MHz to 1 GHz [9].

For tissues and suspensions at frequencies below 1 MHz, eq.(3) can no longer be used for the selection of the optimum capacitance. A numerical analysis of eq.(2) for given uncertainties in measurements of the amplitude and phase of the reflection coefficient has to be done.

Measurement Systems

Schematic diagrams of measurement systems for frequencies between 20 kHz and 100 MHz, and between 100 MHz and 18 GHz are shown in Figures 3 and 4, respectively. Both systems are based on

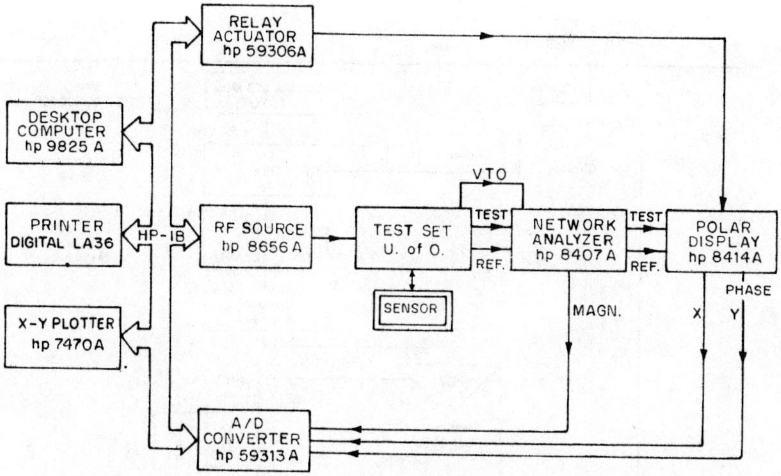

Fig. 3. Measurement system for frequencies between 20 kHz and 100 MHz.

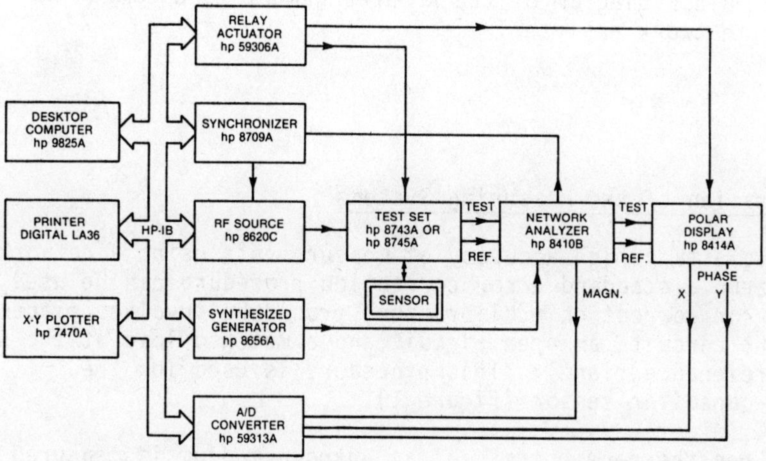

Fig. 4. Measurement system for frequencies between 100 MHz and 18 GHz.

Hewlett-Packard Network Analyzers. To automate the HP 8407A network analyzer for operation between 20 kHz and 100 MHz a special electronic circuity was developed [10]. The circuit utilizes an HP 8656A Synthesized Signal Generator as a main source, while the reference and test signals are obtained by heterodyning the main signal with a signal from a 200-MHz crystal-controlled oscillator [10]. A block diagram of the test set developed at the University of Ottawa is shown in Figure 5. Both systems (Figures 3 and 4) are controlled by an HP 9825 Desktop computer interfaced through the IEEE-488 interface bus.

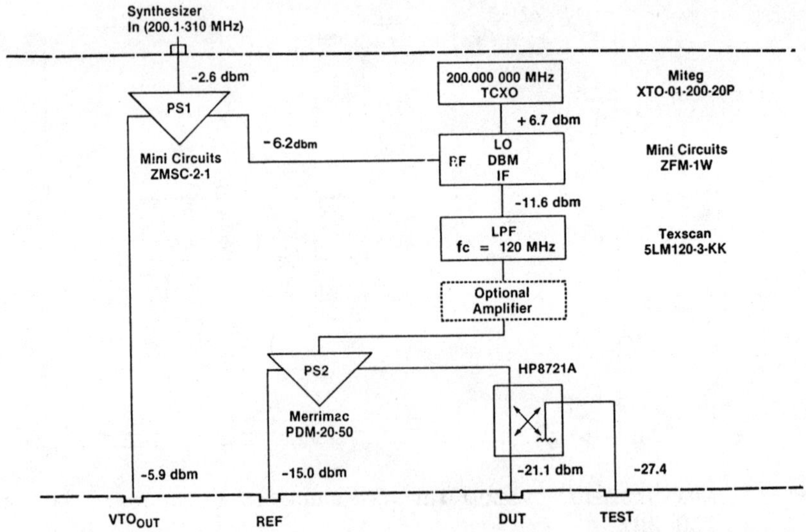

Fig. 5. Block diagram of the RF arrangement to automate the HP 8407A network analyzer.

Calibration of the Measuring Systems

To attain high accuracy of measurements using a network analyzer, a standard error correction procedure can be used [11]. The error correction (calibration) procedure involves placement of a short circuit, an open circuit and a matched load at the sensor end (reference plane). This procedure is used for the shunt-capacitor sensor (Figure 1).

When the permittivity of an unknown medium is measured using an open-ended coaxial line sensor, the input reflection coefficient is affected by the discontinuities between the tip of the sensor and the connector, as well as by the imperfections of the automatic network analyzer system. This effect is particularly pronounced at frequencies above 100 MHz.

For the open-ended coaxial-line sensor, a matched termination or a perfect short-circuit are difficult to realize at the tip of the sensor and therefore, some other "standard" terminations have to be used. It can be shown that the reflection coefficient of the sensor immersed in a liquid of known dielectric properties, such as distilled and deionized water or saline solution, may be used for calibration [12].

The reflection coefficient of the open-ended coaxial-line

sensor can be calculated from

$$\hat{\Gamma}_i = \frac{1-j\omega C_0 Z_0 \hat{\varepsilon}_\ell - j\omega C_f Z_0}{1+j\omega C_0 Z_0 \hat{\varepsilon}_\ell + j\omega C_f Z_0} \tag{4}$$

where the permittivity of the standard liquid is described by

$$\hat{\varepsilon}_\ell = \varepsilon_\infty + \frac{\varepsilon_s - \varepsilon_\infty}{(1+j\omega\tau)^{1-\alpha}} + j\frac{\sigma}{\omega\varepsilon_0} \quad , \tag{5}$$

where ε_s and ε_∞ are the static and high-frequency dielectric constants of the liquid, respectively, τ is the relaxation time, σ is the ionic conductivity in S/m, ε_0 is the permittivity of free space ($8.854.10^{-12}$ F/m), $\omega = 2\pi f$ and f is the frequency in Hz.

The reflection coefficient Γ_i from eq.(4) is then used together with the reflection coefficients corresponding to other standard terminations in the error correction procedure. Generally, up to three different liquids may be used for calibration as required. However, even one standard liquid together with a short-circuit and an open-circuit provide a significant improvement over the standard calibration procedure [11,12].

Water and saline solutions constitute convenient liquids for calibration, as their dielectric properties are well known in a wide range of frequencies.

An improvement in the measurement accuracy as compared with the standard calibration procedure obtained by use of the modified calibration procedure is illustrated in Figure 6. When a 0.7-N saline solution is used in addition to the short and open circuit, errors in the dielectric constant (Figure 6a) and the loss factor (Figure 6b) are very small. This is not the case, though, when a matched load is used, because the matched load cannot be connected at the end of the sensor, so that the imperfections of a sensor section of the coaxial line and a connector are not accounted for.

Measurement Uncertainty

The accuracy of measurement of the dielectric properties depends on the sensor selection (capacitance) and the uncertainty in the reflection coefficient measurements. The effect of the error correction procedure i.e. the system calibration is included in the latter factor.

The uncertainties in the measurement of the magnitude and phase of the reflection coefficient for the modified calibration

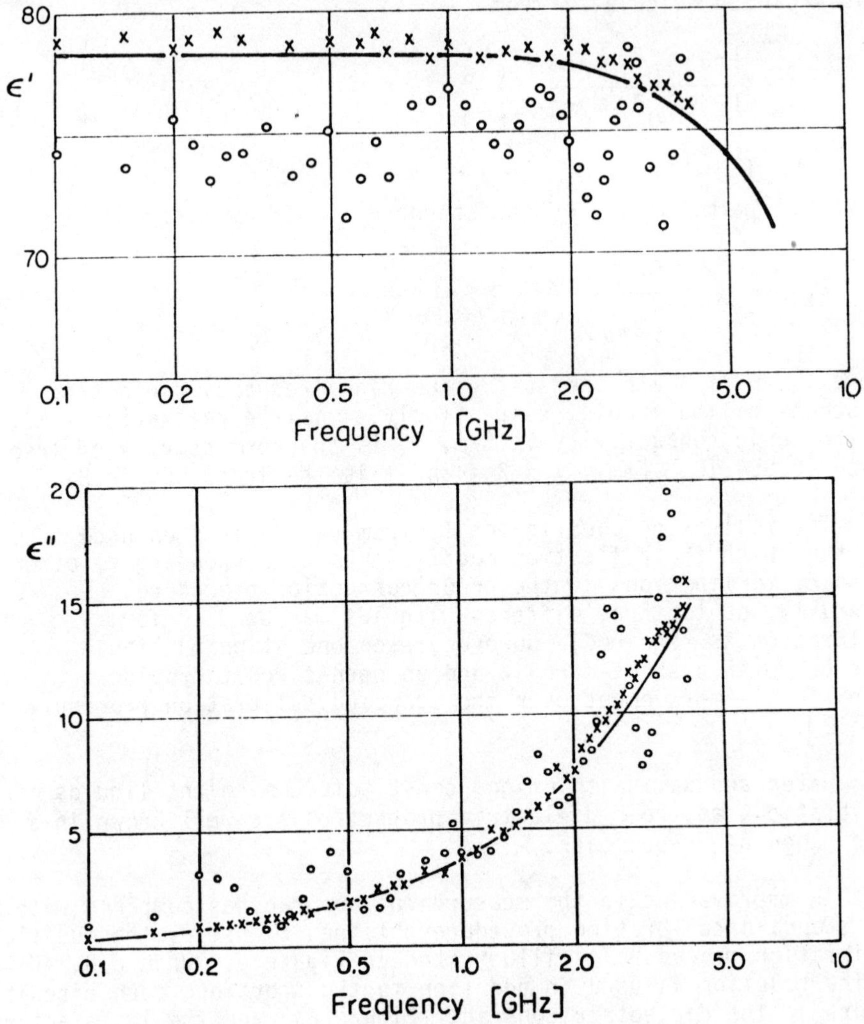

Fig. 6. Dielectric constant and the loss factor of distilled and deionized water at 25°C measured in the same system shown in Figure 4 using the conventional calibration method (O) and the modified method with 0.7-N saline solution (X). Solid line shows the reference data for water.

procedure are $\Delta\Gamma = 0.05$ dB and $\Delta\phi = 0.2^0$, respectively [11]. For these uncertainties the errors in the muscle tissue permittivity were calculated, and are shown in Figure 7 for two coaxial-sensors. The errors in the dielectric constant are below 2% for frequencies below 5 GHz, and increase up to 10% to 10 GHz. The errors in the loss factor remain below 5% between 10 MHz and 10 GHz for any of the two sensors.

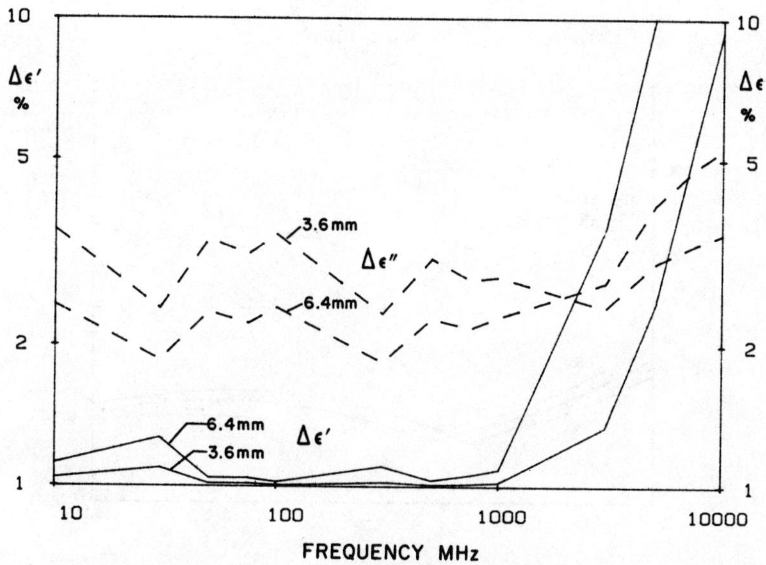

Fig. 7. Errors in the dielectric constant (solid lines) and the loss factor (dashed lines) of muscle tissue in percent, measured with two open-ended coaxial line sensors of diameters 3.6 mm and 6.4 mm. The uncertainties in the measurement of the reflection coefficient were assumed to be $\Delta\Gamma = 0.05$ dB, $\Delta\phi = 0.2^{\circ}$.

At lower frequencies (below 1 MHz) special open coaxial line sensors have to be used to obtain high accuracy of measurements. The capacitance of these sensors has to be relatively high. For example for muscle tissue, the optimum sensor capacitance is about 0.33 pF, as shown in Figure 8. For such a sensor the errors in the dielectric constant are below 2% at frequencies greater than 40 kHz and increase to 8% at 10 kHz. The errors in the loss factor are below 1% in the frequency range from 10 kHz to 100 MHz.

Experimental Results

The measurement system operating from 100 MHz to 10 GHz was used to measure electrical properties of various tissues in vitro and in vivo for different animals [7,8]. Figure 9 illustrates typical results and uncertainty limits obtained in our measurement system and other similar systems.

Recently obtained data at lower frequencies using the system shown in Figure 3 and a shunt capacitor with $C_o = 0.4$ pF are given in Figure 10. This figure shows the changes in the dielectric constant and the loss factor of brain tissue as a function of time after death.

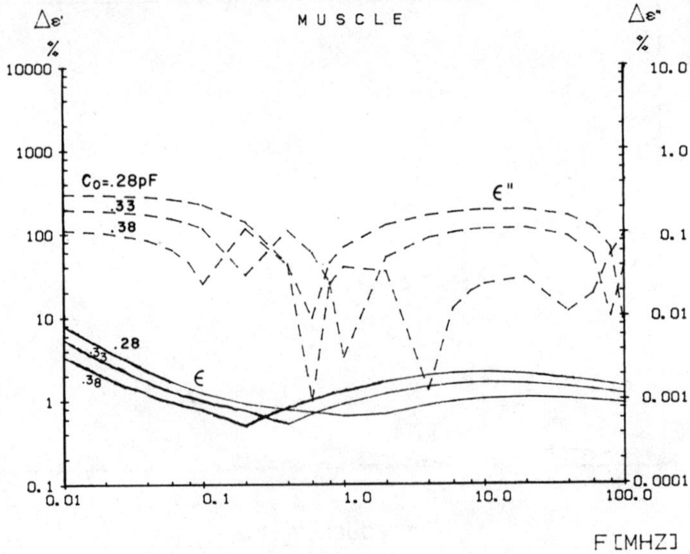

Fig. 8. Errors in the dielectric constant (solid lines) and the loss factor (dashed lines) of muscle tissue in percent, for three different capacitances of the sensor. The uncertainties in the measurement of the reflection coefficient $\Delta\Gamma = 0.05$ dB, $\Delta\phi = 0.2^{0}$.

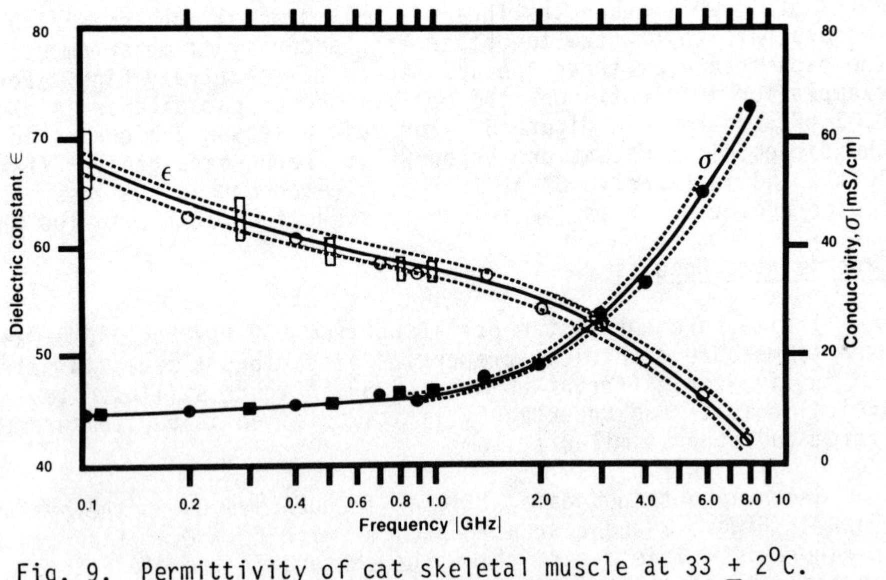

Fig. 9. Permittivity of cat skeletal muscle at 33 $\pm$ 2^{0}C.
Solid lines - average experimental results in vivo;
Dashed lines - uncertainty limits;
Circles - experimental results in vitro;
Rectangles - in vivo data from [9].

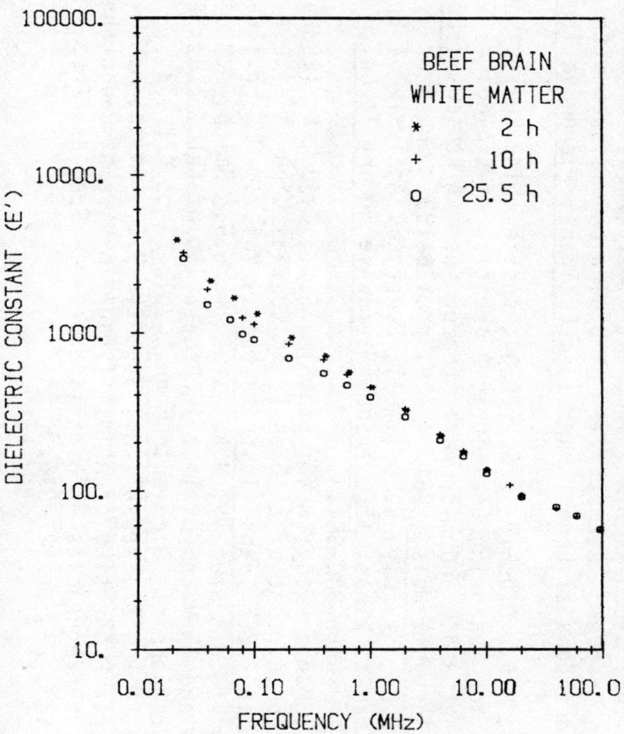

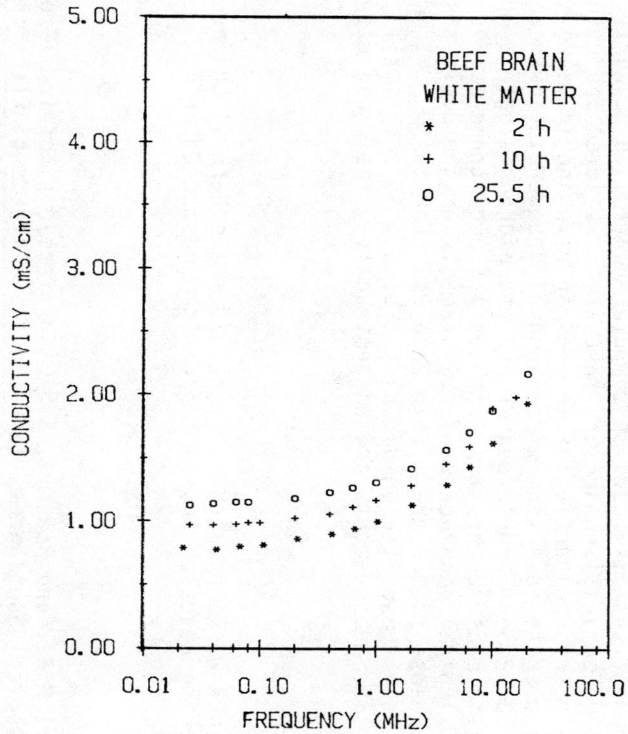

Fig. 10. Dielectric properties of beef brain as a function of time after death, temperature 23°C; (a) dielectric constant, (b) conductivity.

CONCLUSIONS

Measurement systems based on computer-controlled network analyzers and coaxial-line sensors have been developed and used to measure dielectric properties of tissues in vivo and in vitro. The systems operate in a very wide range of frequencies between 20 kHz and 18 GHz. A single sensor and therefore a single sample can be used to measure the tissue properties over three octaves in frequency. Typical accuracies of measurements of the dielectric constant and loss factor are within 2 to 3%. For the shunt-capacitor sensor the volume of the sample is very small (~ 30 μl). The measurements are nondestructive, rapid and easy to perform.

ACKNOWLEDGEMENTS

This work was supported by grants from the Natural Sciences and Engineering Research Council of Canada and the Ontario Ministry of Labour.

REFERENCES

1. Stuchly, M.A., and Stuchly, S.S. (in press), Electrical Properties of Biological Substances, in Biological Effects of Electromagnetic Radiation, O.P. Gandhi, Ed., Academic Press, 1985.
2. Stuchly, S.S., Rzepecka, M.A., and Iskander, M.F. (1974), Permittivity Measurements at Microwave Frequencies Using Lumped Elements, IEEE Trans. Instrum. Meas., Vol. IM-23, pp.56-62.
3. Athey, T.W.; Stuchly, M.A.; Stuchly, S.S., (1982), Measurements of Radio Frequency Permittivity of Biological Tissues with an Open-Ended Coaxial Line. Part I, IEEE Trans. Microwave Theory Techn., Vol. MTT-30, pp.82-86.
4. Stuchly, M.A., Stuchly, S.S. (1980), Coaxial Line Reflection Methods for Measuring Dielectric Properties of Biological Substances at Radio and Microwave Frequencies - A Review, IEEE Trans. Instrum. Meas., Vol. IM-29, pp.176-183.
5. Gajda, G.B.; Stuchly, S.S. (1983), Numerical Analysis of Open-Ended Coaxial Lines, IEEE Trans. Microwave Theory Techn., Vol. MTT-31, pp.380-384.
6. Stuchly, M.A.; Brady, M.M.; Stuchly, S.S.; Gajda, G. (1982), Equivalent Circuit of an Open-Ended Coaxial Line in a Lossy Dielectric, IEEE Trans. Instrum. Meas., Vol. IM-31, pp.116-119.
7. Kraszewski, A.; Stuchly, M.A.; Stuchly, S.S.; Smith, A.M. (1982), In Vivo and In Vitro Dielectric Properties of Cat Tissues at Radio Frequencies, Bioelectromagn., Vol. 3, pp.421-432.
8. Stuchly, M.A.; Kraszewski, A.; Stuchly, S.S.; Smith, A.M. (1982), Dielectric Properties of Animal Tissue In Vivo at Radio and Microwave Frequencies - Comparison Between Species, Phys. Med. Biol., Vol. 27, pp.927-936.

9. Stuchly, M.A.; Athey, T.W.; Stuchly, S.S.; Samaras, G.M.; Taylor, G. (1981), Dielectric Properties of Animal Tissues In Vivo at Frequencies 10 MHz - 1 GHz, J. Bioelectromagn., Vol. 2, pp. 93-103.
10. Symons, S.; Stuchly, S.S. (1983), A Simple Circuit to Automate the HP 8407 Network Analyzer, Microwave J., Vol. 26, No.10, pp.137-138.
11. Kraszewski, A.; Stuchly, S.S.; Stuchly, M.A.; Symons, S. (1983), On Measurement Accuracy of the Tissue Permittivity In Vivo, IEEE Trans. Instrum. Meas., Vol. IM-32, pp.37-42.
12. Kraszewski, A.; Stuchly, M.A.; Stuchly, S.S. (1983), ANA Calibration Method for Measurements of Dielectric Properties, IEEE Trans. Instrum. Meas., Vol. IM-32, pp.385-386.

COHERENCE-SYNCHRONIZATION-CHAOS

COOPERATIVE PROCESSES IN EXCITED BIOLOGICAL SYSTEMS

Friedemann Kaiser*

Institute of Applied Physics, Theory

Technische Hochschule Darmstadt

6100 Darmstadt, FRG

INTRODUCTION

The purpose of the present paper is twofold. On the one hand externally driven self-sustained oscillations are investigated. They exhibit different bifurcation routes from free internal oscillations via a great variety of intermediate states to irregular and to unstable oscillations. On the other hand the possible relevance of this behaviour for certain rhythmic processes in biological systems and the interpretation of experimental results are discussed.

Oscillating processes seem to play a prominent role in physical, chemical and biological systems. They include both, self-sustained internal oscillations and their response to external perturbations (1).

The self-sustained oscillations are established by means of a physical model based on some rather general assumptions with respect to biological order and function. The model is given in some detail in Chapter III.

---

*also: Institute of Theoretical Physics, University of Stuttgart Pfaffenwaldring 57, 7000 Stuttgart 80, FRG

Complex bifurcation routes are well known effects in nonlinear systems. They include nonlinear mechanical and electrical oscillators, chemical and biochemical reactions and a lot of other types of internally or externally forced or of parametrically driven oscillating systems. A common instability found in nonlinear dynamical systems (including nonlinear maps) is the period doubling of a periodic orbit, e.g. of a self-sustained oscillation (limit cycle) (2). In order to present some of the essential features of this type of bifurcational behaviour, we start with a rather simple system.

I. NONLINEAR TORSION PENDULUM

Based on an idea of G. Mayer-Kress (3) we investigated an externally driven torsion pendulum. This <u>linear</u> system exhibits the well-known resonance behaviour (frequency response) of the periodically driven, linearly damped harmonic oscillator. However, an additional small mass fixed on the oscillating wheel causes a nonlinear torque, leading to two new positions of equilibrium for a vanishing external drive (Fig. 1).

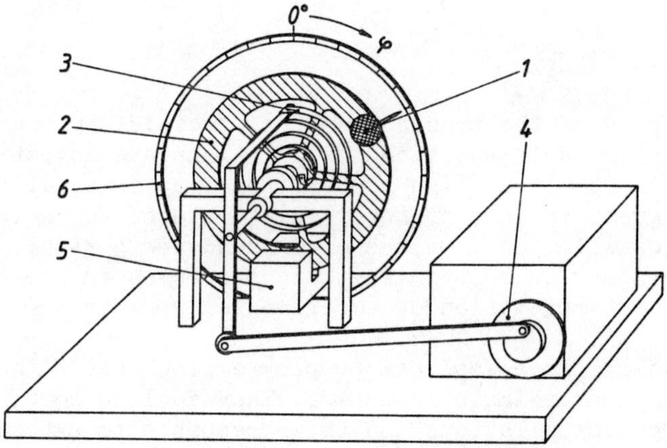

Fig. 1. Simple mechanical system (experimental set-up)
1. additional mass, 2. oscillating wheel, 3. torsional spring, 4. stepper motor, 5. eddy current brake, 6. dial.

The system is periodically driven. Without additional mass it oscillates with the external frequency around its position of equilibrium ($\varphi = 0$). The resulting amplitude depends on both, the strength of the external drive and the magnitude of the damping. In the case of two additional equilibria (i.e. additional mass fixed at the old position of equilibrium), the system exhibits a large number of different oscillating pattern.

Restricting to a frequency of the external drive below that of the free oscillation ($\lambda < \omega_0$) the bifurcations of Fig. 2 have been found.

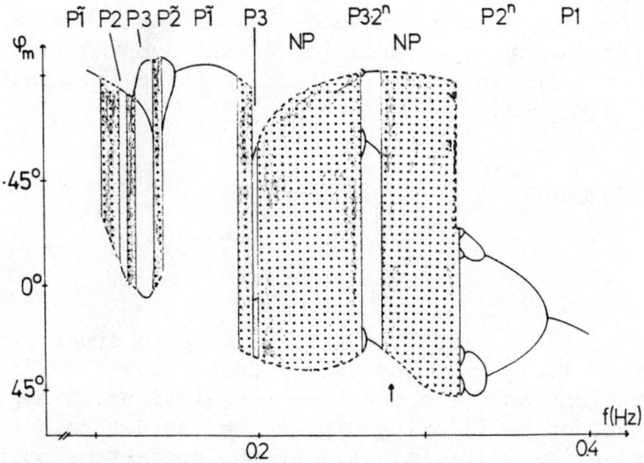

Fig. 2. Bifurcation diagram. Amplitude φ as a function of the external frequency λ (the strength of external drive and the damping current I are fixed, φ_m is the maximum left amplitude).

Near $\lambda = 0.4$ Hz, a small amplitude oscillation around one of the two positions of equilibrium results (periodic state P1). With decreasing external frequency, a sudden split-up of the amplitude occurs. A strongly periodic state is built up, exhibiting a proper sequence of 2 different values for the amplitude (i.e. an alternation of a small and a large amplitude; periodic state P2). A further decrease of the frequency leads to a new bifurcation, a strongly periodic state with 4 different amplitudes is created (periodic state P4). Additional period doubling bifurcations are found, the well-known period doubling sequence is exhibited.

An infinitesimal further decrease of λ leads to an abrupt change of the system's behaviour: a completely irregular sequence of amplitudes suppresses the regular oscillations. In addition, within this chaotic region, the system oscillates around both positions of equilibrium (shaded regions in Fig. 2). However, further periodic regions do exist, e.g. the second period doubling sequence ($P3 \cdot 2^n$, i.e. P3, P6, P12..) and other periodic states. Among these periodic states, the $P\tilde{1}$ should be mentioned. This state is a periodic oscillation with frequency λ around both positions of equilibria (large amplitude oscillation).

The occurence of the various set of bifurcation sequences in such a simple experimental system is highly instructive for the understanding of a similar behaviour found in more complex systems. It should be stressed that extremely small changes of some parameters (in our experiment the external frequency λ) can lead to dramatic changes in both, the quantitative and qualitative behaviour of a driven system. This, in turn requires a very accurate control and stabilization of the frequency and of the other parameters. For a fixed frequency λ, a variation of the damping parameter (by an increase or decrease of the current in the coil) leads to similar bifurcation sequences.

II. GENERAL CONCEPT

Coherence

With coherence we mean a static or dynamic order principle. Some degrees of freedom of the system under investigation exhibit phase correlations and/or space-time correlations. Examples are fixed phases of a set of oscillating dipoles or the determination of a state at space-time point (x_2, t_2) by the space-time point (x_1, t_1).

From a rather fundamental point of view, Fröhlich has promoted the ideas that coherent states can be established in active biological systems (4). Very briefly, the basic results are summarized: Types of coherent excitations

- highly polar metastable states
- coherent vibrations

Types of coherent vibrations

- highly excited single modes
- excitation of highly polar metastable states
- existence of periodic processes

Whereas the establishment of highly polar states requires a certain amount of energy to be put into the system ($E \geqslant E$ min) the excitation of coherent vibrations is only possible, if the rate of energy flux through the system is beyond a certain threshold ($S \geqslant S_0$).

The possible excitation of coherent vibrations has been demonstrated with some simple model calculations (5-7). The coherent vibrations include a very large frequency range (from extreme slow chemical or electrical oscillations to oscillations beyond the GHz range in membranes and proteins).

Questions and problems concerning coherent vibrations are related to their existence and establishment as well as to their link to a biological process. Among them we mention the concept of activation and deactivation of enzymes, the nonchemical storage and transport of energy and, quite generally, the periodicity in biological development.

Modelling Approach

The mathematical description of a certain biological function rises a lot of problems. An extremely complex, many variable system which is open and far beyond thermal equilibrium must be reduced to a few variable system. Necessary prerequisites for an adequate approach are nonlinear dynamics and the inclusion of external influences. The external influences and the internal ones as well can be of a deterministic type or of a stochastic one. Which one is the adequate one? Simplified, the origin of the stochastic behaviour may be classified within three types of statistics:

1. thermodynamic statistics
   (thermal fluctuations caused by large numbers)
2. quantum theoretical statistics
   (quantum fluctuations by Heisenberg's uncertainty principle)
3. turbulent statistics
   (macroscopic fluctuations in strongly deterministic systems).

The occurence of an irregular behaviour (i.e. turbulent, chaotic, nonperiodic) in strongly deterministic systems is surprising and fascinating as well. Within the last few years a lot of both, experimental and theoretical investigations have revealed its existence. Our simple experiment mentioned in the introduction , belongs to this type of systems. Furthermore, a computational check of the corresponding mathematical description yields the same results as in Fig. 2. Consequently, a nonlinear, driven system can exhibit a stochastic behaviour if it is periodically driven, though it is completely deterministic.

For a modelling approach, we choose the following scheme:

<u>general</u>:	<u>realization</u>:
phenomena	periodic processes
↓	↓
dynamical model	set of equations
↓	↓
mathematical image	solutions

A certain phenomenon(behaviour, effect) from physics, chemistry, biology etc. is to be described. This can be done with the aid of a dynamical model. In our case, we restrict to a description of periodic processes, which can be performed by a set of nonlinear differential equations, the solutions of which yield the mathematical image. For a three-dimensional system, the solutions are:

- fixed point (static attractor, steady state)
- limit cycle (periodic attractor, motion on a closed orbit)
- torus (attracting sphere, motion with 2 frequencies)
- chaotic state (strange attractor, irregular motion)

All solutions are structurally stable and attracting ones, which means that all trajectories within the attracting basin run towards these attractors.

Special realization

In relation to the above scheme, we restrict to the more special scheme:

periodic internal processes + external influence

↓ ↓

set of equations: $\dot{\underset{\sim}{x}} = \underset{\sim}{f}(\underset{\sim}{x})$ + $F(t)$

↓ ↓

self-sustained oscillations (limit cycle) | nonlinear superposition of internal + external oscillation

Result: nonlinear response of the driven system.

The whole procedure has to take into account both, a certain theoretical concept and the corresponding experimental basis. Before going into the details of one of our models, a more specified set of steps from the onset of the free internal oscillation to the perturbed one will be presented. This set of steps demonstrates that the concept of perturbed limit cycles is rather universal whereas a specific model includes a series of assumptions and specifications with respect to the mechanism creating the oscillations.

General set of steps:

Phase A : onset of a limit cycle ⟶ LC (ω_{LC}, x_{LC})

requirements: nonlinear and dissipative processes

result : oscillation with a fixed frequency and amplitude, cooperative behaviour (no externally forced order)

Phase B: perturbation of a limit cycle ⟶ LC (λ, x)

restriction : external periodic drive (deterministic)
consequences: competition free internal ↔ external oscillation (nonlinear superposition).
result : entrainment (i.e. synchronized state, oscillation with the external frequency) if the drive is sufficient strong; frequency response shows a sharp resonance, surrounded by quasi-periodic oscillations and beat; sup- and superharmonic entrainment in addition.

Phase C: perturbation of a synchronized limit cycle

requirements: further increase of external drive
result : series of bifurcations; instability of the oscillation (collapse), leading to an onset of propagating pulses; restabilization of periodic or static attractor.
consequences: a relatively small amount of external energy supply (energy input per second) can create a large internal response by applying the internally stored oscillator energy ("trigger action").

Complete result: LC (ω_{LC}, x_{LC}) ⟶ LC (λ, x) ⟶ pulse.

III. MODEL

The concept of chapter II, i.e. the creation of a limit cycle, its bifurcation routes and the subsequent onset of propagating pulses, will be applied to a specific situation. The model is based on both, a series of experimental results and on theoretical considerations with respect to the order and function in biological systems. Very briefly, the basic ideas can be summarized.

Motivation: Extreme sensitivity of certain biological systems to electromagnetic fields of very low intensity (far below thermal sensitivity) (5,8).

Basis: It is assumed that in active biological systems some specific processes can occur (5): coherent electric vibrations, highly polarized states, long range selective interactions.

Model: The model has been proposed by Fröhlich (9) in order to describe slow oscillations in the greater membrane of the brain. Furthermore, the model may be viewed as an example for the creation of limit cycle oscillations and their response to external influences. It thus yields a possibility to describe rhythmic processes and their external perturbation (10).

Assumptions

The model under consideration is mainly concerned with the third type of coherent vibrations: excitation of a limit cycle (vid. Chapter II: coherent excitations). It can be described by a series of events:

nonlinear enzyme-substrate reaction
creation of slow chemical oscillations
creation of slow electric oscillations
↓ interaction with external fields

Mathematical Approach

Details of the model have been given in (9, 11). The resulting set of equations reads

$$\begin{aligned} \dot{\nu} &= \gamma\,\sigma + \alpha\,A\,\sigma\,\nu + (c^2\,e^{-\Gamma^2\nu^2} - d^2)\,\nu + F(t) \\ \dot{\sigma} &= -\,\beta\nu - \alpha\,A\,\sigma\,\nu \end{aligned} \qquad \text{(I)}$$

This system is a superposition of both, a slow Lotka-Volterra dynamics (steming form the nonlinear chemical reaction) and a generalized Van der Pol dynamics (having its origin in "ferroelectric" interactions).

α, β, γ and A are chemical parameters, c, Γ and d belong to the electric interactions. ν and σ are excess concentrations, i.e. derivations from chemical equilibrium values of enzyme and substrate concentrations, respectively.

It should be emphasized that this model is by no means a mathematical one. The basis is completely within profound physical considerations.

IV. RESULTS

1. F(t) = 0

Without external drive the system I exhibits a limit cycle

oscillation for $c^2 > d^2$ with frequency ω_o and amplitude ν_o. For $c^2 < d^2$ stable nonoscillating solutions exist (11). The requirements of "Phase A: onset of a limit cycle " are given. $c^2 > d^2$ corresponds to a preference of a cooperative interaction (ferroelectric tendency) compared to dielectric losses for small values of ν.

2. $F(t) = F_o$

A static external field leads to a reduction of both, the frequency and the amplitude of the oscillation. For $F_o > F_c$, a nonoscillating state gets stabilized (static attractor). A highly polar static state results (for details vid. refs. 11-13).

3. $F(t) = F_o + F_1 \cos \lambda t$

For $F_o = 0$, a purely periodic field drives the system into a synchronized or entrained state (Phase B: Perturbation of a limit cycle). A further increase of the synchronized oscillator leads to a destabilization of the oscillation and subsequently to a collapse and the onset of propagating pulses (Phase C: Perturbation of a synchronized limit cycle).

The combination of both, a static and a periodic field is determined by a competition of two different types of motion: the transition to a static attractor (via F_o) and the bifurcation to other periodic attractors and their subsequent collapse (via F_1). This competition of two different processes yields exciting bifurcation routes. Some of the results will be presented with a series of pictures. In these series, the internal parameters and the external frequency λ are fixed.

Series 1:

Figs. 3-5	: Stability diagrams	$(F_o - F_1)$
Fig. 6	: Stroboscopic diagrams	$(\tilde{\nu} - F_1)$
Fig. 7	: Poincaré-plot	$(\nu_{t+t} - \nu_t)$
Fig. 8	: Maximum Aplitude	$(\nu_{max} - F_1)$
Fig. 9	: Maximum Amplitude	$(\nu_{max} - F_o)$
Table 1	: Bifurcation routes	

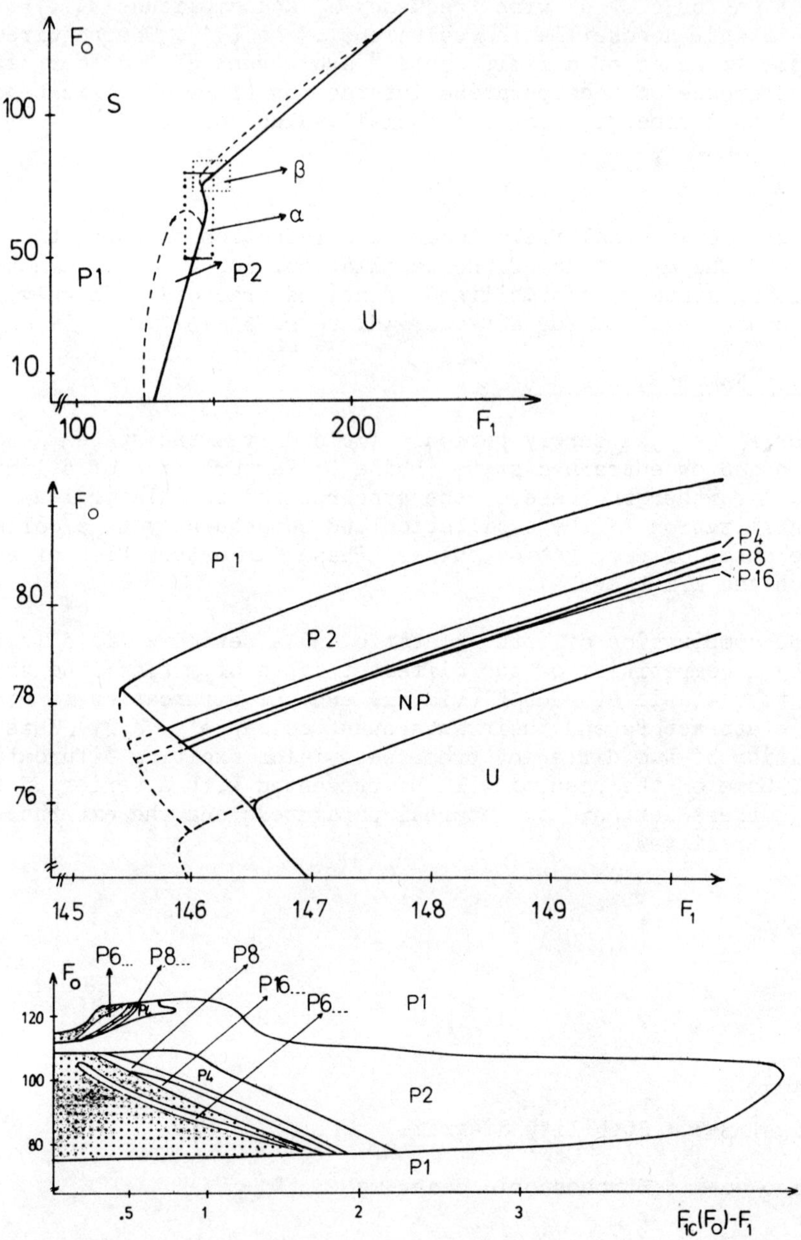

Figs. 3-5: Stability diagrams. Lines, separating regions of different behaviour. S = stable, U = unstable. Pn = periodic state with frequency λ/n; Pn.... : higher periodics (following Pn) are included; shaded regions and NP : nonperiodic states; ....:coexistence of 2 states (in Fig. 4). (Fig. 4 : α-region of Fig. 3; Fig. 5 : β-region of Fig. 3)

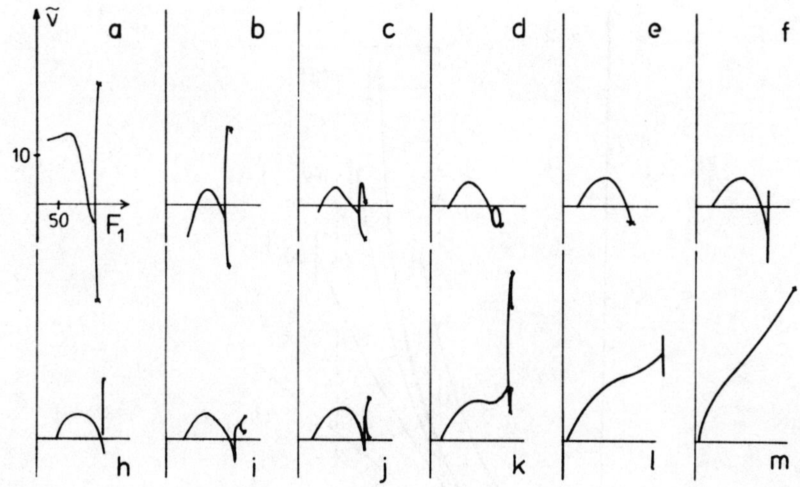

Fig. 6: Stroboscopic amplitude $\tilde{\nu}$ as a function of the periodic field F_1 for 12 different values of F_o increasing from $F_o = 0$ to $F_o = 150$ in a) to m). In i), j) and k) the states beyond P4 are not drawn. The F_o -region corresponds to that of Fig. 3, the whole route from entrainment to collapse is given. * = collapse. ($\tilde{\nu}$: amplitude for $F_1 \cos \lambda t = F_1$)

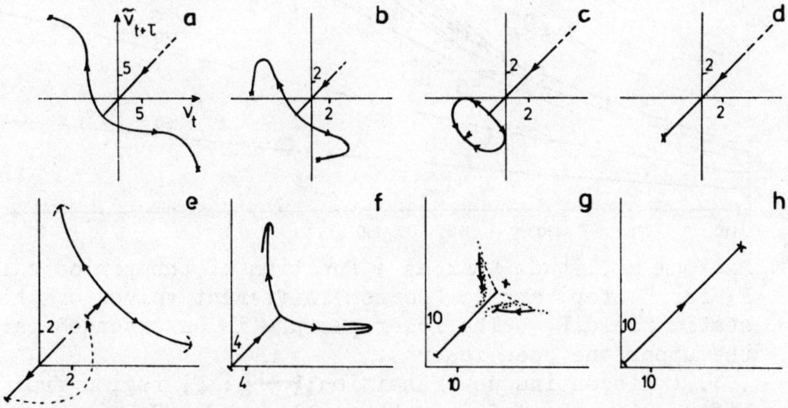

Fig. 7: Poincaré-plots. Stroboscopic amplitude $\tilde{\nu}$ at a time $t + \tau$ as a funktion of $\tilde{\nu}$ taken at time t. $\tau = 2\pi/\lambda$. F_o increases from $F_o = 0$ to $F_o = 150$ in a) to h). The arrows show the development of $\tilde{\nu}$ with increasing F_1, i.e. from entrainment to collapse. * = collapse. Pictures a-h correspond to pictures a, c, d, e, j, k, l and m of Fig. 6, respectively.

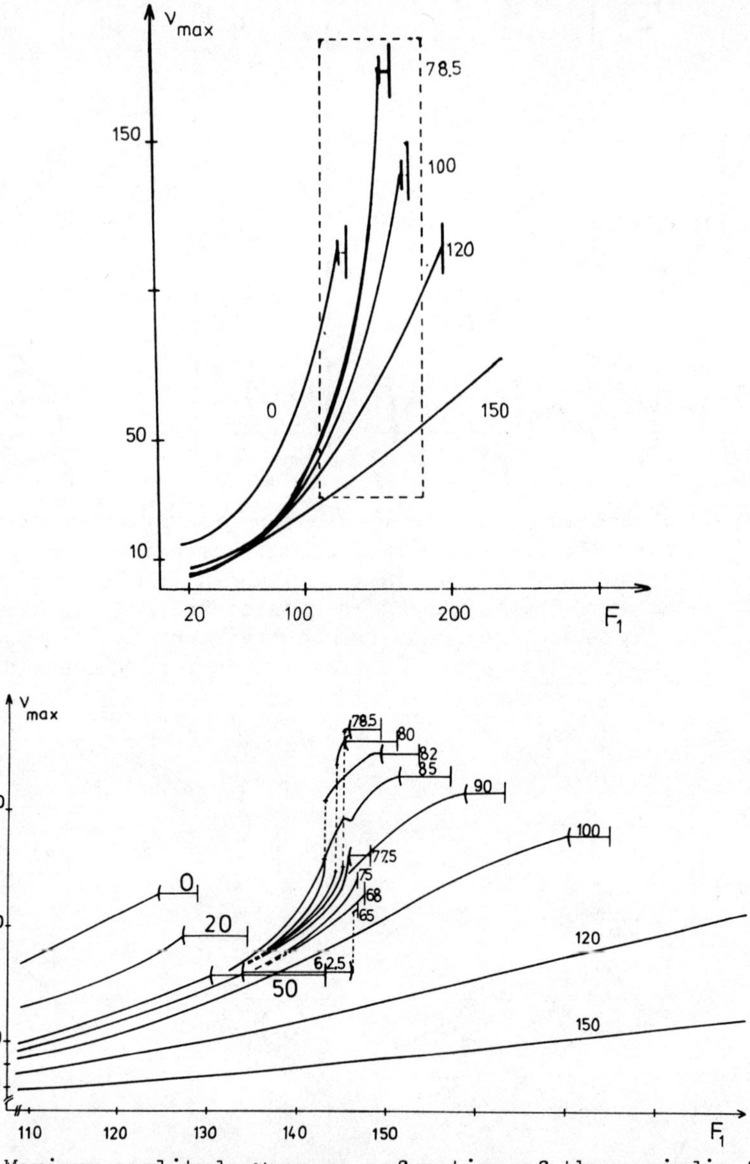

Fig. 8: Maximum amplitude ν_{max} as a function of the periodic field F_1 for 5 (top) and 16 (bottom) different values of the static field F_o. The lower picture is an extended part of the upper one (see insert).
.....: discontinuous transition; $\vdash^{F_o}\dashv$: F_1 region from first bifurcation (P1 → P2) to the collapse for F_o given in the diagram (within this F_1 region, the ν_{max}-values are not drawn).

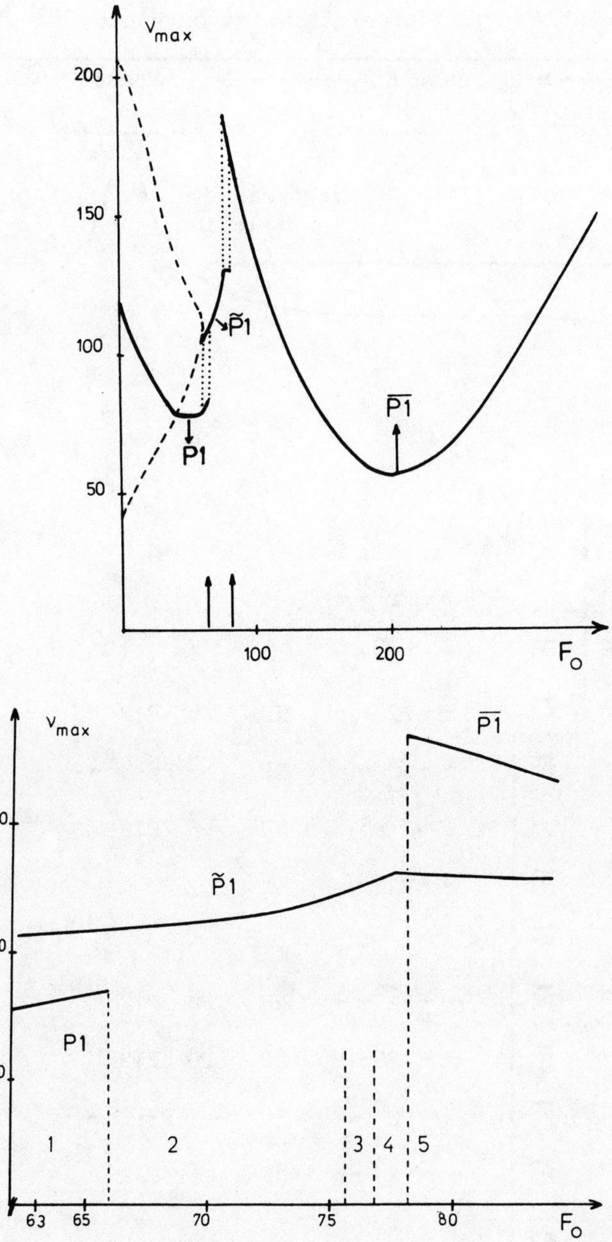

Fig. 9: Maximum amplitude νmax as a function of the static field F_o. Two discontinuous transition regions occur. The first one is given in more detail in the picture at the bottom (region between the 2 arrows). Region 1: P1 → P2 → P1
Region 3: P1 → Pn..NP (n = 2 or 4 or 8, depending on F_o)
Region 4: P1 → P2..NP; Region 5: P1 → P2..NP.

Table 1: Different Bifurcation Routes

Phase A ⟶ Phase B ⟶ Phase C

⟶ LC (ω_o) ⟶ LC (λ) P1 ⟶ C

free oscillation — entrained oscillation — collapse

P1	P2											
P1	P2	$\widetilde{P1}$										
		$\widetilde{P1}$										
		$\widetilde{P1}$									NP	
		$\widetilde{P1}$				--$P2^n$--					NP	
		$\widetilde{P1}$	$\overline{P1}$			--$P2^n$--					NP	
		$\widetilde{P1}$	$\overline{P1}$			--$P2^n$--		NP	P6	P12	NP	
			$\overline{P1}$			--$P2^n$--		NP		--$P3\cdot 2^n$--	NP	
			$\overline{P1}$	P2	P4	P8	P4	NP		P12	NP	
			$\overline{P1}$			--$P2^n$--					NP	
			$\overline{P1}$	P2								
			$\overline{P1}$			--$P2^n$--		NP	P6	P12	NP	
			$\overline{P1}$			--$P2^n$--		NP	P6	P12	NP	$\overline{P1}$
			$\overline{P1}$			--$P2^n$--		NP		--$P3\cdot 2^n$--	NP	$\overline{P1}$
			$\overline{P1}$			--$P2^n$--		NP	P6	P12	NP	$\overline{P1}$
			$\overline{P1}$									

F_o increases from top to bottom, F_1 from left to right. Each route starts with the entrained state (P1, $\widetilde{P1}$, or $\overline{P1}$, respectively)

Series 2: Static field $F_o = 0$ and $F_o = 100$

For $F_o = 0$ only one bifurcation occurs, a transition P1-P2 (vid. Figs 6a, 7a and Table 1, first sequence).
For $F_o = 100$ a whole bifurcation sequence is created (vid. Figs. 6k, 7k and Table 1, sequence 8).

Fig. 10 :	Oscillation diagrams	$(\nu - t)$	$F_o = 0$
Fig. 11	Bifurcation diagram	$(\tilde{\nu} - F_1)$	$F_o = 100$
Fig. 12	Bifurcation diagram	$(\nu_{max} - F_1)$	$F_o = 100$
		$(\Delta\phi - F_1)$	$F_o = 100$
Fig. 13	Poincaré-plot	$(\nu_{t+\tau} - \nu_1)$	$F_o = 100$
Fig. 14	Ljapunov exponents	$(\overline{\lambda} - F_1)$	$F_o = 100$
Fig. 15	Oscillatión diagrams	$(\nu - t)$	$F_o = 100$

Ljapunov-exponent

The Ljapunov-exponent is a measure for the average divergence of nearby trajectories. For its calculation one has to solve numerically the differential equations in order to get the development of 2 trajectories (orginally nearby) for a long time (more than 100 000 oscillations in actual situations). Then the average distance must be calculated. For a three-dimensional system (e.g. model I is of the order $d = 3$) the basic results for the three existing exponents $(\overline{\lambda}_1, \overline{\lambda}_2, \overline{\lambda}_3)$ are:

fixed point	(-, -, -)	limit cycle	(0, -, -)
torus	(0, 0, -)	strange attractor	(0, +, -)

+ (-) : nearby trajectories diverge (converge) on the average,
0 : nearby trajectories remain nearby.

Limit Cycle : one has converging trajectories in two directions and a closed trajectorie in the third one $\longrightarrow$ (0, -, -).

It can be shown that for model I one gets $\overline{\lambda}_1 = 0$, $\overline{\lambda}_3 < 0$ and $\overline{\lambda}_2 \gtrless 0$, i.e. one has transitions

$$(0, -, -) \longleftrightarrow (0, +, -)$$

which are transitions from a periodic state (limit cycle) to a chaotic state (strange attractor) and vice versa (vid. Fig. 14)

Series 3: Specific bifurcations

For certain values of the F_o - field discontinuous transitions occur, if F_1 increases. Furthermore, one can change the F_o -field for fixed values F_1. For appropriate F_1-values, an increasing F_o leads to inverted bifurcations, i.e. inverted period-doubling sequences can occur.

Fig. 16 : Bifurcation diagrams (ν_{max} - F_1), F_o fixed

Fig. 17 : Bifurcation diagrams (ν_{max} - F_o), F_1 fixed.

A more detailed investigation yields additional results:

1. Coexistence: For F_o and F_1 fixed, the state of the system can be determined by initial conditions. This is the case in the dashed part of Fig. 4, where one finds (P1-P2), (P1-P4), (P1-P8) (P1-NP); vid. Fig. 18 for a P1-P8 coexistence.

2. Hysteresis: For F_o fixed, the critical value $F_{1,c}$ can be shifted near the instability or bifurcation lines depending on the way, the system is driven, i.e. if the boundaries are crossed for increasing or decreasing F_1.

3. Transients: Near the bifurcation points, a large number of oscillations is required to drive the system into its stable steady state (vid. Fig. 19).

4. Intermittency: Near some bifurcation points, the system jumps in a nonsystematic way from one state to another one. Irregular bursts are separated by regular oscillations (vid. Fig. 20).

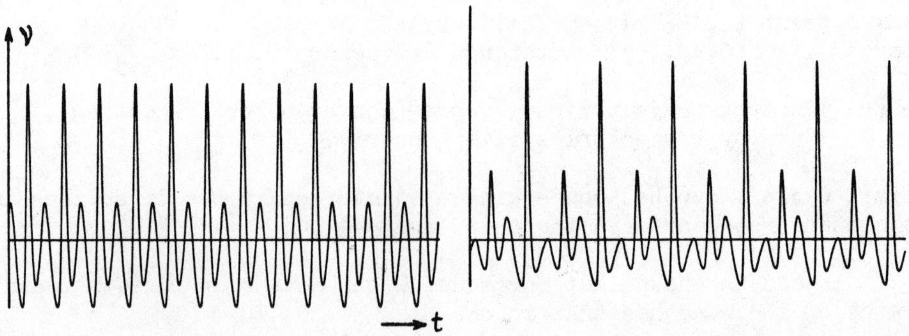

Fig. 10: Oscillation diagram. Amplitude ν as a function of time t for the P1 (l.h.s.) and P2-state. F_o = 0, only P1 and P2 do exist.

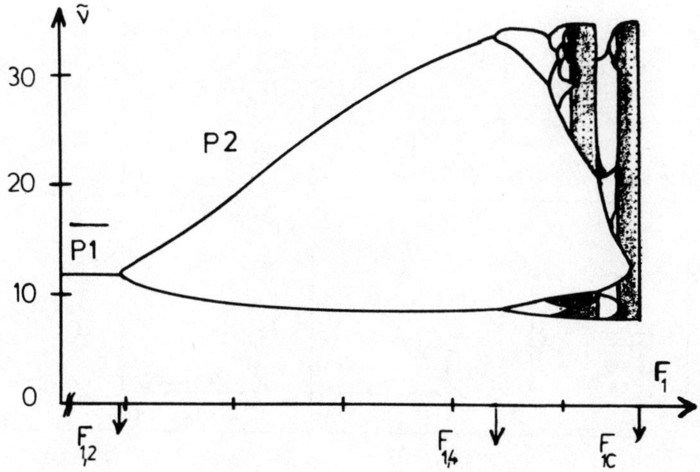

Fig. 11: Bifurcation diagram. Stroboscopic amplitude $\tilde{\nu}$ as a function of the field F_1 for $F_o = 100$. The shaded regions are the 2-band and 1-band chaotic regions. Both period doubling sequences ($P2^n$ and $P3 \cdot 2^n$) are given.

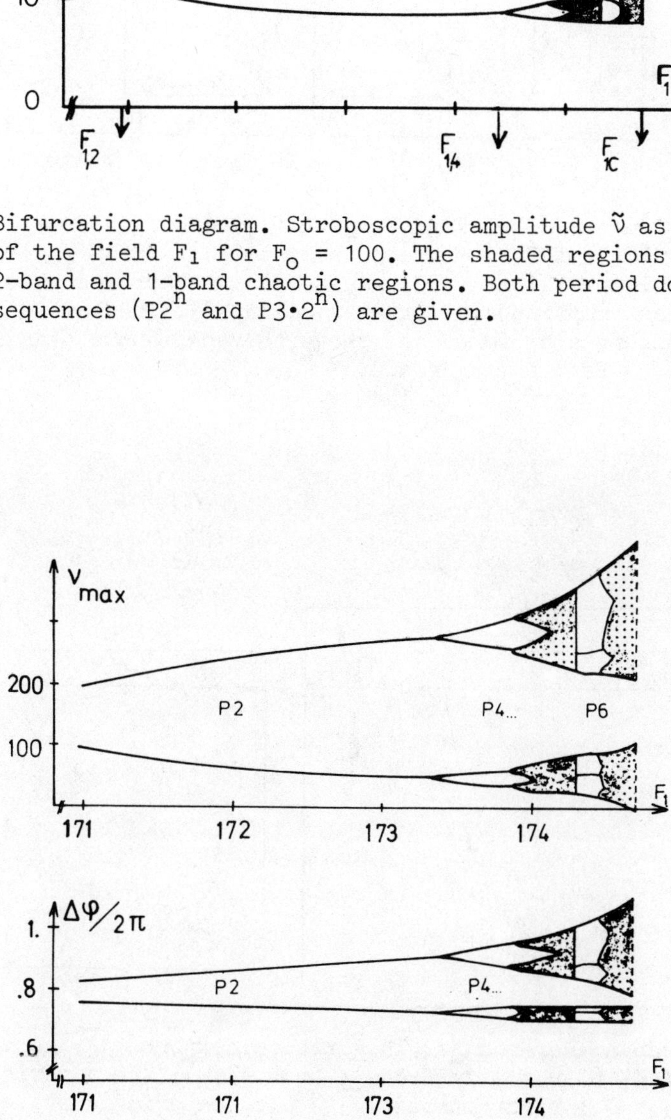

Fig. 12: Bifurcation diagram. Amplitude ν_{max} and phase difference $\Delta \phi$ as a function of F_1 beyond the P1–P2 transition. $F_o = 100$. $\Delta \phi : \nu \rightarrow \nu_{max}$ in units of 2π ($\hat{=}$ frequency λ).

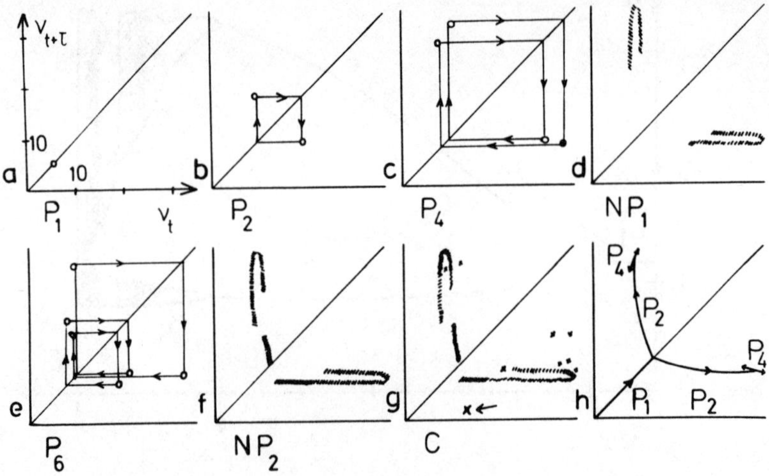

Fig. 13: Poincaré-plot. Fig. 7, part f is redrawn for 7 different values of F_1 (→ pictures a) to g)). Picture g) shows the collapse after 150 transients (the arrowed star is the last amplitude before the collapse). The last diagram, h), combines a) to c) and shows the continuous transitions P1 → P2 → P4. F_o = 100.

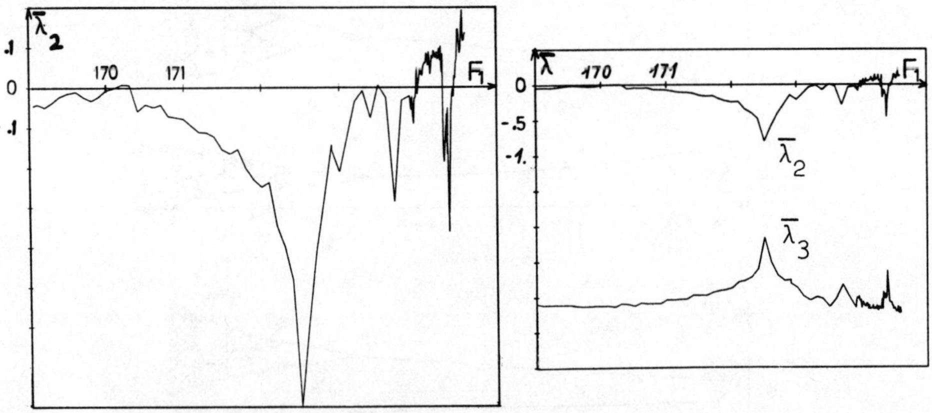

Fig. 14: Ljapunov-exponents. The critical exponent $\bar{\lambda}_2$ as a function of F_1 (l.h.s.), $\bar{\lambda}_2$ and $\bar{\lambda}_3$ (r.h.s.) $\bar{\lambda}_2 < 0$: periodic states; $\bar{\lambda}_2 = 0$: bifurcations; $\bar{\lambda}_2 > 0$: nonperiodic states. The period doubling sequences and some periodic windows can be seen as they develop with F_1.
$\bar{\lambda}_1 = 0$; $\bar{\lambda}_3 < 0$. F_o = 100.

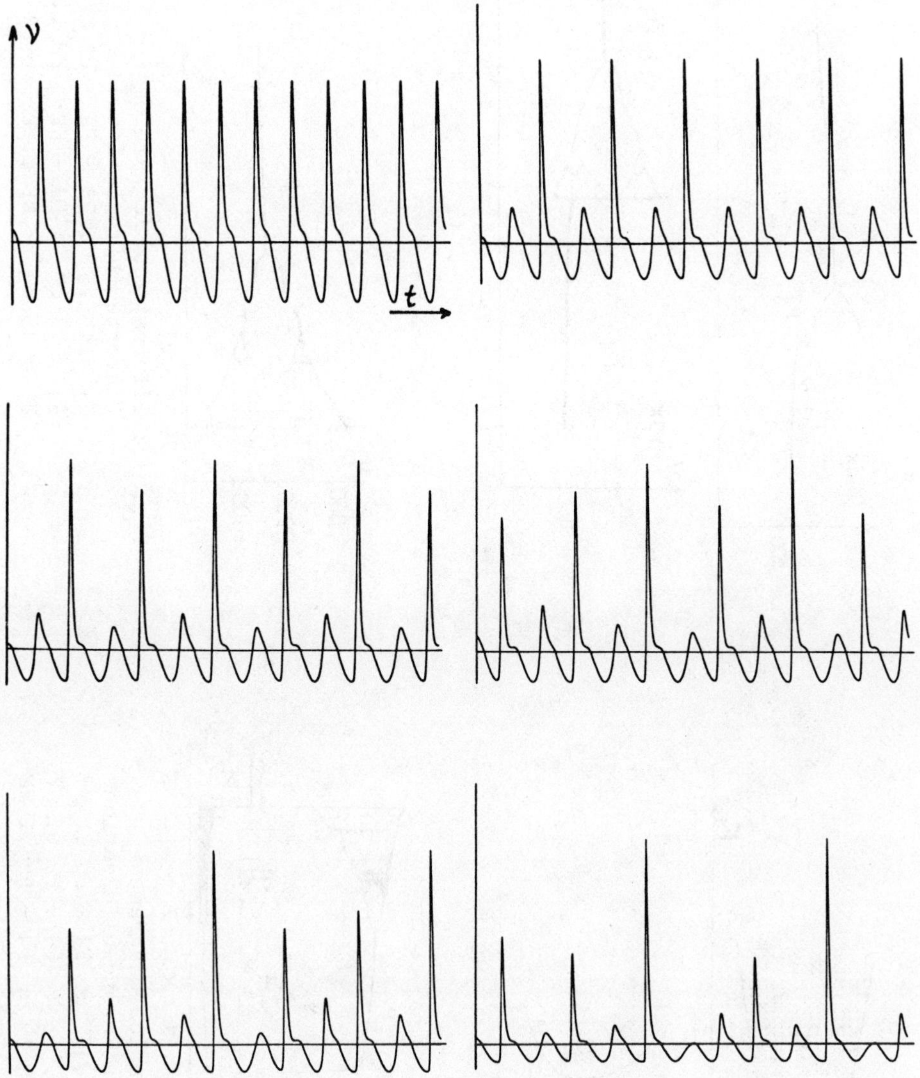

Fig. 15: Oscillation diagrams. Amplitude ν as a function of time for six different F_1 values, increasing from left to right and from top to bottom. The states P1, P2, P4, NP_1, P6 and NP_2 are given. F_o = 100.

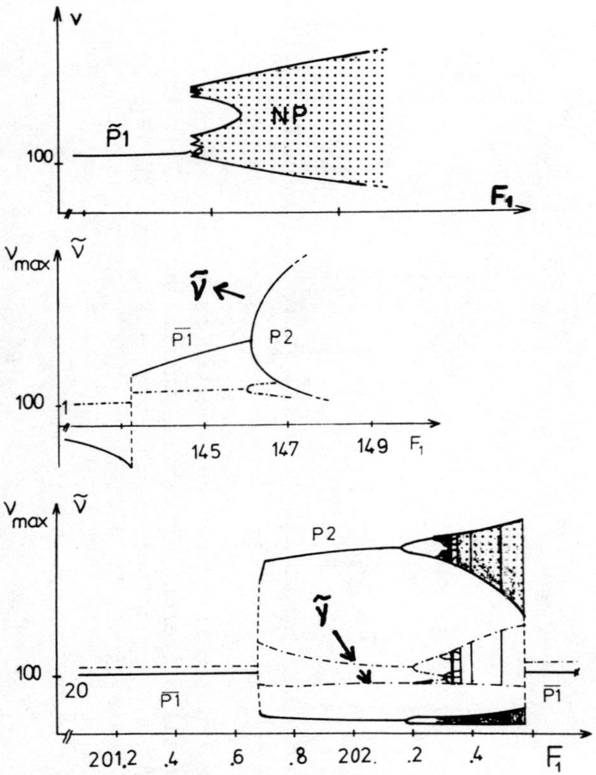

Fig. 16: Bifurcation diagrams
(discontinuous bifurcations)
F_o= 77.15 : P1 - P8 .. NP
F_o= 80 : P1 - P2 .. NP
F_o= 123 : P1 - P8 .. NP

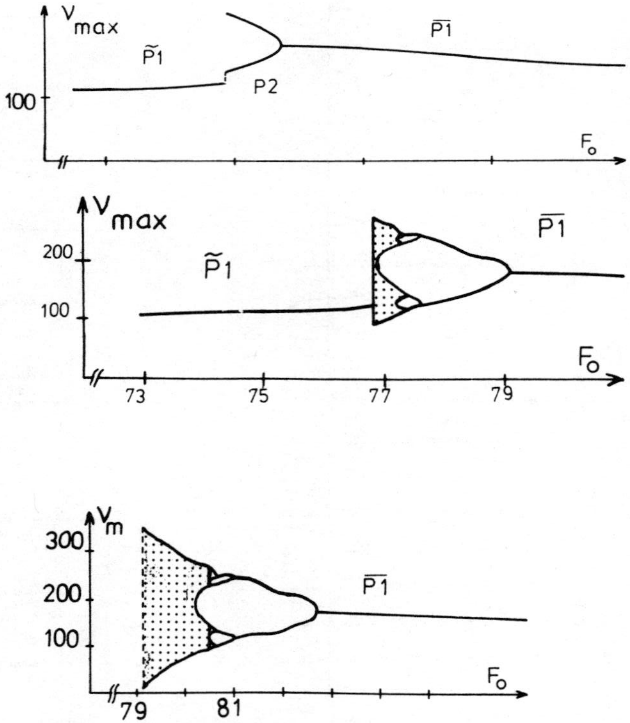

Fig. 17: Bifurcation diagrams
(inverted bifurcations)
F_1 = 145.5 : P1 - P2 - P1
F_1 = 146.0 : P1 - NP - $P2^n$ .. P1
F_1 = 150 : NP .. $P2^n$ - P1

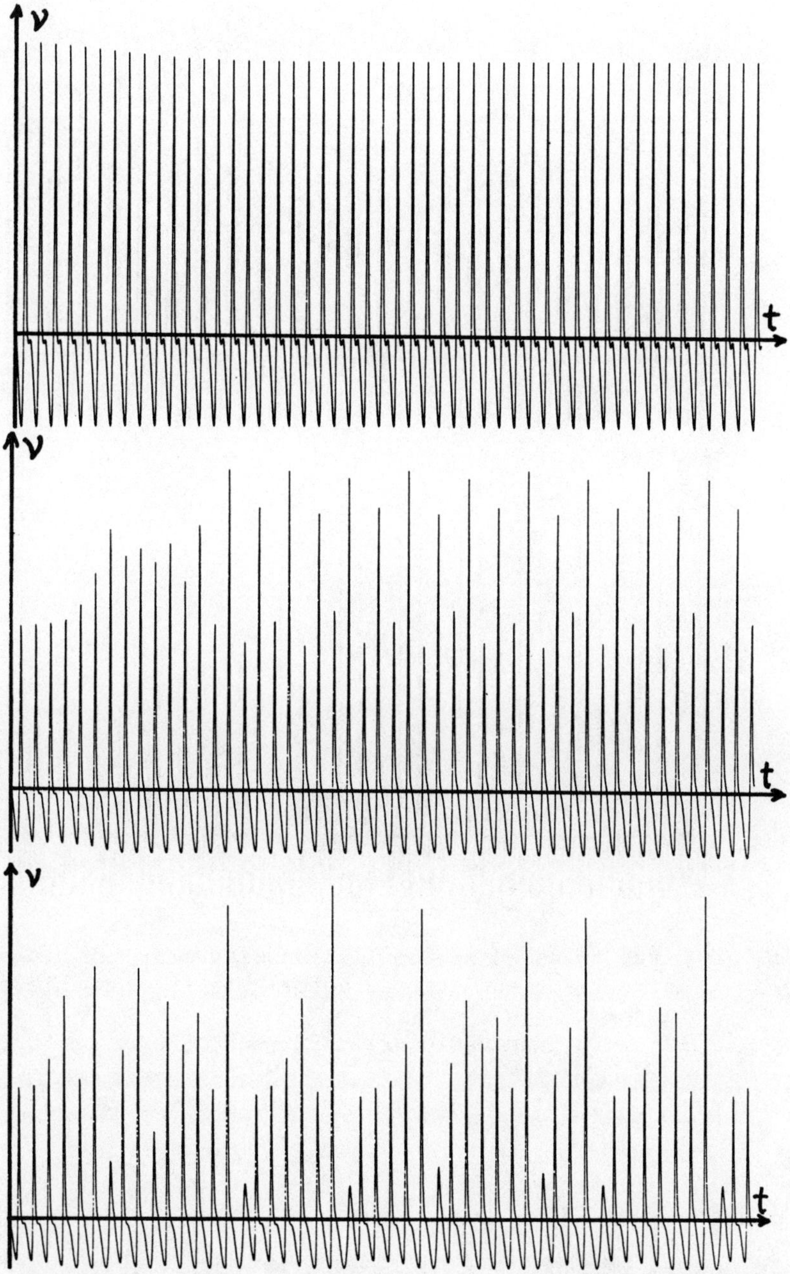

Fig. 18: Coexistence of oscillations. Oscillation diagrams, all parameters (F_0, F_1 included) are fixed. Different initial conditions lead to different oscillations:
P1 (upper picture) P8 (central picture)
NP (lower picture).

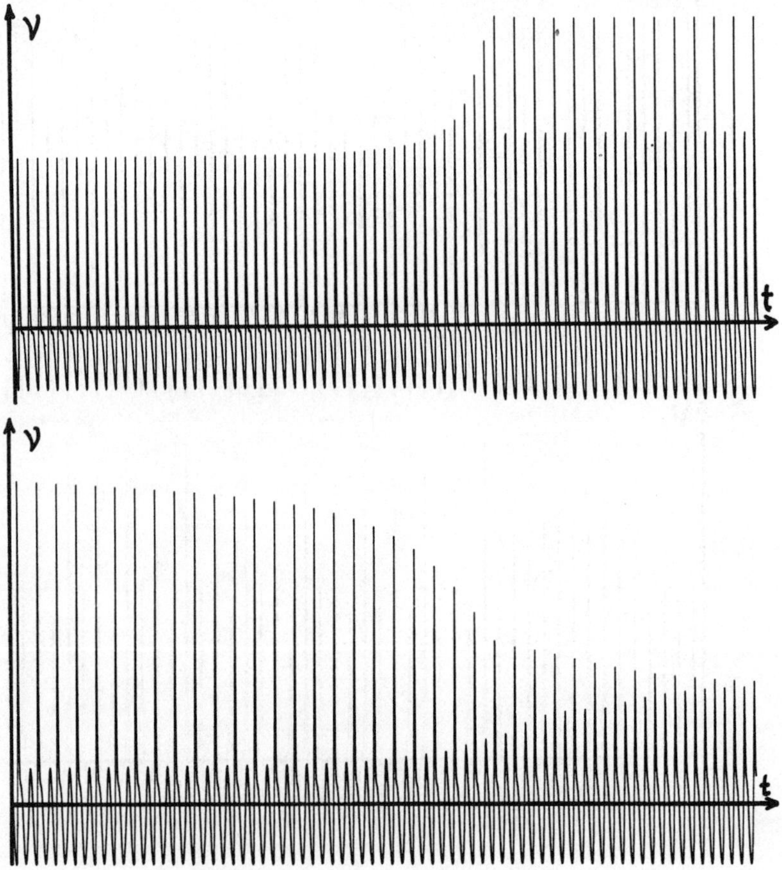

Fig. 19: Large transients. Oscillation diagrams, exhibiting large large transient states, before the final stable state is reached.
P1⟶P2 (upper picture); P2⟶P1 (lower picture).

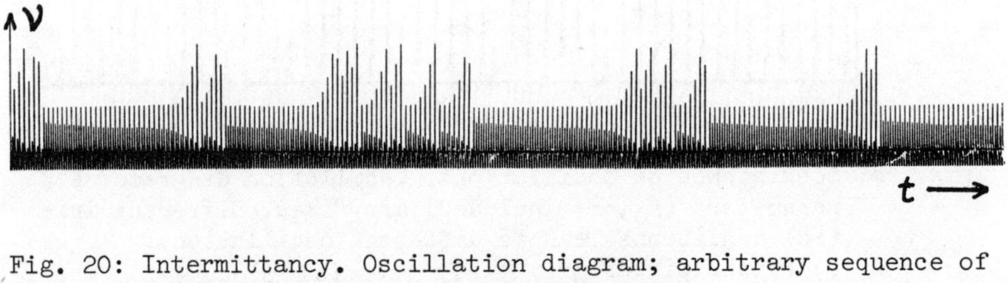

Fig. 20: Intermittancy. Oscillation diagram; arbitrary sequence of regular pattern and irregular bursts.

V. CONCLUSION

Model I exhibits a bifurcation behaviour which, at least in its essential points, is wellknown from a number of systems. The most important ones are nonlinear dynamical systems (14) and nonlinear discrete maps (15). The bifurcation routes occur by a variation of either internal parameters or of external ones (e.g. external periodic fields). Furthermore, a similar behaviour has been found in a number of experiments in different fields of research (14, 15).

Consequences and biological relevance

Model I exhibits the onset of a self-organized state, its synchronization to an external field and the final onset of a propagating pulse. The complexity of this route offers a variety of different behaviour. The occurrence of sharp resonances, of threshold and time of irradiation effects leads to an extreme (frequency dependent) sensitivity of the driven systems (16, 17).

The bifurcation to an irregular (chaotic) behaviour yields additional effects. It should be emphasized that the existence of chaotic states is of great importance for the following reasons:

1. Regular external influences (static plus periodic fields) applied to strongly deterministic systems are able to drive the system into a state of irregular motion (third type of stochastic behaviour).

2. The transition to and the behaviour of this type of stochasticity is determined by a lot of rules (universality and scaling laws; order the different frequencies occur and disappear; contributions of the different peaks to the power spectrum etc. ...).

3. The occurrence of stochastically distributed experimental data does not necessarily point to noise as their origin. The data have to be checked with respect to some details of their irregularity (vid. 2). Possibly, a new interpretation can be adequate, irregular looking results can either possess a deterministic or a noisy basis or a combination of both.

4. Real systems are always exposed to noise (fluctuations). Usually, external noise drives a system from a periodic behaviour into a chaotic one. Noise applied to chaotic systems results in the continuation of chaos. However, there exist some deviations from this behaviour. Noise can lead to order in chaotic systems (noise induced order in chaotic systems). Such a transition is wellknown from the influence of noise onto the chaotic behaviour of discrete maps, 18, 19).

5. Periodic influences can drive a system into a chaotic state. A further periodic stimulus can cause the reappearance of a new periodic behaviour.

The competition of both, periodic and chaotic motion under the combined influence of a deterministic (periodic) and a stochastic (noisy) drive exhibits a variety of additional effects. However, a lot of further investigations remains to be done.

References

1. F.Kaiser, "Nonlinear oscillations (limit cycles) in physical and biological systems", in:"Nonlinear electromagnetics", P.L.E. Uslenghi ed., Academic Press, New York (1980); MIR, Moskwa (1983)
2. P.Collet, J-P.Eckmann, "Iterated maps on the intervall as dynamical systems", Birkhäuser, Boston (1980)
3. G.Mayer-Kress, Thesis, University of Stuttgart (1984)
4. H. Fröhlich, "General theory of coherent excitations on biological systems", in:"Nonlinear electrodynamics in biological systems", Plenum Publ. Corp. New York (1984)
5. H.Fröhlich, "The biological effects of MW and related questions" "Advances in electronics and electron physics", 53: 85-152 (1980)
6. F.Kaiser, "Coherent modes in biological systems, perturbation by external fields", in: "Biological effects of nonionizing radiation", K.Illinger ed., ACS Symp. Ser. 157: 219-241 (1981)
7. F.Kaiser, "Boltzmann equation approach to Fröhlich's vibrational model", Z. Naturforsch. 34a: 134 (1977)
8. W.R.Adey and S.M. Bawin, "Brain interactions with weak electric and magnetic fields", Neurosci. Res. Progr. Bull. 15 (1): 1-107 (1977)
9. H.Fröhlich, in: "Brain interactions with weak electric and magnetic fields", S.M.Bawin and W.R.Adey eds., Neurosci. Res. Progr. Bull. 15: 67-72 (1977)
10. F. Kaiser, "Theory of resonant effects of rf and mw energy" in: "Biological effects and dosimetry of nonionizing radiation", M.Grandolfo, S.M.Michaelson and A.Rindi (eds.) Plenum Publ. Corp., New York (1983)
11. F.Kaiser, "Limit cycle model for brain waves", Biol. Cybernet. 27: 155 (1977)
12. F.Kaiser "Coherent oscillations in biological systems I", Z. Naturforsch. 331: 294 (1978)
13. F.Kaiser, "Coherent oscillations in biological systems II", Z. Naturforsch. 33a: 418 (1978)
14. J.-P.Eckmann, "Roads to turbulence in dissipative dynamical systems", Rev. Mod. Phys. 53: 643 (1981)
15. E.Ott, "Strange attractors and chaotic motions of dynamical systems", Rev. Mod. Phys. 53: 655 (1981)

16. F.Kaiser, "Specific effects in externally driven self-sustained oscillating biophysical model systems", in: "Coherent excitations in biological systems", H.Fröhlich and F.Kremer eds., Springer Verlag, Berlin (1983)

17. F.Kaiser, "Entrainment-quasiperiodicity-chaos-collapse: Bifurcation routes of externally driven self-sustained oscillating systems", in:"Nonlinear electrodynamics in biological systems", W.R.Ross and A.F.Lawrence eds., Plenum Press, New York (1984)

18. G.Mayer-Kress, H.Haken, "Influences of noise on the logistic model", Journ. Stat. Phys. 26: 149 1981)

19. I.Isuda and K.Matsumoto, "Noise induced order-complexity theoretical digression", preprint, to appear in Journ. Stat. Phys. (1984)

ORDERED STRUCTURES AS A RESULT OF THE PROPAGATION OF COHERENT ELECTRIC WAVES IN LIVING SYSTEMS

E. Del Giudice[1], S. Doglia[2], and M. Milani[3]

Dipartimento di Fisica, Università di Milano
Via Celoria 16, I-20133 Milano, Italy

[1]Istituto Nazionale di Fisica Nucleare, Sezione di
Milano

[2]Gruppo Nazionale Struttura della Materia del
C.N.R, Milano

[3]also Istituto Nazionale di Fisica Nucleare
Sezione di Milano

INTRODUCTION

In this paper we discuss the appearance and disappearance of one-dimensional structures - simple filaments or organized networks of filaments - as a consequence of the dynamics of living systems.

Filamentous structures have been detected inside living eukariotic cells (1), where cytoplasm appears as a complex network of filaments immersed in an aqueous solution of biomolecules and electrolytes. These structures change shape and size and eventually disappear when cells die.

The network of filaments in the cytoplasm - usually referred to as "cytoskeleton" - seems instrumental in a large number of biochemical cellular processes and, moreover, seems to play the role of a network of rails. Materials are transported along them to the different parts of the cell (2, 3). Structures of the same kind have been detected also in the extracellular medium. Metabolically active human erytrocytes have been found to form rouleaux, i.e. stacks of red blood cells; in this structure cells appear to be connected by contractil fibrils (4). These fibrils can be observed by electron microscopy when the erythrocytes are pulled apart by a gentle traction and disappear when the traction is removed. When the erythrocytes are metabolically inactive, the rouleaux are produced at a much lower rate and very easily break down under traction; no fibrils appear at all.

One dimensional dynamically sustained microstructures can appear then both at intercellular and at intracellular level, as a quite general feature. Possible mechanisms underlying biological processes will be investigated looking for solutions characterized by a dynamical confinement of the energy, originally flowing in three dimensional domains, into one-dimensional ones. The appearance of such a confinement would be peculiar of particular theories and is a constraint put on the possible dynamics of living processes imposed by the phenomenological evidence.

In the following we will discuss the experimental findings about cell cytoskeleton and fibrils appearing in rouleau formation; then we will discuss the possibility of generating one dimensional structures in living systems.

1. CYTOPLASMIC MICROSTRUCTURES

Cytoplasm of living (i.e., metabolically active) cells has been studied by means of high voltage electron microscopy (1). The main result led to the discovery of a microscopic cell architecture made up by at least three separate networks of branched filaments:

i) microfilament network, whose main component is actin, about 50 A° thick
ii) microtubule network, whose main component is tubulin, about 250 A° thick
iii) microtrabecular network, the most complex one, whose components are a number of proteins including actin; it forms bundles of filaments about 100 A° thick (1, 5).

The interstices of the above networks are filled up by water and solutes. Most biomolecules are carried along the filaments, which are also the privileged site of their chemical reactions. Most of the cell dynamics therefore appear to occur along these filaments. This system has been described as "a vectorial framework upon which some other force generating component is organized" (6). In Ref. 2 the details of the biochemical role of cytoskeleton are discussed.

We wish here to discuss the dynamical features of such

networks. First of all, their shape and size are time-dependent and follow the cell inner dynamics as well as external perturbations (1). It is possible to make the filaments extend or contract by suitably changing chemical and physical parameters of the environment. Their length is the result of a balance between polymerization and depolymerization processes taking place in the solution. When the filament length increases, monomers are extracted from the solution and bound at the ends of the polymer; the opposite occurs when the length decreases. The above process has been studied "in vitro" for a tubulin solution, where microtubules were artificially produced (7). The whole of this process has been termed "treadmilling". It has also been observed a movement of the monomers along the filament associated to the turnover (association and disassociation of the monomers to the filament).

Furthermore, there is a relationship between length and diameter of a filament: a shortening filament becomes thicker and a lengthening one becomes thinner. The above description of the filament evolution - in particular, treadmilling - suggests that a definite polarity can be assigned to each filament. On the other hand, the transport of biomolecules can occur both ways along it. Finally it is highly significant that the cytoskeleton molecular components seem to be universal (3) for all living species. It is intriguing that, among all the cell proteins, only actin, tubulin and some other ones are selected as the components of the cytoplasm microstructures. This selection rule is a further challenge for any dynamical explanation.

Another interesting feature of cytoskeleton is that the system seems not to be directly attached to the membrane or the nucleus or the organelles, but is able to move, e.g. rotate, respect to them (3). Small gaps seem to exist between the ends of filament and the cell sites they refer to. The links between the main cell organs and cytoskeleton are not chemical bonds.

2. CONTRACTIL FIBRILS IN THE ERYTHROCYTE ROULEAUX

Erythrocytes, namely the red blood cells, are very likely to get attached, forming stacks of cells, named rouleaux. Such a process is an essential step in blood clotting. The amount of rouleaux in the blood increases in most diseases and ruling the rouleau formation rate will be an important medical achievement.

Recent experiments have shown (4) that the formation rate cannot be accounted for by a mechanism of random collisions among erythrocytes moving in the suspending medium in a Brownian way. Investigations on human erythrocytes by means of time-lapse cinema-scopy techniques have shown that these cells, whose diameter is about 8 μm, when approaching at a distance of three or four cell diameters "seem to rush together on the time scale of the slow Brownian motion" (4). The measured rate is definitely higher than that expected on the basis of the Brownian motion, when erythrocytes are metabolically active. On the contrary red blood cells starved or damaged by chemical or physical agents follow quite well the Brownian law. Once again the dynamics is ruled by metabolism. These results have been interpreted (8) in terms of a long range attractive force among metabolically active erythrocytes. This force has been shown to be consistent with Fröhlich theory (9) which predicts that metabolically active biological systems generate coherent electric polarization waves and that a long range attraction between systems oscillating coherently with (almost) the same frequency will appear. Actually the observed attraction is species specific; erythrocytes belonging to different species exhibit an attraction rate definitely lower than erythrocytes belonging to the same species.

A very interesting feature is that attraction fails to appear when some particular molecular species is missing in the suspending medium (10). For instance, a Brownian - like attraction rate is recovered when fibrinogen is removed from human blood. It is interesting to note that the role of fibrinogen is not unique, because one gets again anomalously higher attraction when molecules usually not present in human blood, such as polyvinylpyrrolidone, polyethylenoxide or dextran, are added to the suspending medium. Also albumin, which is a blood protein and does not usually transmit the interaction, when preheated (11), can play the role of a transmitter.

When rouleaux are investigated by electron microscopy, a new feature becomes apparent (4). If one pulls apart attached erythrocytes by micromanipulation, a strong resistance is met, which shows actually that an attractive force is at work. Moreover thin fibrils appear, which lie along the line connecting the centres of adjacent erythrocytes. Notice that a similar procedure upon rouleaux of metabolically inactive erythrocytes, whose formation can be assumed to occur under a pure Brownian mechanism, reveals a lower resistance to traction and no fibrils at all. Fibrils appear then as a product of the same dynamics responsible for the anomalous attraction among metabolically active red cells. Under a strong traction these fibrils, named contractils, can reach a length up to 100 μm. When the traction is released, contractils shorten and eventually disappear. They do not behave as viscoelastic threads.

The contractils are not directly attached to the red cell membranes, but end gaps of about 0.1 μm (the same order of magnitude of the diameter) are visible. When the cells are brought to rotate, fibrils do not follow the rotation, but "slide" on the membranes and remain on the straight line between the cell centres. The same pattern as in cytoskeleton case then shows up. The fibrils are dynamically sustained and are not bound to the cells by chemical ties.

3. NONLINEAR EXCITATIONS IN A LIVING SYSTEM

A common property of the filament molecular components seems to be their relevant nonlinear electric response. Such response has been directly checked for actin and fibrinogen. Let us examine actin and fibrinogen cases. Filamentous F-actin is a polymer of G-actin, G-actin is a globular molecule, whose aqueous solutions exhibit positive birefringence. A positive birefringence has been observed for solutions of G- and F-actin at field strengths as low as 10^4 V/cm. (12).Fibrinogen solutions at physiological pH show a remarkable positive birefringence also (13).

In both cases a d.c. voltage has been applied to the solutions and values of 10^{-6} esu in actin case and 10^{-5} esu in fibrinogen case have been obtained for the Kerr constant K_{sp} with a relaxation time of the order of 10^2 μsec. On the contrary globular proteins such as albumin show much lower Kerr constants ($K_{sp} \sim 10^{-8}$ esu) with a relaxation time less than 1μsec.

A biological system, as a particular case of a physical system, is made up of its elementary components (molecules) and <u>their interactions</u> and works as an open system while immersed in an external flow of energy. In these conditions the effect of the dynamics is to shift each component from its ground state to some excited state. The interactions among components appear then as molecule excitations, characterized by the same physical parameters which appear in the interaction mechanism. Living matter can be schematized as a set of electric dipoles supplied with metabolic energy coming from outside. The interactions among biocomponents can be assumed to be electric dipole interactions. These dipole interactions in the open system produce the net effect of putting the dipole of each interacting molecule in an excited configuration. The whole interaction then appears as a dipole excitation sweeping through the system.

For particular systems stable configurations could exist. The living system exhibits indeed a rather steady behaviour. That means that dipole excitations must be rather persistent. On these grounds, Fröhlich (14) has predicted that living systems must exhibit a set of metastable (long-lived) excited states, labelled by an electric polarization number. Long-lived oscillatory modes can be switched on and electric polarization waves could appear as a manifestation of inner dynamics.

It has been shown (15) that a necessary condition for the emergence of such dipole excitations is the appearance of a privileged direction in the system produced just by a mechanism supported by the external flow of energy. Since the dipole interactions are rotationally invariant, the privileged direction compels the system to produce electric dipole waves to restore the lost symmetry. Actually most biomolecules have been shown to produce in the surrounding water a net polarization along a privileged direction (16). It has been argued that the frequency of these waves could range in a very wide interval up to 10^{13} Hz and both longitudinal and transversal components might be present.

The oscillation of the electric polarization vector occurs by the vibration of rather massive particles (the molecular subgroups carrying the dipoles). The velocity of such waves could be expected to be, then, of the same order of magnitude of the sound velocity in the same medium.

Wave velocity depends upon the strength of the coupling among the oscillating dipoles, which is a result of the fundamental dipole dynamics. Dynamics determines the relationship $\omega = \omega(K)$ between frequency ω and wave vector K; no space symmetry is expected so that no particular restriction upon K is expected.

Assuming for the polarization wave velocity an order of magnitude of some Km/sec, a frequency of 10^{10} Hz would correspond to a wavelength of some tenth of a micron, which is an acceptable value on the cell scale.

The basic interaction among the oscillators is mediated by the electric field E. The polarization field is expressed as a function of E:

$$P = P\,(E) \qquad (1)$$

which is in principle quite complicated and must be expected nonlinear.

The dynamics is defined in a formal way by the Maxwell equation:

$$\frac{1}{c^2}\frac{\partial^2}{\partial t^2}E - \nabla^2 E = \frac{1}{c^2}\frac{\partial^2 P}{\partial t^2} \qquad (2)$$

that, when eq. (1) is inserted, becomes a nonlinear equation, whose solutions give the evolution of the system excitation.

We don't know, of course, the explicit form of eq. (1) in a living system which depends on the microstructure of the system. We will use then some general results.

4. SELF-FOCUSING OF THE POLARIZATION WAVE

Consider a wave field X propagating with a velocity v(X). We can define a "refractive index" of the medium

$$n(X) = n_o + \delta n(X) \qquad (3)$$
$$\delta n(0) = 0$$

where n_o is the usual linear "refractive index". We have:

$$v(X) = \frac{v_o}{n(X)} \qquad (4)$$

Suppose that for $|X_1| > |X_2|$, $v(X_1) < v(X_2)$, namely the medium is "optically" denser for increasing fields. In that case

$$n(X_1) > n(X_2) \qquad (5)$$

Inequality (5) implies that a beam of definite profile of X (for instance, a Gaussian profile) would shrink to a very small cross-section. The beam would actually shrink to a zero cross-section, if not counteracted by diffraction or some other mechanism (17). In the case of electromagnetic waves it has been shown that the beam reaches a limiting diameter beyond a path R_{nl} - the self-focusing length -. Such a limiting diameter holds for a while, after which the beam can spread out or collapse or have a more complex behaviour. Such a phenomenon is called self-focusing and has been observed in the case of electromagnetic waves, plasma waves and acoustical waves.

When a beam self-focuses, its amplitude increases because of the cross-section shrinking, so that in presence of constant absorption, i.e. absorption independence of wave field intensity, its penetration in the medium is correspondingly enhanced.

The nonlinearity of the Maxwell equation (2) in the case of electric dipole waves we are concerned with implies a nonlinear refractive index. How we can garantee the occurrence of condition (5)? A conclusive answer will of course depend on the specific form of eq. (1).

Experimental results pointing to nonlinear electric response of solution of biomolecules such as t-RNA, actin and other molecules of biological interest have been achieved, either under the influence of DC or AC electric fields (up to optical frequencies)(18). Furthermore impressive results have been obtained when propagation properties of different wavelength is considered (19, 20): a very low attenuation, totally unexpected, has been obtained for selected wavelengths, independently of the origin of the signal, in living systems.

In principle two different, but not mutually exclusive mechanisms may be at work to provide this effective propagation.

The first one (21) is directly connected with the model we are proposing here and is based on the ability of the living system of building a network of privileged paths for AC electric fields (a network of labile "optical" fibers). In this case the whole mechanism can be traced back to the nonlinear response of the medium to externally applied or self-generated alternate electric or electromagnetic fields and the eventual appearance of structures in an otherwise homogeneous medium. Attenuation can attain very low levels since propagation of electric waves along self-generated built in "optical fibers" is practically lossless.

In the second mechanism, as it is figured out in ref. 19, the medium (i.e. the living system) traversed by an electric field can be thought of as an active medium, since when metabolism is at work, the living system is far from thermal equilibrium.

The medium then plays the same role as the active material has in a laser system: in particular it might act as an amplifier for the incoming wave, transforming the excitation energy supplied by metabolism into electric waves of the same frequency as the incoming wave.

This situation then could provide lossless propagation of signals too, in the sense that usual absorption processes take place in living systems as they do in non living ones, but losses are compensated in living systems by gain, i.e.by amplification of the propagating electric wave.

As we were anticipating the two mechanisms could take place simultaneously.

Here we restrict our analysis only to mechanisms of the first type: we will discuss the consequences of propagation of AC electric fields in a nonlinear material on the medium itself.

Biological media usually show positive birefringence, which is a necessary condition for self-focusing. We expect then that the electric dipole waves would self-focus whenever a positive birefringence is observed, namely, for instance, for actin and fibrinogen solutions and a threshold is overcome. Self-focusing of a polarization beam produces high density fields with high boundary dishomogeneities. Forces are generated by these field gradients. A "particle" present in the medium would be acted upon by the force (22):

$$F = \Lambda \sum_K \frac{\omega_K^2 - \omega^2}{(\omega_K^2 - \omega^2)^2 + \Gamma_K^2} \, \mathrm{grad}\, E^2 + \frac{1}{c} \int_V \frac{\partial \vec{P}}{\partial t} \times \vec{H} \, dV \qquad (6)$$

In eq. (6) ω is the frequency of the traveling wave, Γ_k is the damping of the k -th mode of the "particle", Λ is a positive constant and the integral is performed on the volume V of the "particle".

Relevant frequency selective forces develop across the boundaries of the beam when a molecule is present able to oscillate on a frequency ω_k near to the frequency ω of the polarization wave. The force is attractive when ω is slightly less than ω_k and repulsive in the opposite case. Small changes either of ω or ω_k would change the sign of the force. Eq. (6) provides, then, a flexible mechanism able to produce time-dependent condensations or rarefactions of selected molecular species around the beam axis.

Eq. (6) shows also that the gradient force is smaller at the ends of the self-focused beam where grad E^2 has not yet acquired a large value. When the limiting diameter is reached, the field inhomogeneity becomes relevant and quite large forces can be produced. The gradient force must overcome friction in the medium. That could not be possible at the ends of the beam where the gradient force is smaller. Significant condensations or rarefactions of selected molecular species are then expected to appear only at some distance from the beam filament ends. It is interesting to observe that this feature of the gradient force matches with the observed feature of filamentous microstructures in living systems.

The second term in eq. (6) is a sort of "radiation pressure force". In particular, the transversal component of the polarization field is able to produce a force along the beam axis, which can provide kinetic energy to the molecules trapped in the filament.

When molecules with appropriate vibrational frequencies are present in the medium and can get attracted on the filament boundaries, the beam acquires a material coating, that behaves as a wave guide for it.

5. A QUALITATIVE MODEL FOR THE FILAMENTOUS MICROSTRUCTURES

The above considerations on the propagation of polarization waves in a nonlinear medium and on the induced dynamical consequences allow a qualitative understanding of the main features of filamentous microstructures appearing in living systems (23, 24, 25).

The starting point is the existence of polarization waves in living (i.e., metabolically active) systems. A comprehensive evaluation of evidence so far accumulated in favor of this proposal can be found in ref. 26. In the present analysis we assume, then, that a living system can generate a set of electric polarization waves, whose frequencies are typical of the particular system. We assume also, on the basis of the Kerr properties of most biomolecules, that these waves could self-focus and attract, furthermore, specific molecules. The universality of the cytoskeleton protein components would suggest that some of the frequencies could be the same in all cells and correspond to vibrational frequencies of those protein components (actin, tubulin). On the other hand, filaments are not produced in the erythrocyte rouleaux when particular molecules are missing. That means, in our scheme, that the wave emitted by the red cell does not find in the traversed medium any molecule providing a nonlinear electric response of the medium and thus the wave cannot self-focus and give rise to ponderomotive forces.In conclusion, this selective mechanism allow to understand both the specificity of the cytoskeleton components and the role of the "interaction transmitters" in erythrocyte rouleaux.

The material coating of the polarization beam depends upon the existence of the forces given in eq. (6). These forces are a function of the underlying waves and follow their evolution. Consequently we expect that the molecule condensations around the beam axis modify their shape and size according to self-trapped beam evolution. Disappearance of the wave implies disappearance of the molecule condensation. There is again an agreement with the observed facts.

The problem could be raised about the interactions among the attracted molecules, once aligned along the beam axis. One could ask about possible polymerization mechanisms. It is reasonable to expect that the attracted molecules could get oriented by the propagating polarization field. Nonequilibrium concentrations and ordering of the molecules then appear in the solution and could ease a subsequent polymerization. The push produced by the "radiation pressure force" is also useful in that purpose. Additional conditions must be, of course, fulfilled in order to get polymerization (presence of ions, pH value and so on). Anyway we expect that the above mechanism would play an important role in polymer formation inside biological structures. Further research is needed to clarify completely the problem.

Filaments produced through the present dynamical mechanism obviously are not viscoelastic threads. The gradient force (6) does not depend upon the length of the trapped beam, so that the number of attracted molecules per unit length (which depends on grad E^2 only) does not change.

The existence of filament end gaps is also taken into account in the present model, as we have said above. The property of treadmilling can be also understood in terms of the pushing provided by the "radiation pressure force". Since the wave travels in a definite direction along its axis, the existence of a polarity in cytoskeletal elements becomes clear.

We have so far outlined a scheme in qualitative agreement with experimental evidence. We wish to stress that the order of magnitude of the involved physical quantities may be correct. Reported measurements (27) of the frequency of the electric field around human erythrocytes give the value of about 10 GHz. The measured diameter of the contractil fibril between adjacent erythrocytes - which can be assumed of the same order of magnitude of the wavelength - is 0.1 μm. Consequently the velocity of the electric dipole wave responsible for the fibril would be 1Km/sec. The length of a fibril - up to 100 μm - would then contain a thousand of wavelengths.

Investigation on the quantitative features of this model is under way.

REFERENCES

1. Wolosewick, I.I. and Porter, K.R., The Journal of Cell Biology 82 (1979) 114.

2. Clegg, J.S., Collective Phenomena, 3 (1981) 289.

3. Clegg, J.S., in: "Coherent Excitations in Biological Systems" (eds. H. Frohlich and F. Kremer), Springer, Berlin (1983) p. 189.

4. Rowlands, S., in: "Coherent Excitations in Biological Systems" (eds. H. Frohlich and F. Kremer), Springer, Berlin (1983) p. 145.

5. Hameroff, S.R., Smith, S.A. and Watt, R.C., in: "Nonlinear Electrodynamics in Biological Systems" (eds. W.R. Adey and A.F. Lawrence), Plenum Press, New York (1984) p.567.

6. Hyams, J., Nature 291 (1981) 107.

7. Margolis, R.L. and Wilson, L., Nature 293 (1981) 705.

8. Rowlands, S., Sewchand, L.S., Lovlin, R.E., Beck, J.S. and Enns, E.G., Phys. Lett. A 82 (1981) 436.

9. Frohlich, H., Adv. in Electronics and Electron Physics (eds. L. Marton and C. Marton) 53 (1980) 85.

10. Sewchand, L.S., Roberts, D. and Rowlands, S., Cell Biophysics 4 (1982) 253.

11. Rowlands, S., Sewchand, L.S. and Skibo, L., Cell Biophysics 5 (1983) 197.

12. Kobayasi, S., Asai, H. and Oosawa, F., Biochim. Biophys. Acta 88 (1964) 528.

13. Yoshioka, K., in Molecular Electro-Optics (ed. C.T. O'Konski), Dekker, New York and Basel (1976) p. 601.

14. Frohlich, H., Proc. Natl. Acad. Sci. USA 72 (1975) 4211.

15. Del Giudice, E., Doglia, S., Milani, M. and Vitiello, G., Phys. Lett. A 95 (1983) 508; Nuclear Physics B (FS), in press.

16. Hasted, J.B., Millany, H.M. and Rosen, D., J. Chem. Soc. Faraday Trans. 77 (1981) 2289;
Celaschi, S. and Mascarenhas, S., Biophysical Journal 20 (1977) 273.

17. Shen, Y.R., Prog. Quant. Electr. 4 (1975) 1;
Marburger, J.H., Prog. Quant. Electr. 4 (1975) 35.

18. Fredericq, E. and Houssier, C. "Electric Dichroism and Electric Birefringence", Clarendon, Oxford (1975).

19. Popp, F.A., Nagl, W, Li, K.H., Scholz, W., Weingartner, O. and Wolf, R., Cell Biophysics 6 (1984) 33.

20. Mandoli, D.F. and Briggs, W., Proc. Natl. Acad. Sci. USA 79 (1982) 2902.

21. Del Giudice, E., Doglia, S. and Milani, M., Phys. Lett. A 90 (1982) 104.

22. Askar'yan, G.A., Sov. Phys. JETP 15 (1962) 1088.

23. Del Giudice, E., Doglia, S. and Milani, M. in: "Coherent Excitations in Biological Systems" (eds. H. Frohlich and F. Kremer), Springer, Berlin (1983) p. 123.

24. Del Giudice, E., Doglia, S. and Milani, M. in: "Nonlinear Electrodynamics in Biological Systems" (eds. W.R. Adey and A.F. Lawrence) Plenum Press, New York (1984) p. 477.

25. Del Giudice, E., Doglia, S. and Milani, M., Journal of Biol. Physics, in press.

26. Coherent Excitations in Biological Systems (eds. H. Frohlich and F. Kremer), Springer, Berlin (1983).

27. Pohl, H., in: "Coherent Excitations in Biological Systems" (eds. H. Frohlich and F. Kremer), Springer, Berlin (1983) p. 199.

ELECTRIC FIELD EFFECTS ON RATE PROCESSES AND ON COOPERATIVE TRANSITIONS OF MACROMOLECULAR STRUCTURES

Gerhard Schwarz

Department of Biophysical Chemistry
Biocenter of the University
CH 4056 Basel, Switzerland

INTRODUCTION

It is well-known that the state of a material system depends on certain physical conditions imposed on it. These are described by appropriate variable parameters such as the pressure, p, and particularly the temperature, T. In principle, a possibly existing electric field strength, E, also falls in this category of state variables[1] (for the sake of simplicity we consider a homogeneous field with E being a scalar quantity).

Accordingly any physico-chemical properties must generally be subject to electric field effects. In most ordinary cases of practical interest these can be safely disregarded because of their really insignificant magnitude. However, the situation may be quite different for comparatively complex systems as they are encountered in a biological context. On one hand, they could involve variations of extremely strong electric fields (especially in cell membranes). On the other side, the underlying macromolecular structures can give rise to cooperative interactions which substantially intensify the effects of (even small) electric fields.

MASS ACTION AND RATE LAWS

Thermodynamic Basis

Quantitative expressions to do pertinent calculations in a given special case can be derived from some fundamental relations implied by the principles of thermodynamics. We note that for differential changes of state in a closed material system

$$dU = \delta q + \delta w \tag{1}$$

$$dq \leqslant TdS \tag{2}$$

(U: internal energy, S: entropy, T: absolute temperature of the surroundings, δq: heat input, δw: work done on the system) where the < and = signs apply to changes under non-equilibrium and equilibrium conditions, respectively. Taking into account externally applied pressure and electric field strength, one has

$$\delta w = -\, pdV + EdM \tag{3}$$

(V: volume, M: component of the overall electric dipole moment parallel to the field). Inserting this in (1), defining a (modified) free enthalpy

$$G = U + pV - ME - TS \tag{4}$$

and combining (1) with (2) eventually gives

$$dG \leqslant -SdT + Vdp - MdE \tag{5}$$

for any closed system. Apparently the present G-function coincides with the conventional one at zero-field.

Changes of G due to exchange of matter with the surroundings are expressed in terms of appropriate chemical potentials

$$\mu_i = (\partial G/\partial n_i)_{T,p,E,n_{j \neq i}} \tag{6}$$

related to a pure chemical component, B_i, whose amount in the system is measured by its mole number, n_i. Extending the equilibrium version of (5), the total differential of G then reads

$$dG = -SdT + VdP - MdE + \Sigma_i \mu_i dn_i \tag{7}$$

when all relevant variables of state are considered.

Following the ordinary practice we define an activity, a_i, according to the relation

$$\mu_i = \mu_i^{\ominus} + RT \ln a_i \tag{8}$$

(R: gas constant) with a suitable standard state value $\mu_i^{\ominus}$. In the case of a solute this $\mu_i^{\ominus}$ is chosen so that in the limit of infinite dilution $a_i \to c_i$ (i.e. the molar concentration of B_i). In addition we generally set $a_i = y_i c_i$ by means of introducing the proper activity coefficient, y_i. Under the present circumstances, $\mu_i^{\ominus}$ and y_i must naturally be functions of the electric field strength.

Electric Field Effect on Chemical Equilibrium[2]

A chemical reaction process causes a transformation of matter which is formally described by a corresponding stoichiometric equation, generally formulated

$$\Sigma_i \nu_i B_i = 0 \tag{9}$$

(e.g., $-2H_2 - O_2 + 2H_2O = 0$ stands for $2H_2 + O_2 \to 2H_2O$). The stoichiometric factor, ν_i (taken negative for a reaction partner on the left side of the conventional chemical equation), determines the relative turn-over of B_i. Because of mass conservation we have in a closed reaction system

$$n_i = n_i^o + \nu_i \xi \qquad dn_i = \nu_i d\xi \tag{10}$$

(n_i^o being the respective mole number in some arbitrarily chosen reference state of the reaction) with only one variable, ξ, the extent of reaction. According to (7) any possible variation of state thus implies a change of free enthalpy given by

$$dG = -SdT + Vdp - MdE + \Delta G\, d\xi \tag{11}$$

involving a molar free enthalpy of reaction

$$\Delta G \equiv (\partial G/\partial \xi)_{T,p,E} = \Sigma_i \nu_i \mu_i \tag{12}$$

When (11) is compared with (5) we see that the reaction tends to go off spontaneously if

$$\Delta G \cdot d\xi < 0 \tag{13}$$

It would therefore proceed from the left to the right (corresponding to $d\xi > 0$) provided $\Delta G < 0$ (and vice versa). Naturally, chemical equilibrium is established once $\Delta G = 0$.

Because of (8) we can always write

$$\Delta G = -RT \left\{ \ln K^{\bullet} - \Sigma_i \nu_i \ln a_i \right\} \tag{14}$$

with

$$K^{\bullet} = \exp\left(-\Delta G^{\ominus}/RT \right) \tag{15}$$

($\Delta G^{\ominus} = \Sigma_i \nu_i \mu_i^{\ominus}$) being the chemical equilibrium constant under the mass action law. Owing to the multiple-differentiation rule for state functions applied to (11) we find

$$-(\partial \Delta G/\partial E)_{T,p,\xi} = (\partial M/\partial \xi)_{T,p,\xi} \equiv \Delta M \tag{16}$$

which is the so-called reaction dipole moment, i.e. the change of M due to one molar chemical turn-over (at constant T,p,E). Inserting (14) in (16) yields

$$\Delta M = RT \left\{ (\partial \ln K^{\bullet}/\partial E)_{T,p} - (\partial \Sigma_i \nu_i \ln y_i/\partial E)_{T,p,\xi} \right\} \tag{17}$$

In most cases of interest the term involving the activity coefficients does practically not depend on ξ. Then it proves more suitable to use the apparent equilibrium constant

$$K = K^{\bullet}/(y_1^{\nu_1}\, y_2^{\nu_2} \ldots) \tag{18}$$

It permits a formulation of the mass action law using real concentrations instead of activities. Under these circumstances (17) leads to the "van't Hoff" equation

$$(\partial \ln K / \partial E)_{T,p} = \Delta M/RT \tag{19}$$

In other words, an increase of the electric field strength would displace the chemical equilibrium towards that side of the reaction which makes the greater contribution to M.

Elementary Rate Constants

Turning to the rate problem we consider first the basic case of elementary reactions. The underlying mechanism is most simple in so far as just one activation barrier has to be overcome in the course of the reaction path between the initial and the final states of the process. In condensed phases this implies either first- or second-order kinetics depending on the number of molecules involved (which is one or two, respectively). The appropriate rate constant has been derived from the transition state theory as

$$k = (k_B T/h) \exp\left(-\Delta G^{\ddagger}/RT\right) \tag{20}$$

(k_B: Boltzmann's constant, h: Planck's constant) where $\Delta G^{\ddagger}$ stands for the standard free enthalpy of activation, i.e. of the formation of the "activated" state on top of the free enthalpy barrier observing the condition that the transition towards the reaction product is "frozen". In principle, $\Delta G^{\ddagger}$ must be field dependent. By application of (16) ee obtain here

$$\Delta G^{\ddagger} = \Delta G_o^{\ddagger} - \Delta M^{\ddagger} \cdot E \qquad (21)$$

(the subscript o is to indicate values taken at zero-field). The dipole moment of activation, $\Delta M^{\ddagger}$, must possibly be averaged over an orientational distribution of the individual molecular dipole moments. Anyway, inserting (21) in (20) gives

$$k = k_o \exp \left\{ (\Delta M^{\ddagger}/RT) \cdot E \right\} \qquad (22)$$

Structural transitions occurring in the course of an elementary reaction can hardly produce a $\Delta M^{\ddagger}$ larger than a few Debye units per activated molecule. Even 100 kV/cm would therefore cause only a few percent change of k at the most (ordinary temperatures presumed). Thus a single elementary rate constant does usually not exhibit a significant field effect.

Complex Rate Processes

The mechanism of a certain chemical transformation of interest may actually be composed of a great number of elementary reaction steps. This is especially be expected when conversions of macromolecular structures are involved. Under such circumstances the combined action of all these steps may very well result in appreciable sensibility to an electric field. A theoretical treatment turns out to be comparatively easy if one of the steps is clearly rate limiting. This is indeed encountered quite frequently in practice.

As an example, we shall examine a conformational transition, $R \rightarrow P$, of a macromolecule. Among the many intermediate elementary steps there may be one, $R^* \rightarrow P^*$, which has an exceptionally high activation barrier. The essential elements of the whole kinetics can then be formulated

$$R \underset{}{\overset{(fast)}{\rightleftharpoons}} R^* \overset{k^*}{\rightleftharpoons} P^* \overset{(fast)}{\rightleftharpoons} P \qquad (23)$$

implying that the overall rate becomes equal to $k^*[R^*]$. Since equilibrium must be established between R^* and R, one has $[R^*] = K^* [R]$ where K^* stands for the appropriate equilibrium constant. Accordingly R → P adopts first-order kinetics with a rate constant of $k = k^*\cdot K^*$. A possible electric field dependence of K^* must generally be taken into account. It can be calculated from (19) leading to

$$k = k_o \exp\left\{ (\overline{\Delta M}^*/RT)\cdot E \right\} \tag{24a}$$

which involves an effective (orientationally averaged) reaction dipole moment

$$\overline{\Delta M}^* = \frac{1}{E}\cdot\int_0^E \Delta M^*\, dE \tag{24b}$$

related to R → R*, i.e. the formation of the rate limiting intermediate state. This $\overline{\Delta M}^*$ could in fact assume any value when the structure in question is only large enough

A completely analogous result is likewise obtained for other macromolecular processes provided a pronounced rate limiting step exists. The latter condition, however, does not naturally apply to all complex reaction mechanisms. Hence a more sophisticated theoretical treatment of the rate process may be required in certain cases.

SPECIAL MACROMOLECULAR SYSTEMS

Cooperative Transition of a Membrane Protein

The electric potential across a biological membrane generally gives rise to an unusually strong electric field inside (with field strengths of the order of 100 kV/cm). The observed changes of this membrane potential may therefore cause significant field effects on the functional properties of a protein incorporated in the membrane. Basically, the problem can be dealt with quite readily if a cooperative all-or-none mechanism between two states, $P_1 \rightleftharpoons P_2$, is assumed. The fraction of protein in the P_2 state is to be called the degree of transition, Θ.

Under equilibrium conditions, application of the mass action law immediately yields

$$\Theta = K/(1 + K) \tag{25a}$$

with the proper equilibrium constant K. The latter quantity will apparently depend on the electric field in the membrane if different dipole moments, m_1 and m_2, are contributed to M in the two molecular states. We have $\Delta M = N_A m_{12}$ (N_A: Avogadro's number,

$m_{12} = m_2 - m_1$) so that because of (19)

$$K = K_o \exp\left\{(m_{12}/k_BT)\cdot E\right\} \qquad (25b)$$

Applying a field strength of the order of (m_{12}/k_BT) would thus tend to induce appreciable changes of Θ (provided a reasonable value of K_o is given). As a numerical example we consider a case where m_{12} = -100D, K_o = 1 (equivalent to 50% transition at E = 0). Reversal of a membrane field strength of 120 kV/cm would displace this transition from 27% to 73%.

Assuming a rate limiting step in $P_1 \rightarrow P_2$, both the forward and reverse reactions must exhibit a first-order rate law. The relevant kinetic equation then simply reads

$$d\Theta/dt = k_{12}(1-\Theta) - k_{21}\Theta \qquad (26)$$

(t: time). The field effect on the rate constants, k_{12} and k_{21}, can be expressed according to (24). This gives

$$k_{12} = (k_{12})_o \exp\left\{(m_1^*/k_BT)\cdot E\right\} \qquad (27)$$

where $m_1^* = m^* - m_1$ (m^* being the effective dipole moment parallel to the field, contributed by the rate determining intermediate structure of the protein). An analogous expression applies to k_{21} with $m_2^* = m^* - m_2$ instead of m_1^*. We note that the forward and reverse rates are usually influenced differently by the field.

<u>Helix-Coil Transition in an Isotropic Solvent</u>

The conversion between the random coil and α-helix forms of polypeptides, particularly synthetic polyamino acids, in homogeneous solution may be considered as a basic model system for a cooperative conformational transition of a protein. The experimental as well as theoretical aspects of this phenomenon have therefore been studied to a very great extent.[3]

Usually the polymer molecules in question do not undergo a simple all-or-none transition but assume a large number of different structural states composed of alternating randomly coiled and α-helical fragments. The relevant reactions of the individual peptide subunits (with regard to helix formation) can be classified as either a growth step (at the end of an already existing helical fragment) or a nucleation step (within a coiled fragment). For entropic reasons, the latter step is much less likely than the former one. The equilibrium constants for growth and nucleation are usually introduced as s and σs, respectively, where $\sigma \ll 1$. This preference for growth determines the cooperative nature of the overall conversion process. Its equilibrium properties can be calculated on the basis of mathematical procedures developed for the linear Ising model. The degree of transition (i.e. the fraction of peptide units in the helical form) becomes

$$\Theta = (1/2) \left\{ 1 + (s-1)/ \sqrt{(s-1)^2+4\sigma s} \right\} \qquad (28)$$

provided the polymer chain is sufficiently long (i.e. the degree of polymerization is $\gg 1/\sqrt{\sigma}$). Very small values of the cooperativity parameter, σ ($\lesssim 10^{-4}$ as ordinarily encountered in practice), imply remarkably sharp transitions associated with the occurrence of mostly rather long helical fragments.

The α-helix has a substantial permanent electric dipole moment (amounting to about 5 D per subunit when side groups are disregarded).[4] On the other hand, the random coil largely cancels out all the individual group dipoles. Consequently an appreciable field effect on the helix-coil transition must be expected. It essentially arises from the inherent field dependence of the growth parameter s. Due to (19) we have

$$s = s_o \exp \left\{ (\delta m/k_B T) \cos \psi \cdot E \right\} \qquad (29)$$

involving δm, the helix dipole moment per peptide unit (including side group), and the angle ψ between the dipole (of the respective helix fragment) and the field. This dependence on orientation implies a complication of the theoretical treatment because generally a distribution of helix directions has to be considered. A detailed analysis related to free rotation in solution has been given elsewhere.[5]

Apparently field induced changes of Θ may be calculated, however, in a quite straight-forward way provided a largely uniform direction of the individual helical fragments exists. This applies for instance to a parallel alignment of helices brought about by incorporation in a membrane or forced orientation in a sufficiently strong field. For illustration we give a numerical example: With $\sigma = 10^{-4}$, $\delta m = 4$ D, and $\psi = 0^o$, a primarily coiled system of $\Theta_o = 5\%$ (s = 0.98) is converted to an almost completely helical one of $\Theta = 95\%$ (s = 1.02) when a field strength of E = 120 kV/cm is switched on.

These helix-coil transitions can proceed quite rapidly. In ordinary solvents the reaction times fall in the microsecond range.[6] They are naturally expected to be longer in a membrane environment where the viscosity is much greater. Nevertheless, values clearly below a millisecond would still be possible.

REFERENCES

1. F.O. Koenig, Thermodynamics of the electric field, J. Phys. Chem. 41: 597 (1937).
2. G. Schwarz, On dielectric relaxation due to chemical rate processes, J. Phys. Chem. 71: 4021 (1967).
3. J. Engel and G. Schwarz, Cooperative conformational transitions of linear biopolymers. Angew. Chem. Int. Ed. 9: 389 (1970).
4. A. Wada, The α-helix as an electric macro-dipole, Adv. Biophys. 9: 1 (1976).
5. G. Schwarz, On the displacement of cooperative transitions of linear biopolymers by an electric field, Biopolymers 14: 1173 (1975).
6. B. Gruenewald, C.U. Nicola, A. Lustig, G. Schwarz and H. Klump, Kinetics of the helix-coil transition of a polypeptide with non-ionic side groups, derived from ultrasonic relaxation measurements, Biophys. Chem. 9: 137 (1979).

INTERACTION-FORCES BETWEEN MICROSCOPIC PARTICLES IN AN EXTERNAL ELECTROMAGNETIC FIELD

Friedrich A. Sauer

Max-Planck-Institut für Biophysik
D-6000 Frankfurt 70
Federal Republic of Germany

1. INTRODUCTION

Force effects on microscopic particles in an external electromagnetic field are of interest for the discussion of the interaction of such fields with biological materials. Because of the complexity of the materials a general approach seems to be impossible. Therefore we restrict ourselves to a discussion of the phenomenon of pearl chain formation. Here a suspension of microscopic particles is exposed to an electromagnetic field. Above a critical field strength chains of particles are formed. One might look at this effect being a kind of phase transition under the influence of the external field. Below the critical field strength the particle phase is disordered and above the critical field strength it is ordered. Krasny-Ergen[1] and Saito and Schwan[2] have given a qualitative and semi-quantitative analysis of the effect for the special case of lossfree particles and medium. This analysis is based on relations for the body forces, which were derived from the Helmholtz energy principle of electrostatics. As Landau and Lifschitz[3] pointed out this principle fails and cannot be applied for media or particles with losses (dielectric losses, conducting media). In this case one has to calculate the forces with the help of the electromagnetic momentum balance via the Maxwell stress tensor. This leads to other complications because of the intimate connection of the Maxwell stress tensor with the pressure tensor (see Meixner[4]). One finally has to deal with a combined hydrodynamical and electromagnetic problem. We will give an approximative solution of the problem and justify it for our special case.

2. THE ELECTROMAGNETIC ENERGY AND MOMENTUM BALANCE: CALCULATION OF PONDEROMOTIVE FORCES

Two balance equations can be derived with the help of the time dependent Maxwell equations; one for the electromagnetic field energy density

$$u_e = \frac{1}{2}\left(\vec{E}^T\cdot\vec{D} + \vec{H}^T\cdot\vec{B}\right)$$

and the other for the electromagnetic momentum density

$$\vec{p}_e = \frac{1}{c^2}\left(\vec{E}\times\vec{H}\right)$$

This will be done for small hydrodynamic velocities $\vec{v}$ only. We assume that $|\vec{v}|$ is small compared to the phase velocity of light c, which is the case in our applications. The balance equations we have in mind are of the form

$$\frac{\partial A}{\partial t} = - \text{divergence (Flow of A)} + \text{(Production of A)} \quad .$$

In the following we make use of the nabla notation and $\vec{E}^T$ for example means the transposed vector. The symbols are explained in the list of the symbols. We follow here the derivations given by Meixner[4] and De Groot and Mazur[5]. With the space-time dependent Maxwell equations (see Stratton[6]).

$$\nabla\times\vec{H} = \vec{j} + \frac{\partial}{\partial t}\vec{D} \quad ; \quad \nabla\times\vec{E} = -\frac{\partial}{\partial t}\vec{B}$$
$$\nabla^T\cdot\vec{D} = \rho \quad ; \quad \nabla^T\cdot\vec{B} = 0 \tag{2.1}$$

we obtain for the time change of the field energy density

$$\frac{\partial u_e}{\partial t} = -\nabla^T\cdot\left(\vec{E}\times\vec{H}\right) + \frac{1}{2}\left(\vec{D}^T\cdot\frac{\partial\vec{E}}{\partial t} - \vec{E}^T\cdot\frac{\partial\vec{D}}{\partial t}\right) + \frac{1}{2}\left(\vec{B}^T\cdot\frac{\partial\vec{H}}{\partial t} - \vec{H}^T\cdot\frac{\partial\vec{B}}{\partial t}\right) - \vec{E}^T\cdot\vec{j} \quad . \tag{2.2}$$

This is in general no conservation law for the field energy and therefore an energy principle cannot be used to calculate body forces in media with losses. This is possible only when the dissipated power

$$Q = \vec{E}^T\cdot\vec{j} - \frac{1}{2}\left(\vec{D}^T\cdot\frac{\partial\vec{E}}{\partial t} - \vec{E}^T\cdot\frac{\partial\vec{D}}{\partial t} + \vec{B}^T\cdot\frac{\partial\vec{H}}{\partial t} - \vec{H}^T\cdot\frac{\partial\vec{B}}{\partial t}\right) \tag{2.3}$$

becomes zero. For periodic fields of the frequency ω the time average of Q is of the form (we assume time dependence $\sim e^{-i\omega t}$ and $\vec{D} = \varepsilon\vec{E}$, $\vec{B} = \mu\vec{H}$)

$$\bar{Q} = \frac{1}{2}\left(\sigma' + \omega\varepsilon''\right)|\vec{E}_o|^2 + \frac{1}{2}\omega\mu''|\vec{H}_o|^2 \tag{2.4}$$

where

$$\varepsilon'' = \text{Im}\{\varepsilon\} \quad ; \quad \mu'' = \text{Im}\{\mu\} \quad ; \quad \sigma' = \text{Re}\{\sigma\} \quad .$$

For dissipative media or particles $(\varepsilon'' \neq 0 \text{ or } \mu'' \neq 0 \text{ or } \sigma' \neq 0)$ $\bar{Q}$ will be unequal zero and the force calculation from the variation of the field energy must lead to meaningless results. Therefore the energy principle of electrostatics, which states that the variation of the field energy $\int u_e dV$ with respect to certain coordinates (e.g. center of mass coordinates) is equal to the work done by the ponderomotive forces ($F_\alpha d\alpha$) cannot be applied. On the other hand there exists the electromagnetic momentum balance which can be used to derive meaninful expressions for the body force. For the local time change of $\vec{p}_e$ one obtains with the help of Maxwell's equation

$$\frac{\partial}{\partial t}\vec{p}_e = (\Pi\cdot\nabla) - \vec{f}_e \quad ; \quad (\Pi\cdot\nabla)_i = \sum_{k=1}^{3}\frac{\partial\Pi_{ik}}{\partial x_k} \tag{2.5}$$

where Π is the Maxwell stress tensor and $\vec{f}_e$ the electromagnetic body force. There are infinite many ways to write the right hand side of (2.5) in the form of a balance equation. We observe that the definition of Π determines $\vec{f}_e$. With that in mind we makes the following choice for Π

$$\Pi = \frac{1}{2}\left(\vec{E}\vec{D}^T + \vec{D}\vec{E}^T + \vec{H}\vec{B}^T + \vec{B}\vec{H}^T - \left(\vec{D}^T\cdot\vec{E} + \vec{B}^T\cdot\vec{H}\right)U\right) \tag{2.6}$$

which is a symmetric tensor. U means the unit tensor, products with a dot are scalar products and without a dot dyadic products of vectors. With (2.6) for Π one gets for the body force

$$\begin{aligned} f_e = \frac{1}{2}\Big[&\left(\nabla^T\cdot\vec{E}\right)\vec{D} - \left(\nabla^T\cdot\vec{D}\right)\vec{E} + \vec{D}\times\nabla\times\vec{E} - \vec{E}\times\nabla\times\vec{D} \\ &+ \left(\nabla^T\cdot\vec{H}\right)\vec{B} + \vec{B}\times\nabla\times\vec{H} - \vec{H}\times\nabla\times\vec{B}\Big] \\ &+ \rho\vec{E} + \vec{j}\times\vec{B} + \frac{\partial}{\partial t}\left(\vec{D}\times\vec{B} - \frac{1}{c^2}\vec{E}\times\vec{H}\right) \quad . \end{aligned} \tag{2.7}$$

This expression contains the well known forces on charges and currents. The additional terms are related to dielectric and magnetic inhomogeneities and the radiation pressure. For the vacuum $\vec{f}_e$ becomes zero.

For the total momentum, the sum of mechanical momentum $\tau\vec{v}$ and $\vec{p}_e$ there exists a conservation law (in the nonrelativistic approximation)[5]

$$\frac{\partial}{\partial t}\left(\tau\vec{v} + \vec{p}_e\right) = \left(-\tau\vec{v}\vec{v}^T + \Pi - P\right)\cdot\nabla \tag{2.8}$$

if no other external forces are present. P is the pressure tensor and τ the mass density. Only the difference $\Pi - P$ has a physical meaning[4]. Combination of (2.5) and (2.8) leads to a material momentum balance

$$\frac{\partial}{\partial t}\left(\tau\vec{v}\right) = -\left(\tau\vec{v}\vec{v}^T + P\right)\cdot\nabla + \vec{f}_e \tag{2.9}$$

where the pressure tensor P depends on the field quantities and the thermodynamic state. This dependence is determined by the choice of $\vec{f}_e$ and is not yet specified. If the electromagnetic field becomes zero, P becomes the usual mechanical pressure tensor.

For periodic fields $\vec{E}(\vec{r}, t) = \mathrm{Re}\{\vec{E}(\vec{r})e^{-i\omega t}\}$ of the frequency ω

we obtain for the time average of the body force

$$\overline{\vec{f}_e} = -\frac{1}{4}\,\mathrm{Re}\{|\vec{E}|^2\nabla\varepsilon + |\vec{H}|^2\nabla\mu - \varepsilon^*\vec{E}\vec{E}^{T*}\,\nabla\left(\frac{\varepsilon}{\varepsilon^*}\right) - \mu^*\vec{H}\vec{H}^{T*}\,\nabla\left(\frac{\mu}{\varepsilon^*}\right)$$

$$+ i\omega\left(\vec{E}^*\times\vec{H} + \vec{E}\times\vec{H}^*\right)\varepsilon\mu\} \quad . \qquad (2.10)$$

Here we assumed $\rho = 0$; $\vec{j} = 0$; $\vec{D} = \varepsilon\vec{E}$; $\vec{B} = \mu\vec{H}$ with complex and frequency dependent ε and μ.
Use has been made of Maxwell's equations for periodic fields in the complex notation

$$\nabla\times\vec{H} = \vec{j} - i\omega\vec{D} \quad ; \quad \nabla\times\vec{E} = i\omega\vec{B}$$
$$\nabla\cdot\vec{D} = \rho \quad ; \quad \nabla^T\cdot\vec{B} = 0 \qquad (2.11)$$

* means the complex conjugated quantity. In (2.10) we have terms proportional to the gradient of ε and μ well known from the body force in electro and magneto statics. For media with losses we get additional terms which depend on the gradients of $\varepsilon/\varepsilon^*$ and μ/μ^*. Furthermore, there are terms with $\vec{E}^*\times\vec{H}$, which contribute to the body force at higher frequencies.
In the time average the electromagnetic momentum balance equation changes into

$$\overline{(\Pi\cdot\nabla)} = \overline{\vec{f}_e} \qquad (2.12)$$

with

$$\bar{\Pi} = \frac{1}{4}\,\mathrm{Re}\,\{\varepsilon\}\cdot\left(\vec{E}^*\vec{E}^T + \vec{E}\vec{E}^{T*} - |\vec{E}|^2\,U\right)$$
$$+ \frac{1}{4}\,\mathrm{Re}\,\{\mu\}\left(\vec{H}^*\vec{H}^T + \vec{H}\vec{H}^{T*} - |\vec{H}|^2\,U\right) \quad . \qquad (2.13)$$

For the material momentum balance we obtain in the time average, if no other external forces are present

$$\overline{(P\cdot\nabla)} = \overline{\vec{f}_e} \qquad (2.14)$$

Here we neglect the convective contribution $\tau\vec{v}\vec{v}^T$ as small at not too high field strength.
With an external force $\vec{f}_{ex}$, which keeps the body at rest, we get

$$\bar{P}\cdot\nabla - \overline{\vec{f}_e} = \vec{f}_{ex} \qquad (2.15)$$

and because of (2.12)

$$\int \vec{f}_{ex}\,dv_s = \int \left(\bar{P}_s - \bar{\Pi}_s\right)\cdot\vec{n}_s\,da_s \qquad (2.16)$$

where v_s is the volume of a solid body, a_s its surface and $\vec{n}_s$ the surface normal vector. Use has been made of Gauss' theorem.
At a surface of discontinuity between a solid body (s) and a liquid (l) we get a surface force $\vec{t}_{sl}$. For a surface at rest this surface force must be compensated by an external force and we have

$$\vec{t}_{ex} = -\vec{t}_{sl} \quad . \qquad (2.17)$$

From limit considerations of (2.12) and (2.14) applied on a small volume including the surface of discontinuity we obtain

$$\vec{t}_{sl} = \left(\bar{\Pi}_l - \bar{P}_l - \bar{\Pi}_s + \bar{P}_s\right)\cdot\vec{n}_s \quad . \tag{2.18}$$

The total external force, which keeps the body at rest in the time average is given by

$$\vec{F}_{ex} = \int \vec{t}_{ex}\, da_s + \int \vec{f}_{ex}\, dv_s \quad . \tag{2.19}$$

With the help of the equations (2.16) - (2.18) we derive

$$\vec{F}_{ex} = \int \left(\bar{P}_l - \bar{\Pi}_l\right)\cdot\vec{n}_s\, da_s \quad . \tag{2.20}$$

Equation (2.20) enables one to calculate the force acting on the center of mass of a solid body immersed in a liquid if the pressure tensor and the Maxwell stress tensor are known in the liquid at the surface of the body. An external force $\vec{F}_{ex}$ prevents the movement of the center of mass. Similar considerations allow us to calculate the external torque, which prevents the body from rotation. This will be done in another article of this volume.
In the following we discuss (2.20) for special cases. For this we must solve the Maxwell equations for special boundary conditions and take into account the material momentum balance (2.14) for the liquid. Because the general problem is too complex, we cannot give a general solution of the problem. Many approximations must be done to arrive finally at applicable formulas.

3. FORCES ON A SINGLE SPHERICAL PARTICLE IN A FLUID THE GRADIENT APPROXIMATION WITH DIELECTRIC LOSSES

We start with equation (2.20) for the calculation of the force on a spherical particle in an external RF field (RF = radio frequency). Under the conditions that the radius a of the particle is small compared to the distance d of the electrodes and this distance is small compared to the vacuum wave length λ_o we can neglect the terms with the magnetic field strength $\vec{H}$ in the Maxwell stress tensor (2.13) and the body force (2.10). These terms are small of the order $\frac{d}{\lambda_o}$. The Maxwell equations change into

$$\nabla\times\vec{E} = 0 \quad \text{or} \quad \vec{E} = -\nabla\Phi \quad \text{for} \quad a < d < \lambda_o = \frac{2\pi c}{\omega} \tag{3.1}$$

This means for the field calculation we have to solve an electrostatic problem.
For the determination of the pressure tensor $\bar{P}$ we have to solve (2.14) with the given expression for $\vec{f}_e$. In the special case we discuss here, (2.14) changes into

$$(\bar{P}\cdot\nabla) = -\frac{1}{4}\,\mathrm{Re}\left\{ |\vec{E}|^2\nabla\varepsilon - \varepsilon^*\,\vec{E}\vec{E}^{T*}\,\nabla\left(\frac{\varepsilon}{\varepsilon^*}\right) \right\} \quad . \tag{3.2}$$

In genral, equation (3.2) does not lead to hydrostatic equilibrium with $\bar{P} = pU$ where p is a scalar pressure. Therefore, one can ex-

pect the appearance of hydrodynamic motion in media with dielectric losses at higher field strength.
For lower field strength we neglect the contributions of the terms with $\nabla\varepsilon$ and $\nabla \varepsilon/\varepsilon^*$ and get a constant pressure tensor P if there are no other sources for a hydrodynamic motion. This is justified for slightly compressible liquids at low field strength if $|\kappa\, \varepsilon\, E^2| \ll 1$, with κ being the isothermal compressibility. As a typical example we have for $\kappa = 10^{-4}\mathrm{atm}^{-1}$, $E = 10^3 \mathrm{Vcm}^{-1}$ and $\varepsilon = 7\cdot 10^{-12}\mathrm{Asec\ V}^{-1}\ \mathrm{cm}^{-1}$

$$|\kappa\, \varepsilon\, E^2| \simeq 7\cdot 10^{-9} \quad .$$

In such a case the pressure tensor, which we have chosen in accordance with (2.13) becomes constant. A constant pressure tensor does not contribute to the surface integral (2.20) and one obtains for the external force, which keeps the particle at rest

$$\vec{F}_{ex} = -\frac{1}{4}\,\mathrm{Re}\,\{\varepsilon_1\} \int \left(\vec{E}_1^{*}\vec{E}_1^{T} + \vec{E}_1\vec{E}_1^{T*} - |\vec{E}_1|^2\, U\right)\cdot\vec{n}_s\, da_s \ . \quad (3.3)$$

This generalizes a result of Stratton[6] for a dielectric particle in a static field for the case of a periodic field. If the particle is rigid and homogeneous, the surface integral (3.3) can be transformed into

$$\vec{F}_{ex} = -\frac{1}{4}\,\mathrm{Re}\,\{\varepsilon_1\} \int \Big[(b^* - b)\left(\vec{E}_s^{T*}\cdot\vec{n}_s\right)\vec{E}_s + b|\vec{E}_s|^2\, \vec{n}_s + |b|^2\, |\vec{E}_s^{T}\cdot\vec{n}_s|^2\, \vec{n}_s\Big]\, da_s \quad (3.4)$$

with $b = \dfrac{\varepsilon_s - \varepsilon_1}{\varepsilon_1}$.

The surface integral (3.4) contains only field quantities at the solid side of the liquid-solid interface. Use has been made of the transition conditions

$$\varepsilon_1\vec{E}_1^{T}\cdot\vec{n}_s = \varepsilon_s\vec{E}_s^{T}\cdot\vec{n}_s \quad \text{and} \quad \vec{E}_1\times\vec{n}_s = \vec{E}_s\times\vec{n}_s \quad (3.5)$$

at the solid-liquid interface and of the fact that we assumed $\nabla\varepsilon_s = 0$ (homogeneous particle).
If one applies (3.4) for a special case, various approximations must be made to evaluate the surface integral in equation (3.4). For a homogeneous rigid sphere in a slightly inhomogeneous field we make use of the gradient approximation. This means, we express the field at the surface by a Taylor expansion around the center of the sphere up to first order terms. This expansion reads

$$\vec{E}_s(a\,\vec{n}_s) = \vec{E}_s(0) + a\left(\vec{E}_s\nabla^T\right)_o\cdot\vec{n}_s \quad (3.6)$$

where $\vec{E}(0)$ is the electric field strength at the center of the sphere, $(\vec{E}_s\nabla^T)_o$ the gradient of $\vec{E}$ at the center and a the radius of the sphere. For the integral (3.4) we derive

$$\vec{F}_{ex} = -\frac{V}{20}\left[b^{*}(2b+5)\left(\nabla\vec{E}^{T}\right)\cdot\vec{E}^{*} + b(2b^{*}+5)\left(\nabla\vec{E}^{T*}\right)\cdot\vec{E}\right]\cdot \mathrm{Re}\,\{\varepsilon_1\} \tag{3.7}$$

V is the volume of the sphere. The field and the gradient must be taken at the center of the sphere. A similar integration has been performed by Denner and Pohl[7] in the electrostatic case.
One can now introduce the external field $\vec{E}_o$ instead of the actual field $\vec{E}$ inside the sphere. The external field is the field that exists at the point $\vec{r}$ inside the sphere, if $\varepsilon_s = \varepsilon_1$. In general, there is a functional relation between the actual field and the external field. At the center of the sphere we obtain

$$\vec{E} = \frac{3}{b+3}\vec{E}_o \tag{3.8}$$

and

$$\nabla\vec{E}^{T} = \frac{5}{2b+5}\nabla\vec{E}_o^{T} \quad . \tag{3.9}$$

These relations can be calculated from a multipole expansion. They are valid only at the center of the sphere. For the external force one gets

$$\vec{F}_{ex} = -\frac{3}{16}V\left(\frac{b^{*}}{3+b^{*}} + \frac{b}{3+b}\right)(\varepsilon_1^{*}+\varepsilon_1)\nabla|\vec{E}_o|^2 \quad . \tag{3.10}$$

This expression is valid for slightly inhomogeneous fields. It changes into the well known result of electrostatics in the case $b^{*}=b$ and $\varepsilon_1 = \varepsilon_1^{*}$ when one observes that $\vec{E}_o$ is the amplitude of the external field.
We compare equation (3.10) with a formula for the external force derived by Sher[8] from an energy principle. Sher's result is given by

$$\vec{F}_{ex}(\mathrm{Sher}) = -\frac{3}{8}V\left(\frac{\varepsilon_1^{*}\,b}{b+3} + \frac{\varepsilon_1\,b^{*}}{b^{*}+3}\right)\nabla|\vec{E}_o|^2 \tag{3.11}$$

which is different from (3.10) for complex dielectric constants ε_1 and ε_s. The difference between the two expressions has the form

$$\begin{aligned}\vec{F}_{ex} - \vec{F}_{ex}(\mathrm{Sher}) &= -\frac{9}{16}V\frac{(b-b^{*})(\varepsilon_1-\varepsilon_1^{*})}{|3+b|^2}\nabla|\vec{E}_o|^2 \\ &= \frac{9}{4}V\frac{(\varepsilon_s''\varepsilon_1' - \varepsilon_s'\varepsilon_1'')\varepsilon_1''}{|\varepsilon_1|^2\,|3+b|^2}\nabla|\vec{E}_o|^2 \quad .\end{aligned} \tag{3.12}$$

Here we have introduced $\varepsilon_1 = \varepsilon_1' + i\,\varepsilon_1''$ and $\varepsilon_s = \varepsilon_s' + i\,\varepsilon_s''$.

The difference becomes zero if $\varepsilon_1'' = 0$ or if $\varepsilon_s'/\varepsilon_s'' = \varepsilon_1'/\varepsilon_1''$.

This means the loss angle of liquid and solid must be the same. For $\varepsilon''_s = 0$ the difference between the force expressions becomes proportional to ε''^2_1. So we come to the conclusion that Sher's formula cannot be applied for media with large losses when the deviations (3.12) become large. This is the case for conducting media at low frequencies where the imaginary part of ε_1 becomes equal to σ/ω with σ being the conductivity.
The gradient approximation breaks down for stronger inhomogeneous fields. In this case one must go back to the original expression (2.20) and find other ways to evaluate the surface integral. This situation occurs for example when another particle is in the close vicinity of the particle of interest. The gradient approximation in the form (3.10) is invalid for inhomogeneous particles.

4. ELECTROMAGNETIC AND HYDRODYNAMIC FORCES BETWEEN TWO IDENTICAL SPHERES IN A HOMOGENOUS EXTERNAL FIELD

The field inside a single homogeneous sphere in an external homogeneous field is also homogeneous (Stratton[6]). Therefore, the integral in equation (3.4) vanishes and the resulting force on the center of mass becomes zero. If there is a second sphere in the vicinity of the other, the field inside the sphere is perturbed and the perturbation leads to a resulting force. The electrostatic problem for two spheres in an external homogeneous field cannot be solved in a closed form in general. We rely here on asymptotic solutions. Therefore, one expands the potential in a power series of $\gamma^{-1} = \frac{a}{\xi}$ where a is the radius of the sphere and ξ the distance of the centers. The result of an elaborate calculation leads to the force of interaction between the two spheres. The leading term is the dipole-dipole interaction between the induced dipoles of the spheres.

The time-averaged force acting on a spherical particle 1 in a homogeneous electric high frequency field if another particle 2 of the same volume V and same dielectric properties is located in the vicinity of a particle 1 is given by (neglecting terms $\sim \frac{1}{\xi^9}$)

$$\vec{F}_1 = -\frac{\alpha}{\xi^4}\Big\{ 2\left(\vec{E}_o^T\cdot\vec{\xi}_o\right)\left(1+\frac{s}{\gamma^3}\right)\vec{E}_o - \left[\left(5\left(\vec{E}_o^T\cdot\vec{\xi}_o\right)^2 - |\vec{E}_o|^2\right) + \frac{2s}{\gamma^3}\left(|\vec{E}_o|^2 + 4\left(\vec{E}_o^T\cdot\vec{\xi}_o\right)^2\right)\right]\vec{\xi}_o\Big\} + O\left(\frac{1}{\xi^9}\right) = -\vec{F}_2 \qquad (4.1)$$

with

$$\alpha = \frac{27\,V^2\,|b|^2\,\mathrm{Re}\,\{\varepsilon_1\}}{8\pi\,|b+3|^2} \quad ; \quad b = \frac{\varepsilon_s - \varepsilon_1}{\varepsilon_1} \qquad (4.2)$$

$$s = \mathrm{Re}\left\{ \frac{b}{3+b} \right\} \tag{4.3}$$

$$\gamma = \frac{\xi}{a} \tag{4.4}$$

where V is the volume of the sphere of radius a, ξ the distance of the centers of the spheres, $\vec{\xi}_o$ the unit vector in the direction of the connection line and $\vec{E}_o$ the field vector amplitude of the homogeneous electric field. This amplitude can be chosen to be real without loss of generality. With $\vec{r}_1$ and $\vec{r}_2$ being the position vectors for the centers of the particles, we have

$$\vec{\xi} = \vec{r}_2 - \vec{r}_1 \tag{4.5}$$

and

$$\vec{\xi} = \xi\, \vec{\xi}_o \quad . \tag{4.6}$$

The equations of motion for the centers of two identical particles under the influence of friction read

$$\begin{aligned} m \frac{d^2\vec{r}_1}{dt^2} + R \frac{d\,\vec{r}_1}{dt} &= \vec{F}_1 \quad ; \\ m \frac{d^2\vec{r}_2}{dt^2} + R \frac{d\,\vec{r}_2}{dt} &= \vec{F}_2 \quad . \end{aligned} \tag{4.7}$$

Here we neglect the influence of Brownian motion. This is justified for larger field strength

$$E_o \gg \left(\frac{3}{2}\frac{kT}{\alpha}\xi^3\right)^{1/2}$$

as will be shown below. Equations (4.7) are valid only for larger distances $\xi \gg a$ because we neglect hydrodynamic interaction between the particles. For R, the hydrodynamic resistance, we assume the validity of Stokes'law

$$R = 6\pi\, \eta_1\, a \tag{4.8}$$

with η_1 being the viscosity of the liquid. We come back to this point later and give corrections for Stokes'law in special cases. Equations (4.5) and (4.7) together lead to

$$m \frac{d^2\,\vec{\xi}}{dt^2} + R \frac{d\,\vec{\xi}}{dt} = -\,2\,\vec{F}_1 \tag{4.9}$$

the equation of motion for the relative movement of the particles. Here we made use of $\vec{F}_2 = -\vec{F}_1$. Introduction of a special coordinate system with the field vector $\vec{E}_o$ parallel to the x-direction leads to

$$\vec{F}_1 = -\frac{\alpha E_o^2}{\xi^4}\left[2\xi_{ox}\left(1+\frac{s\,a^3}{\xi^3}\right)\vec{e}_x + \left[\left(1-5\xi_{ox}^2\right) - \frac{2a^3 s}{\xi^3}\left(1+4\xi_{ox}^2\right)\right]\vec{\xi}_o\right] \quad . \tag{4.10}$$

If one introduces dimensionless variables

$$\tau \equiv \frac{2\alpha E_o^2}{Ra^5} t = \frac{2 E_o^2 \operatorname{Re}\{\varepsilon_1\}}{\eta_1}\left|\frac{b}{b+3}\right|^2 \cdot t$$

$$\vec{\gamma} \equiv \frac{\vec{\xi}}{a} \tag{4.11}$$

the equation of motion (4.9) changes into

$$\beta\frac{d^2\vec{\gamma}}{d\tau^2} + \frac{d\vec{\gamma}}{d\tau} = \tag{4.12}$$

$$\frac{1}{\gamma^4}\left[2\gamma_{ox}\left(1+\frac{s}{\gamma^3}\right)\vec{e}_x + \left[\left(1-5\gamma_{ox}^2\right) - \frac{2s}{\gamma^3}\left(1+4\gamma_{ox}^2\right)\right]\vec{\xi}_o\right]$$

with $\gamma_{ox} = \xi_{ox}$ and so on. The dimensionless parameter β is given by

$$\beta = \frac{4\rho\, a^2 s^2 \varepsilon_1 E_o^2}{9\,\eta_1^2} \tag{4.13}$$

where ρ is the mass density of the particle material. For water as liquid, a field strength of

$$100\ \frac{\text{V rms}}{\text{cm}}$$

and the radius $a = 10^{-4}$ cm the parameter β becomes very small $(\beta \simeq 1.5 \cdot 10^{-5})$. Under these circumstances we neglect the acceleration term in (4.12) and discuss the quasistationary motion of the particles, which is governed by the equation of motion

$$\frac{d\vec{\gamma}}{d\tau} = \frac{1}{\gamma^4}\left[2\gamma_{ox}\left(1+\frac{s}{\gamma^3}\right)\vec{e}_x + \left[\left(1-5\gamma_{ox}^2\right) - \frac{2s}{\gamma^3}\left(1+4\gamma_{ox}^2\right)\right]\vec{\xi}_o\right] \quad . \tag{4.14}$$

The condition $\beta \ll 1$ which leads to equation (4.14), is consistent with the assumption of a Reynolds number small compared to unity which one needs for the calculation of the hydrodynamic resistance according to equation (2.8).

Before we try to solve the system of differential equations (4.14), it should be mentioned that the force $\vec{F}_1$ in equation (4.10) can be derived from a potential U, which should not be confused with the field energy

$$\vec{F}_1 = -\nabla_\xi U \tag{4.15}$$

where U is given by

$$\begin{aligned} U &= \frac{\alpha E_o^2}{3}\left[\frac{1}{\xi^5}\left(3\xi_x^2 - \xi^2\right) + \frac{sa^3}{\xi^8}\left(\xi^2 + 3\xi_x^2\right)\right] + 0\left(\frac{1}{\xi^8}\right) \\ &= \frac{2\alpha E_o^2}{3\,\xi^3}\left[P_2(\cos\theta_o) + \frac{sa^3}{\xi^3}\left[P_2\left(\cos\theta_o\right) + 1\right]\right] + 0\left(\frac{1}{\xi^8}\right) \end{aligned} \tag{4.16}$$

with $\xi^2 = \xi_x^2 + \xi_y^2 + \xi_z^2$. This result neglects terms of the order $\frac{1}{\xi^8}$. The first term in (4.16) with $P_2\left(\cos\theta_o\right)$ (Legendre polynomials of second degree) was already discussed by Saito and Schwan[2]. The term proportional $\frac{sa^3}{\xi^3}$ is a correction which becomes effective for closer distances.
For U as a whole we have $\Delta U \neq 0$. That the force is a potential force is due to the fact that the approximation (4.1) coincides with the gradient approximation. The system of differential equations (4.14) cannot be solved in general in closed form. We leave this for numerical analysis. In the following we discuss only special cases. In the first case we assume that $\vec{\xi}_o$ is parallel or perpendicular to the field. In the second case we stay with the dipole approximation for larger distances.
One special property of (4.14) is the fact that the solution vector $\vec{\gamma}(\tau)$ lies in the plane defined by the x-direction and $\vec{\gamma}(0)$. Scalar multiplication of (14) with $\vec{\xi}_o$, $\vec{e}_x$, $\vec{e}_y$ and $\vec{e}_z$ leads to the following scalar differential equations

$$\frac{d\gamma}{d\tau} = \frac{1}{\gamma^4}\left[1 - 3\gamma_{ox}^2 - \frac{2s}{\gamma^3}\left(1 + 3\gamma_{ox}^2\right)\right] \tag{4.17}$$

$$\frac{d\gamma_{ox}}{d\tau} = \frac{2\gamma_{ox}(1 - \gamma_{ox}^2)}{\gamma^5}\left(1 + \frac{s}{\gamma^3}\right) \tag{4.18}$$

$$\frac{d\gamma_{oy}}{d\tau} = -\frac{2\gamma_{oy}\,\gamma_{ox}^2}{\gamma^5}\left(1 + \frac{s}{\gamma^3}\right) \tag{4.19}$$

$$\frac{d\gamma_{oz}}{d\tau} = -\frac{2\gamma_{oz}\,\gamma_{ox}^2}{\gamma^5}\left(1 + \frac{s}{\gamma^3}\right) \tag{4.20}$$

Because $(\vec{\gamma}_o \cdot \vec{\gamma}_o) = 1$ ($\vec{\gamma}_o$ is a unit vector) equations (4.18), (4.19) and (4.20) are not independent.
For the initial condition

$$\gamma_{ox} = 1 \;;\quad \gamma_{oy} = \gamma_{oz} = 0 \;;\quad \gamma = \gamma_o \quad \text{at} \quad \tau = 0 \tag{4.21}$$

which means the field is parallel to $\vec{\gamma}_o$, we have

$$\gamma_{oy}(\tau) = \gamma_{oz}(\tau) = 0 \;;\qquad \gamma_{ox}(\tau) = 1 \tag{4.22}$$

and

$$\frac{d\gamma}{d\tau} = -\frac{2}{\gamma^4}\left(1 + \frac{4s}{\gamma^3}\right) \quad . \tag{4.23}$$

The solution of (4.23) is given by

$$\tau = -\frac{1}{2}\int_{\gamma_o}^{\gamma} \frac{x^7\,dx}{x^3 + 4s}$$

$$= \frac{1}{10}\left(\gamma_o^5 - \gamma^5\right) - s\left(\gamma_o^2 - \gamma^2\right) - 8s^2\int_{\gamma_o}^{\gamma} \frac{x\,dx}{x^3 + 4s} \quad . \tag{4.24}$$

The integral in (4.24) can be expressed by

$$\int_{\gamma_o}^{\gamma} \frac{x\,dx}{x^3 + 4s} = \frac{1}{3A}\left(\log\left[\frac{(\gamma^2 - A\gamma + A^2)^{1/2}}{(\gamma_o^2 - A\gamma_o + A^2)^{1/2}} \cdot \left(\frac{\gamma_o + A}{\gamma + A}\right)\right] + \sqrt{3}\,\arctan\left(\frac{Z - Z_o}{1 + Z\,Z_o}\right)\right) \tag{4.25}$$

with

$$A^3 = 4s \;;\quad Z = \frac{2\gamma - A}{A\sqrt{3}} \;;\quad Z_o = \frac{2\gamma_o - A}{A\sqrt{3}} \quad . \tag{4.26}$$

For the initial conditions

$$\gamma_{ox} = \gamma_{oy} = 0 \;;\quad \gamma_{oz} = 1 \;;\quad \gamma = \gamma_o \quad \text{at} \quad \tau = 0 \tag{4.27}$$

which means the field is perpendicular to $\vec{\gamma}_o$, we have

$$\gamma_{ox}(\tau) = \gamma_{oy}(\tau) = 0 \quad ; \quad \gamma_{oz}(\tau) = 1 \tag{4.28}$$

and

$$\frac{d\gamma}{d\tau} = \frac{1}{\gamma^4}\left(1 - \frac{2s}{\gamma^3}\right) \quad . \tag{4.29}$$

The solution of equation (29) can be given in a closed form.

$$\tau = \int_{\gamma_o}^{\gamma} \frac{x^7\,dx}{x^3 - 2s}$$

$$= \frac{1}{5}\left(\gamma^5 - \gamma_o^5\right) + s\left(\gamma^2 - \gamma_o\right) + 4(s)^2 \int_{\gamma_o}^{\gamma} \frac{x\,dx}{x^3 - 2s} \tag{4.30}$$

The value of the integral in (4.30) can be taken from (4.25) and (4.26) replacing $A^3 = 4s$ by $A^3 = -2s$.
Equation (4.24) can be used to calculate the approach time of two particles when the field is parallel to $\vec{\xi}_o$. But one must keep in mind that the Stoke's law (4.9) might fail at closer distances.
For the initial conditions

$$\gamma_{oy}(0) = 0 \quad ; \quad \gamma_{ox}^2 = 1 - \gamma_{oz}^2 = \gamma_{ox}^2(0) \quad ;$$
$$\gamma = \gamma_o \quad \text{at} \quad \tau = 0 \tag{4.31}$$

equations (4.17)-(4.20) can be integrated, if s << 1 . One obtains

$$\frac{\gamma(\tau)}{\gamma_o} = \left[\frac{\gamma_{ox}(\tau)\left(1-\gamma_{ox}^2(\tau)\right)}{\gamma_{ox}(0)\left(1-\gamma_{ox}^2(0)\right)}\right]^{1/2} \tag{4.32}$$

and

$$2\tau = \frac{\gamma_o^5}{\left(\gamma_{ox}(0)\{1-\gamma_{ox}^2(0)\}\right)^{5/2}} \int_{\gamma_{ox}(0)}^{\gamma_{ox}(\tau)} \left(x(1 - x^2)\right)^{3/2} dx \tag{4.33}$$

where the integral in (4.33) has the value

$$\int \left(x(1 - x^2)\right)^{3/2} dx$$

$$= \frac{2}{77}\left[\left(-7x^4 + 13x^2 - 4\right)y - 2\sqrt{2}\,F\left(\arccos\sqrt{x}\,,\ \frac{1}{\sqrt{2}}\right)\right]$$

with $y^2 = x(1-x^2)$. F(u, k) is the incomplete elliptic integral of the first kind.

Equation (4.32) is a parameter representation of the trajectories in the x-z plane. The curved trajectory of the relative motion is parallel to the x-axis for $\theta = 63.4^{\circ}$; the particle distance ξ reaches a maximum for $\theta = 54.7^{\circ}$ and the trajectory is parallel to the z-axis for $\theta = 39.2^{\circ}$. The special angles are independent of the dielectric properties as long as $\gamma > 2.5$. For details see fig.1. This behaviour of the particles is mainly due to the complicated angle- and distance dependence of the force (4.10), which is purely repulsive for $\theta = 90^{\circ}$ and attractive for $\theta = 0^{\circ}$.

So far we neglected the influence of Brownian motion. This can be done for large field strength E_o. The following considerations lead to an estimate of the field strength from which on we can neglect Brownian motion. The potential U in equation (4.16) is the potential energy of interaction between two identical particles in an external field E_o. This is correct as long as we restrict ourselves to the gradient approximation. If $|U|$ is large

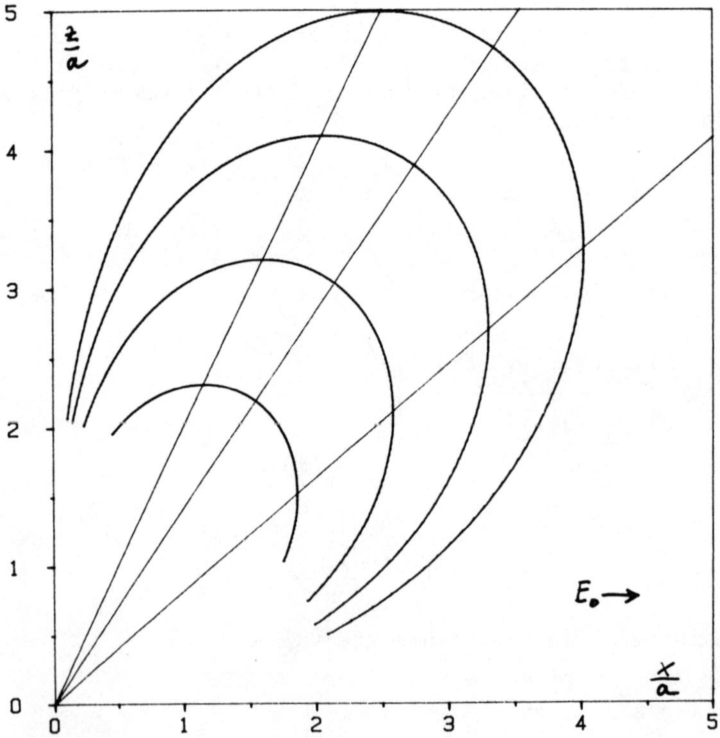

Fig.1. Trajectories of the relative motion of two identical particles according to equ.(4.31) and (4.32), neglecting hydrodynamical interaction and putting $s = 0$. External field in x-direction.

compared to kT (k = Boltzmann constant, T = absolute temperature) we can use equation (4.7) as the equation of motion. From

$$|U| \gg kT \tag{4.35}$$

we derive the condition for the field strength amplitude

$$E_o \gg \left(\frac{3kT\,\xi^3}{\alpha} \left| \frac{1}{3\gamma_{ox}^2 - 1 + s/_{\gamma^3}\,(3\gamma_{ox}^2 + 1)} \right| \right)^{1/_2} . \tag{4.36}$$

For polystyrene particles of radius $a = 10^{-4}$ cm in water, $\xi/_a = 3$ and $\gamma_{ox} = 1$ (field parallel to $\vec{\xi}_o$) the field strength amplitude must be larger than 70 V/cm. For that kind of particles at a distance $\xi = 4a$ and a field strength of 500 Vrms/cm the atractive force (two particles parallel to the field) becomes

$$F_1 \approx 2 \cdot 10^{-13} \text{ Newton} \quad .$$

One can use the right hand side of the inequality (4.36) also for an estimate of the threshold field strength for pearl chain formation in dilute suspensions. For two particles with $\vec{\xi}_o$ parallel to the field one finds

$$E^c_{eff} = \left(\frac{3kT\,d^3}{4\alpha} \right)^{1/_2} \tag{4.37}$$

where d is the average distance between the particles in the suspension. Introducing equation (4.2) for α into (4.37) one obtains

$$E^c_{eff} = \left[\frac{kT}{8\pi s^2\, a^3\, \varepsilon_1} \left(\frac{d}{a}\right)^3 \right]^{1/_2} = \left(\frac{kT}{6s^2\, a^3\, \varepsilon_1} \right)^{1/_2} \frac{1}{p^{1/2}}$$

$$= \left(\frac{kT}{6s^2\, p\, \varepsilon_1} \right)^{1/_2} \frac{1}{a^{3/_2}} \tag{4.38}$$

where p is the volume fraction of the particles in the suspension with

$$p = \frac{4\pi}{3} \left(\frac{a}{d} \right)^3 \quad . \tag{4.39}$$

From equation (4.38) one concludes

$$E^c_{eff} \cdot a^{3/_2} = \text{const.} \qquad \text{for fixed p and } \omega$$

and

$$E^c_{eff} \cdot p^{1/_2} = \text{const.} \qquad \text{for fixed a and } \omega \quad . \tag{4.40}$$

The relations (4.38) and (4.40) may be not valid for concentrated suspensions, because many-partial interaction and the p-dependence of ε_1 must be taken into account.
For distances ξ of the particles with $\xi/_a < 8$ hydrodynamic interactions between the particles occur and lead to a $\xi/_a$ dependence of the hydrodynamic resistance. Faxén[9] calculated correction terms for Stokes'law up to the 9th order in $a/_\xi$ in the special case that the speed of both particles is parallel to $\vec{\xi}_o$. This applies to our initial conditions (4.21) and (4.27), the direct approach and the direct repulsion of the particles.
Then the equations of motion (4.7) have the form

$$m \frac{d^2\vec{r}_1}{dt^2} + R_{11} \frac{d\vec{r}_1}{dt} + R_{12} \frac{d\vec{r}_2}{dt} = \vec{F}_1$$

$$m \frac{d^2\vec{r}_2}{dt^2} + R_{12} \frac{d\vec{r}_1}{dt} + R_{11} \frac{d\vec{r}_2}{dt} = \vec{F}_2 \; . \tag{4.41}$$

This means that a hydrodynamic force $R_{12} \frac{d\vec{r}_2}{dt}$ which has its origin in the movement of the second particle, acts on the first particle and vice versa. Faxén[9] derives the following power series in γ^{-1} for R_{11} and R_{12} from an asymptotic expansion

$$R_{11} = 6\pi\,\eta_1 a \left(1 + \sum_{v=1}^{4} \frac{\beta_{2v}}{2^{2v}} \gamma^{-2v}\right)$$

$$R_{12} = -\,6\pi\,\eta_1 a \sum_{v=0}^{4} \frac{\beta_{2v+1}}{2^{2v+1}} \gamma^{-2v-1} \tag{4.42}$$

with $\beta_1 = 3$; $\beta_2 = 9$; $\beta_3 = 19$; $\beta_4 = 93$; $\beta_5 = 387$

$\beta_6 = 1197$; $\beta_7 = 5331$; $\beta_8 = 19821$; $\beta_9 = 76115$.

This is a good approximation for all values of $\gamma \geq 2.5$. Because of the asymptotic nature of the expansion it cannot be used for γ values close to 2.
The equation of motion for the relative movement of the particles is given by

$$m \frac{d^2\vec{\xi}}{dt^2} + \left(R_{11} - R_{12}\right) \frac{d\vec{\xi}}{dt} = -\,2\vec{F}_1 \; . \tag{4.43}$$

Introduction of dimensionless variables according to (4.11) leads to the equation of motion for the quasi stationary movement of the particles

$$\left(1 + \sum_{v=1}^{9} \frac{\beta_v}{2^v} \gamma^{-v}\right) \frac{d\gamma}{d\tau} = -\frac{2}{\gamma^7} \left(\gamma^3 + 4s\right) \quad . \tag{4.44}$$

Here we neglected the acceleration term and assumed that $\vec{\xi}_o$ is parallel to $\vec{E}_o$. If $\vec{\xi}_o$ is perpendicular to $\vec{E}_o$ one obtains

$$\left(1 + \sum_{v=1}^{9} \frac{\beta_v}{2^v} \gamma^{-v}\right) \frac{d\gamma}{d\tau} = \frac{1}{\gamma^7} \left(\gamma^3 - 2s\right) \quad . \tag{4.45}$$

Equation (4.44) can be integrated in a closed form. The result of the integration is given by

$$\tau(\gamma, \gamma_o) = f_o(\gamma, \gamma_o) + \sum_{v=1}^{9} \frac{\beta_v}{2^{v+1}} f_v(\gamma, \gamma_o) \tag{4.46}$$

with $\gamma = \gamma(\tau)$ and $\gamma_o = \gamma(0)$.

For the f_v one gets

$$\begin{aligned}
f_o &= \frac{1}{10}\left(\gamma_o^5 - \gamma^5\right) - \frac{A^3}{4}\left(\gamma_o^2 - \gamma^2\right) - \frac{A^6 \cdot I_1}{2} \\
f_1 &= \frac{1}{4}\left(\gamma_o^4 - \gamma^4\right) - A^3\left(\gamma_o - \gamma\right) - A^6 \cdot I_o \\
f_2 &= \frac{1}{3}\left(\gamma_o^3 - \gamma^3\right) + A^3 I_2 \\
f_3 &= \frac{1}{2}\left(\gamma_o^2 - \gamma^2\right) + A^3 I_1 \\
f_4 &= \gamma_o - \gamma + A^3 I_o \\
f_5 &= - I_2 \;; \quad f_6 = - I_1 \;; \quad f_7 = - I_o \\
f_8 &= \frac{1}{A^3}\left[\log\left(\frac{\gamma_o}{\gamma}\right) + I_2\right] \\
f_9 &= \frac{1}{A^3}\left[\frac{1}{\gamma} - \frac{1}{\gamma_o} + I_1\right]
\end{aligned} \tag{4.47}$$

with $A^3 = 4s$ and

$$I_o \equiv \int_{\gamma_o}^{\gamma} \frac{dx}{x^3 + A^3} = \frac{1}{3A^2} \left\{ \log \left(\frac{(\gamma + A)}{(\gamma_o + A)} \frac{(\gamma_o^2 - A\gamma_o + A^2)^{1/2}}{(\gamma - A\gamma + A^2)^{1/2}} \right) + \sqrt{3} \arctan \left(\frac{Z - Z_o}{1 + ZZ_o} \right) \right\}$$

$$Z = \frac{2\gamma - A}{A\sqrt{3}} \quad ; \qquad Z_o = \frac{2\gamma_o - A}{A\sqrt{3}} \tag{4.48}$$

$$I_1 \equiv \int_{\gamma_o}^{\gamma} \frac{xdx}{x^3 + A^3} = A \cdot I_o - \frac{2}{3A} \log \left(\frac{(\gamma + A)(\gamma_o^2 - A\gamma_o + A^2)^{1/2}}{(\gamma_o + A)(\gamma^2 - A\gamma + A^2)^{1/2}} \right)$$

$$I_2 = \frac{1}{3} \log \left(\frac{\gamma^3 + A^3}{\gamma_o^3 + A^3} \right) \quad .$$

A numerical comparison of equation (4.24), the result of integration without hydrodynamic correction and (4.46) shows that the hydrodynamic interaction becomes important for $\gamma = \xi/_a < 8$.
If $4 \geq \gamma$, $\gamma_o \geq 2.5$ the time τ calculated from (4.46) is twice the time from calculated (4.24).
In fig. 2 the contribution of the hydrodynamic interaction is shown. The case s = 0 means that we neglect terms in the electromagnetic interaction force, which go beyond the dipole-dipole interaction.
One gets the solution of (4.45) quite easily out of (4.46) by making the following substitutions:

$$\tau \rightarrow -\frac{\tau}{2} \quad \text{and} \quad s \rightarrow -\frac{s}{2} \quad .$$

From equation (4.11) we find for the actual approach time t_a of two particles which start with γ_o at t=0 and have γ at t_a

$$t_a = \frac{R\,a^5}{2\alpha\, E_o^2}\, \tau(\gamma, \gamma_o) \tag{4.49}$$

where $\tau(\gamma, \gamma_o)$ is given by equation (4.46). Substituting equation (4.2) for α leads to

$$t_a = \frac{\eta_1}{2s^2\, \varepsilon_1\, E_o^2}\, \tau(\gamma, \gamma_o) \quad . \tag{4.50}$$

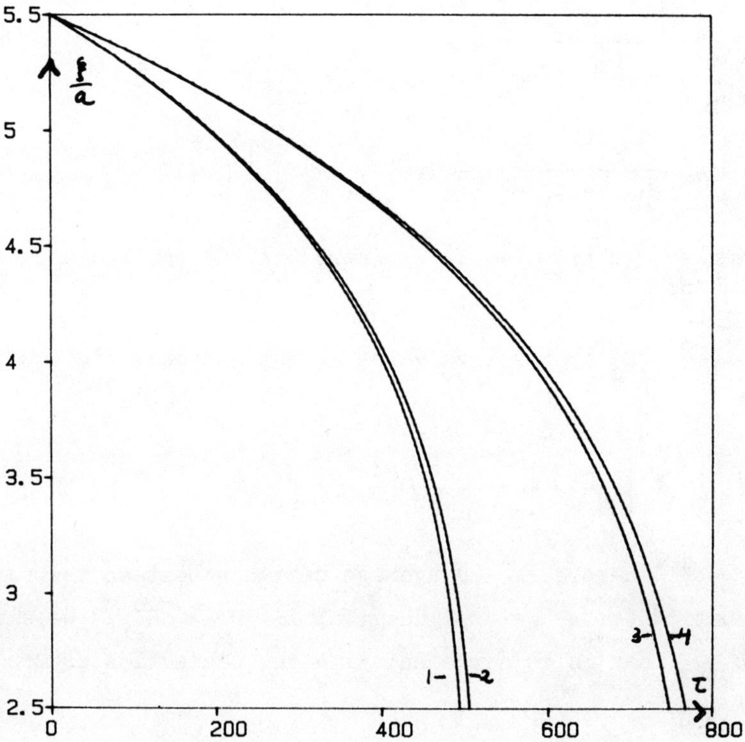

Fig. 2. Direct approach of two particles. External field in $\vec{\xi}_o$ direction. $\gamma = \gamma(\tau)$.

1: no hydrodynamical interaction, according to (4.24) $s = 0$,

2: no hydrodynamical interaction, according to (4.24), $s = -0.48$

3. with hydrodynamical interaction, according to (4.46), $s = 0$,

4: with hydrodynamical interaction, according to (4.46). $s = -0.48$

This means

$$t_a \cdot E_o^2 = \text{const.} \quad \text{for fixed } \gamma \text{ and } \gamma_o \quad . \tag{4.51}$$

If one compares t_a with the characteristic diffusion time t_d of noninteracting particles (no electrical and no hydrodynamic interaction) over a distance d which is given by

$$t_D = \frac{6\pi \eta_1 a^3}{kT} \left(\frac{d}{a} \right)^2 \tag{4.52}$$

one gets

$$\frac{t_a}{t_D} = \frac{kT}{12\pi s^2 \varepsilon_1 E_o^2 a^3} \tau(\gamma, \gamma_o) \cdot \left(\frac{a}{d} \right)^2 \quad . \tag{4.53}$$

According to equation (4.38) we introduce the threshold field strength and

$$E_e^2 = \frac{E_o^2}{2} \quad (E_e = \text{effective value of the external field strength}).$$

Then we get

$$\frac{t_a}{t_D} = \frac{1}{3} \left(\frac{E_{eff}^c}{E_e} \right)^2 \cdot \frac{\tau(\gamma, \gamma_o)}{\bar{\gamma}^5} \tag{4.54}$$

with $\bar{\gamma} = \frac{d}{a}$ where d is the average distance between the particles. From equation (4.54) we conclude that if $E_e >> E_{eff}^c$ we also have $t_a << t_D$. That is in agreement with the neglection of Brownian motion.

5. SUMMARY AND CONCLUSIONS

Interaction forces between two identical spherical particles in an external high frequency field were calculated when the medium and the particles have dielectric losses. The results can also be applied to conducting media (electrolyte solutions) if the radius a of the particle is large compared to the Debye-length δ_D . In this case we observe that the normal component of the total current (conduction current plus displacement current) is continuous at a surface of discontinuity. Equation (4.1) for the interaction force can be generalized in case that the particles have different sizes and different dielectric properties. The α becomes proportional $V_1 \cdot V_2$. If the ratio of the radii deviates largely from unity the expansion we used breaks down and a completely different approach must be used.

The calculated two particle-interaction can be applied to dilute suspensions of particles to obtain results for the frequency dependence of the critical field strength. A mean field approach leads to the result that

$$E_c \cdot \left(\mathrm{Re}\{\varepsilon_1\}\right)^{1/2} \left|\frac{b}{b+3}\right| = \text{const} \quad \text{for fixed p and a} ,$$

where one can take for ε_1 the effective dielectric constant of the suspension, which varies strongly with the frequency.

ACKNOWLEDGEMENTS

I thank Prof. R.Schlögl and Prof. H.P.Schwan for many valuable discussions.

LIST OF SYMBOLS

$\vec{D}$ = vector of dielectric displacement

$\vec{E}$ = vector of electric field strength

$\vec{B}$ = vector of magnetic induction

$\vec{H}$ = vector of magnetic field strength

ρ = lectric charge density

$\vec{j}$ = vector of conductive current density

τ = mass density

c = phase velocity of light in vacuum

$\vec{v}$ = vector of center of mass velocity

ε = dielectric constant = (relative dielectr.const.)$\cdot\varepsilon_o$

μ = magnetic permittivity = (relative perm.)$\cdot\mu_o$

∇ = Nabla operator, a column vector

A quantity $\vec{A}$ is a column vector

$\vec{A}^T$ is a row vector, the transposed of $\vec{A}$.

Re {a} = real part of complex number a

Im {a} = imaginary part of complex number a .

REFERENCES

1. W. Krasny-Ergen, Nicht-thermische Wirkungen elektrischer Schwingungen auf Kolloide, Hochfreq. u. Elektroak.,48:126 (1936).
2. M. Saito and H.P. Schwan, The Time Constants of Pearl-Chain Formation, in: "Biological Effects of Microwave Radiation" Mary Fouse Peyton ed., Plenum Press, New York 1961.

3. L. D. Landau and E. M. Lifshitz, "Electrodynamics of Continuous Media". Course of Theoretical Physics Vol.8, page 253-256, Pergamon Press, New York 1960.
4. J. Meixner, Zur Thermodynamik des elektromagnetischen Feldes, Z.Physik 229:352(1969).
5. S. R. De Groot and P. Mazur, "Nonequilibrium Thermodynamics", page 376-404, North Holland, Amsterdam 1962.
6. J. A. Stratton, "Electromagnetic Theory", McGRAW-HILL, New York 1941.
7. V. Denner and H. A. Pohl, Dielectric Force in Electrostatic Fields, Research Note 127, Quantum Theoretical Research Group, Oklahoma State University, Stillwater, OK (1982).
8. L. D. Sher, Dielectrophoresis in Lossy Dielectric Media, Nature, 220:695(1968).
9. H. Faxén, Gegenseitige Einwirkung zweier Kugeln, die in einer zähen Flüssigkeit fallen, Ark. Mat. Astronom. Fys., 19 A, No.13:1-8(1925/27).

TORQUES EXERTED ON CYLINDERS AND SPHERES BY EXTERNAL ELECTROMAGNETIC FIELDS: A CONTRIBUTION TO THE THEORY OF FIELD INDUCED CELL ROTATION

Friedrich A. Sauer and Reinhard W. Schlögl

Max-Planck-Institut für Biophysik
D-6000 Frankfurt 70
Federal Republic of Germany

1. INTRODUCTION

The interpretation of observed cell rotations, induced by external electromagnetic fields at various frequencies (Pohl[1,2]), led to a theoretical analysis of the effect in terms of the electric cell properties (Holzapfel et al.[3]). The experimental method could be improved by the employment of rotating electric fields (Arnold and Zimmermann[4]).

Because these effects occur in media and with particles both of which suffer dielectric losses and have finite electric conductivity, any theory of the phenomena faces a fundamental difficulty. The energy considerations, usually used for an electrostatic field to calculate ponderomotive forces and the moments of this forces, cannot be applied (Landau and Lifshitz[5], Sauer[6]). Therefore, the theory must be reexamined. Calculations of the forces and the moments of the forces via the electromagnetic momentum balance help to overcome the difficulties. This naturally leads to the introduction of Maxwell's stress tensor. The earlier theories of Lampa[7] and Lertes[8] give correct results only if the surrounding medium suffers no losses and has a dielectric constant of unity. Fürth[9,10] was one of the first investigators to introduce the Maxwell'stress tensor in the problem. He calculated the torque exerted on a metallic ellipsoid in a conducting medium by an external field at low frequencies. This led to the method for the measurement of the dielectric constant of electrolyte solutions, if the conductivity of the medium and the metal were known.

Our treatment generalizes Fürth's approach, in that it can be applied over a wider frequency domain. The theory enables us then to calculate torques exerted on bodies of simple shape such as cylinders and spheres with a shell structure. From the calculated frequency response of the rotation, conclusions may be drawn as to the magnitude of the effect, the direction of the rotation in various fields and the dependence of the rotation on the dielectric properties of the medium and the particles.

2. THE TORQUE EXERTED ON A BODY IN A DISSIPATIVE SYSTEM

Forces and torques exerted by electric fields in a non-conducting medium, are usually calculated from a variational principle as applied to the electric field energy (see Stratton[11]). In a conducting medium or a medium with dielectric losses the field energy is not conserved (see Landau and Lifshitz[5], Sauer[6]). Therefore, a variational principle cannot be applied. Body forces and torques must be calculated by means of the momentum balance equations. A general balance equation for a quantity A has the form

$\frac{\partial A}{\partial t}$ = - divergence (flow of A) + (production of A) (see Meixner and Reik[12]). For the electromagnetic momentum density

$$\vec{p}_e = \frac{1}{c^2} (\vec{E}\times\vec{H})$$

a balance equation can be derived by utilizing the time dependent Maxwell equations. They have the form (see Stratton[11])

$$\nabla\times\vec{H} = \vec{j} + \frac{\partial\vec{D}}{\partial t} \quad ; \quad \nabla\times\vec{E} = -\frac{\partial}{\partial t}\vec{B} \tag{2.1}$$
$$\nabla^T\cdot\vec{D} = \rho_{el} \quad ; \quad \nabla^T\cdot\vec{B} = 0 \quad .$$

The vector $\vec{E}$ is the electric field strength, $\vec{H}$ the magnetic field strength, $\vec{D}$ the dielectric displacement, $\vec{B}$ the magnetic induction, $\vec{j}$ the conduction current density and the scalar quantity ρ_{el} the electric charge density. Use will be made of the formal nabla notation throughout this article. The derivation will be given for small hydrodynamic velocities $\vec{v}$. We assume that $|\vec{v}| << c$, c being the phase velocity of light. We follow here the derivations given by Meixner[13] and De Groot and Mazur[14]. For the local time change of $\vec{p}_e$ one obtains

$$\frac{\partial\vec{p}_e}{\partial t} = \frac{1}{c^2}\left(\frac{\partial\vec{E}}{\partial t}\times\vec{H}\right) + \frac{1}{c^2}\left(\vec{E}\times\frac{\partial\vec{H}}{\partial t}\right) \quad . \tag{2.2}$$

Using Maxwell's equations the right hand side of (2.2) can be cast in the form below

$$\frac{\partial\vec{p}_e}{\partial t} = (\Pi\cdot\nabla) - \vec{f}_e \quad ; \quad (\Pi\cdot\nabla)_i \equiv \sum_{k=1}^{3}\frac{\partial\Pi_{ik}}{\partial X_k} \tag{2.3}$$

Π is the Maxwell stress tensor and $\vec{f}_e$ the electromagnetic body force density. There are infinitely many ways of writing (2.2) in the form of a balance equation. We observe that the definition of Π determines $\vec{f}_e$. With this in mind we make the following choice for Π

$$\Pi = \frac{1}{2}\left(\vec{E}\vec{D}^T + \vec{D}\vec{E}^T + \vec{H}\vec{B}^T + \vec{B}\vec{H}^T - \left(\vec{D}^T\cdot\vec{E} + \vec{B}^T\cdot\vec{H}\right)U\right) \quad . \qquad (2.4)$$

This is a symmetric tensor. U is the unit tensor, products with a dot are scalar products and those without a dot dyadic products of two vectors. Using Maxwell's equations and the definition of Π one obtains for $\vec{f}_e$

$$\begin{aligned}\vec{f}_e = \frac{1}{2}\Big(&\left(\nabla^T\cdot\vec{E}\right)\vec{D} - \left(\nabla^T\cdot\vec{D}\right)\vec{E} + \vec{D}\times\left(\nabla\times\vec{E}\right) \\ &- \vec{E}\times\left(\nabla\times\vec{D}\right) + \left(\nabla^T\cdot\vec{H}\right)\vec{B} + \vec{B}\times\left(\nabla\times\vec{H}\right) \qquad (2.5) \\ &- \vec{H}\times\left(\nabla\times\vec{B}\right)\Big) + \rho_{el}\vec{E} + \vec{j}\times\vec{B} + \frac{\partial}{\partial t}\left(\vec{D}\times\vec{B} - \frac{1}{c^2}\vec{E}\times\vec{H}\right) \quad .\end{aligned}$$

The body force density $\vec{f}_e$ contains the well known forces on charges and currents. The additional terms are contributions to the body force due to inhomogeneities in the dielectric and magnetic properties. For the vacuum $\vec{f}_e$ becomes zero.
The total momentum density, the sum of mechanical momentum $\rho\vec{v}$ and $\vec{p}_e$ obeys a conservation law. In the non-relativistic approximation this balance equation has the form

$$\frac{\partial}{\partial t}\left(\rho\vec{v} + \vec{p}_e\right) = \left(-\rho\vec{v}\vec{v}^T + \Pi - P\right)\cdot\nabla \qquad (2.6)$$

in absence of other external forces. P is the pressure tensor and ρ the mass density. Only the difference $\Pi - P$ has a physical meaning and is thus a measurable quantity (see Meixner[13]). Combination of (2.3) and (2.6) leads to the material momentum balance

$$\frac{\partial}{\partial t}\left(\rho\vec{v}\right) = -\left(\rho\vec{v}\vec{v}^T + P\right)\cdot\nabla + \vec{f}_e \quad . \qquad (2.7)$$

The pressure tensor P depends on the field quantities and the thermodynamic state. This dependence is determined by the choice of $\vec{f}_e$ as yet unspecified. If the electromagnetic field vanishes, P reduces to the usual mechanical pressure tensor. The form of (2.7) justifies the interpretation of the quantity $\vec{f}_e$ as the electromagnetic body force density. In general, to each conceivable choice of Π and hence of $\vec{f}_e$ there corresponds a different P.
For periodic fields of frequency ν ($\omega = 2\pi\nu$), we adopt the complex notation (see Stratton[11]). Each field quantity $A(\vec{r}, t)$ is then written as

$$A(r, t) = \text{Re}\,\{A(\vec{r})\,e^{-i\omega t}\}$$

with $\text{Re}\{A\}$ = real part of A. $A^*(\vec{r})$ and $\text{Im}\{A(\vec{r})\}$ stands for the complex conjugate and the imaginary part of $A(\vec{r})$ respectively.

The Maxwell equations for periodic fields have the form

$$\nabla\times\vec{H} = \vec{j} - i\omega\vec{D} \quad ; \quad \nabla\times\vec{E} = i\omega\vec{B}$$
$$\nabla^T\cdot\vec{D} = \rho_{el} \quad ; \quad \nabla^T\cdot\vec{B} = 0 \quad . \tag{2.8}$$

In the sequel we postulate at low field strengths the linear relations

$$\vec{D} = \varepsilon(\vec{r},\omega)\vec{E} \quad ; \quad \vec{B} = \mu(\vec{r},\omega)\vec{H} \tag{2.9}$$

and (under given conditions) that

$$\vec{j} = \sigma(r,\omega)\vec{E} \quad . \tag{2.10}$$

In general, the quantities ε, μ, σ are complex, and the dependence on the space coordinates can be explicit (as for an inhomogeneous medium) or implicit (as the dependence on the mass density). In our notation ε, μ and σ have dimensions. ε contains the factor ε_o (permittivity of free space) and μ contains the factor μ_o (permeability of free space).
One finds the time average of the body force density (2.5) (see Appendix A1)

$$\begin{aligned}\overline{\vec{f}_e} = \frac{1}{4}\,\mathrm{Re}\,\{ &- |\vec{E}|^2\,\nabla\varepsilon - |\vec{H}|^2\,\nabla\mu + \varepsilon^*\vec{E}\vec{E}^{T*}\cdot\nabla\left({}^{\varepsilon}/\varepsilon^*\right) \\ &+ \mu^*\vec{H}\vec{H}^{T*}\cdot\nabla\left(\frac{\mu}{\mu^*}\right) + i\omega\left(\vec{H}\times\vec{E}^* + \vec{H}^*\times\vec{E}\right)\cdot\varepsilon\cdot\mu \\ &+ \left(1 + {}^{\varepsilon}/\varepsilon^*\right)\rho_{el}^*\,\vec{E} + \left(1 + \frac{\mu}{\mu^*}\right)\vec{j}\times\vec{B}^* \} \quad .\end{aligned} \tag{2.11}$$

This is different from the force density in static fields in several respects: we have additional terms in (2.11) containing $\nabla({}^{\varepsilon}/\varepsilon^*)$ and $\nabla({}^{\mu}/\mu^*)$, and the forces on charges and currents are also different. One notes that terms of the form $\vec{H}\times\vec{E}^*$ may contribute at higher frequencies. One observes further that the dependence on ε and σ does not occur in the combination of a single variable $\varepsilon + \frac{i\sigma}{\omega}$. This means that the force density distinguishes between the conduction current $\vec{j}$ and the displacement current $-i\omega\varepsilon\vec{E}$.
On taking the time average, the electromagnetic momentum balance (2.3) has the form

$$\overline{\Pi}\cdot\nabla = \overline{\vec{f}_e} \tag{2.12}$$

with

$$\begin{aligned}\overline{\Pi} = &\frac{1}{4}\,\mathrm{Re}\{\varepsilon\}\left(\vec{E}^*\vec{E}^T + \vec{E}\vec{E}^{T*} - |\vec{E}|^2\,U\right) \\ &+ \frac{1}{4}\,\mathrm{Re}\{\mu\}\left(\vec{H}^*\vec{H}^T + \vec{H}\vec{H}^{T*} - |\vec{H}|^2\,U\right) \quad .\end{aligned} \tag{2.13}$$

Time averaging of the material momentum balance leads to

$$\overline{P}\cdot\nabla - \overline{\vec{f}}_e = \vec{f}_{ex} \tag{2.14}$$

or because of (2.12)

$$(\overline{P} - \overline{\Pi})\cdot\nabla = \vec{f}_{ex} \quad . \tag{2.15}$$

Here we have introduced an additional external force $\vec{f}_{ex}$ and neglected the convective contribution $\rho\vec{v}\vec{v}^T$. At a surface of discontinuity between two phases (s = solid, l = liquid) with different properties a surface force $\vec{t}_{sl}$ develops. To keep this surface at rest, we must apply an external surface force $\vec{t}_{ex}$ given by

$$\vec{t}_{ex} + \vec{t}_{sl} = 0 \quad . \tag{2.16}$$

Taking a small volume including the surface of discontinuity and subjecting (2.15) to a limiting procedure we obtain

$$\vec{t}_{sl} = -\left(\overline{P}_l - \overline{\Pi}_l - \overline{P}_s + \overline{\Pi}_s\right)\cdot\vec{n}_s \tag{2.17}$$

and

$$\vec{t}_{ex} = \left(\overline{P}_l - \overline{\Pi}_l - \overline{P}_s + \overline{\Pi}_s\right)\cdot\vec{n}_s \tag{2.18}$$

$\vec{n}_s$ being the surface normal vector of the solid body.
We now proceed to calculate the total external torque $\vec{L}_{ex}$ exerted on a body immersed in a liquid by an external electromagnetic field. This torque is the sum of the moment of the surface forces and that of the volume forces. The torque with respect to the center of mass of the body is

$$\vec{L}_{ex} = \int\left(\vec{r}_s\times\vec{t}_{ex}\right)da_s + \int\left(\vec{r}_s\times\vec{f}_{ex}\right)dv_s \tag{2.19}$$

$\vec{r}_s$ being the distance vector from the center of mass, v_s the volume and a_s the surface area of the body. In view of equations (2.15) and (2.18) we get

$$\vec{L}_{ex} = \int \vec{r}_s\times\left(\overline{P}_l - \overline{\Pi}_l - \overline{P}_s + \overline{\Pi}_s\right)\cdot\vec{n}_s\, da_s + \int\vec{r}_s\times\left[\left(\overline{P}_s - \overline{\Pi}_s\right)\cdot\nabla\right]dv_s \tag{2.20}$$

On applying Gauss' theorem to the volume integral in (2.20), one obtains

$$\vec{L}_{ex} = \int\vec{r}_s\times\left(\overline{P}_l - \overline{\Pi}_l\right)\cdot\vec{n}_s\, da_s \tag{2.21}$$

and the torque $\vec{L}_s$ exerted on the body is evidently

$$\vec{L}_s = \int\vec{r}_s\times\left(\overline{\Pi}_l - \overline{P}_l\right)\cdot\vec{n}_s\, da_s \quad . \tag{2.22}$$

Here we have assumed that $\overline{P}_s$ is a symmetric tensor. As in the case of the body forces, the measurable quantity $\overline{\Pi}_l - \overline{P}_l$ appears under the integral sign. Equation (2.22) enables one to determine the torque if the Maxwell stress tensor and the pressure tensor on the liquid side of the solid-liquid interface are known. The solutions of the Maxwell equations (2.8) and the mechanical equations (2.15) for the liquid and solid phase provide precisely this information. This task cannot be performed, however, in its full generality. For our special problem, we shall have to make drastic simplifications to make the problem tractable.

If we deal with periodic external electric fields of frequency ν generated by an electrode system in which the average distance d between the electrodes is small compared to vacuum wavelength $\lambda_o = \frac{c}{\nu}$, we may neglect all contributions from the variable magnetic field in the Maxwell equations and arrive at a quasi electrostatic problem (see Stratton[11]). At low field strengths, the hydrodynamical problem also can be simplified. Because we deal with small hydrodynamic velocities, we calculate at first the torque exerted on a body at rest. This torque then results in a rotation of the body with an angular velocity $\vec{\omega}$, which is supposed to be governed by the equation of motion (angular momentum balance)

$$\theta \cdot \frac{d\vec{\omega}}{dt} + R_r \cdot \vec{\omega} = \vec{L}_s \quad . \tag{2.23}$$

θ is the tensor of the moment of inertia and R_r the tensor characterizing the hydrodynamic resistance to rotation. The rotation governed by equation (2.23) occurs in quite a different time domain as the time variation of the field.

For a body at rest, the pressure tensor P_1, for our choice (2.13) of Π_1 becomes a constant tensor if we neglect space charge effects and $|E|$ is small compared to $(\kappa \cdot \varepsilon)^{-1/2}$ where κ is the isothermal compressibility. For slightly compressible liquids $(\kappa \cdot \varepsilon)^{-1/2}$ is of the order of 10^7 V/cm (see A1). Because a constant tensor does not contribute to the surface integral (2.22), we get for the torque

$$\vec{L}_s = \int \vec{r}_s \times \left(\overline{\Pi}_1 \cdot \vec{n}_s\right) da_s \tag{2.24}$$

and with (2.13)

$$\vec{L}_s = \frac{1}{4} \operatorname{Re} \{\varepsilon_1\} \int \vec{r}_s \times \left[\vec{E}_1 \left(\vec{E}_1^{*T} \cdot \vec{n}_s\right) + \vec{E}_1^{*} \left(\vec{E}_1 \cdot \vec{n}_s\right) - |E_1|^2 \, \vec{n}_s\right] da_s \quad . \tag{2.25}$$

Any other possible choice for Π leads to the same result (2.25) if one observes that the new pressure tensor is then not a constant. Because in equation (2.25) appears the actual field $\vec{E}_1$ and not the external field $\vec{E}_o$, we face another difficulty. For the calculation of the field one must allow for possible elastic deformation of the body. Depending on the elastic properties of the material and deformations might occur at low field strengths. Therefore, we restrict ourselves to rigid bodies. In this case we must solve the Laplace equation with appropriate transition conditions at the

geometrical phase boundary (solid-liquid) and proper conditions at surfaces of discontinuity inside the solid body. From this solution we caculate the field $\vec{E}_1$ at the solid-liquid interface, perform the integration in (2.25) and solve (2.23). It need hardly be said that this can be done in a closed form only for simple geometric shapes of the body.
In general, the field can be evaluated only if we neglect space charges and if the characteristic linear dimension a (e.g. radius of the sphere) is small compared to the electrode distance d. Therefore the following conditions must be met

$$\nabla\times\vec{E} = 0 \quad ; \quad (\nabla^T\cdot\varepsilon\vec{E}) = 0 \quad ; \quad a \ll d \ll \lambda_o = \frac{c}{\nu} \quad . \tag{2.26}$$

This implies for a microscopic body of radius $a = 10^{-4}$cm, $d \geqq 10^{-2}$cm and $\lambda \geqq 1$ cm, while for a macroscopic body of radius $a = 1$ cm, $d \gtrsim 10^2$cm and $\lambda_o \gtrsim 100$ m.
We now derive an upper bound for the magnitudes of the torque and the stationary angular velocity. From (2.25) we conclude that for a single body

$$|\vec{L}_s| < \frac{3}{4} V_s \,\mathrm{Re}\{\varepsilon_1\}\, |\vec{E}_o|^2 \quad . \tag{2.27}$$

With $R_r \simeq V_s \eta_1$ (see Kirchhoff[15]) we get

$$|\vec{\omega}| = \frac{|\vec{L}_s|}{R_r} < \frac{3\,\mathrm{Re}\{\varepsilon_1\}\cdot|\vec{E}_o|^2}{4\,\eta_1} \tag{2.28}$$

where V_s is the volume of the solid body, η_1 the viscosity of the liquid, $|\vec{E}_o|$ the amplitude of the applied external field and ε_1 the permittivity of the liquid. For a body immersed in water at a field strength of 100 V/cm, the angular velocity should not exceed the value

52.5 Rad/sec.

In formula (2.28) the volume V_s cancels out and the upper bound is independent of the volume. More refined calculations for bodies of simple geometry should confirm this result.
In the equation of motion (2.23) we have neglected Brownian rotation. This is justified (see Einstein[16]) if

$$2\pi|\vec{L}_s| \gg kT \quad . \tag{2.29}$$

Here k is the Boltzmann constant and T the absolute temperature. Introduction of (2.27) into (2.29) leads to

$$\pi/2\cdot V_s \,\mathrm{Re}\{\varepsilon_1\}\, |\vec{E}_o|^2 \gg kT \tag{2.30}$$

yielding a lower bound for the electric field strength of the external field, namely

$$|\vec{E}_o| \gg \left(\frac{2\ kT}{\pi\ V_s\ \mathrm{Re}\{\varepsilon_1\}}\right)^{1/2} \quad . \tag{2.31}$$

For a spherical particle of radius $a = 10^{-4}$ cm in water at $T = 300^{\circ}$K the right hand side has the value 9.5 V/cm. Thus the external electric field strength should be well above 9.5 V/cm

$$E_o \gg 9.5\ \text{V/cm} \quad (\text{for a radius} \quad a = 10^{-4}\ \text{cm}) \quad .$$

This preliminary result will sharpened by means of those obtained for simple geometries.

3. A SINGLE CYLINDER: TORQUE AND ROTATION IN A ROTATING FIELD

We now consider a single circular cylinder of radius R and height h, its axis pointing in z-direction exposed to an external rotating electric field $\vec{E}_o$. The field is assumed to rotate in the x-y plane, perpendicular to the cylinder axis. The material of the cylinder has the permittivity ε_s and the conductivity σ_s and in the surrounding liquid we have ε_1 and σ_1. Because the external field is independent of z, we have a two dimensional electrostatic problem if R is small compared to h. The actual fields in the cylinder and in the liquid can be derived from potentials, if R is small compared to $\frac{c}{\nu}$. We seek solutions of the Laplace equation

$$\Delta\Phi = 0$$

in the interior of the cylinder and in the liquid. At the cylinder wall the field must obey the transition conditions

$$\vec{n}_s \times \vec{E}_s = \vec{n}_s \times \vec{E}_1 \tag{3.1}$$

and

$$\left(\varepsilon_s + \frac{i\sigma_s}{\omega}\right) \vec{n}_s^T \cdot \vec{E}_s = \left(\varepsilon_1 + \frac{i\sigma_1}{\omega}\right) \vec{n}_s^T \cdot \vec{E}_1 \tag{3.2}$$

$\vec{n}_s$ being the normal vector. Condition (3.2) is equivalent to the condition of continuity of the normal component of the total current $\vec{j}_t = \vec{j} + \frac{\partial \vec{D}}{\partial t}$. This condition can be applied to electrolyte solutions when we neglect diffusion. This in turn requires the Debeye length δ_D be small compared to R. (Debye length $\delta_D = \left(\frac{\varepsilon\ RT}{F^2\ I}\right)^{1/2}$ with R = gas constant; T = absolute temperature; F = Faraday constant; I = ionic strength).

In addition to (3.1) and (3.2) we demand that the actual field $\vec{E}$ tends to $\vec{E}_o$ at infinity. The solution of the electrostatic problem is given by

$$\Phi_s = -\frac{2}{b+2}\left(\vec{r}^T\cdot\vec{E}_o\right) \quad ; \quad |\vec{r}| < R \tag{3.3}$$

and

$$\Phi_1 = \left[\frac{b}{b+2}\left(\frac{R}{r}\right)^2 - 1\right]\left(\vec{r}^T\cdot\vec{E}_o\right) \quad ; \quad |\vec{r}| > R \tag{3.4}$$

with

$$b = \frac{\varepsilon_s + \frac{i\sigma_s}{\omega}}{\varepsilon_1 + \frac{i\sigma_1}{\omega}} - 1 \quad . \tag{3.5}$$

Here $\vec{r}$ is the two-dimensional position vector. The field $\vec{E}_1$ in the liquid is then

$$\vec{E}_1 = -\nabla\Phi_1 = \left[1 - w\left(\frac{R}{r}\right)^2\right]\vec{E}_o + \frac{2w\,R^2}{r^4}\left(\vec{r}^T\cdot\vec{E}_o\right)\vec{r} \tag{3.6}$$

with

$$w = \frac{b}{b+2} \quad .$$

At the cylinder wall we get

$$E^*_{1n} \equiv \vec{E}^{*T}_1\cdot\vec{n}_s = \left(1 + w^*\right)E^*_{on} \tag{3.7}$$

with

$$E^*_{on} \equiv \vec{E}^{*T}_o\cdot\vec{n}_s \quad \text{and} \quad \left(\vec{r}\times\vec{E}_1\right) = R\cdot(1-w)\left(\vec{n}_s\times\vec{E}_o\right) \quad . \tag{3.8}$$

Because the z-components of $\vec{n}_s$ and $\vec{E}_o$ are zero, only the z-component of $(\vec{r}\times\vec{E}_1)$. Introduction of (3.7) and (3.8) into (2.25) gives for the z-component of the torque $\vec{L}_s$

$$L_{sz} = \frac{1}{4}R^2\cdot h\cdot \mathrm{Re}\{\varepsilon_1\}\int_0^{2\pi}\Big[(1+w^*)(1-w)\left(E^*_{ox}\cdot n_x + E^*_{oy}n_y\right)\left(n_xE_{oy} - n_yE_{ox}\right) + \text{c.c.}\Big]\,d\phi \tag{3.9}$$

where c.c. means the complex conjugate. In deriving (3.9) we used the fact that $\vec{n}_s\times\vec{r} = 0$ at the wall of the cylinder. Noting that $n_x = \cos\phi$ and $n_y = \sin\phi$ we obtain for L_{sz}

$$L_{sz} = \frac{\pi}{4}R^2\cdot l\cdot \mathrm{Re}\{\varepsilon_1\}\cdot\Big[(1+w^*)(1-w)\left(E^*_{ox}E_{oy} - E_{ox}E^*_{oy}\right) + \text{c.c.}\Big] \tag{3.10}$$

If there is no phase difference between E_{ox} and E_{oy}, the right hand side of (3.10) vanishes. There is thus no torque on a cylinder in a non-rotating field. In a rotating field there is a phase difference of $\pi/2$ between E_{ox} and E_{oy}

$$E_{oy} = i\,E_{ox} \tag{3.11}$$

for a field of frequency ν that rotates with a constant field strength. This field rotates in the positive sense about the z-axis.

For a rotating field the torque becomes

$$L_{sz} = 2V \cdot \mathrm{Re}\{\varepsilon_1\}\ \mathrm{Im}\{w\}\ |E_{ox}|^2 \tag{3.12}$$

where $V = \pi R^2 h$ is the volume of the cylinder. For a cylinder the resistance R_r against rotation is given by

$$R_r = 4V \cdot \eta_1 \tag{3.13}$$

η_1 being the viscosity of the liquid. For the stationary angular velocity $\omega_{r\infty}$ one gets

$$\omega_{r\infty} \equiv \frac{L_{sz}}{R_r} = \frac{1}{2\eta_1} \mathrm{Re}\{\varepsilon_1\}\ \mathrm{Im}\{w\}\ |E_{ox}|^2 \tag{3.14}$$

which is independent of the radius R of the cylinder for $R << h$. For a nonconducting system without losses, Im{w} vanishes and $\omega_{r\infty}$ becomes zero. Since L_{sz} is independent of the coordinates of the center of mass, the solution of the equation of motion (2.23) can be found in the form

$$\omega_{rz}(t) = \omega_{r\infty} \left(1 - \exp\left(- \frac{t}{\tau r}\right)\right) \tag{3.15}$$

with

$$\tau_r = \frac{\rho_s R^2}{8\ \eta_1} \tag{3.16}$$

ρ_s being the mass density of the cylinder material. For R = 1 cm and $\rho_s = 1\ g/cm^3$, $\tau_r \simeq 12$ sec. This has consequences for the observation of the stationary rotation of macroscopic bodies. As the example given above shows one must wait for nearly half a minute before the rotation becomes stationary.
The sign and magnitude of the rotation in a given liquid at a given field strength depend on the sign and magnitude of Im {w} . As w is given by

$$w = \frac{b}{b + 2} \tag{3.17}$$

we have a mapping of the complex b plane into the complex w plane. From the definition (3.5) of b we conclude that the physical domain of admissible b values is given by

$$-1 < \mathrm{Re}\ \{b\} < +\infty \quad ; \quad -\infty < \mathrm{Im}\{b\} < +\infty \tag{3.18}$$

because $\mathrm{Re}\{\varepsilon\}$, $\mathrm{Im}\{\varepsilon\}$ and σ must be positive. The domain (3.18) is mapped into the interior of the unit circle in the complex w-plane by (3.17). Hence, the physical domain of w is

$$-1 < \mathrm{Re}\{w\} < 1 \quad ; \quad -1 < \mathrm{Im}\{w\} < 1 \quad . \tag{3.19}$$

This result leads to a sharper upper bound for $|\omega_{r\infty}|$.

Noting that $|\mathrm{Im}\{w\}| < 1$ by virtue of (3.19), we get

$$|\omega_{r\infty}| < \frac{\mathrm{Re}\{\varepsilon_1\}\ |E_{ox}|^2}{2\eta_1} \quad . \tag{3.20}$$

If one compares (3.20) with the rough estimate (2.28), it is seen that the upper bound for a cylinder is well below that given in the latter. For a given field strength and a given viscosity η_1 of the liquid, the product

$$\mathrm{Re}\{\varepsilon_1\} \cdot \mathrm{Im}\{w\} \tag{3.21}$$

determines the frequency dependence of the stationary rotation. The sense of the rotation (with the field or against it) is determined by the sign of $\mathrm{Im}\{w\}$. $\mathrm{Im}\{w\} > 0$ means "positive rotation" of the body and $\mathrm{Im}\{w\} < 0$ "negative rotation". Choosing the simple situation

$$\mathrm{Re}\{\varepsilon_1\} = \varepsilon_1' \ ; \ \mathrm{Im}\{\varepsilon_1\} = 0 \ ; \ \mathrm{Re}\{\varepsilon_s\} = \varepsilon_s' \ ; \ \mathrm{Im}\{\varepsilon_s\} = 0$$

with frequency independent ε_1' and ε_s', $\mathrm{Im}\{w\}$ can be cast in the form

$$\mathrm{Im}\{w\} = \frac{W \cdot \omega\tau}{1 + (\omega\tau)^2} \tag{3.22}$$

with

$$W = \frac{2\,(\sigma_s\varepsilon_1' - \sigma_1\varepsilon_s')}{(\varepsilon_s' + \varepsilon_1')\ (\sigma_s + \sigma_1)} \quad ; \quad \tau = \frac{\varepsilon_s' + \varepsilon_1'}{\sigma_s + \sigma_1} \quad . \tag{3.23}$$

In this special case the sign of $\mathrm{Im}\{w\}$ is determined by

$$\mathrm{sgn}\,\{\sigma_s\varepsilon_1' - \sigma_1\varepsilon_s'\} \quad .$$

The knowledge of W and τ enables us to determine ε_s' and σ_s if ε_1' and σ_1 are also known.

The frequency dependence of $\mathrm{Im}\{w\}$ can become very complex if we allow for dielectric losses ε''. Even so, since $w(\omega)$ is related to the linear response function that relates the potential Φ_1 with the external field $\vec{E}_o$ (3.4), it must obey the Kramers-Kronig relations (see Landau-Lifshitz[5]).

A more refined assessment of the influence of Brownian rotation can now be made. From (2.29) we conclude that if $|E_{ox}|$ is large compared to E_c (the critical field strength) defined by

$$E_c \equiv \left(\frac{kT}{4\pi\ V\ \mathrm{Re}\{\varepsilon_1\}\ |\mathrm{Im}(w)|} \right)^{1/2} \quad , \tag{3.24}$$

and if

$$|E_{ox}| \gg E_1 \tag{3.25}$$

one can neglect the influence of Brownian rotation. Equations (3.24) and (3.25) improve the estimate we have given in (2.31) by taking into account the frequency dependence of the effect.

4. TWO IDENTICAL CYLINDERS: TORQUES AND ROTATIONS DUE TO INTERACTION WITH A HOMOGENEOUS EXTERNAL FIELD

We now consider two identical circular cylinders of radius R and height h, with identical dielectric and conductive properties. The homogeneous external field $\vec{E}_o$ is assumed to be in the x-y plane and has the components

$$E_{ox} = E_o \cos\theta \quad \text{and} \quad E_{oy} = E_o \sin\theta \quad . \qquad (4.1)$$

(For details see Fig. 1)

For the calculation of the field inside and outside the cylinders one must solve the Laplace equation. The solution must obey the transition conditions (3.1) and (3.2) at both cylinder walls. The actual field $\vec{E}$ must also tend to $\vec{E}_o$ at infinity.
Fortunately the solution can be found in a closed form (see Appendix A2). Likewise, the torque also may be obtained in a closed form on performing the relevant integrations.

a

b

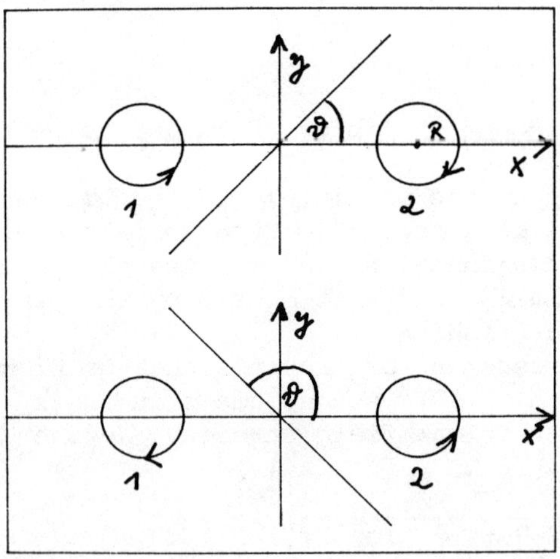

Fig. 1. Two identical cylinders or spheres. The coordinate system. a) $0 < \theta < \pi/2$. b) $\frac{\pi}{2} < \theta < \pi$. The arrows at the surfaces indicate the direction of the rotation.

The torque exerted on cylinder 2 according to (A2. 21) can be written in the form

$$L_z = - \frac{2V \cdot \mathrm{Re}\{\varepsilon_1\} \left(\mathrm{Im}\{w\}\right)^2 \cdot \sin 2\theta \cdot E_o^2}{\gamma^2} \cdot \left(1 + g(\gamma, |w|, \chi)\right), \tag{4.2}$$

where V is the volume of the cylinder, $\gamma = \frac{\xi}{R}$ (ξ = the distance between the centers of the cylinders) and w is given by (3.17). The "loss angle" χ is defined by

$$w = |w| \, e^{i\chi} \quad . \tag{4.3}$$

L_z vanishes for external fields in the x and y directions.
For larger distances ξ and $\gamma >> 2$, noting that g is small compared to unity, we get the asymptotic formula

$$L_z \simeq L_{z\infty} = - \frac{2V \cdot \mathrm{Re}\{\varepsilon_1\} \left(\mathrm{Im}\{w\}\right)^2 \sin 2\theta \; E_o^2}{\gamma^2} \qquad \gamma >> 2 \tag{4.4}$$

Taking the ratio of $L_{z\infty}$ to L_{sz}, the latter being the torque on the same cylinders in a rotating field of the same field strength, we have

$$k \equiv \left| \frac{L_{z\infty}}{L_{sz}} \right| = \frac{|\mathrm{Im}\{w\} \cdot \sin 2\theta|}{\gamma^2} \quad . \tag{4.5}$$

Since $|\mathrm{Im}\{w\}| < 1$, $|\sin 2\theta| \leq 1$ and $\gamma^2 \geq 4$,

$$k < \frac{1}{4} \tag{4.6}$$

holds. Taking into account the proportionality of the torques to the angular velocities at large distances, the same inequality holds for the ratio of the angular velocities of interacting cylinders as that in a rotating field. Angular velocities may not satisfy inequality (4.6), should the cylinders be in close proximity ($\gamma \simeq 2$).
The stationary angular velocity of an interacting cylinder at larger distances is given by

$$\omega_{r\infty} = \frac{L_{z\omega}}{R_r} = - \frac{\mathrm{Re}\{\varepsilon_1\} \left(\mathrm{Im}\{w\}\right)^2 \sin 2\theta \; E_o^2}{2\gamma^2 \, \eta_1} \quad . \tag{4.7}$$

The angular velocity vanishes if the external field is in the x or y direction.
Equation (4.7) ceases to be valid for $\gamma \simeq 2$, since equation (3.13) for R_r pertains to a single cylinder and its derivation does not take into account hydrodynamic interaction (see Sauer[17]).

Thanks to the quadratic dependence of $\omega_{r\infty}$ on Im{w}, the direction of rotation is independent of the sign of the latter. As seen from equation (4.7), the direction of rotation of cylinder 2 is solely determined by the sign of sin 2θ. One can show that cylinders 1 and 2 rotate in opposite directions as illustrated in Fig. 1 (see also Appendix A2). Clearly this is true only for an external homogeneaous field.

5. A SINGLE SPHERE: TORQUE AND ROTATION IN A ROTATING FIELD

In view of the fact that spherical bodies play an important role in the biological applications of rotating fields, we shall present the results of the theory in some more detail. We consider a single sphere of radius R acted on by an external rotating field $\vec{E}_o$ rotating in the x-y plane. In order to determine the actual field $\vec{E}$, we seek solutions of the Laplace equation

$$\Delta\Phi = 0$$

in the interior of the sphere and in the liquid. At infinity the actual field must tend to $\vec{E}_o$. At surfaces of discontinuity we demand continuity of the tangential component of the field strength

$$\vec{n}_s \times \vec{E}_1 = \vec{n}_s \times \vec{E}_s \quad , \tag{5.1}$$

$\vec{n}_s$ being the normal vector. The conditions on the normal components of $\vec{E}$ are of a different kind and related to the properties of the medium and the body. If the medium and the body have simple dielectric properties describable by complex valued ε_s and ε_1, the imaginary part Im{ε} of which is finite and positive

$$0 \le \text{Im}\{\varepsilon\} < \infty \quad ,$$

we have the transition condition

$$\varepsilon_1\left(\vec{n}_s^T \cdot \vec{E}_1\right) = \varepsilon_s\left(\vec{n}_s^T \cdot \vec{E}_s\right) \quad . \tag{5.2}$$

For a medium and a body characterized by finite conductivities σ_s and σ_1, by virtue of the continuity of the normal component of the total current (sum of conduction current and displacement current), we get the transition condition

$$\tilde{\varepsilon}_1\left(\vec{n}_s^T \cdot \vec{E}_1\right) = \tilde{\varepsilon}_s\left(\vec{n}_s^T \cdot \vec{E}_s\right) \tag{5.3}$$

with

$$\tilde{\varepsilon} \equiv \varepsilon + \frac{i\sigma}{\omega} \quad . \tag{5.4}$$

The transition condition (5.3) can be applied to electrolyte solutions, if the Debye length δ_D is small compared to R. A complex situation arises when the surface of discontinuity carries mobile surface charges of surface charge density Ω_a. This gives rise to a surface current of density $\vec{J}_a$ which is tangential to the surface of discontinuity. Now the normal component of the total current is discontinuous because of the surface current. Limiting arguments applied to the charge balance equation lead to the condition

$$\tilde{\varepsilon}_1 \left(\vec{n}_s^T \cdot \vec{E}_1\right) - \tilde{\varepsilon}_s \left(\vec{n}_s^T \cdot \vec{E}_s\right) = \Omega_a \quad . \tag{5.5}$$

Ω_a in (5.5) should not be confused with the total surface charge density Ω defined by

$$\Omega \equiv \varepsilon_1 \left(\vec{n}_s^T \cdot \vec{E}_1\right) - \varepsilon_s \left(\vec{n}_s^T \cdot \vec{E}_s\right) \quad . \tag{5.6}$$

In fact, one gets from (5.5) and (5.6)

$$\Omega = \Omega_a - \frac{i}{\omega} \left[\sigma_1 \left(\vec{n}_s^T \cdot \vec{E}_1\right) - \sigma_s \left(\vec{n}_s^T \cdot \vec{E}_s\right)\right] \quad . \tag{5.8}$$

Thus, Ω is the sum of Ω_a and the "Maxwell-Wagner" surface charge density. For an isolated system carrying surface charges (no exchange with the surrounding media), Schwarz[18] has shown that the charge balance equation

$$- i\omega\Omega_a = - \nabla_t \cdot \vec{J}_a \tag{5.9}$$

holds. ∇_t is the tangential component of the nabla operator. Furthermore, Schwarz[18] was able to show that for spherical surfaces, Ω_a has the form

$$\Omega_a = \frac{- a\, \Phi_a}{1 - i\omega\, \tau_a} \quad , \tag{5.10}$$

where Φ_a is the surface potential, τ_a the relaxation time of the surface charge system and a is related to the unperturbed surface charge density. Ω_a, as given by (5.10), leads to dispersion effects of the low frequency dielectric constant for a suspension of particles. (For more details see Schwarz' paper[18]). Regardless of the transition conditions for the normal component of $\vec{E}$, the field in the liquid can be derived from a potential

$$\Phi_1 = \left[u \cdot \left(\frac{R}{r}\right)^3 - 1\right] \left(\vec{r}^T \cdot \vec{E}_o\right) \qquad r \geq R \tag{5.11}$$

as long as the sphere has spherically symmetric (internal structure). u is a function of all the physical properties (dielectric constant, conductivity, surface charge density) of the body and the medium and the internal structure. For the problems of interest here (a homogeneous sphere, or a sphere with a membrane shell), u is given by

$$u = \frac{b}{b + 3} \tag{5.12}$$

with

$$b = \frac{\hat{\tilde{\varepsilon}}_s}{\tilde{\varepsilon}_1} - 1 \quad . \tag{5.13}$$

Here $\hat{\varepsilon}_s$ is the "effective dielectric constant" of the solid sphere. We shall give formulas for $\hat{\varepsilon}_s$ for the problems of interest.
As u is given by (5.12) we have a mapping of the complex b plane into the complex u plane. From the definition (5.13) of b, we conclude that the physically admissible b values lie in the domain

$$-1 < \mathrm{Re}\{b\} < +\infty \quad ; \quad -\infty < \mathrm{Im}\{b\} < +\infty \tag{5.14}$$

because $\mathrm{Re}\{\hat{\varepsilon}\}$ and $\mathrm{Im}\{\hat{\varepsilon}\}$ must be positive. The domain (5.14) is mapped into the interior of a circle of radius 3/4, whose center is located at 1/4 on the positive real axis of the complex u-plane. Hence the physical domain of u is

$$-\frac{1}{2} < \mathrm{Re}\{u\} < 1 \quad ; \quad -\frac{3}{4} < \mathrm{Im}\{u\} < \frac{3}{4} \quad . \tag{5.15}$$

Because $u(\omega)$ is related to a linear response-function, it must obey the Kramers-Kronig relation

$$u(\omega) - u(\infty) = \frac{P}{\pi i}\int_{-\infty}^{+\infty} \frac{u(k) - u(\infty)}{(k - \omega)}\, dk \ .$$

P stands for the principal value of the integral (for details see Landau-Lifshitz[5]).
The field $\vec{E}_1$ in the liquid is

$$\vec{E}_1 = -\nabla\Phi_1 = \left[1 - u\left(\frac{R}{r}\right)^3\right]\vec{E}_o + 3u\left(\frac{R}{r}\right)^3 \frac{(\vec{E}_o^T \cdot \vec{r})}{r^2}\vec{r} \tag{5.16}$$

with $|\vec{r}| \equiv r \geq R$. At the surface of the sphere we get

$$E_{1n}^* \equiv \vec{E}_1^{T*} \cdot \vec{n}_s = \left(1 + 2u^*\right)\left(E_{ox}^* n_x + E_{oy}^* n_y\right) \tag{5.17}$$

and

$$\left(\vec{r}\times\vec{E}\right) = R\cdot(1 - u)\left(\vec{n}_s \times \vec{E}_o\right) \quad . \tag{5.18}$$

Substitution of (5.17) and (5.18) in (2.25) gives for the torque $\vec{L}_s$

$$\vec{L}_s = \frac{1}{4} R\, \mathrm{Re}\{\varepsilon_1\}\cdot \int \left[(1-u)(1+2u^*)\left(E_{ox}^* n_x + E_{oy}^* n_y\right)\left(\vec{n}_s \times \vec{E}_o\right) + \text{c.c.}\right] da_s \tag{5.19}$$

where c.c. means the complex conjugate. In deriving (5.19) we have used the fact that $\vec{n}_s \times \vec{r} = 0$ at the surface of the sphere. Noting that

$$\int n_x n_y\, da_s = \int n_x n_z\, da_s = \int n_y n_z\, da_s = 0$$

and that

$$R\int n_x^2\, da_s = R\int n_y^2\, da_s \quad R\int n_z^2\, da_s = V \ ,$$

V being the volume of the sphere, we obtain for the components of $\vec{L}_s$

$$L_{sz} = \frac{V}{4} \operatorname{Re}\{\varepsilon_1\} \left[(1-u)(1+2u^*) \left(E^*_{ox} E_{oy} - E^*_{oy} E_{ox}\right) + c.c. \right] \qquad (5.20)$$

and

$$L_{sx} = L_{sy} = 0 \quad . \qquad (5.21)$$

If there is no phase difference between E_{ox} and E_{oy}, the right hand side of (5.20) vanishes. There is thus no torque on a single sphere in a constant homogeneous field. In a rotating field there is a phase difference of $\pi/2$ between E_{ox} and E_{oy}

$$E_{oy} = i\, E_{ox}$$

for a field of frequency ν that rotates with a constant field strength. This field rotates in the positive sense about the z-axis.
In a rotating field the torque is given by

$$L_{sz} = 3V \cdot \operatorname{Re}\{\varepsilon_1\} \cdot \operatorname{Im}\{u\} \cdot |E_{ox}|^2 \quad . \qquad (5.22)$$

For a sphere the resistance against rotation is (valid for small Reynolds number)

$$R_r = 6V\, \eta_1 \qquad (5.23)$$

η_1 being the viscosity of the liquid. For the stationary angular velocity $\omega_{r\infty}$ one gets

$$\omega_{r\infty} \equiv \frac{L_{sz}}{R_r} = \frac{1}{2\eta_1} \operatorname{Re}\{\varepsilon_1\} \cdot \operatorname{Im}\{u\} |E_{ox}|^2 \quad , \qquad (5.24)$$

which is independent of the radius R of the sphere. We write (5.24) in the form

$$\omega_{r\infty} = N_r(\omega) \cdot \frac{\operatorname{Re}\{\varepsilon_1(0)\}}{2\eta_1} |E_{ox}|^2 \qquad (5.25)$$

with

$$N_r(\omega) = \frac{\operatorname{Re}\{\varepsilon_1(\omega)\} \cdot \operatorname{Im}\{u\}}{\operatorname{Re}\{\varepsilon_1(0)\}} \qquad (5.26)$$

$N_r(\omega)$ determines the frequency dependence of $\omega_{r\infty}$. Noting that $|\operatorname{Im}(u)| < {}^3/4$ by virtue of (5.15), we get

$$|N_r(\omega)| < {}^3/4 \quad , \qquad (5.27)$$

which leads to the inequality

$$|\omega_{r\infty}| < \frac{3}{8} \frac{\operatorname{Re}\{\varepsilon_1(0)\}}{\eta_1} |E_{ox}|^2 \quad . \qquad (5.28)$$

This is sharper than the rough estimate (2.28). For a spherical body immersed in water at a field strength of 100 V/cm, $|\omega_{r\infty}|$ should not exceed the value of

26 Rad/sec.

Clearly, the relations (5.25) and (5.26) for $\omega_{r\infty}$ are valid only if the distances of the sphere from the electrodes and the walls of the chamber, in which the measurements are performed, are large compared to R. This restriction stems from the fact that in evaluating the field and the resistance against rotation, we have used solutions, which hold only for an unbounded fluid medium. Besides, the influence of Brownian rotation and hydrodynamic scale invariance also give rise to severe restrictions. We define a critical field strength by

$$2\pi \cdot |L_{sz}| = kT \quad . \tag{5.29}$$

Substituting (5.22) we obtain

$$|E_c| = \left(\frac{kT}{6\pi\ V \cdot \mathrm{Re}\{\varepsilon_1\} \cdot |\mathrm{Im}(u)|}\right)^{1/2} \quad . \tag{5.30}$$

If now (see (2.29))

$$|E_{ox}| \gg E_c \quad ,$$

one can neglect the effects of Brownian rotation. For spheres of radius $R = 10^{-4}$ cm in water and $|\mathrm{Im}(u)| = 3/4$ (the largest possible value), we get

$$E_c = 3.16 \text{ V/cm} \quad .$$

Formula (5.23) was derived for small Reynolds numbers $\dfrac{\rho_s \cdot R^2\ \omega_{r\infty}}{\eta_1}$.
For microscopic spheres in water with radius $R = 10^{-4}$ cm and $\omega_{r\infty} = 10\ \dfrac{\text{Rad}}{\text{sec}}$, the Reynolds number is 10^{-5} and (5.25) can be applied. For macroscopic spheres in water with $R = 1$ cm and for the same value for $\omega_{r\infty}$, the Reynolds number is 10^3 and the result (5.25) does not apply. For a sphere the equation of motion (2.23) has the solution

$$\omega_{rz}(t) = \omega_{r\infty}\left(1 - \exp - \frac{t}{\tau_r}\right) \tag{5.1}$$

with

$$\tau_r = \frac{\rho_s\ R^2}{15 \cdot \eta_1} \quad . \tag{5.32}$$

For microscopic spheres τ_r is very small, while for macroscopic spheres of radius R = 1 cm in water, $\tau_r \simeq 6$ sec.
Expressions for the torque and the angular velocity, similar to (5.22) and (5.25), describing stationary rotations of spheres, may be found in the earlier works of Lampa[7], Lertes[8] and Arnold and

Zimmermann[4]. Unfortunately their results differ from ours by a factor of 80 with regard to the magnitude of the effect, since they use a factor ε_o in the place of $Re\{\varepsilon_\omega\}$. Indeed this has already been noted by Lovelace et al.[19]. Therefore, their results can only be used for the calculation of the torque exerted on a sphere in vacuum. In addition, the work of Arnold and Zimmermann[4] does not give the correct frequency dependence of the torque because these authors assume a frequency dependence of the dipole-response that is incorrect in general. We show this below by means of some examples. The theory presented by Lovelace et al.[19] differs from ours in the choice of the Maxwell stress tensor. Instead of (2.13), they use one of the form

$$\frac{1}{2}\cdot Re\{\varepsilon_1\left(\vec{E}^*\vec{E}^T - \frac{1}{2}|E|^2 U\right)\} \quad .$$

In the above, contributions from the magnetic field have been neglected and justifiably so. This choice of the stress tensor, the same as that used in electrostatics, is permissible if one incorporates a change in the pressure tensor; but this has not been done by these authors. The different expression for N_r may be attributed to this fact. Their result (in our notation)

$$\tilde{N}_r = \frac{Im\{\varepsilon_1(\omega)\cdot u(\omega)\}}{Re\{\varepsilon_1(0)\}}$$

differs from ours (5.26) if ε_1 is complex and frequency dependent, for example, as in polar liquids at high frequencies. The relative difference between the two expression is given by

$$\left|\frac{N_r - \tilde{N}_r}{N_r}\right| = \left|\frac{Re\{u\}\cdot Im\{\varepsilon_1\}}{Re\{\varepsilon_1\}\cdot Im\{u\}}\right| \quad .$$

It vanishes if either $Re\{u\}$ or $Im\{\varepsilon_1\}$ is zero. At low frequencies the result of Lovelace et al.[19] agrees with ours.
As pointed out by Arnold and Zimmermann[4], lovelace et al.[19] and Glaser et al.[20], a study of the rotation of biological cells may serve as a useful tool for the determination of dielectric properties and conductivities of cell membranes and those of the cell interior. The frequency dependence of N_r contains precisely this information. In the following we shall discuss this dependence in more detail in a few cases of interest. Let us suppose that with the currently available techniques angular velocities of 10^{-2} Rad/sec and more are observable and that the field strength is below 50 V/cm. This provides us with a lower bound for the observable N_r

$$|N_r| > 1.5\cdot 10^{-3} \quad . \tag{5.33}$$

We now turn to some specific examples of homogeneous spheres. To begin consider a dielectric particle in a polar liquid. Since we exclude conductivity, the transition condition (5.2) applies and u is given by (see Stratton[11])

$$u = \frac{\varepsilon_s - \varepsilon_1}{\varepsilon_s + 2\varepsilon_1} \quad . \tag{5.34}$$

For the polar liquid with one relaxation time τ_o we have

$$\varepsilon_1 = \varepsilon'_{1\infty} + \frac{\Delta\varepsilon_1}{1 - i\omega\tau_o} \tag{5.35}$$

and for the material of the sphere

$$\varepsilon_s = \varepsilon'_s \quad . \tag{5.36}$$

This leads to

$$N_r = - 3A\cdot U\cdot X\cdot\left(1 + X^2\right)^{-1}\cdot\left[\left(U + 2A + 2\right)^2 + X^2(U + 2)^2\right]^{-1} \tag{5.37}$$

with

$$X = {}^{\omega}/\omega_o \quad ; \quad A = \frac{\Delta\varepsilon_1}{\varepsilon'_{1\infty}} \quad ; \quad U = \frac{\varepsilon'_1}{\varepsilon'_{1\infty}} \quad ; \quad \omega_o = \frac{1}{\tau_o} \quad .$$

The sphere thus rotates against the field with a frequency dependence that is more complex than that envisaged by Arnold and Zimmermann[4] and Lovelace et al.[19] in simpler situations. N_r, calculated from (5.37) for special values of the parameters, is displayed in a semilogarithmic plot in Fig.2. The curve has three extrema. Use has been made of the scaling properties of N_r. Only ratios of frequencies and dielectric constants occur in (5.37).

As a second example we consider a conducting sphere in a conducting liquid without dielectric losses. The transition condition (5.3) applies. This gives for u

$$u = \frac{\tilde{\varepsilon}_s - \tilde{\varepsilon}_1}{\tilde{\varepsilon}_s + 2\tilde{\varepsilon}_s} \quad . \tag{5.38}$$

We assume that

$$\tilde{\varepsilon}_s = \varepsilon'_s + \frac{i\sigma_s}{\omega}$$

and that (5.39)

$$\tilde{\varepsilon}_1 = \varepsilon'_1 + \frac{i\sigma_1}{\omega} \quad .$$

Introduction of (5.39) into (5.38) leads to

$$N_r = \frac{3\ (K-U)}{(K+2)\ (U+2)} \left(\frac{\frac{X(U+2)}{(K+2)}}{1 + \left(\frac{U+2}{K+2}\right)^2 X^2} \right) \tag{5.40}$$

with

$$K = \frac{\sigma_s}{\sigma_1} \quad ; \quad U = \frac{\varepsilon'_s}{\varepsilon'_1} \quad ; \quad X = \frac{\omega}{\omega_o} \quad ; \quad \omega_o = \frac{\sigma_1}{\varepsilon'_1} \; .$$

In this case the frequency dependence is quite simple. $|N_r|$ has a maximum at

$$X_r = \frac{K+2}{U+2} \quad .$$

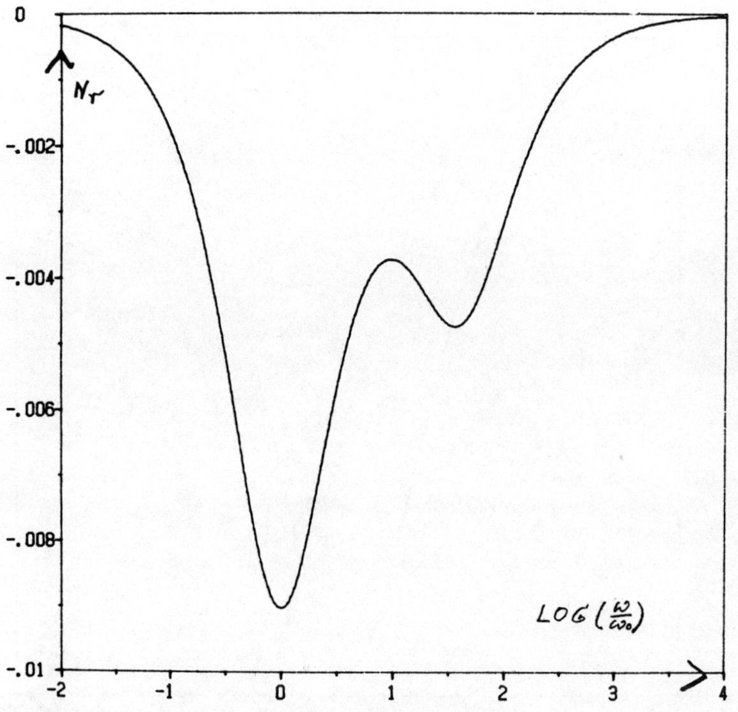

Fig. 2. N_r for a polar liquid . Parameter values ("water") :

$\frac{\varepsilon'_s}{\varepsilon'_{1\infty}} = 2$, $\frac{\Delta\varepsilon_1}{\varepsilon'_{1\infty}} = 80$ and $\omega_o = \frac{1}{\tau_o}$. According to (5.37)

In general, this value will be different from unity. Measurements of X_r and the maximum value of $|N_r|$ enables us to determine K and U and hence the ratios of the dielectric constants and the conductivities.

Fig.3 shows N_r as a function of the frequency for a nonconducting sphere in a conducting liquid. Fig.4 shows N_r for a highly conducting sphere in a weakly conducting medium (ion exchange material). In this case, it is best to employ the present method to determine dielectric constants and conductivities as these cannot easily be obtained by other methods. As seen from Fig.4, the theory predicts a large effect ($\bar{N}_r \propto 0.45$) for frequencies around 30 MHz and for $\sigma_1 = 10^{-4}(\Omega\ \text{cm})^{-1}$.

Using
$$X_r = \frac{K+2}{U+2} \tag{5.41}$$

and the known maximum value of $|N_r|$ (paying due attention to the sign), we find

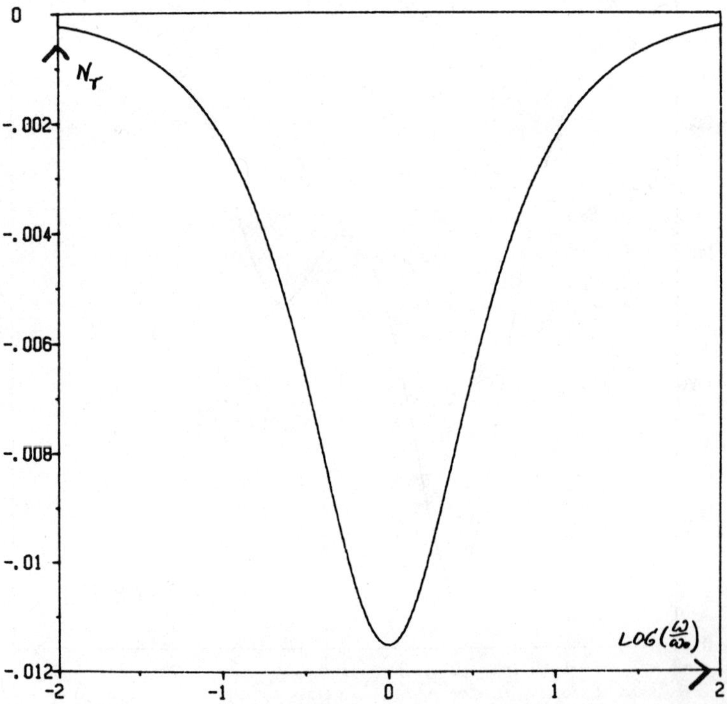

Fig. 3. N_r for a nonconducting sphere in a conducting liquid according to (5.40). Values of the parameters:
$$\frac{\varepsilon'_s}{\varepsilon'_1} = 0.03 \quad , \quad \frac{\sigma_s}{\sigma_1} = 0 \quad \text{and} \quad \omega_o = \frac{\sigma_1}{\varepsilon'_1} .$$

$$\bar{N}_r = \frac{3}{2}\,\frac{K - U}{(K+2)(U+2)} \quad , \tag{5.42}$$

$$\frac{\sigma_s}{\sigma_1} \equiv K = \frac{3\cdot(X_r - 1)}{2\cdot\bar{N}_r} - 2 \quad ,$$

and

$$\frac{\varepsilon'_s}{\varepsilon'_1} \equiv U = \frac{3\cdot(X_r - 1)}{2\cdot X_r\,\bar{N}_r} - 2 \quad . \tag{5.43}$$

We draw attention to the fact that the direction of rotation is determined by the sign of K - U.

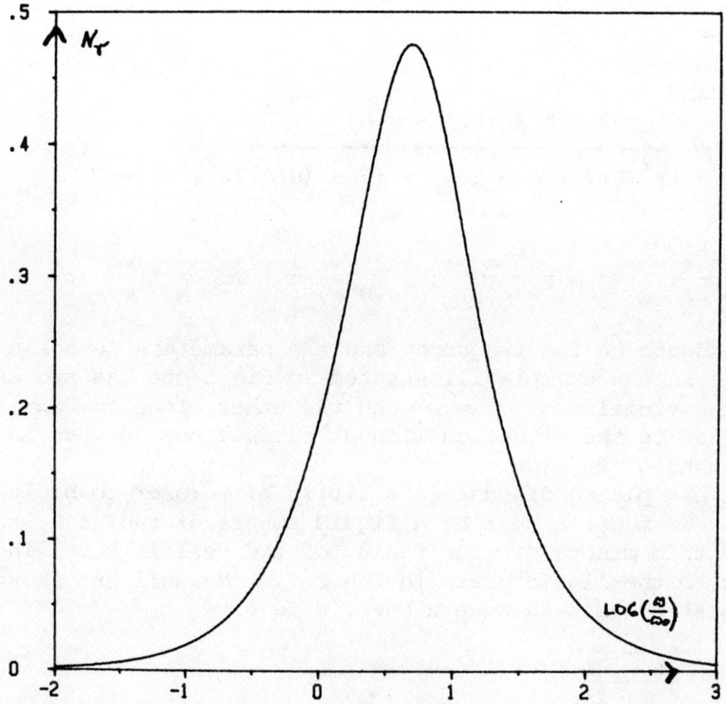

Fig. 4. N_r for a highly conducting sphere in a weakly conducting liquid according to (5.40). Values of the parameters: $\frac{\varepsilon'_s}{\varepsilon'_1} = 0.5$, $\frac{\sigma_s}{\sigma_1} = 10$ and $\omega_o = \frac{\sigma_1}{\varepsilon'_1}$.

Next we turn to inhomogeneous spheres. Let us first consider a homogeneous nonconductive sphere with mobile surface charges in a conducting liquid. In this case the transition condition (5.5) together with (5.10) apply, and as Schwarz[18] has shown, one obtains for u

$$u = \frac{\hat{\tilde{\varepsilon}}_s - \tilde{\varepsilon}_1}{\hat{\tilde{\varepsilon}}_s + 2\tilde{\varepsilon}_1} \tag{5.44}$$

with

$$\hat{\tilde{\varepsilon}}_s = \varepsilon'_s + \frac{\Delta\varepsilon_a}{1 - i\omega\tau_a} \tag{5.45}$$

$\Delta\varepsilon_a$ and τ_a depend on the radius R of the sphere. For spheres of radius $R = 10^{-4}$cm, $\frac{1}{2\pi\tau_a} \simeq 500$ Hz and $\frac{\Delta\varepsilon_a}{\varepsilon'_1}$ can attain values up to 10^3. $\tilde{\varepsilon}_1$ is given by

$$\tilde{\varepsilon}_1 = \varepsilon'_1 + i\,\frac{\sigma_1}{\omega} \quad . \tag{5.46}$$

We then find

$$N_r = \frac{3X\left(PX^2(A-PU) - U-A\right)}{X^2(U+2 + A + 2P)^2 + \left(2 - (U+2)PX^2\right)^2} \tag{5.47}$$

with

$$U = \frac{\varepsilon'_s}{\varepsilon'_1} \; ; \quad A = \frac{\Delta\varepsilon_a}{\varepsilon'_1} \; ; \quad \omega_o = \frac{\sigma_1}{\varepsilon'_1} \; ; \quad X = \frac{\omega}{\omega_o} \; ; \quad P = \omega_o\tau_a .$$

The dependence on the frequency and the parameters is again very complex. In the example illustrated in Fig.5 one has two extrema, one in the vicinity of $X = \frac{1}{P}$ and the other close to $X = 1$. In contrast to the situation without surface charges, as in Fig.3, N_r has changed its sign.
The next example we discuss is a little bit closer to biological reality. We model a cell by a liquid sphere of radius R_2 surrounded by a thin membrane. The radius of the cell is R_1. The index 2 refers to the liquid phase in the cell. Maxwell has shown in electrostatics that in such a case, u is given by

$$u = \frac{\hat{\varepsilon}_s - \varepsilon_1}{\hat{\varepsilon}_s + 2\varepsilon_1} \tag{5.48}$$

with an effective dielectric constant $\hat{\varepsilon}_s$,

$$\hat{\varepsilon}_s = \frac{\varepsilon_m\left(2(1 - 1^3)\,\varepsilon_m + (1 + 2\cdot 1^3)\,\varepsilon_2\right)}{(2 + 1^3)\,\varepsilon_m + (1 - 1^3)\,\varepsilon_2} \quad , \tag{5.49}$$

where ε_m is the dielectric constant of the membrane, ε_2 that of the inner sphere and $l = \frac{R_2}{R_1}$. The field for $|\vec{r}| > R_1$ is then given by (5.16). For a conducting system, equations (5.48) and (5.49) modify to (see Schwan[21])

$$u = \frac{\hat{\tilde{\varepsilon}}_s - \tilde{\varepsilon}_1}{\hat{\tilde{\varepsilon}}_s + 2\varepsilon_1} \tag{5.50}$$

with

$$\hat{\tilde{\varepsilon}}_s = \frac{\tilde{\varepsilon}_m\left(2(1 - l^3)\ \tilde{\varepsilon}_m + (1 + 2l^3)\ \tilde{\varepsilon}_2\right)}{(2 + l^3)\ \tilde{\varepsilon}_m + (1 - l^3)\ \tilde{\varepsilon}_2} \quad . \tag{5.51}$$

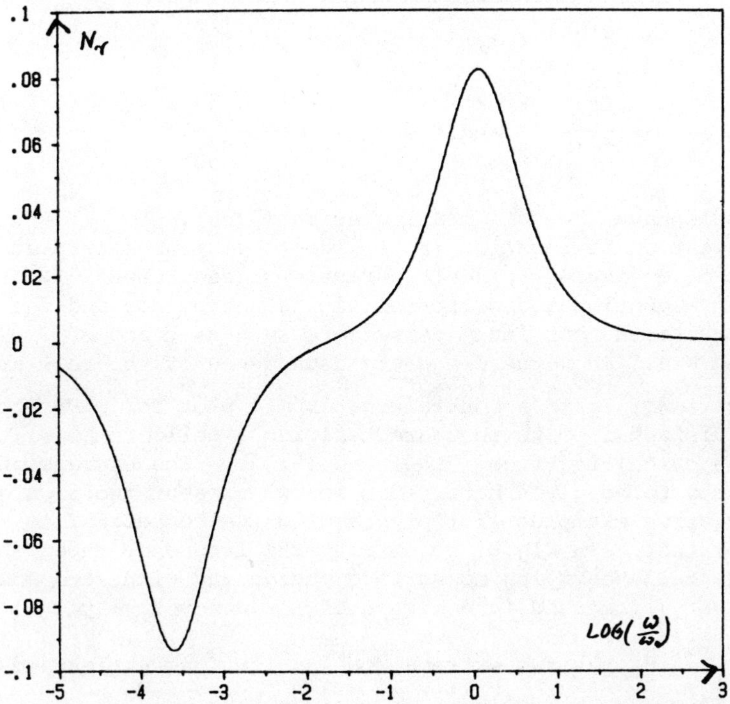

Fig. 5. N_r for a nonconducting sphere with surface charges in a conducting liquid according to (5.47). Values of the parameters:

$$\frac{\varepsilon_s'}{\varepsilon_1'} = 0.03 \ ; \quad \frac{\sigma_s}{\sigma_1} = 0 \ ; \quad \omega_o = \frac{\sigma_1}{\varepsilon_1'} \ ;$$

$$\frac{\Delta\varepsilon_a}{\varepsilon_1'} = 10^3 \ ; \quad \omega_o\tau_a = 3.5 \cdot 10^3 \quad .$$

Here one assumes the validity of the transition conditions (5.3) at the inner and outer membrane surfaces. A fact that is often overlooked is that for electrolyte solutions, the validity of (5.50) and (5.51) presupposes small values of the Debye-length and absence of surfaces charges. If the membrane exhibits surface conduction on both sides, the solution of the Laplace equation with the transition conditions (5.5) and (5.9) can be found. We quote the results without proof. One obtains for u under the same conditions as above

$$u = \frac{\hat{\tilde{\varepsilon}}_s - \tilde{\varepsilon}_1}{\hat{\tilde{\varepsilon}}_s + 2\tilde{\varepsilon}_1} \qquad (5.52)$$

with

$$\hat{\tilde{\varepsilon}}_s = \frac{\tilde{\varepsilon}_m\left(2(1 - l^3)\tilde{\varepsilon}_m + (1 + 2l^3)(\tilde{\varepsilon}_2 + ls_2)\right)}{(2 + l^3)\,\tilde{\varepsilon}_m + (1 - l^3)(\tilde{\varepsilon}_2 + ls_2)} + s_1 \qquad (5.53)$$

and

$$s_1 = \frac{\Delta\varepsilon_1}{1 - i\omega\tau_{a1}} \quad \text{and} \quad s_2 = \frac{\Delta\varepsilon_2}{1 - i\omega\tau_{a2}} \qquad (5.54)$$

The symbols have the same meaning as in (5.45).
The application of equation (5.51) led to various interpretations in terms of equivalent circuit parameters (see Schwan[21]). Here we deal with a continuum description of the phenomenon and shall therefore stick to the continuum parameters such as ε and σ.
Fig.6 and Fig.7 show the frequency dependence of the real and imaginary parts of $\hat{\tilde{\varepsilon}}_s$ in a double logarithmic plot for parameter values not too distant from those characterizing a cell.
N_r can be calculated from (5.50) and (5.52). The expressions are too clumpsy to be given here; even so, we present two figures as representative examples (for programming the computer Im{u} had to be calculated). In Fig.8, we display the frequency dependence of N_r for a "cell" carrying no surface charge and with zero membrane conductivity, immersed in a liquid of low conductivity $\left(\frac{\sigma_2}{\sigma_1} = 60\right)$. N_r has two extrema, one at low frequencies and the other close to ω_o (slightly beyond ω_o). The first extrema owes its existence to the relatively high value of $\hat{\tilde{\varepsilon}}'_s$ at low frequencies and to the low conductivity, a situation comparable to that in Fig.3. At higher frequencies we approach the situation delineated in Fig.4.
In Fig.9 the influence of a mobile surface charge and of a finite membrane conductivity is shown. The surface conductivity changes the behaviour of N_r at very low frequencies (part a of Fig.9).
It has been reported in the literature that one observes deviations

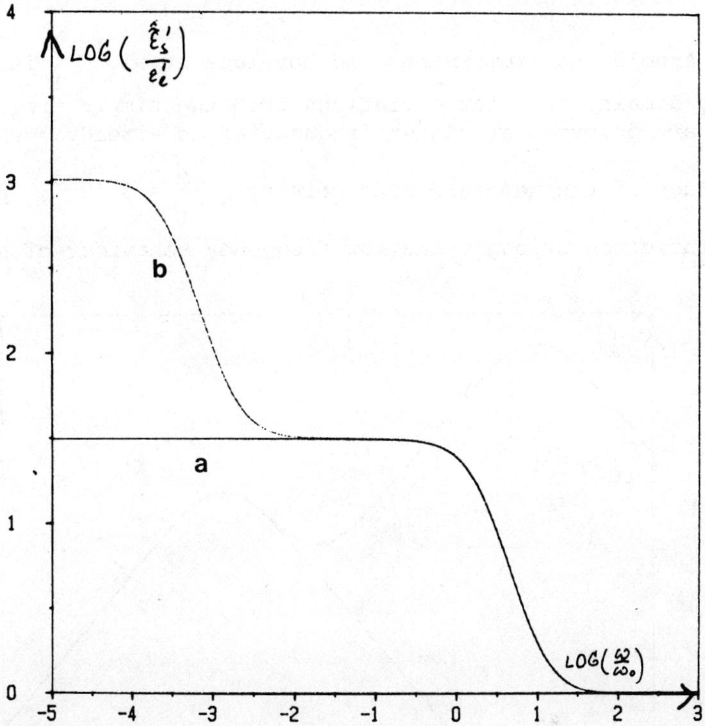

Fig. 6. $\frac{\hat{\tilde{\varepsilon}}'_s}{\varepsilon'_1} \equiv \frac{\mathrm{Re}\{\hat{\tilde{\varepsilon}}_s\}}{\varepsilon'_1}$ for a "cell" according to (5.51) and (5.53)

a) Without surface charge. Values of the parameters:

$\frac{\varepsilon'_2}{\varepsilon'_1} = 1$; $\frac{\varepsilon'_m}{\varepsilon'_1} = 0.03$; $\frac{\sigma_m}{\sigma_1} = 0$; $\frac{\sigma_2}{\sigma_1} = 60$;

$\omega_o = \frac{\sigma_1}{\varepsilon'_1}$; $\frac{R_2}{R_1} = 0.999$.

b) With surface charge. Values of the parameters:

$\frac{\Delta\varepsilon_1}{\varepsilon'_1} = 10^3$; $\omega_o \tau_{a1} = 3.5 \cdot 10^3$; $\frac{\Delta\varepsilon_2}{\varepsilon'_1} = 0$;

all other parameters as in a).

from the simple frequency dependence $\left(\sim \frac{\nu/\nu_x}{1 + (\nu/\nu_x)^2} \right.$ as proposed by Arnold and Zimmermann[4] and Lovelace et al.[19] $\left. \right)$ in the low frequency domain. Similar deviations from the simple frequency dependence are observed at higher frequencies as already seen in Fig.8. Small values of the membrane conductivity $\left(\frac{\sigma_m}{\sigma_1} = 3 \cdot 10^{-3}, \right.$ part b of Fig.9 $\left. \right)$ influence strongly the low frequency behaviour of N_r .

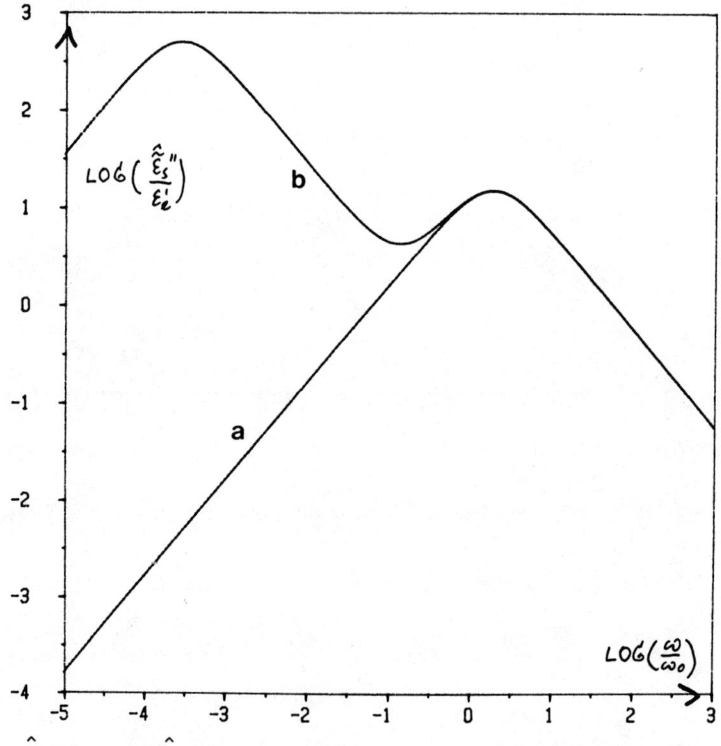

Fig. 7. $\frac{\hat{\tilde{\varepsilon}}_s''}{\varepsilon_1'} \equiv \frac{\mathrm{Im}\{\hat{\tilde{\varepsilon}}_s\}}{\varepsilon_1'}$ for a "cell" according to (5.51) and (5.53)

a) Without surface charge

b) With surface charge

Values of the parameters as in Fig.6.

The surface charge effect practically disappears and the low frequency extremum is reduced to a third of its original value. More detailed studies are needed to attain our final objectice, a complete description of cell-rotation in terms of cell properties. Since a number of parameters determine all rotation, high precision experiments are called for. Nevertheless, the method can be used for a categorization of the effect and to estimate the order of magnitude of the parameters. The experimental results of Arnold and Zimmermann[4] and Lovelace et al.[19] are in accord with the order of magnitude of the effect as predicted by (5.25).

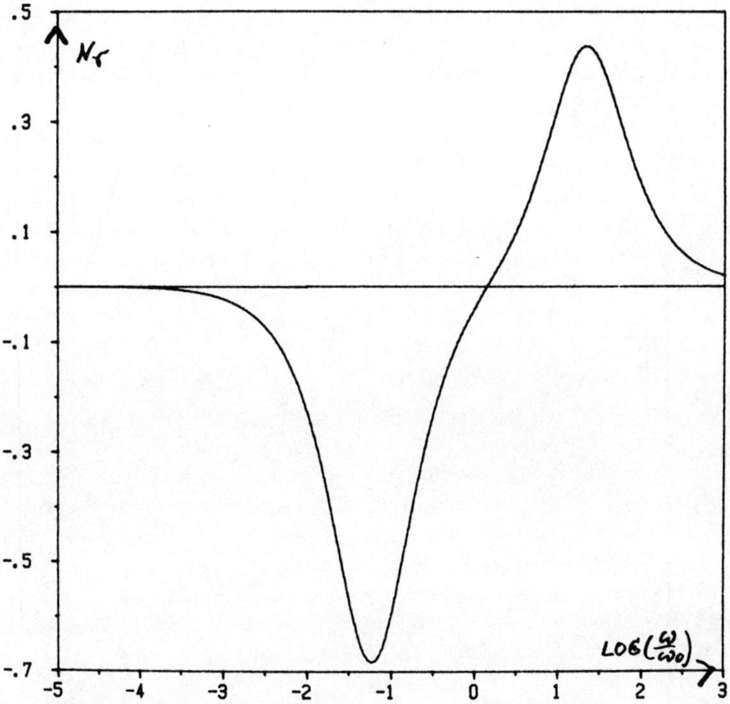

Fig. 8. N_r for a "cell" without surface charge. Values of the parameters:

$$\frac{\varepsilon_m}{\varepsilon_1'} = 0.03 \quad ; \quad \frac{\varepsilon_2'}{\varepsilon_1'} = 1 \quad ; \quad \frac{\sigma_m}{\sigma_1} = 0 \quad ; \quad \frac{\sigma_2}{\sigma_1} = 60 \quad ;$$

$$\omega_o = \frac{\sigma_1}{\varepsilon_1'} \quad ; \quad \frac{R_2}{R_1} = 0.999 \quad .$$

6. TWO IDENTICAL SPHERES: TORQUES AND ROTATION DUE TO INTERACTION WITH A HOMOGENEOUS EXTERNAL FIELD

We now consider two identical spheres both of Radius R, identical with regard to their dielectric and conductive properties, the centers of which are separated by a distance ξ. The homogeneous external field E_o is assumed to be in the x-y plane and has the components

$$E_{ox} = E_o \cos\theta \quad \text{and} \quad E_{oy} = E_o \sin\theta \quad . \tag{6.1}$$

(For details see Fig.1)

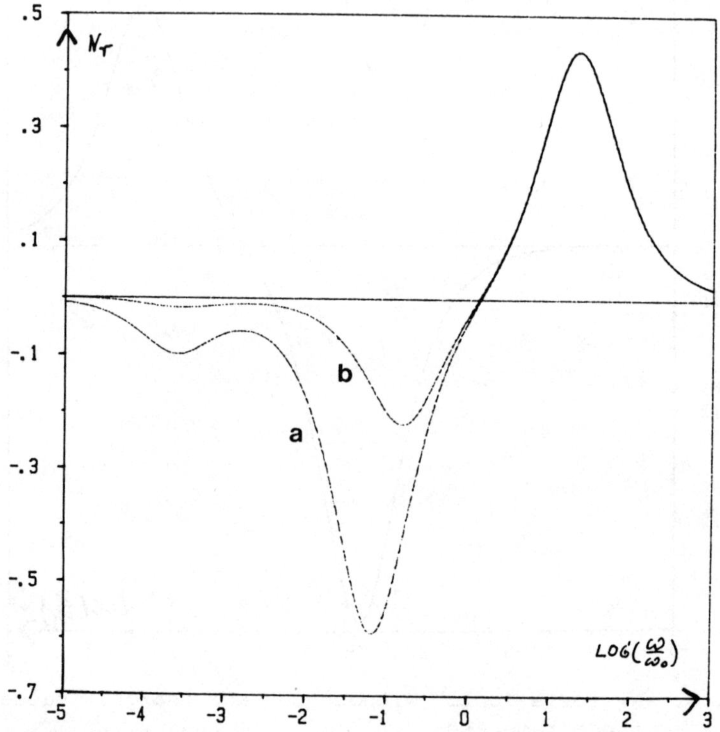

Fig. 9. N_r for a cell with surface charges. Values of the parameters:

a) The same values as in Fig.8 and

$$\frac{\Delta\varepsilon_1}{\varepsilon_1'} = 10^3 \quad ; \quad \omega_o \tau_{a_1} = 3.5 \cdot 10^3 \quad ; \quad \frac{\Delta\varepsilon_2}{\varepsilon_1'} = 0$$

b) The same values as in a) and $\frac{\sigma_m}{\sigma_1} = 3 \cdot 10^{-3}$

For the calculation of the fields within and outside the spheres one must solve the Laplace equation. The solution must necessarily obey the transition condisions (5.1) and one of (5.2), (5.3) and (5.5), depending on the dielectric and conductive properties of the media. The actual field $\vec{E}$ must also tend to $\vec{E}_o$ at infinity. Unfortunately the solution cannot be given in a closed form. Therefore we are forced to make approximations. An approximate solution can be given in the form of a power series in $\frac{1}{\gamma}$ ($\gamma = \xi/R$). Since this series is asymptotic in nature, it does not converge for γ values close to 2 (sphers in contact). We get as a zeroth order approximation for the field in the liquid (valid for $\gamma \to \infty$)

$$\vec{E}_1 = \vec{E}_o + \vec{E}_{d1} + \vec{E}_{d2} \quad . \tag{6.2}$$

This is a superposition of the original field $\vec{E}_o$ and two dipole fields $\vec{E}_{d1}$ and $\vec{E}_{d2}$. The latter, due to the induced dipoles in the spheres 1 and 2, are given by (see 5.16)

$$\vec{E}_{d1} = -u\left(\frac{R}{r_1}\right)^3 \vec{E}_o + 3u\left(\frac{R}{r_1}\right)^3 \frac{(\vec{E}_o^T \cdot \vec{r}_1)}{r_1^2} \vec{r}_1 \tag{6.3}$$

and

$$\vec{E}_{d2} = -u\left(\frac{R}{r_2}\right) \vec{E}_o + 3u\left(\frac{R}{r_2}\right)^3 \frac{(\vec{E}_o^T \cdot \vec{r}_2)}{r_2^2} \vec{r}_2 \tag{6.4}$$

$\vec{r}_1$ and $\vec{r}_2$ being the position vectors originating from the centers of the spheres 1 and 2 respectively.
For small values of $\frac{1}{\gamma}$, $\vec{E}_{d2}$ on the surface of sphere 1 can be approximated by

$$\vec{E}_{d2} \simeq \frac{u}{\gamma^3}\left(3E_{ox}\vec{e}_x - \vec{E}_o\right) \quad ; \quad r_1 = R \quad ; \quad \frac{1}{\gamma} << \frac{1}{2} \quad , \tag{6.5}$$

$\vec{e}_x$ being the unit vector in the x-direction. We replace in (6.3) $\vec{E}_o$ by $\vec{E}_o + E_{d2}$ at $r_1 = R$ and obtain for the field on the surface of the sphere 1 (in the liquid)

$$\vec{E}_1 \simeq \vec{E}_o + \vec{E}_{d2} - u\left(\vec{E}_o + \vec{E}_{d2}\right) + 3u\left[\left(\vec{E}_o^T + \vec{E}_{d2}^T\right)\cdot\vec{n}_s\right]\vec{n}_s \tag{6.6}$$

$\vec{E}_{d2}$ being given by (6.5). Here we have neglected terms of the order of $\left(\frac{1}{\gamma}\right)^5$.

On substituting (6.6) in (2.25) and performing the integration, we obtain for the torque exerted on sphere 1

$$L_{s_1 z} = \frac{V}{4}\cdot\mathrm{Re}\{\varepsilon_1\}\left[(1-u)(1+2u^*)\left(\tilde{E}^*_{ox}\tilde{E}_{oy} - \tilde{E}^*_{oy}\tilde{E}_{ox}\right) + \text{c.c.}\right] \tag{6.7}$$

with

$$\tilde{E}_{ox} = \left(1 + \frac{2u}{\gamma^3}\right) E_{ox} \quad \text{and} \quad \tilde{E}_{oy} = \left(1 - \frac{u}{\gamma^3}\right) E_{oy} . \tag{6.8}$$

Introducting (6.8) in (6.7) and again neglecting terms of the order $\left(\frac{1}{\gamma}\right)^5$, we have

$$L_{s_1z} = \frac{9}{2} \cdot \frac{V}{\gamma^3} \, Re\{\varepsilon_1\}\left(Im\{u\}\right)^2 \sin(2\,\theta) \cdot E_o^2 . \tag{6.9}$$

The x and y-components of the torque are given by

$$L_{s_1x} = L_{s_1y} = 0 ,$$

in every order of approximation.
The stationary angular velocity given by

$$\omega_{r\infty} = \frac{3\,Re\{\varepsilon_1\}}{4 \cdot \eta_1 \cdot \gamma^3} \left(Im\{u\}\right)^2 \cdot \sin(2\,\theta)\, E_o^2 \tag{6.10}$$

is again independent of the volume of the sphere. The direction of rotation is determined by the sign of $\sin 2\theta$ ({see Fig.1} for illustration). Here we have used (5.23), (valid only for small values of $\frac{1}{\gamma}$), for the resistance against rotation. If one compares the expression (6.10) for the angular velocity with that derived by Holzapfel et al.[3] one sees agreement with respect to the dependence on E_o, $\sin 2\theta$ and η_1. Since they replace the factor $Re\{\varepsilon_1\}$ by ε_o, their result is in error by a factor of 80. The dependence on $Im\{u\}$ is obscure in the treatment of Holzapfel et al.[3]. The same may be said of the dependence on γ, since these authors use the approximation (6.5) for $\gamma=2$, which is clearly inadmissible. Furthermore they have used (5.23) which is invalid for $\gamma=2$. Because of the condition $\frac{1}{\gamma} \ll \frac{1}{2}$, the formula (6.10) is of limited value. It does not cover the situation of the spheres in contact. Greater insight into this problem may be gained by a closer scruting of the case of cylinders in contact, where one obtains a result in a closed form. Considerations similar to those in chapter 5 lead to an expression for the critical field strength for interacting spheres in the form

$$E_c \equiv \left(\frac{\gamma^3\, kT}{9\pi V\, Re\{\varepsilon_1\}\left(Im\{u\}\right)^2 \sin 2\theta}\right)^{1/2} . \tag{6.11}$$

In general the E_c for interacting spheres will be larger than that for a single sphere in a rotating field.
The ratio of the angular velocity given by (6.10) to that in a rotating field of the same field strength is

$$\frac{3}{2} \frac{|\mathrm{Im}\{u\}|}{\gamma 3} \sin 2\theta \quad ,$$

which is smaller than $^9/_8 \frac{1}{\gamma^3}$.

Since two spheres in a homogeneous external field tend to change the angle θ and the distance between the centers under the influence of the body forces, the time available for the observation of a stationary rotation is necessarily limited. Indeed, we require $\frac{d\theta}{dt}$ to be small compared to $\omega_{r\infty}$ for such an observation to be possible at all. According to Sauer[17] $\frac{d\theta}{dt}$ is given by

$$\frac{d\theta}{dt} = - \frac{2 \cdot \mathrm{Re}\{\varepsilon_1\}|u|^2 \sin (2\theta) E_o^2}{\gamma^5 \eta_1} \quad . \tag{6.12}$$

The ratio of $\omega_{r\infty}$ to $\frac{d\theta}{dt}$ is

$$\left| \frac{\omega_{r\infty}}{\frac{d\theta}{dt}} \right| = \frac{3}{2} \gamma^2 \frac{(\mathrm{Im}\{u\})^2}{|u|^2} \tag{6.13}$$

which is smaller than $\frac{3}{2} \gamma^2$. If one intends to observe n rotations as θ changes from $\pi/3$ to $\pi/6$, this ratio becomes 12 n and hence

$$12\ n < \frac{3}{2} \gamma^2$$

or

$$\gamma > (8\ n)^{1/2} \quad .$$

Thus for 10 rotations, we must have a γ larger than 9, and this by (6.10) represents quite a small rotation.

All that has been said about the frequency dependence in chapter 5 can now be taken over.

In the case of two spheres of different radii R_1 and R_2 and different u_1 and u_2, it is quite easy to derive asymptotic formulae for the torque and the stationary angular velocity of rotation at large distances (γ_1 and $\gamma_2 >> 2$). Skipping this derivation the result is

$$L_{sz}^{(1)} = - L_{sz}^{(2)} = \frac{9}{2} \frac{V_1}{\gamma_2^3} \mathrm{Im}\{u_1\} \cdot \mathrm{Im}\{u_2\} \mathrm{Re}\{\varepsilon_1\} E_o^2 \sin 2\theta \tag{6.14}$$

with $\gamma_1 = {}^\xi/R_1$ and $\gamma_2 = {}^\xi/R_2$.

For the stationary angular velocity of rotation we get

$$\omega^{(1)}_{r\infty} = -\left(\frac{\gamma_1}{\gamma_2}\right)^3 \omega^{(2)}_{r\infty}$$

$$= \frac{3}{4} \frac{\mathrm{Re}\{\varepsilon_1\}}{\eta_1 \gamma_2^3} \mathrm{Im}\{u_1\} \mathrm{Im}\{u_2\} E_o^2 \sin 2\theta \qquad (6.15)$$

provided that $\gamma_1 , \gamma_2 >> 2$.

The direction of rotation is now determined by the sign of $\mathrm{Im}\{u_1\}\cdot\mathrm{Im}\{u_2\}\cdot\sin 2\theta$.

7. SUMMARY

The subject of field induced cell rotation has made rapid strides lately and the interpretation of the data in terms of the electrical cell parameters calls for a reexamination of the underlying theories. In the case of spheres and long cylinders, it is possible to derive results in a closed form in some special cases. As one deals with dissipative media, the formulation must start with the Maxwell stress tensor to calculate the forces and torques exerted on the cell (averaged over time) by the field. A certain care must be exercised in the choice of the mechanical pressure tensor, since its form depends upon that used for the Maxwell stress tensor. Fortunately, electrostrictive effects in slightly compressible liquids may be neglected at field strengths below 10^6 V/cm. This facilitates the derivation leading to our final results. Our results point to errors in earlier theories (Holzapfel et al.[3], Arnold and Zimmermann[4]). The existence of dielectric losses or nonzero conductivities for at least part of the system turn out to be, in the light of our results, a necessary condition for cell rotation. Hydrodynamic scaling properties impose restrictions on the application of our results to macroscopic bodies, as their validity is limited to small Reynolds numbers. For interacting cells the theory presented here gives reliable results only for relatively large separations between the cells. A more comprehensive theory must include electric and hydrodynamic interactions to higher orders of approximation. Reports in the literature (Pohl[2]) that single cells rotate in a constant homogeneous external field do not find corroboration in our theory. There may be several reasons for this discrepancy. One may visualize that another cell finds itself in the vicinity of the one rotating, or that the single cell is close to one of the electrodes. Another conceivable explanation (a very speculative one) invokes symmetry breaking: Due to elastic deformations (neglected in our treatment) in the external field, the cell assumes a shape that corresponds to an unstable equilibrium of the torques. Under these circumstances, a rotation may indeed result; but its direction cannot be predicted, since it depends on random disturbances or hidden asymmetrics.

ACKNOWLEDGEMENT

The authors thank Prof. H.P.Schwan for bringing the problem of cell rotation to our attention and for many valuable discussions. Thanks are also due to Prof. R.A.Spangler for illuminating discussions on the pressure tensor and the Maxwell stress tensor. Finally we thank Mrs. O.Woinoff for her excellent typing of complicated formulas.

APPENDIX

A1. THE TIME AVERAGE OF THE BODY FORCE DENSITY AND THE CALCULATION OF THE PRESSURE DISTRIBUTION

We use a complex representation of quantitaties with a periodic time dependence in the form

$$a(t) = \mathrm{Re}\{A\, e^{-i\omega t}\} \quad \text{and} \quad b(t) = \mathrm{Re}\{B\, e^{-i\omega t}\} \quad ,$$

A and B being complex quantities. We recall that the time average of the product at a(t) and b(t) is given by

$$\overline{a(t)\cdot b(t)} = \frac{1}{2}\,\mathrm{Re}\{AB^*\} = \frac{1}{2}\,\mathrm{Re}\{A^*B\}$$

and that the time average of the time derivative of the product $a(t)\cdot b(t)$ vanishes

$$\overline{\frac{d}{dt}\big(a(t)\cdot b(t)\big)} = 0 \quad .$$

Taking the time average of the body force density (2.5) we obtain

$$\begin{aligned}\overline{\vec{f}_e} = \frac{1}{4}\,\mathrm{Re}\{ & \left(\nabla^T\cdot\frac{\vec{D}^*}{\varepsilon^*}\right)\vec{D} - \left(\nabla^T\cdot\vec{D}^*\right)\vec{E} \\ & + \vec{D}^*\times\left(\nabla\times\vec{E}\right) - \vec{E}^*\times\left(\nabla\times\varepsilon\vec{E}\right) \\ & + \left(\nabla^T\cdot\frac{\vec{B}^*}{\mu^*}\right)\vec{B} \\ & + \vec{B}^*\times\left(\nabla\times\vec{H}\right) - \vec{H}^*\times\left(\nabla\times\mu\vec{H}\right) \\ & + 2\rho^*_{el}\,\vec{E} + 2\vec{j}\times\vec{B}^* \;\} \quad .\end{aligned} \tag{A1.1}$$

On using the following relations

$$\left(\nabla^T \cdot a\vec{A}\right) = a\left(\nabla^T \cdot \vec{A}\right) + \vec{A}^T \cdot \nabla a$$

$$\nabla \times (s\vec{A}) = a\nabla \times \vec{A} + (\nabla a) \times \vec{A}$$

$$\vec{A} \times \left(\vec{B} \times \vec{C}\right) = \left(\vec{A}^T \cdot \vec{C}\right) \vec{B} - \left(\vec{A}^T \cdot \vec{B}\right) \vec{C}$$

and the Maxwell equations (2.8), one obtains equation (2.11) of the text.
According to (2.14) the pressure distribution in the liquid is given by

$$\overline{P} \cdot \nabla = \overline{\vec{f}}_e \quad . \tag{A1.3}$$

Neglecting space charge effects and the contribution from the $\vec{H}$ field we get for $\overline{\vec{f}}_e$

$$\overline{\vec{f}}_e = \frac{1}{4} \text{Re} \{ - |\vec{E}|^2 \nabla \varepsilon + \varepsilon^* \vec{E}^* \vec{E}^{T*} \cdot \nabla \left({}^{\varepsilon}/\varepsilon^* \right) \} \quad . \tag{A1.3}$$

For a liquid without dielectric losses we find

$$\nabla \overline{p} = - \frac{1}{4} \text{Re} \{ |\vec{E}|^2 \frac{\partial \varepsilon}{\partial \rho} \nabla \rho \} \quad . \tag{A1.4}$$

or

$$\nabla \left(\overline{p} + \frac{1}{4} \text{Re} \{ \rho \frac{\partial \varepsilon}{\partial \rho} |\vec{E}|^2 \} \right) = \frac{1}{4} \text{Re} \{ \rho \nabla \left(\frac{\partial \varepsilon}{\partial \rho} |\vec{E}|^2 \right) \} \quad . \tag{A1.5}$$

It follows from (A1.5) that $\overline{p} + \frac{1}{4} \text{Re} \{ \rho \frac{\partial \varepsilon}{\partial \rho} |\vec{E}|^2 \}$ is a function of ρ only, namely the thermodynamic pressure p_o corresponding to the density ρ :

$$\overline{p} + \frac{1}{4} \text{Re} \{ \rho \frac{\partial \varepsilon}{\partial \rho} |\vec{E}|^2 \} = p_o(\rho) \quad . \tag{A1.6}$$

For slightly compressible liquids we have

$$p_o(\rho) = p_o(\rho_\infty) + \frac{(\rho - \rho_\infty)}{\kappa \rho_\infty} \quad , \tag{A1.7}$$

κ being the isothermal compressibility. Integration of (A1.5) leads to

$$\frac{\rho - \rho_\infty}{\kappa \rho \rho_\infty} = \frac{1}{4} \text{Re} \{ \frac{\partial \varepsilon}{\partial \rho} |\vec{E}|^2 - \left(\frac{\partial \varepsilon}{\partial \rho} |E|^2 \right)_\infty \} \tag{A1.8}$$

and in view of (A1.6)

$$\overline{p} = p_o(\rho_\infty) - \frac{1}{4} \text{Re} \{ \left(\rho \frac{\partial \varepsilon}{\partial \rho} |\vec{E}|^2 \right)_\infty \} - \frac{1}{4} \text{Re} \{ (\rho - \rho_\infty) \left(\frac{\partial \varepsilon}{\partial \rho} |\vec{E}\}^2 \right)_\infty \} \quad . \tag{A1.10}$$

Substitution of (A1.8) yields

$$\bar{p} = p_\infty - \frac{\kappa\,\rho\,\rho_\infty}{16}\,\mathrm{Re}\,\{\left(\frac{\partial\varepsilon}{\partial\rho}\,|\vec{E}|^2\right)_\infty\}\cdot\mathrm{Re}\,\{\frac{\partial\varepsilon}{\partial\rho}\,|\vec{E}|^2 - \left(\frac{\partial\varepsilon}{\partial\rho}\,|\vec{E}|^2\right)_\infty\} \quad \text{(A1.10)}$$

with

$$p_\infty = p_o(\rho_\infty) - \frac{1}{4}\,\mathrm{Re}\,\{\left(\rho\,\frac{\partial\varepsilon}{\partial\rho}\,|\vec{E}|^2\right)_\infty\} \quad .$$

The second term in (A1.10) is of the order of

$$\kappa\,\left(\mathrm{Re}\,\{\varepsilon|\vec{E}|^2\}\right)^2 \quad .$$

This must be compared with the contributions of the Maxwell stress tensor to the integral (2.22). These are of the order of

$$\mathrm{Re}\,\{\varepsilon|\vec{E}|^2\} \quad .$$

Since κ is small, we may neglect the second term in (A1.10) for field strengths not exceeding 10^6 V/cm. Therefore we use just

$$\bar{p} \simeq p_\infty$$

what does not contribute to the integral (2.22). For liquids with losses obeying equation (A1.3), similar reasoning leads to the approximation

$$P \simeq P_\infty \quad , \quad \text{(A1.11)}$$

where again terms of the order of

$$\kappa\,\left(\mathrm{Re}\,\{\,\varepsilon|\vec{E}|^2\}\,\right)^2$$

for the components of the pressure tensor have been neglected. The difference between the pressure $\tilde{P}$ which belongs to the choice of the Maxwell stress tensor used by Lovelace et al.[19] and our pressure tensor P is given by

$$\tilde{P} - P \equiv \tilde{\Pi} - \Pi = \frac{1}{4}\,\mathrm{Re}\,\{(\varepsilon - \varepsilon^*)\,\vec{E}^*\vec{E}^T\} \quad .$$

Using the approximation (A1.11) one obtains

$$\tilde{P} = P_\infty + \frac{1}{4}\,\mathrm{Re}\,\{(\varepsilon - \varepsilon^*)\,\vec{E}^*\vec{E}^T\} \quad \text{(A1.12)}$$

If $\mathrm{Im}\{\varepsilon\} \neq 0$, $\tilde{P}$ cannot be approximated by a constant tensor, since the second term in (A1.12) is of the same order as the contributions from the components of the Maxwell stress tensor to the integral (2.22).

A2. TWO PARALLEL DIELECTRIC CYLINDERS IN A TRANSVERSE PERIODIC FIELD. TORQUES RESULTING FROM DIELECTRIC LOSSES

Both cylinders are assumed to have the same radii as R and the same dielectric constants. Their axes are oriented parallel to the z-axis of a cartesian system and the shortest line connecting the cylinder axes defines the x-axis. An externally imposed spatially homogeneous field can be written in the form

$$\mathrm{Re}\ \{ \left(E_{ox}\vec{e}_x + E_{oy}\vec{e}_y\right) e^{-i\omega t} \}$$

where $\vec{e}_x$ and $\vec{e}_y$ are the unit vectors in the x and y-directions. E_{ox} and E_{oy} represent the amplitudes in the x and y-directions respectively. E_{ox} and E_{oy} may be complex (as for example in a rotating field). (For illustration see Fig.1).

We seek to determine the time averaged torque acting upon an individual cylinder. From (2.25) we obtain for the z-component L_{sz} of the torque

$$L_{sz} = \frac{1}{4}\ \mathrm{Re}\ \{\varepsilon_1\}\cdot R\cdot \int \left(E_{1t}E^{*}_{1n} + E^{*}_{1t}\ E_{1n}\right) da_s \quad . \qquad \text{(A2.1)}$$

For a homogeneous rigid body the surface integral

$$\int \vec{r}_s\times\left(\bar{\Pi}_s\cdot\vec{n}_s\right) da_s \qquad \text{(A2.2)}$$

vanishes, since $\bar{\Pi}_s = \bar{\Pi}_s^{\ T}$ and $\left(\bar{\Pi}_s\cdot\nabla\right) = 0$. $\bar{\Pi}_s$ is the Maxwell stress tensor in the solid. Taking the z-component of (A2.2) for a long cylinder one obtains the relation

$$\int \left(E^{*}_{st}E_{sn} + E_{st}E^{*}_{sn}\right) da_s = 0$$

which, together with the transition conditions (3.2), leads to

$$\int E_{1t}E^{*}_{1n}\ da_s = -\frac{1+b}{1+b^{*}} \int E_{1t}E^{*}_{1n}\ da_s \quad .$$

Substituting this in (A2.1) one obtains

$$L_z = -\,i\,\frac{\mathrm{Re}\{\varepsilon_1\}\ \mathrm{Im}\{w\}\cdot R\cdot h}{(1+w^{*})\,(1-w)} \int E_{1t}E^{*}_{1n}\ ds \quad . \qquad \text{(A2.3)}$$

Here use has been made of

$$b = \frac{\varepsilon_s}{\varepsilon_1} - 1 \quad \text{and} \quad w = \frac{b}{b+2} \qquad \text{(A2.4)}$$

ε_s and ε_1 are respectively the dielectric constants of the solid body and the liquid. We have written h ds for the surface element d a_s , where ds is a line element of the cylinder periphery and h an arbitrary chosen height of a portion of the cylinder.

E_{sn} and E_{st} are respectively the normal and the tangential components of the E-field in the solid body; E_{ln} and E_{lt} are the corresponding components in the liquid phase.
We take the values of the components of the $\vec{E}$-field from a solution of the potential problem in bipolar coordinates.* The transformation relating the cartesian and bipolar coordinates is

$$x = a \frac{\sinh \mu}{\cosh \mu - \cos \eta} \quad ; \quad y = a \frac{\sin \eta}{\cosh \mu - \cos \eta} . \qquad \text{(A2.5)}$$

The cylinder surfaces correspond to the circles

$$(x \mp a \coth \mu_o)^2 + y^2 = \frac{a^2}{\sinh^2 \mu_o} .$$

It is useful to introduce an auxiliary parameter

$$p = e^{2\mu_o} .$$

Then the distance between the two "source-points" turns out to be

$$2\,a = R \left(p^{1/2} - p^{-1/2}\right) . \qquad \text{(A2.7)}$$

The following relation holds

$$\frac{\xi}{R} = 2 \cosh \mu_o = p^{1/2} + p^{-1/2} . \qquad \text{(A2.8)}$$

η is a periodic variable, which increases by 2π on going around the right source-point at $x = a$ in the counter-clockwise direction.
A line element on the circle $\mu = \mu_o$ is given by

$$d\,s_\eta = \frac{a}{\cosh \mu_o - \cos \eta} d\eta .$$

The gradients of the potentials are given by

$$\text{grad}_\mu \psi \equiv E_n = \frac{1}{a} (\cosh \mu - \cos \eta) \frac{\partial \psi}{\partial \mu}$$

$$- \text{grad}_\eta \psi \equiv E_t = -\frac{1}{a} (\cosh \mu - \cos \eta) \frac{\partial \psi}{\partial \eta} .$$

These yield on substitution in (A2.3) for the torque on the right cylinder

$$L_z = \frac{i\,\mathrm{Re}\{\varepsilon_1\} \cdot \mathrm{Im}\{w\}}{(1+w^*)(1-w)} \times \frac{Rh}{a} \int_0^{2\pi} (\cosh \mu_o - \cos \eta) \left(\frac{\partial \phi^*}{\partial \mu} \frac{\partial \phi}{\partial \eta}\right)_{\mu_o} d\eta . \qquad \text{(A2.9)}$$

---

* (see Schlögl[22]. This paper is referred to as I in the following. Note the different labelling $\xi \rightarrow \mu$ and $\phi \rightarrow \psi$)

The solution of the potential problem (Eq.25,I) given in I leads to

$$\left(\frac{\partial\psi}{\partial\mu}\right)_o = (1+w)\, E_{ox}\, 2a \sum_1^\infty a_m \cos m\,\eta + (1+w)\, E_{oy}\, 2a \sum_1^\infty b_m \sin m\,\eta \qquad \text{(A2.10)}$$

$$\left(\frac{\partial\psi}{\partial\eta}\right)_o = (1-w)\, E_{ox}\, 2a \sum_1^\infty a_m \sin m\,\eta - (1-w)\, E_{oy}\, 2a \sum_1^\infty b_m \cos m\,\eta \qquad \text{(A2.11)}$$

for the outer fluid phase ($-\mu_o < \mu < \mu_o$). Here the coeeficients had the values (Eq.26.I)

$$a_m = \frac{m\,p^{m/2}}{p^m - w} \quad ; \qquad b_m = \frac{m\,p^{m/2}}{p^m + w} \qquad \text{(A2.12)}$$

(12 follows from 26,I in the symmetrical case, $q = p$, $\kappa_1 = \kappa_2$. We term here $\kappa_i \to w_i$.)
Taking into account losses, one can write generally

$$w = |w|\, e^{i\chi} \quad . \qquad \text{(A2.13)}$$

By exploiting the orthogonality relations of the trigonometric functions we find by substituting for

$$\left(\frac{\partial\psi^*}{\partial\mu}\right)_{\mu_o} \quad \text{and} \quad \left(\frac{\partial\psi}{\partial\eta}\right)_{\mu_o}$$

from (2.10) and (2.11) in the integral in (2.9)

$$
\begin{aligned}
\int_0^{2\pi} &= 4\pi\, a^2\, (1+w^*)(1-w) \\
&\times \Bigg(\cosh \mu_o \Big(- E^*_{ox} E_{oy} \sum_1^\infty a^*_m b_m + E_{ox} E^*_{oy} \sum_1^\infty a_m b^*_m \Big) \\
&\quad + \frac{1}{2} E^*_{ox} E_{oy} \sum_1^\infty a^*_m \left(b_{m+1} + b_{m-1}\right) \\
&\quad - \frac{1}{2} E_{ox} E^*_{oy} b^*_m \left(a_{m+1} + a_{m-1}\right) \Bigg) \qquad \text{(A2.14)} \\
&= 4\pi\, a^2\, (1+w^*)(1-w) \\
&\times \Bigg(- i \left(p^{1/2} + p^{-1/2}\right) \mathrm{Im}\{E^*_{ox} E_{oy} \sum_1^\infty a^*_m b_m\} \\
&\quad + i\, \mathrm{Im}\{E^*_{ox} E_{oy} \left(a^*_m b_{m+1} + a^*_{m+1} b_m\right)\} \Bigg) \quad ,
\end{aligned}
$$

where we have used the relation $z - z^* = 2\, i\, \mathrm{Im}\{z\}$. For the series that appear in (A2.14), we now find

$$
\begin{aligned}
\sum_1^\infty a^*_m b_m &= \sum_1^\infty \frac{m^2\, p^m}{(p^m - w^*)(p^m + w)} = \sum_1^\infty \frac{m^2}{w+w^*} \left(\frac{p^m}{p^m - w^*} - \frac{p^m}{p^m + w} \right) \\
&= \frac{1}{w+w^*} \sum_1^\infty m^2 \sum_{n=0}^\infty \left(\left(w^* p^{-m}\right)^n - \left(-wp^{-m}\right)^n \right) \\
&= \frac{1}{w+w^*} \sum_{n=0}^\infty \left(w^{*n} - (-)^n w^n\right) \sum_1^\infty m^2 \left(p^{-n}\right)^m \qquad \text{(A2.15)} \\
&= \frac{1}{w+w^*} \sum_{n=1}^\infty \left(w^{*n} - (-)^n w^n\right) \frac{p^n\, (p^n+1)}{(p^n - 1)^3} \\
&= \frac{1}{w+w^*} \sum_{n=1}^\infty p^n\, |w|^n\, \frac{p^n+1}{(p^n-1)^3} \left(e^{-in\chi} - (-)^n e^{in\chi}\right) .
\end{aligned}
$$

In the above we have used the form (A2.13) for w and the summation formula (28,I). Thanks to the symmetry relation $b_m(w) = a_m(-w)$, it suffices to evaluate the series

$$\sum_{1}^{\infty} a_m^* b_{m+1} = \sum_{1}^{\infty} m(m+1) \frac{p^{m+1/2}}{(p^m - w^*)(p^{m+1} + w)}$$

$$= \sum_{1}^{\infty} \frac{m(m+1)\, p^{1/2}}{w + p\, w^*} \left(\frac{p^m}{p^m - w^*} - \frac{p^{m+1}}{p^{m+1} + w} \right)$$

$$= \frac{p^{1/2}}{w + p\, w^*} \sum_{1}^{\infty} m\,(m+1) \sum_{n=0}^{\infty} \left[\left(w^*\, p^{-m}\right)^n - \left(-\, w\, p^{-m-1}\right)^n \right]$$

$$= \frac{p^{1/2}}{w + p\, w^*} \sum_{n=0}^{\infty} \left[(w^*)^n - (-)^n \left(wp^{-1}\right)^n \right] \sum_{1}^{\infty} m(m+1) \left(p^{-n}\right)^m$$

$$= \frac{2\, p^{1/2}}{w + p\, w^*} \sum_{n=1}^{\infty} \frac{p^{2n}}{(p^n - 1)^3} \left[(w^*)^n - (-)^n \left(w\, p^{-1}\right)^n \right] .$$

In the above derivation we invoke the summation formula (29,I). By virtue of the symmetry relation we find the series

$$\sum_{1}^{\infty} a_m^* b_{m+1} = \sum_{1}^{\infty} m(m+1) \frac{p^{m+1/2}}{(p^m - w^*)(p^{m+1} + w)}$$

$$= \sum_{1}^{\infty} \frac{m(m+1)\, p^{1/2}}{w + p\, w^*} \left(\frac{p^m}{p^m - w^*} - \frac{p^{m+1}}{p^{m+1} + w} \right)$$

$$= \frac{p^{1/2}}{w+pw^*} \sum_{1}^{\infty} m(m+1) \sum_{n=0}^{\infty} \left[\left(w^*\, p^{-m}\right)^n - \left(-\, w\, p^{-m-1}\right)^n \right]$$

$$= \frac{p^{1/2}}{w+pw^*} \sum_{n=0}^{\infty} \left[(w^*)^n - (-)^n \left(wp^{-1}\right)^n \right] \sum_{1}^{\infty} m(m+1) \left(p^{-n}\right)^m$$

$$= \frac{2p^{1/2}}{w+pw^*} \sum_{n=1}^{\infty} \frac{p^{2n}}{(p^n - 1)^3} \left[(w^*)^n - (-)^n \left(w\, p^{-1}\right)^n \right] .$$

In the above derivation we invoke the summation formula (29,I). By virtue of the symmetry relation we find

$$\sum_1^\infty \left(a_m^* b_{m+1} + a_{m+1}^* b_m\right) = 2\, p^{1/2} \sum_1^\infty \frac{p^n}{(p^n - 1)^3}$$

$$\times \left(\frac{(w^* p)^n - (-)^n w}{w + p\, w^*} - \frac{(-\,wp)^n - (w^*)^n}{w^* + p\, w} \right)$$

$$= \frac{2\, p^{1/2}}{|w + p\, w^*|^2} \sum_1^\infty \frac{p^n}{(p^n - 1)^3} \tag{A2.16}$$

$$\times \left(\left(p^{n+1} + 1\right)\left(w\, w^{*n} - (-)^n w^* w^n\right) + \left(p^n + p\right)\left(w^{*n+1} - (-)^n w^{n+1}\right) \right)$$

$$= \frac{2\, p^{1/2}}{|w + p\, w^*|^2} \sum_1^\infty \frac{p^n\, |w|^{n+1}}{(p^n - 1)^3}$$

$$\times \left(\left(p^{n+1} + 1\right)\left(e^{-i(n-1)\chi} - (-)^n e^{i(n-1)\chi}\right) + \left(p^n + p\right) \right.$$

$$\left. \times \left(e^{-i(n+1)\chi} - (-)^n e^{i(n+1)\chi}\right) \right) .$$

We write for the complex dielectric constants

$$\varepsilon_s = \varepsilon_s' + i\, \varepsilon_s'' \quad ; \quad \varepsilon_1 = \varepsilon_1' + i\, \varepsilon_1'' \quad . \tag{A2.17}$$

Then the "loss angle" χ is determined from (A2.13):

$$\mathrm{tg}\, \chi = -\, 2\, \frac{\varepsilon_1' \varepsilon_s'' - \varepsilon_1'' \varepsilon_s'}{|\varepsilon_1|^2 - |\varepsilon_s|^2} \quad . \tag{A2.18}$$

$|w|$ is given by

$$|w|^2 = \frac{(\varepsilon_1' - \varepsilon_s')^2 + (\varepsilon_1'' - \varepsilon_s'')^2}{(\varepsilon_1' + \varepsilon_s')^2 + (\varepsilon_1'' + \varepsilon_s'')^2} \quad . \tag{A2.19}$$

Substituting of (A2.14) in (A2.9) yields

$$L_z = 4\pi a \cdot R \cdot h \cdot \mathrm{Re}\{\varepsilon_1\} \cdot \mathrm{Im}\{w\} \left(\left(p^{1/2} + p^{-\,1/2}\right) \mathrm{Im}\left\{E_{ox}^* E_{oy} \sum_1^\infty a_m^* b_m\right\} \right.$$

$$\left. -\ \mathrm{Im}\left\{E_{ox}^* E_{oy} \sum_1^\infty \left(a_m^* b_{m+1} + a_{m+1}^* b_m\right)\right\} \right) . \tag{A2.20}$$

Two cases are to be examined here more closely.

a) E_{ox} and E_{oy} are in phase and can therefore be taken to be real. We obtain then from terms of the form $\mathrm{Im}\left(e^{-i\alpha} - e^{i\alpha}\right) = -2\sin\alpha$ to which the terms with even n in the series appearing in (2.15) and (2.16) reduce

$$L_z = 2\pi R^2 h\, E_{ox} \cdot E_{oy} \cdot \mathrm{Re}\{\varepsilon_1\} \cdot \mathrm{Im}\{w\} \sum_{n=2,4,..}^{\infty} \frac{p^n\,|w|^{n-1}}{(p^n-1)^3}$$

$$\times \left\{ -\frac{(p^n+1)(p^2-1)}{p\cos\chi} \sin n\chi + \frac{4\,(p-1)}{p^2+2p\cos 2\chi+1} \right. \tag{A2.21}$$

$$\left. \times \left((p^{n+1}+1)\sin(n-1)\chi + (p^n+p)\sin(n+1)\chi\right) \right\} .$$

Here we have condensed the expression for 2a from (2.7) by absorbing $p^{1/2}$ and rewritten $|w|^2 / |w + pw^*|^2$ with the help of (A2.13).

b) The field rotates with the angular frequency ω. We set $E_{ox} = E$, $E_{oy} = iE$. In this case, the real parts of the series have to be chosen: We obtain an expression analogous to (A2.21), however, with the difference that the odd numbered n are summed.

$$L_z = 2\pi R^2 h\, E^2\, \mathrm{Re}\{\varepsilon_1\}\, \mathrm{Im}\{w\} \sum_{n=1,3,..}^{\infty} \frac{p^n\,|w|^{n-1}}{(p^n-1)^3}$$

$$\times \left\{ \frac{(p^n+1)(p^2-1)}{p\cos\chi} \cos n\chi + \frac{4(p-1)}{p^2+2p\cos 2\chi+1} \right. \tag{A2.21a}$$

$$\left. \times \left((p^{n+1}+1)\cos(n-1)\chi + (p^n+p)\cos(n+1)\chi\right) \right\} .$$

Limiting behaviour for large separations ξ :

For $p \to \infty$, $(\xi/R)^2 \to p$.

Passage to this limit in (A2.21) yields for cylinder 2 in Fig.(1),

$$L_z = -\frac{2V}{\gamma^2} \mathrm{Re}\{\varepsilon_1\} \left(\mathrm{Im}\{w\}\right)^2 \sin 2\theta\, E_o^2 .$$

V being the volume of the cylinder, $E_o = |\vec{E}|$, $\gamma = \frac{R}{\xi}$.

Limiting case of contact :

We introduce a new coordinate system by (Schlögl[23],... cited as II in what follows. Note the different labelling of the coordinates in this paper.)

$$\mu = \frac{x \cdot l}{x^2 + y^2} \quad ; \quad \eta = \frac{y \cdot l}{x^2 + y^2} \quad .$$

One has to evaluate in this case Fourier integrals instead of Fourier series. For the torque ($\mu = \mu_o > 0$) acting upon the right cylinder one finds

$$L_z = 2\pi \, Rh \, l \, \mathrm{Re}\{\varepsilon_1\} \cdot \mathrm{Im}\{w\} \tag{A2.22}$$

$$\times \left(\mu_o^2 \, \mathrm{Im}\{E_{ox}^* \, E_{oy} \int_0^\infty a_k^* b_k dk\} + \mathrm{Im}\{E_{ox}^* \, E_{oy} \int_0^\infty \frac{\partial a_k^*}{\partial_k} \frac{\partial b_k}{\partial \, k} dk\} \right) \; .$$

(We have introduced here k as an integration variable in the place of ω). As we restrict ourselves to cylinders of equal radii R, the coefficients simplify to

$$a_k^* = \frac{k \, p^{k/2}}{p^k - w^*} \quad ; \quad b_k = \frac{k \, p^{k/2}}{p^k + w} = a_k(-w) \quad . \tag{A2.23}$$

For the length parameter l still free at our disposal, we set

$$l = R \quad ; \quad \text{hence} \quad \mu_o = \frac{1}{2} \; ; \quad p = e \quad . \text{(Compare Eqs.10,II and 7,II)}.$$

We next transform $a_k^* = k \, \alpha_k^*$, $b_k = k \, \beta_k$ and hence obtain

$$\frac{\partial \, a_k^*}{\partial k} \frac{\partial b_k}{\partial k} = k^2 \frac{\partial \alpha_k^*}{\partial k} \frac{\partial \beta_k}{\partial k} + \frac{\partial}{\partial k} \left(k \, \alpha_k^* \, \beta_k\right) \quad .$$

The last term does not contribute to the integral. We may therefore set

$$\int_0^\infty \frac{\partial a_k^*}{\partial k} \frac{\partial b_k}{\partial k} dk = \int_0^\infty k^2 \frac{\partial \alpha_k^*}{\partial k} \frac{\beta_k}{\partial \, k} dk \quad . \tag{A2.25}$$

We substitute (2.23) in the integral

$$\int_0^\infty a_k^* \, b_k \, dk = \int_0^\infty \frac{k^2 \, e^k \, dk}{(e^k - w^*)(e^k + w)}$$

where use has been made of (A2.24).
By decomposition into partial fractions

$$\frac{e^k}{(e^k - w^*)(e^k + w)} = \frac{w^*}{w + w^*} \cdot \frac{1}{e^k - w^*} + \frac{w}{w + w^*} \cdot \frac{1}{e^k + w}$$

we obtain

$$I_1 \equiv \int_0^\infty \frac{k^2 e^k dk}{(e^k - w^*)(e^k + w)} = \frac{2}{w+w^*} \sum_{n=0}^\infty \frac{w^{*(n+1)} + (-)^n w^{n+1}}{(n+1)^3}$$

$$= \frac{2}{w+w^*} \sum_{n=1}^\infty \frac{w^{*n} - (-)^n w^n}{n^3} \quad . \tag{A2.26}$$

In the above w and w^* were treated as independent parameters. We may therefore carry out the partial differentiations

$$I_2 = -\frac{\partial}{\partial w} I_1 = \int_0^\infty \frac{k^2 e^k dk}{(e^k - w^*)(e^k + w)^2}$$

$$= \frac{1}{w+w^*} I_1 + \frac{2}{w+w^*} \sum_{n=1}^\infty \frac{(-)^n w^{n-1}}{n^2} \tag{A2.27}$$

$$I_3 = \frac{\partial}{\partial w^*} I_1 = \int_0^\infty \frac{k^2 e^k dk}{(e^k - w^*)^2 (e^k + w)}$$

$$= \frac{-1}{w+w^*} I_1 + \frac{2}{w+w^*} \sum_{n=1}^\infty \frac{w^{*(n-1)}}{n^2} \tag{A2.28}$$

$$I_4 = -\frac{\partial}{\partial w} I_3 = \int_0^\infty \frac{k^2 e^k dk}{(e^k - w^*)^2 (e^k + w)^2}$$

$$= \frac{-1}{(w+w^*)^2} I_1 - \frac{1}{w+w^*} I_2 + \frac{2}{(w+w^*)^2} \sum_{n=1}^\infty \frac{w^{*(n-1)}}{n^2} \tag{A2.29}$$

$$= \frac{-2}{(w+w^*)^2} I_1 + \frac{2}{(w+w^*)^2} \sum_{n=1}^\infty \frac{w^{*n-1} - (-)^n w^{n-1}}{n^2} \quad .$$

One finds after a brief calculation

$$\frac{\partial \alpha_k^*}{\partial k} = -\frac{1}{2} \frac{e^{k/2}}{e^k - w^*} - \frac{w^* e^{k/2}}{(e^k - w^*)^2}$$

$$\frac{\partial \beta_k}{\partial k} = -\frac{1}{2} \frac{e^{k/2}}{e^k + w} + \frac{w e^{k/2}}{(e^k + w)^2} \quad .$$

On substituting these in (A2.25), we obtain with the use of the integrals introduced in (A2.26) - (A2.29) for the sums occurring in (A2.22) $\left(\mu_o^2 = \frac{1}{4}\right)$

$$\int_0^\infty \left(\frac{\partial a_k^*}{\partial k}\frac{\partial b_k}{\partial k} + \frac{1}{4} a_k^* b_k\right) dk = \frac{1}{2} I_1 - \frac{w}{2} I_2 + \frac{w^*}{2} I_3 - |w|^2 I_4$$

$$= \frac{4\,|w|^2}{(w+w^*)^3} \sum_{n=1}^\infty \frac{w^{*n} - (-)^n w^n}{n^3} - \frac{2\,|w|^3}{(w+w^*)^2}$$

$$\times \sum_{n=1}^\infty \frac{w^{*(n-1)} - (-)^n w^{n-1}}{n^2} \tag{A2.30}$$

$$+ \frac{1}{w + w^*} \sum_{n=1}^\infty \frac{w^{*n} - (n)^n w^n}{n^2}$$

$$= \frac{1}{2} \sum_{n=1}^\infty \frac{|w|^{n-1}}{(n \cos \chi)^3} \left[\left(e^{-in\chi} - (-)^n e^{in\chi}\right)\left(1 + n \cos^2\chi\right)\right.$$

$$\left. - n \cos \chi \left(e^{-i(n-1)\chi} - (-)^n e^{i(n-1)\chi} \right)\right] .$$

We again consider both the cases

a) E_{ox} and E_{oy} are in phase, both are real.

The even-numbered n values are sorted out. They lead to terms with sin n χ and sin (n-1) χ . After minor simplifications we obtain

$$L_z = - 2V E_{ox} E_{oy} \operatorname{Re}\{\varepsilon_1\}\cdot\operatorname{Im}\{w\} \sum_{n=2,4..}^\infty \frac{|w|^{n-1}}{(n \cos \chi)^3}$$

$$\times\left(\sin n \chi + \frac{n}{2} \sin 2\chi \cdot\cos n \chi\right) \tag{A.2.31}$$

b) The E-field rotates with the angular frequency ω, $E_{ox} = E$, $E_{oy} = iE$

$$L_z = 2V E^2 \operatorname{Re}\{\varepsilon_1\}\cdot\operatorname{Im}\{w\} \sum_{n=1,3..}^\infty \frac{|w|^{n-1}}{(n \cos \chi)^3}$$

$$\times\left(\cos n \chi - \frac{n}{2} \sin 2\chi \cdot\sin n\chi \right) .$$

χ passes through a zero when $\frac{\varepsilon_s'}{\varepsilon_s''} = \frac{\varepsilon_1'}{\varepsilon_1''}$. L_z vanishes in this case.

But, together with χ, Im{w} also becomes zero. Should this condition be fulfilled while varying the ε-values, it still does not imply a change of sign of L_z: For, Im{w} and χ change signs together.

LIST OF SYMBOLS

$\vec{D}$ = vector of dielectric displacement

$\vec{E}$ = vector of electric field strength

$\vec{B}$ = vector of magnetic induction

$\vec{H}$ = vector of magnetic field strength

ρ_{el} = electric charge density

$\vec{j}$ = vector of conductive current density

ρ = mass density

c = phase velocity of light in vacuum

$\vec{v}$ = vector of center of mass velocity

ε = dielectric constant = (relative dielectr.const.)$\cdot\varepsilon_o$

μ = magnetic permittivity = (relative perm.)$\cdot\mu_o$

σ = electric conductivity

∇ = Nabla operator, a column vector

A quantitaty $\vec{A}$ is a column vector

$\vec{A}^T$ is a row vector, the transposed of $\vec{A}$

Re{a} = real part of complex number a

Im{a} = imaginary part of complex number a .

REFERENCES

1. H. A. Pohl, "Dielectrophoresis", Cambridge University Press, Cambridge (1978).
2. H. A. Pohl and I. S. Crane, Dielectrophoresis of Cells, Biophys. J. 11:711 (1971).
3. C. Holzapfel, J. Vienken and U. Zimmermann, Rotation of cells in an alternating electric field: Theory and experimental proof, J. Membrane Biol. 67:13 (1982).
4. W. M. Arnold and U. Zimmermann, Rotating-field induced rotation and measurement of membrane capacitance of single mesophyll cells of Avena sativa. Z. Naturforsch. 37:908 (1982).
5. L. D. Landau and E. M. Lifshitz, "Electrodynamics of Continuous Media". Course of Theoretical Physics Vol.8, Pergamon Press, New York (1960).

6. F. A. Sauer, Forces on Suspended Particles in the Electromagnetic Field, in: "Coherent Excitations in Biological Systems", H. Fröhlich and F. Kremer ed., Springer-Verlag, Berlin Heidelberg (1983).
7. A. Lampa, Über Rotation im Elektrostatischen Drehfelde, Wien. Ber. 115(2a):1659 (1906).
8. P. Lertes, Untersuchungen über Rotationen von dielektrischen Flüssigkeiten im elektrostatischen Drehfeld, Zs. f. Physik 4:315 (1921).
9. R. Fürth, Eine neue Methode zur Bestimmung der Dielektrizitätskonstanten guter Leiter, Zs. f. Physik, 22:98 (1924).
10. R. Fürth, Die absolute Bestimmung von Dielektrizitätskonstanten mit der Ellipsoidmethode, Zs. f. Physik 44:256 (1927).
11. J. A. Stratton, "Electromagnetic Theory", McGraw-Hill, New York (1941).
12. J. Meixner and H. G. Reik, Thermodynamik irreversibler Prozesse, in: "Encyclopedia of Physics, Vol III/2, Principles of Thermodynamics and Statistics", S. Flügge ed., Springer Verlag Berlin, Göttingen, Heidelberg (1959).
13. J. Meixner, Zur Thermodynamik des elektromagnetischen Feldes, Zs. f. Physik 229:352 (1969).
14. S. R. De Groot and P. Mazur, "Nonequilibrium Thermodynamics", page 376-404, North Holland, Amsterdam (1962).
15. G. Kirchhoff, "Vorlesungen über mathematische Physik. Mechanik", Leipzig (1883).
16. A. Einstein, Über die von der molekularkinetischen Theorie der Wärme geforderte Bewegung von in ruhenden Flüssigkeiten suspendierten Teilchen, Ann. Phys. 17:549 (1905).
17. F. A. Sauer, Interaction-Forces between microscopic particles in an external electromagnetic field, this volume.
18. G. Schwarz, A theory of low-frequency dielectric dispersion of colloidal particles in electrolyte solutions, J. physical Chemistry 66:2636 (1962).
19. R. V. E. Lovelace, D. G. Stout and P. L. Steponkus, Protoplast Rotation in a rotating electric field: the influence of cold acclimation, J. Membrane Biol. 82:157 (1984).
20. R. Glaser, G. Fuhr and J. Gimsa, Rotation of erythrocythes, plant cells and protoplasts in an outside rotating field, Studia biophysica 96:11 (1983).
21. H. P. Schwan, Dielectric properties of cells and tissues, this volume.
22. R. W. Schlögl, Potential distribution and interaction between two parallel dielectric cylinders in a static, transverse field, Ber. Bunsenges. Phys. Chem., in press (1985).
23. R. W. Schlögl, Potential distribution and interaction between two parallel cylinders in a static, transverse field, limiting case of contacting cylinders. Ber. Bunsenges. Phys. Chem., in press (1985).

ELECTRIC AND MAGNETIC FIELD EFFECTS ON LIGAND BINDING TO THE CELL MEMBRANE

Alessandro Chiabrera, Bruno Bianco,
Franco Caratozzolo, Giulio Giannetti,
Massimo Grattarola, and Rinaldo Viviani

Biophysical and Electronic Engineering Department
University of Genoa
Via all'Opera Pia 11a, 16145 Genoa, Italy

INTRODUCTION

The interactions between electromagnetic (em) fields and cellular processes have been experimentally proved by many authors. Nevertheless, the only quantitative theory so far proposed deals with the microelectrophoretic effects of a low-frequency electric exposure on lectins and their surface receptors [1]. We aim to extend such a theory to a more general case, considering how electric and magnetic signals can affect the aggregation process of a ligand to its binding site.

We deal mainly with the effect of the electric field as a putative transduction mechanism of an em stimulus into a biochemical signal. Furthermore, some specific effects of the magnetic induction field are considered and discussed. The typical signal considered in this paper is characterized by the following features: the signal spectrum has a bandwidth ranging from dc to about 10MHz; the peak electric field intensity is smaller than 100mV/cm; the peak magnetic-induction field intensity is smaller than 10mT, and its maximum rate of change is less than 100T/s.

We assume that a first external messenger (ion, lectin, hormone) acts at the level of a cell binding site, e.g., on the cell membrane, inducing a change in the concentration of a second internal messenger. In the case of an ion (e.g., Ca++) the site can be the mouth of a passive channel, the binding site of an active pump, or, generally speaking, any membrane binding site. The corresponding second messenger can be the ion itself [2], or another ion. If the messenger is a lectin (e.g., phytohemagglutinin, PHA), the binding site is the lectin receptor. The second messenger can be an ion (Ca++, for PHA) whose concentration inside the cell can be enhanced by the lectin-induced receptor aggregation [1] on the cell membrane. If the first messenger is a hormone (e.g., parathyroid hormone, PTH), the binding site is the hormone receptor. The second messenger can be an enzymatic product (cyclic adenosine monophosphate, cAMP, for PTH) [3].

The basic idea is the following: the ligand, which has a net charge, can diffuse and drift by a three-dimensional walk close to its binding site. The site itself can diffuse and drift by a two-dimensional walk over the cell membrane, or it can be considered immobile. The electric and magnetic Lorentz forces affect the values of the dissociation and association rate constants which model the interaction process between the ligand and its binding site. Such an effect may induce a change in the second messenger's concentration.

AC ELECTRIC FIELD EFFECTS

The evaluation of the electric field effects on the binding between a ligand L (ion, lectin, hormone) and its site S closely follows the approach developed in [1].

The aggregation between L and S occurs when their mutual distance δ_{LS} is less than a critical distance R_A, i.e., when

$$(1) \quad \langle \delta_{LS}^2 \rangle \leq R_A^2$$

in the mean-square sense.

The time interval, t_A, necessary for L and S to travel the distance R_A by diffusion and drift, in the mean-square sense (starting from $\delta_{LS}=0$ at time zero), is a measure of the macroscopic rate K^- of the dissociation events, letting

$$(2) \quad K^- = t_A^{-1}$$

Finally, we assume, as a simplifying hypothesis, that the ratio between the association and dissociation rate constants of the binding process is given by

$$(3) \quad a = K^-/K^+ = [(4/3)\pi R_A^3]^{-1}$$

in accordance with [4].

The dynamics of the binding process between L and S can be

studied by solving two Langevin-type equations for L and S, including the effects of the viscous and drag forces, in the presence of a periodic ac electric field $E(t)=\sum_{-\infty}^{+\infty}{}_n E_n \exp(jn\omega_0 t)$, where E_n are the Fourier components, $f_o=\omega_o/(2\pi)$ is the field fundamental frequency, and $E_o=0$:

$$(4) \quad d\bar{v}_j/d\theta+\beta_j\bar{v}_j=\bar{n}_j+\gamma_j E(\theta-\tau) \qquad (j=L,S)$$

where θ is a dummy time variable, β_j is a coefficient related to the viscosity [1], $\bar{n}_j$ is the random force, and the vector $\bar{\gamma}_j$ includes drift and drag effects as well as the effects of the nonuniform spatial distribution of the electric field over the cell surface [1]. The electric field, E(t), is the intensity of the far field. The time shift $\tau=t-\theta$ is a random variable which accounts for the shift between the true time variable for E(t) and the dummy time variable θ. The approximations used for the integration of eq.(4) and the computation of the displacements $\bar{\delta}_L$ and $\bar{\delta}_S$ are given in [1]. Accordingly, the evaluation of $\delta^2_{LS}(t_A)=R^2_A$ in the mean-square sense gives:

$$(5) \quad 1/t_A=1/t_{A,o}+Kt_A\ 4\{\textstyle\sum_1^{\infty}{}_n[|E_n|^2/(2\pi n f_o t_A)^2][1-\cos(2\pi n f_o t_A)]\}$$

where $t_{A,o}$ is the value of t_A for E(t)=0, and K is a positive constant. The procedure for finding relationship (5) has been fully outlined in [1], where $1/t_A$ and $1/t_{A,o}$ have been labelled r_- and r_-^o, respectively. Also the value of K is given in [1]. Eq.(5) reveals that, for a given E(t), the value of t_A depends on two parameters only, i.e., $t_{A,o}$ and K, so that a simple normalizing procedure can be devised. Letting $E^*=(2t_{A,o}K^{1/2})^{-1}$, $y=t_{A,o}/t_A=K^-/K^-_o$, $x=f_o t_{A,o}$ and $e_n=E_n/E^*$, relationship (5) becomes:

$$(6) \quad y=1+y^{-1}\{\textstyle\sum_1^{\infty}{}_n[|e_n|^2 y^2/(2\pi n x)^2][1-\cos(2\pi n x/y)]\}$$

which is a rather general relationship applicable to any process modelled by eqs.(4). The characteristic values of each process are the critical field E^* and the value $t_{A,o}$. The anticipated values of E^* and $t_{A,o}$ for PHA and human lymphocytes are $E^*<25$mV/cm and $t_{A,o}<0.25$ms. A crude evaluation of E^* is offered in [1] as a function of the value assumed by the effective diffusivity of the ligand near its binding site, and as a function of the value of the effective local viscosity. These values should be considered as almost independent parameters because the existence of a local field of forces, due to the binding site, implies that the effective diffusivity and viscosity are no longer inversely proportional, as happens in the classical case of a particle travelling in bulk water. Roughly speaking, E^* is proportional to the product of the effective diffusivity times the viscosity [1] so that, if the endogenous field of forces produces both low-diffusivity and low-viscosity values, the corresponding critical field E^* could become smaller than a few mV/cm. We point out that the value of E^* is not constant on a spherical cell surface because it depends on the nonuniform spatial distribution of E(t) over the cell surface, via the contribution of $\bar{\gamma}_j$. A simple inspection of eq.(6) gives $y \geq 1$, i.e., $t_A \leq t_{A,o}$, irrespective of the signal waveform, so that we conclude that eqs.(4) always imply a decrease of t_A with respect to $t_{A,o}$, irrespective of the type of em exposure used. Other special features of eq.(6) can be brought into evidence by discussing three types of em signals of pratical interest. The first case deals with a sinusoidal electric field

$$(7) \quad E(t)=\sqrt{2}\, E_{rms} \cos \omega_o t = E_1 \left[\exp(-j\omega_o t)+\exp(j\omega_o t)\right]$$

where E_{rms} is the mean-square root amplitude.

Letting $e_{1,rms}=E_{rms}/E^*=\sqrt{2}\, E_1/E^*=\sqrt{2}\, e_1$, the plot of y versus x for different values of $e_{1,rms}$ is shown in Fig. 1. The plot of y versus $e_{1,rms}$ is shown in Fig.2, where $0 \leq x \leq 1$, and in Fig.3, where $1 < x \leq 2$. The second case deals with a periodic train of Dirac's delta functions

$$(8) \quad E(t)=\sum_{m=-\infty}^{+\infty} \phi_o \left[\delta(t-m/f_o)-f_o\right]=\sum_{n=-\infty}^{+\infty} (\phi_o f_o)\exp(jn\omega_o t) \quad (n \neq 0)$$

where ϕ_o is the area of the Dirac delta $\delta(t)$. The amplitudes E_n assume the common value $E_\phi=\phi_o f_o$ for every value of n. We take $e_\phi=E_\phi/E^*$ as a normalized amplitude parameter. Accordingly, if e_ϕ is kept constant and f_o decreases, the area ϕ_o increases. The plot of y versus x for different values of e_ϕ is shown in Fig.4. The plot of y versus e_ϕ for different values of x is shown in Figs.5 ($0 < x \leq 1$) and 6 ($1 < x \leq 2$).

The third case deals with the same periodic train of Dirac's delta functions considered in the second case (eq.(8)). We have already pointed out that, if the value of E_ϕ is kept constant, a change in f_o also affects ϕ_o. Consequently, it is worth considering as a normalized amplitude parameter, the value $e_D=(\phi_o/t_{A,o})/E^*=e_\phi/x$, which is independent of x. By doing so, and keeping e_D constant, less and less power is delivered to the system as f_o (i.e., x) decreases. The plot of y versus x for different values of e_D is shown in Fig.7. The plot of y versus e_D for different values of x is shown in Figs.8 ($0 < x \leq 1$) and 9 ($1 < x \leq 2$).

Even over the low-frequency range ($0 \leq x \leq 1$), y can be a multivalued function of $e_{1,rms}$, or e_ϕ, or e_D. Such a behaviour strongly suggests the possibility of the onset of field-induced instabilities which may occur in the case of time-varying exposure only. In the limit $x \rightarrow 0$ for a sinusoidal field, the variable y

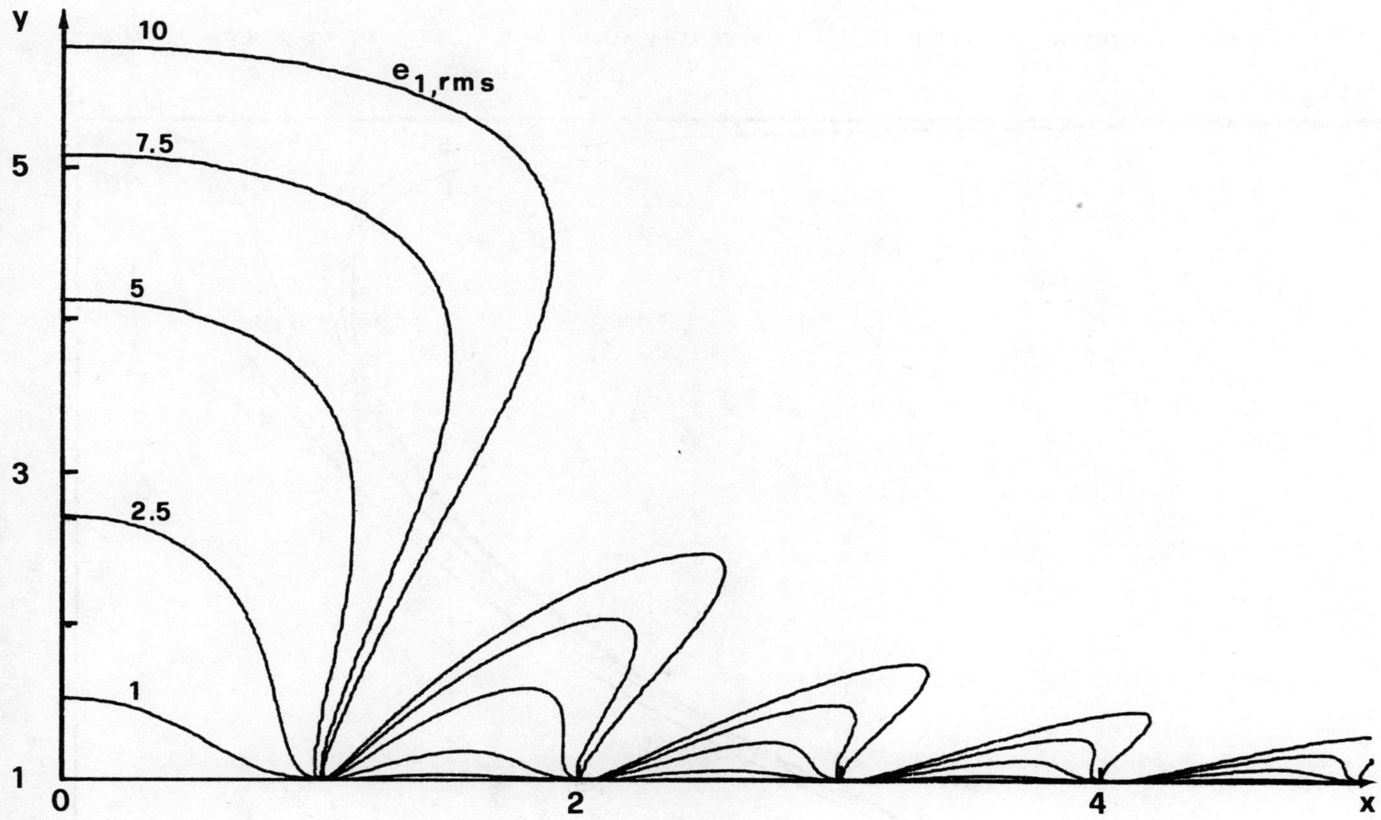

Figure 1 - Effect of sinusoidal electric field on K^-. The field frequency is f_o and its amplitude is $\sqrt{2}\ E_{rms}$. Plot of $y=K^-/K_o^-$ versus $x=f_o/K_o^-$, at different values of $e_{1,rms} = E_{rms}/E^*$.

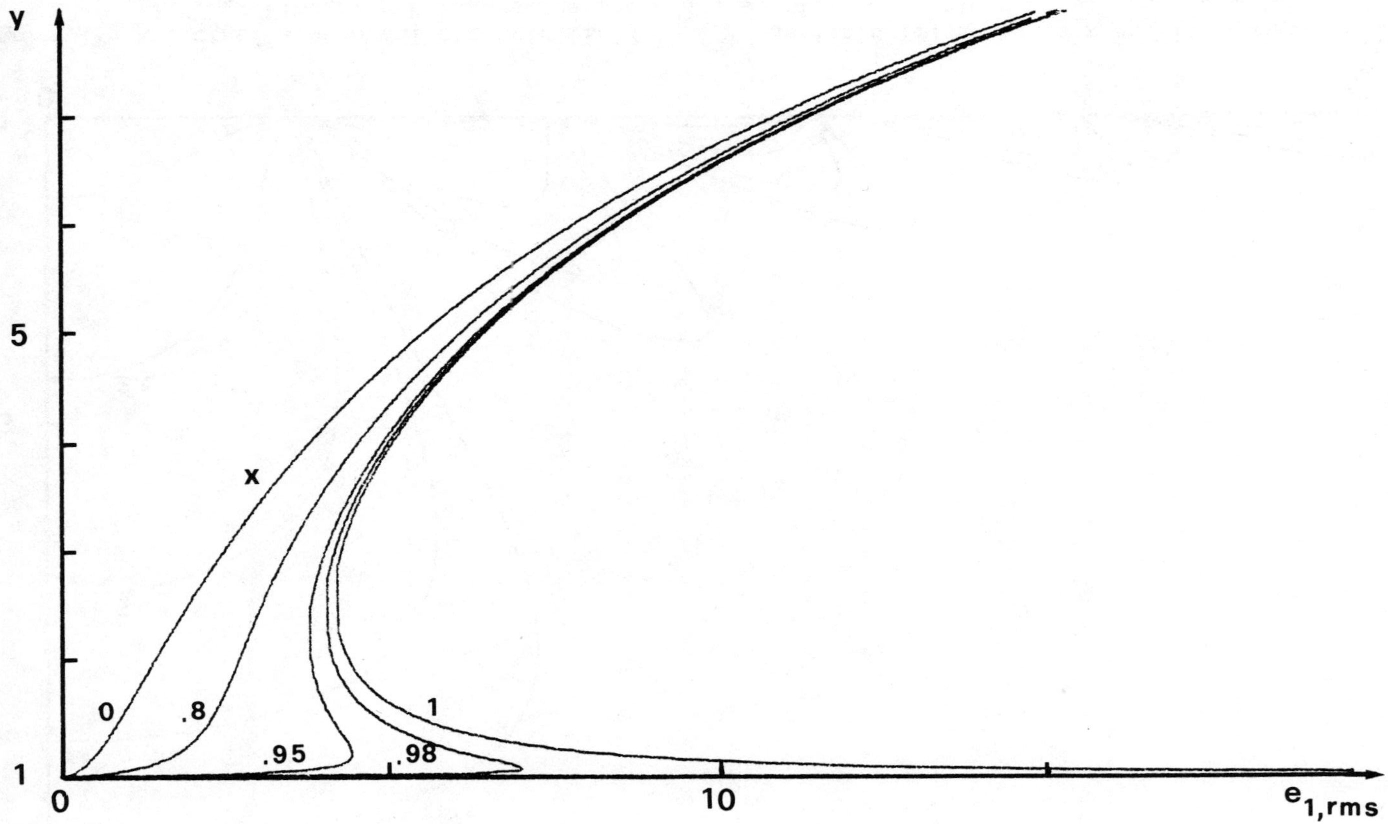

Figure 2 - Same as in Fig. 1. Plot of y versus $e_{1,rms}$, at different values of $0 \leqslant x \leqslant 1$

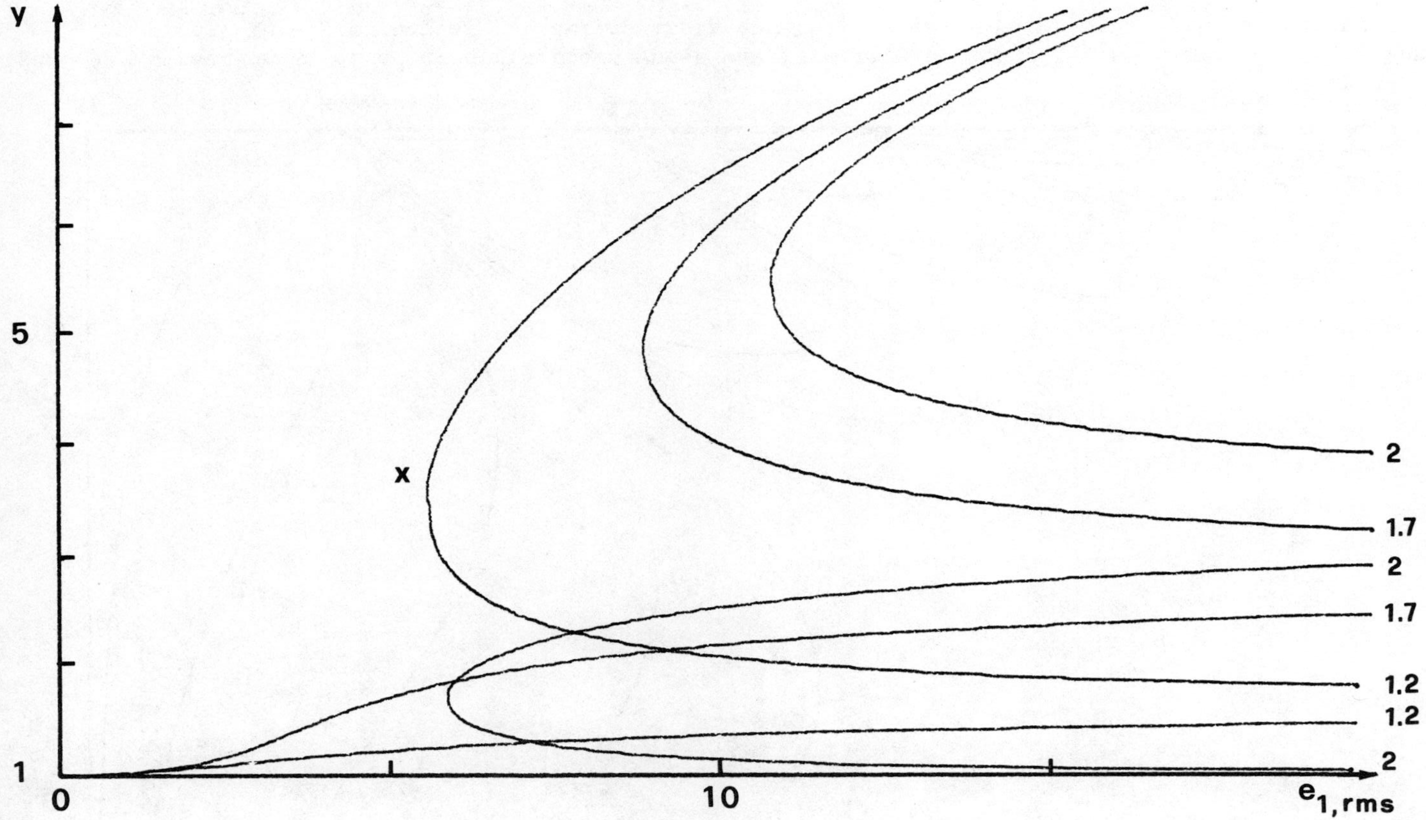

Figure 3 - Same as in Fig. 1. Plot of y versus $e_{1,rms}$, at different values of $1 < x \leqslant 2$.

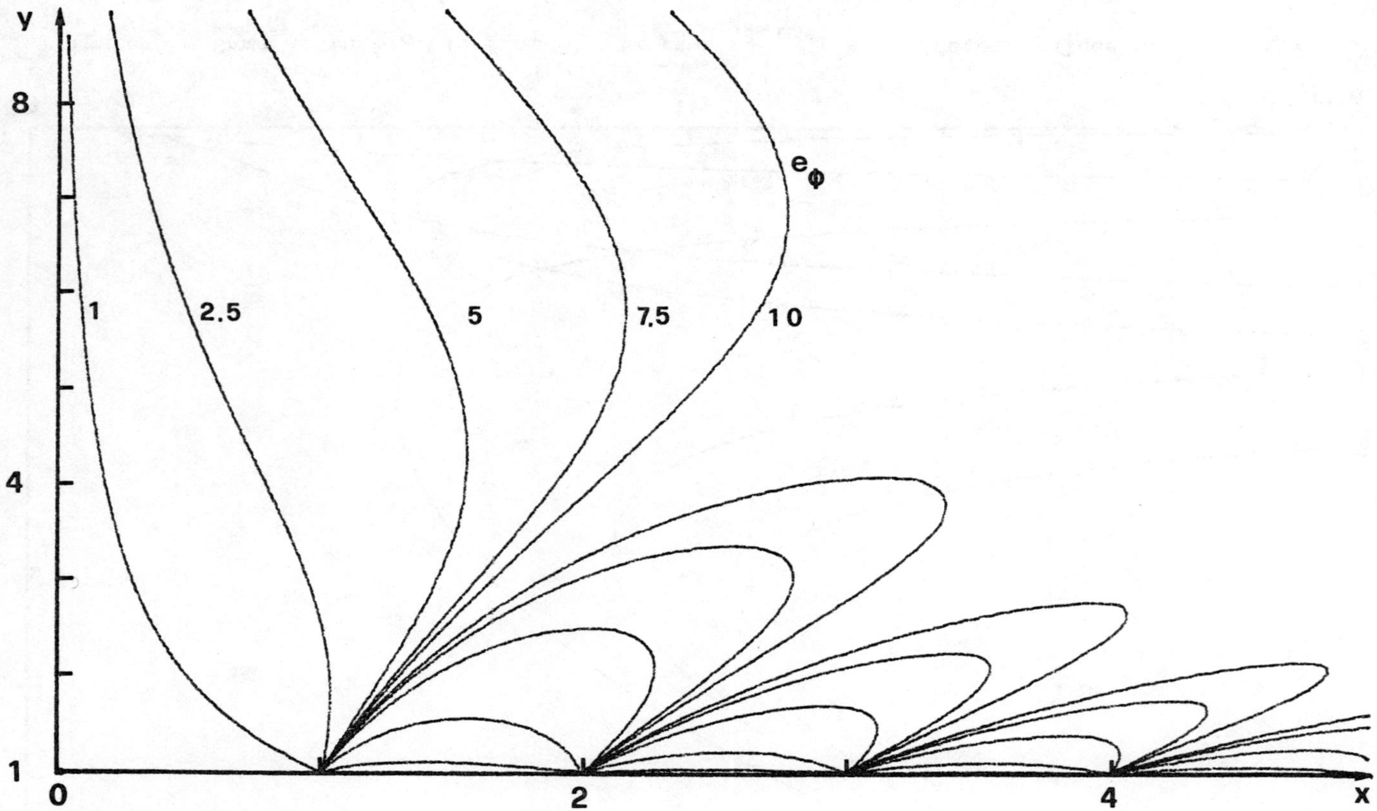

Figure 4 - Effect on K^- of an electric field whose waveform is a periodic train of Dirac's delta functions. The frequency of the deltas is f_o and their area is $\emptyset_o$. Plot of $y=K^-/K_o^-$ versus $x=f_o/K_o^-$, at different values of $e_\emptyset=\emptyset_o f_o/E^*$

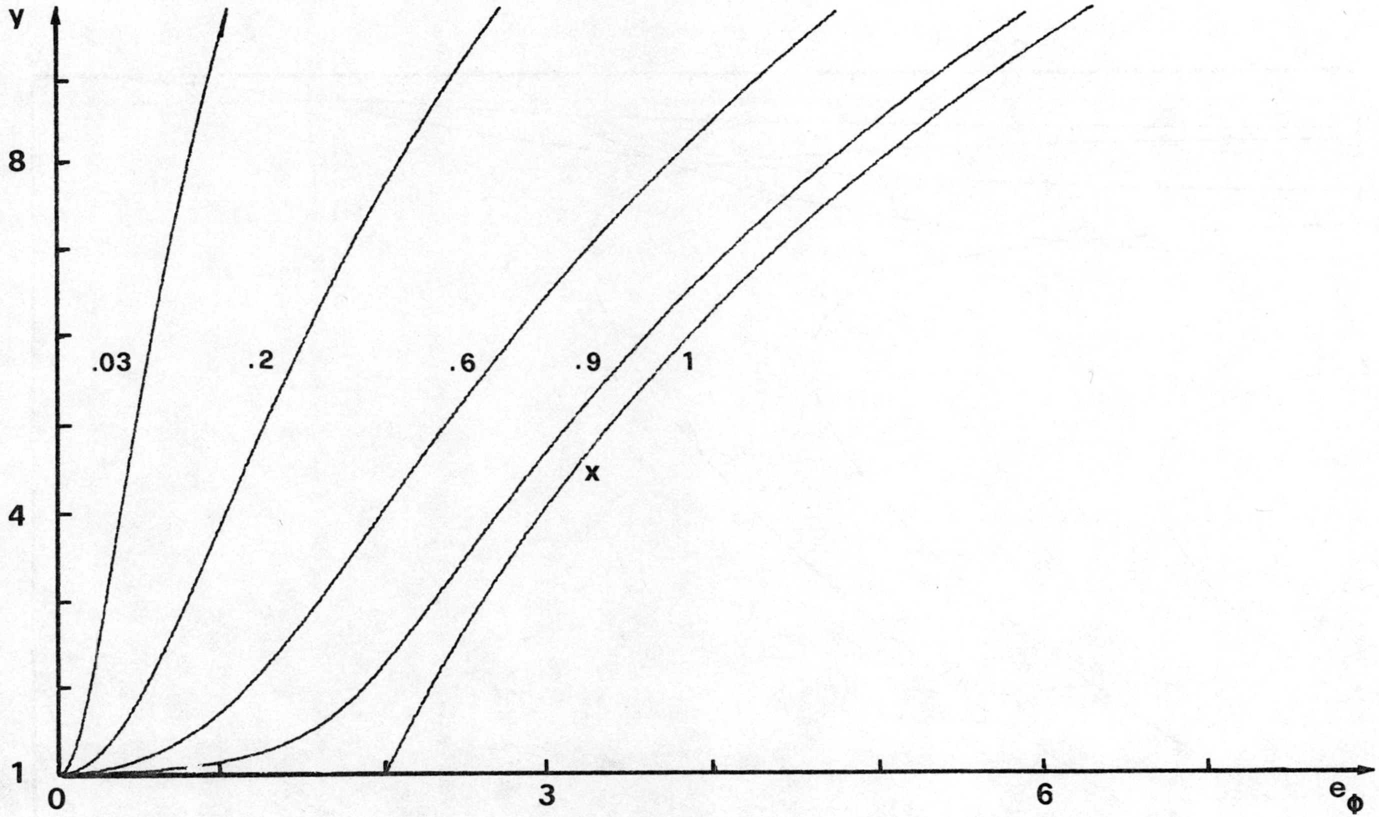

Figure 5 - Same as in Fig. 4. Plot of y versus e_ϕ, at different values of $0 < x \leq 1$

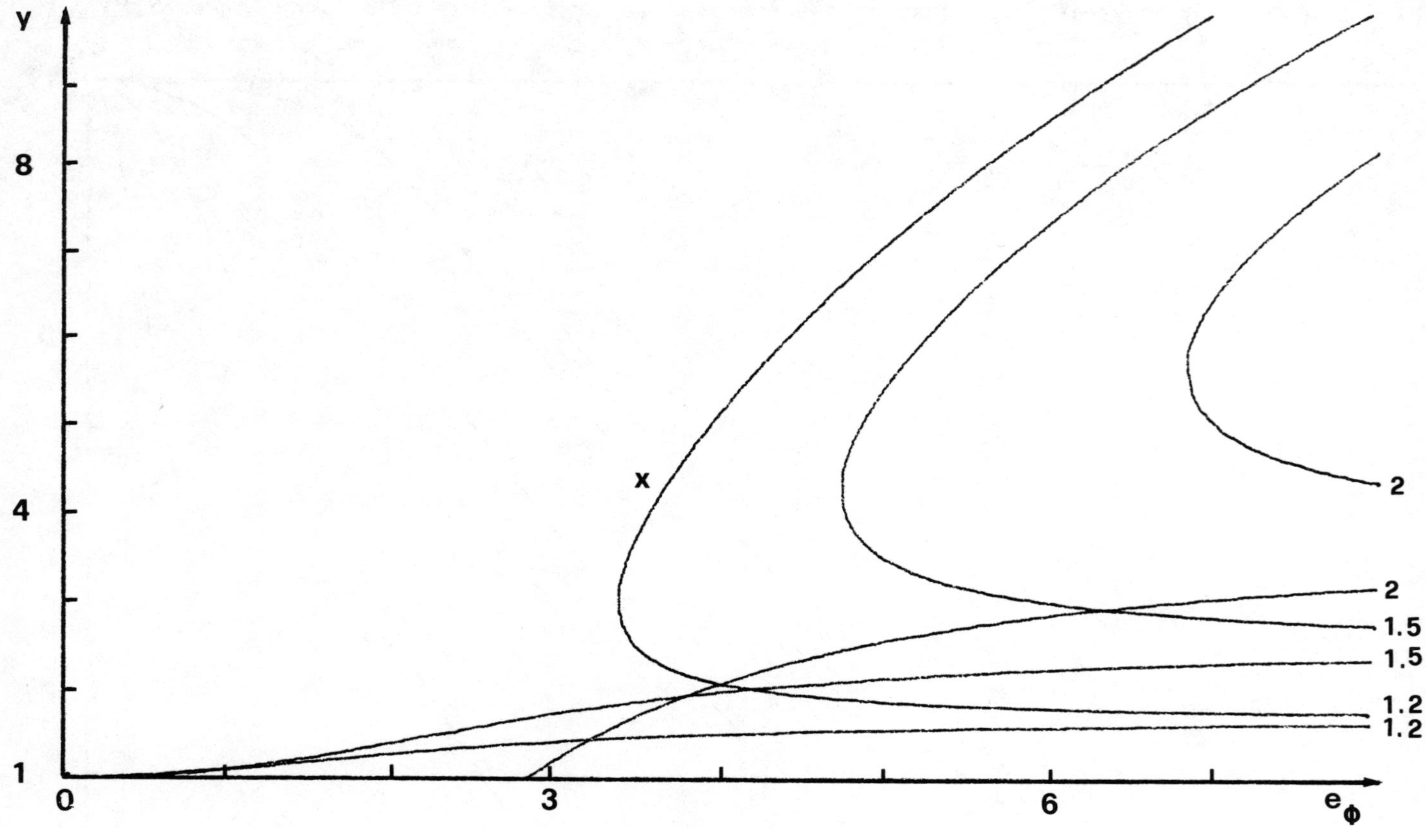

Figure 6 - Same as in Fig. 4. Plot of y versus e_ϕ, at different values of $1 < x \leqslant 2$

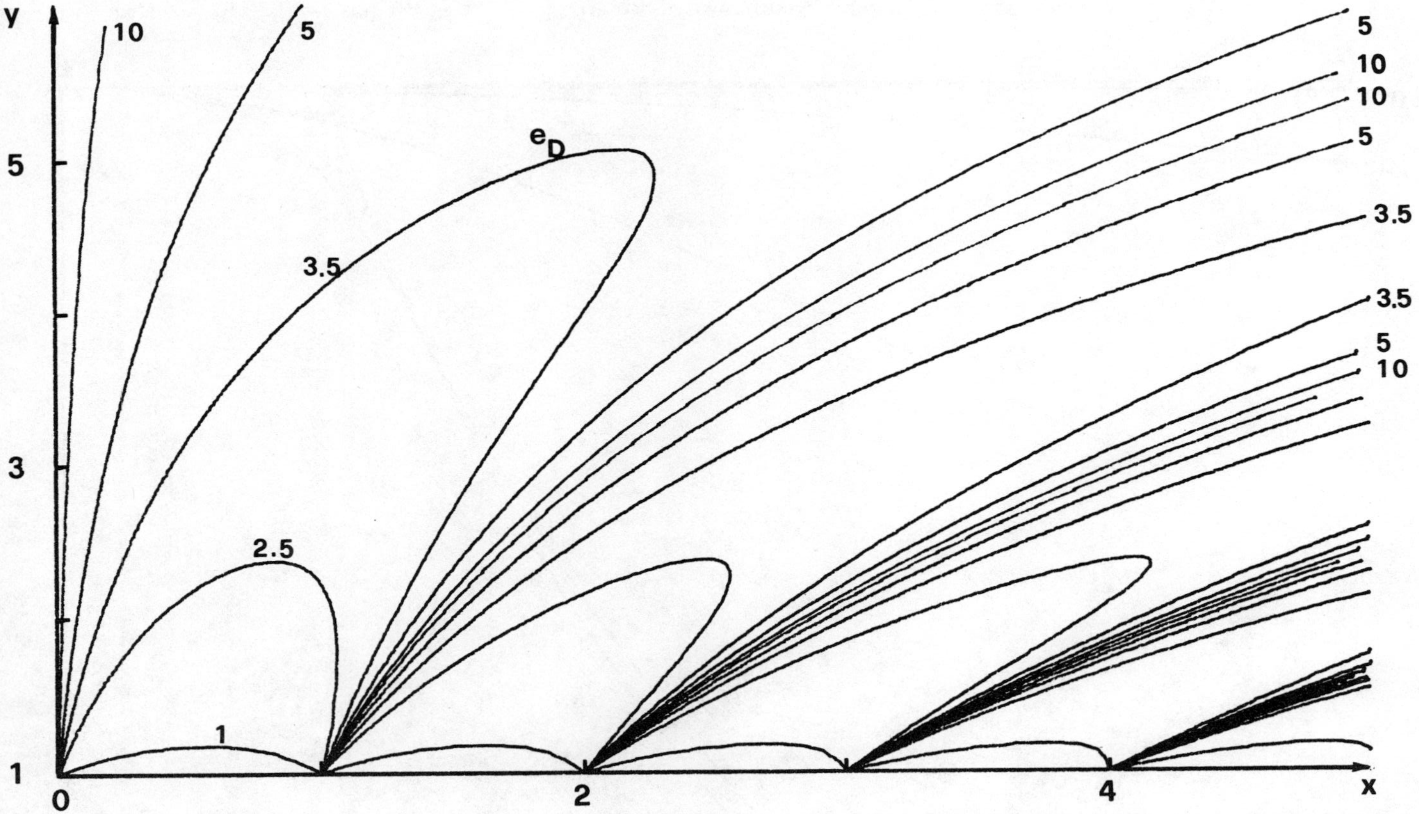

Figure 7 - Same as in Fig. 4. Plot of $y=K^{-}/K_{o}^{-}$ versus $x=f_{o}/K_{o}^{-}$, at different valus of $e_{D}=\emptyset_{o}K_{o}^{-}/E^{*}$

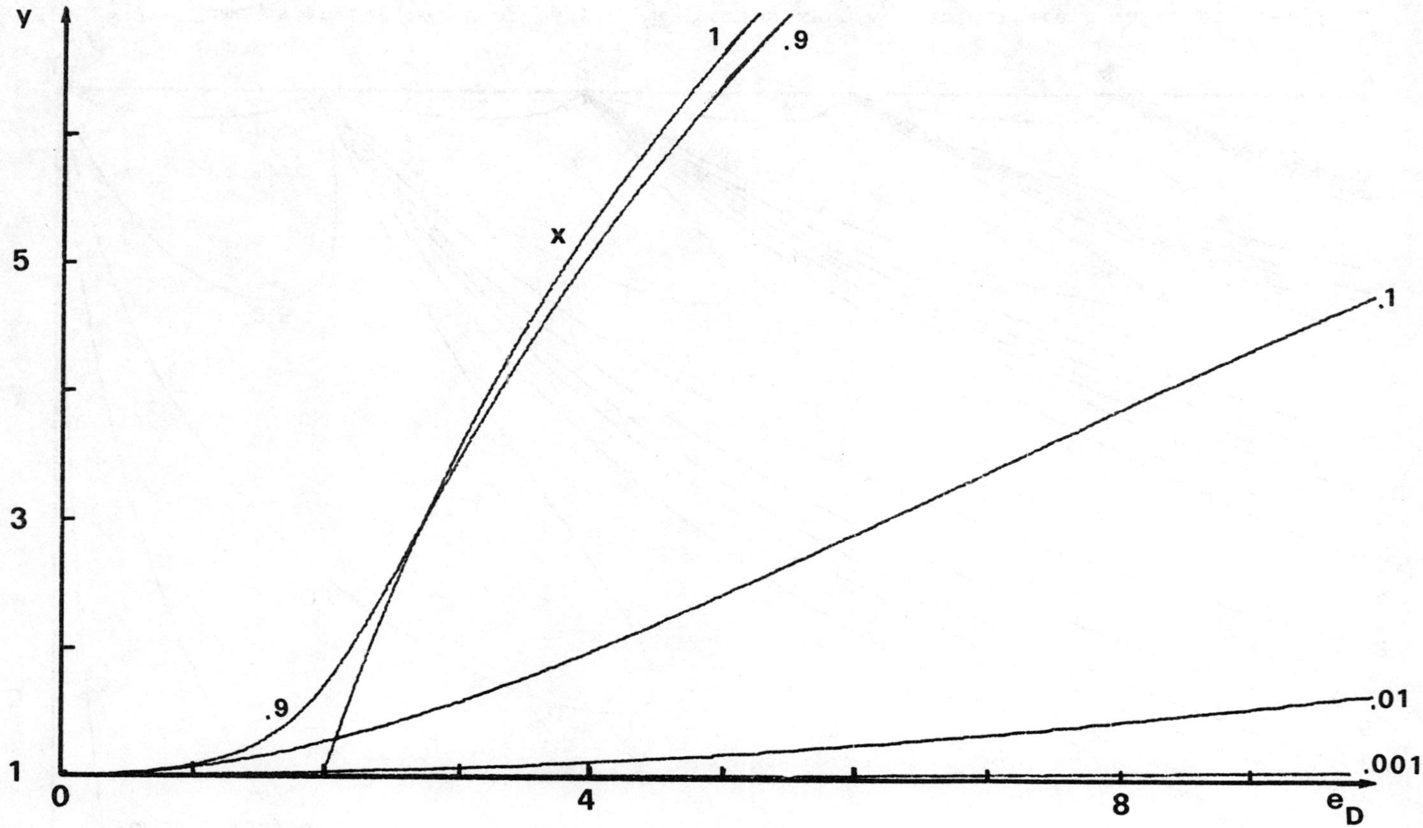

Figure 8 - Same as in Fig. 7. Plot of y versus e_D, at different values of $0 < x \leqslant 1$

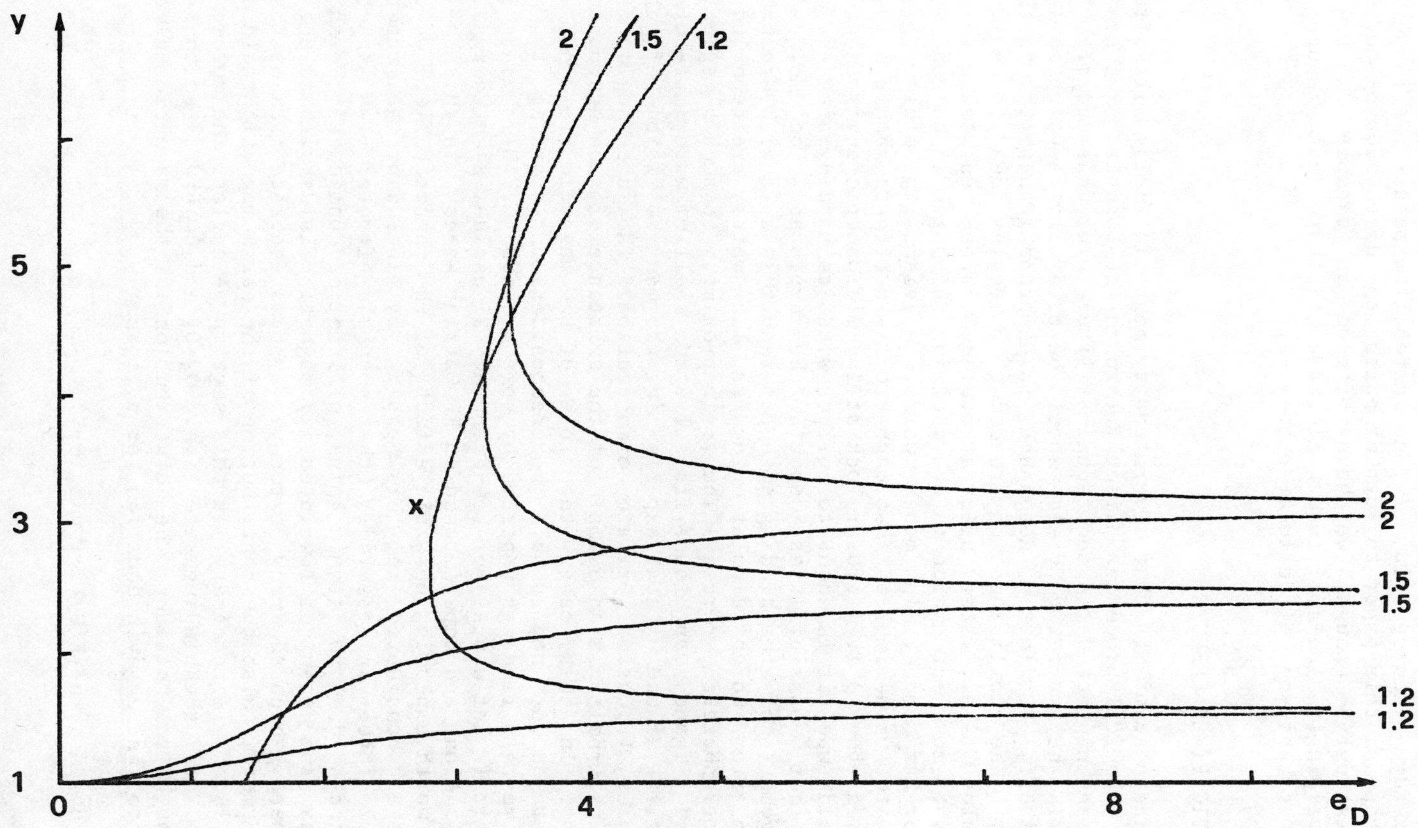

Figure 9 - Same as in Fig. 7. Plot of y versus e_D, at different values of $1 < x \leqslant 2$

becomes a single-valued monotonic function, as shown in Fig.2, and, as already discussed in [1], no instability occurs. It is beyond the scope of this paper to further pursue the discussion of eq.(6). Interestingly enough, we have offered an example of how a time-varying, low-amplitude electric exposure can induce an unstable behaviour of cellular processes.

MAGNETIC FIELD EFFECTS

The possibility that low-intensity magnetic fields will affect electrochemical processes directly through the Lorentz force has so far been neglected in the technical literature. Consequently, the role of the magnetic field should be carefully reevaluated (see A.R.Liboff, "Cyclotron Resonance in Membrane Transport", in this book). The collisions on reacting molecules due to solvent molecules (i.e., water) offer an average viscous resistance to the reactant movement which is overwhelming the magnetic component of the Lorentz force. In other words, a viscosity close to the bulk-water value implies that magnetic field effects should only be observed above 1 T. A closer look at the microscopic environment of a ligand when it is about to collide with its binding site reveals that other possibilities should be explored. The molecular arrangement of the binding site can result in a local hydrophobic region from which dipolar molecules, like water, are repelled by dielectrophoretic forces. Then, in proximity to such a region, a charged ligand can loose its solvation shell of water molecules, and drift around without experiencing too many collisions of water molecules (F.Oosawa, private communication). The inner region of a protein α-helix is an example of the existence of such hydrophobic regions. In this case, the role of the magnetic field per se, whether dc or ac, should be reconsidered. According to the experimental findings reported by some authors (C.F.Blackman, "The Biological Influences of Low-Frequency Sinusoidal Electromagnetic Signals Alone and Superimposed on RF Carrier Waves", in this book), it is worth exploring possible effects of the magnetic field on the processes considered in this paper, by including the magnetic Lorentz force in eqs.(4). Other possible effects, like magnetic resonances due to atomic spins, will not be considered. We shall discuss the case of low-intensity magnetic fields with dc and ac components because many exposure systems generate both of them. Furthermore, also the contribution of the earth magnetic field can be included, if necessary. In this section we neglect the movement, if any, of the binding site (i.e., $\bar{\delta}_S=0$, and $\delta_{LS}=|\bar{\delta}_L|$) so that it is enough to consider the contribution of the magnetic Lorentz force to the Langevin equation for a ligand:

$$(9) \quad d\bar{v}/d\theta+\beta\bar{v}=\bar{n}+\bar{\gamma}E(\theta-\tau)+\gamma_B\bar{v}\wedge\bar{B}_b(\theta-\tau)$$

where γ_B is the ratio of the charge Q to the mass m of the ligand, and $\bar{B}_t$ is the total magnetic induction vector. The subfix L has been dropped for the sake of simplicity. The dc magnetic component is $\bar{B}_o$ and we assume a sinusoidal ac magnetic component $\bar{B}_1 \sin[\omega_o(\theta-\tau)]$, so that $\bar{B}_t=\bar{B}_o+\bar{B}_1 \sin[\omega_o(\theta-\tau)]$. A study of eq.(9) / to evaluate $\delta_{LS}^2(t_A)=R_A^2$ would be rather cumbersome so that we restrict our analysis to the discussion of the solution of eq.(9) in the case $\bar{n}=0$ and $\tau=0$.

Eq.(9) is a linear vectorial equation with time-varying coefficients. We introduce the dc and ac cyclotron frequencies $f_{c,o}=\gamma_B B_o/(2\pi)$ and $f_{c,1}=\gamma_B B_1/(2\pi)$, corresponding to the dc and ac magnetic fields. Furthermore, we assume that both the dc and ac magnetic fields are aligned with the z axis, i.e. $\bar{B}_t=[B_o+B_1 \sin(\omega_o\theta)]\bar{1}_z$, where $\bar{1}_z$ is the unit z vector. The electric field is induced by the time variation of the magnetic field, so that $\bar{\gamma}\cdot\bar{1}_z=0$. Accordingly we let

$$\bar{\gamma}E(\theta)=(\alpha_x\bar{1}_x+\alpha_y\bar{1}_y)f_o f_{c,1}\cos(\omega_o t) \tag{10}$$

where $\bar{1}_{x(y)}$ is the unit x(y) vector, $\alpha_{x(y)}$ is a suitable constant, and $f_o=\omega_o/(2\pi)$.

The eigenfunctions of the homogeneous equation associated to (9) are:

$$\exp(-\beta\theta)\,[S(\theta)\bar{1}_x+C(\theta)\bar{1}_y] \tag{11}$$

$$\exp(-\beta\theta)\,[C(\theta)\bar{1}_x-S(\theta)\bar{1}_y] \tag{12}$$

where it is

$$S(\theta)=\sin\{2\pi f_{c,o}\theta+(f_{c,1}/f_o)[1-\cos(\omega_o\theta)]\} \tag{13}$$

$$C(\theta)=\cos\{2\pi f_{c,o}\theta+(f_{c,1}/f_o)[1-\cos(\omega_o\theta)]\} \tag{14}$$

A solution of eq.(9) in the case $\alpha_y=0$ can be obtained from eqs.(11-14) in the following form:

$$\frac{\bar{v}\cdot\bar{1}_x}{\alpha_x f_o f_{c,1}}=\exp(-\beta\theta)\left\{S(\theta)\int_0^\theta \exp(\beta t)S(t)\cos(\omega_o t)dt+ +C(\theta)\int_0^\theta \exp(\beta t)C(t)\cos(\omega_o t)dt\right\} \tag{15}$$

$$\frac{\bar{v}\cdot\bar{1}_y}{\alpha_x f_o f_{c,1}}=\exp(-\beta\theta)\left\{C(\theta)\int_0^\theta \exp(\beta t)S(t)\cos(\omega_o t)dt+ +S(\theta)\int_0^\theta \exp(\beta t)C(t)\cos(\omega_o t)dt\right\} \tag{16}$$

We choose, as a practical criterion of biological significance, the onset of a dc value of $\bar{v}\cdot\bar{i}_x$ and/or $\bar{v}\cdot\bar{i}_y$ sustained by the sinusoidal em perturbation. In fact such a dc velocity components should affect the macroscopic dissociation and association rate constants of the binding process of the ligand. After evaluating the time average:

$$(17)\quad \langle\bar{v}\rangle = \lim_{T\to\infty}\int_0^T \bar{v}\,dt$$

we conclude that the time average of the velocity can be easily computed if the dc cyclotron frequency is a multiple of the em field frequency (resonance condition):

$$(18)\quad f_{c,o} = n\, f_o$$

After some computations, we obtain

$$(19)\quad \langle\bar{v}\cdot\bar{i}_x\rangle \cong (\alpha_x/\beta) f_o\, f_{c,1}\left[\langle S\cos(\omega_o t)\rangle\langle S\rangle + \right.$$

$$\left. + \langle C\cos(\omega_o t)\rangle\langle C\rangle\right]$$

$$(20)\quad \langle\bar{v}\cdot\bar{i}_y\rangle \cong (\alpha_x/\beta) f_o\, f_{c,1}\left[\langle S\cos(\omega_o t)\rangle\langle C\rangle - \right.$$

$$\left. - \langle C\cos(\omega_o t)\rangle\langle S\rangle\right]$$

in the limit $\beta/f_o<1$. Similar (dual) expressions holds if one lets $\alpha_x=0$ and $\alpha_y\neq0$. The evaluation of eqs.(19,20) in the general case ($\alpha_x\neq0$, $\alpha_y\neq0$) leads to the following result:

$$(21)\quad \langle\bar{v}\cdot\bar{i}_{x(y)}\rangle \cong (\alpha_{y(x)}/\beta) f_o\, f_{c,1}\, J_n(f_{c,1}/f_o)\times$$

$$\times\left[J_{n-1}(f_{c,1}/f_o) - J_{n+1}(f_{c,1}/f_o)\right]/2$$

where $J_n(\cdot)$ is the Bessel function of order n-th. Eqs.(21) prove that an ac magnetic field can induce a dc velocity component of the ligand, irrespective of the electric field being purely ac. This result offers, for the first time, an explanation of the possible role played by an ac magnetic field on a cell process. The multiple maxima (zeros) of the absolute value of

$$(22)\quad \mathcal{J}_n = \left[f_o\, f_{c,1}\, J_n(f_{c,1}/f_o)\right]\left[J_{n-1}(f_{c,1}/f_o) - J_{n+1}(f_{c,1}/f_o)\right]/2$$

give the relationships among f_o, B_o and B_1 in order to (not) observe a putative effect of the em field at resonance (eq.18). The behaviour of the function (22) versus $X=f_{c,1}/f_o$ is shown in Fig.10 for n=0,1,2,3, in the case $f_o\, f_{c,1}$=const., which should apply to the quoted results of C.Blackmann. As an example, the cyclotron resonance of Ca++ at $B_o=38\mu T$ is about $f_{c,o}$=29.4Hz. Then an em field

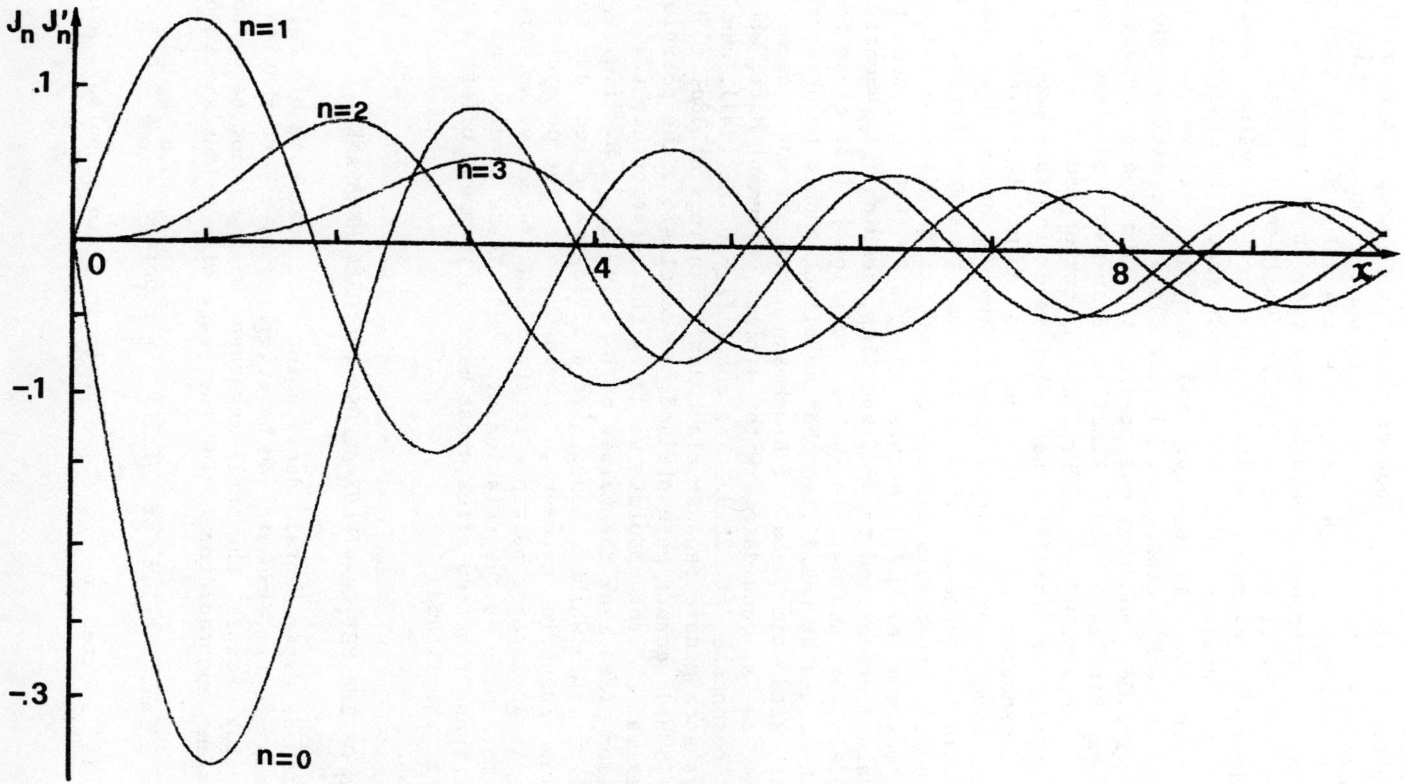

Figure 10 - Plot of $\mathfrak{J}_n/(f_o f_{C,1}) = J_n(\mathcal{X}) J'_n(\mathcal{X})$ versus $\mathcal{X} = f_{C,1}/f_o$, in the case of sinusoidal electric and magnetic exposures, being $J'_n = (J_{n-1} - J_{n+1})/2$

of f =30Hz is about at resonance for n=1 (see eq. 18). The first zero of $|J_1|$ occurs at X=1.85 (Fig. 10). We offer such a condition as an explanation why Blackmann did not observe any effect. Then, if f_o is halved to 15Hz, $f_{c,1}$ doubles in order to keep $f_o f_{c,1}$=const., so that $f_{c,1}/f_o$=4×1.85=7.4 and n=2. In such a condition, $|J_2|$ is close to a maximum, which occurs at the value X=7.53. Then we predict that a response should be observed, as Blackmann did. At B_o=19µT and f =15Hz it is n=1 and $f_{c,1}/f_o$=7.4. In such a condition, $|J_1|$ is close to a zero, which occurs at the value X=7.2. Accordingly we predict that no response should be observed, as Blackmann did not. At B_o=76µT and f =30Hz, it is n=2 and $f_{c,1}/f_o$=1.85. In such a condition, $|J_2|$ is close to a maximum, which occurs at X=2.06, so that we predict that a response should be observed, as Blackmann did. Finally, Blackmann observed some response for B_o=25.3µT at 30Hz, which corresponds to a non resonance condition (n=2/3), not discussed in this paper. The cyclotron resonance of K+ at 76µT is about $f_{c,o}$=29.8Hz. In Blackmann's experiment, a larger Ca++ efflux was observed at 30Hz, so that A.Liboff suggests that potassium ions are involved. We tested Liboff's hypothesis against our model. In this case n=1 and the first maximum of $|J_1|$ occurs at X=0.91 (Fig. 10). When the previous values are halved to 38µT and 15Hz, keeping $f_o f_{c,1}$=const., $f_{c,1}/f_o$=4×0.91=3.64 and n=1. In such a condition $|J_1|$ is close to a zero, which occurs at x=3.76, so that no effect should be observed, in contrast with the data of Blackmann. Then our model seems to exclude K+ as a candidate. With similar argument Mg++, whose cyclotron resonance at 25.3µT is about $f_{c,o}$=30.7Hz (n=1), can be excluded as a candidate because Blackmann's findings at 38µT (15Hz) and 76µT (30Hz) cannot be explained. In conclusion, the positions of the maxima of the absolute value of the relationship (21) and the condition (18) link the values of B_o (i.e. $f_{c,o}$), B_1 (i.e. $f_{c,1}$) and ω_o (i.e. f_o) which should give a biological effect due to a time-varying magnetic exposure, provided that the dc cyclotron frequency is an integer multiple of the field frequency. No effect occurs at the zeros of the relationship (21). These predictions are valid for those reaction sites whose microenvironment offers a low viscosity to the ligand.

MODULATION OF THE NET CA++ FLUX ACROSS THE CELL MEMBRANE

Published experimental data point out that some biological effects due to em exposure can be triggered by a perturbation of the Ca++ flux across the cell membrane. On the other hand, some results seem contradictory. In the case of a 6-16Hz sinusoidal exposure, a net influx or efflux is reported [5,6]. A bimodal response is reported in [7], for a periodic train of pulses. A net influx is reported for 500Hz periodic sequences of pulses [8].

Finally, a net influx is postulated in [9,10,11,12].

In this paper, in order to explore how a low-frequency electric field can affect the Ca++ flux across the cell membrane, a simplified model is proposed (Fig.11). Only the onset of an electrically induced perturbation is considered.

On the outer aspect of the membrane, we consider a compartment of positive counter ions, σ_e, opposed to the negative charges, S_ε, of the glycocalices of integral proteins, which protrude outside the layer of phospholipid heads. The external negative charges of the ionized membrane lipids are neglected, in accordance with available data on erythrocytes [13]. Polyelectrolyte theory predicts that divalent cations are much more effective than monovalent cations in neutralizing these negative charges, so that we postulate that the compartment σ_e is flooded by Ca++ ions (M.Parodi et al., "Ion Interactions with DNA: Results Based on Polyelectrolytes Theory", in this book), which can reversibly bind to the membrane negative sites by a dissociation rate constant K_ε^-.

The passive transport of Ca++ across the membrane is divided into two steps:

i) Ca++ ions approach the passive channel mouths, whose populations of free Ca++ ions are σ_μ on the external side and σ_η on the internal one. The number of available sites in the external (internal) mouths is S_μ (S_η).

ii) Ca++ ions are transported across the membrane through a passive transport mechanism ϕ_m [14].

On the inner aspect of the membrane, we postulate a compartment, σ_i, of free Ca++ ions, which can bind to the negative charges of ionized membrane lipids, S_I. The active transport of Ca++ across the membrane is modelled by Ca-Mg ATPase pumps. The pump-efflux term has been expressed as:

$$(23) \quad A_m = a_m K_e^- \sigma_i / (\sigma_i + K_i^- / K_i^+)$$

where a_m is a suitable positive constant, K_e^- is the release rate constant of Ca++ ions on the outer sides of the pumps, and K_i^- (K_i^+) is the dissociation (association) rate constant of Ca++ ions at the internal surfaces of the pumps. This expression is in agreement with [15]. The concentration of Mg++ ions, which are dominant in the inner compartment, is assumed to remain unchanged at the onset of the perturbation of Ca++ ions. Fig.11 summarizes the model features considered in this section. The frequency value of the induced electric field is low enough so that it only affects the external processes, i.e. the rate constants of the binding process (K_ε^-), of the channel-mouth escaping process (K_μ^-), and of the

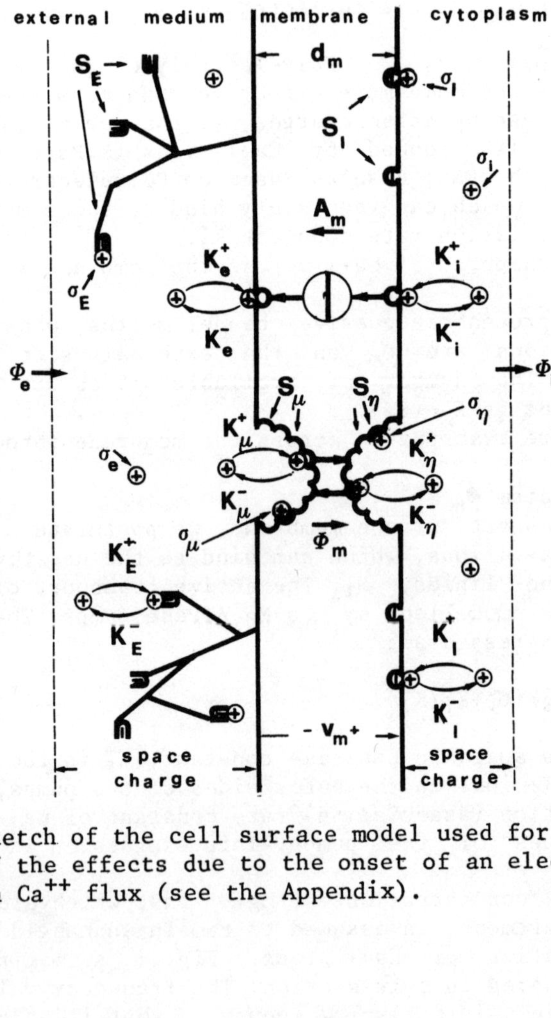

Figure 11 - Sketch of the cell surface model used for the evaluation of the effects due to the onset of an electric exposure on Ca^{++} flux (see the Appendix).

pump-release process (K_e^-). The theory developed in the previous section allows us to evaluate the dependence of any rate constant K_j^- ($j=E,\mu,e$) on the electric field. We have already proved that the relationship between K_j^- and the amplitude of the field harmonics only depends on two parameters, i.e., on the value $K_{j,o}^-$ of K_j^- in the absence of any em exposure, and on the critical value E_j^*, which is defined for each process j. From eq.(6), letting $y_j=K_j^-/K_{j,o}^-$, we obtain :

$$(24)\qquad \Delta y_j = y_j - 1 \geq 0 \qquad (j=E,\mu,e))$$

irrespective of the process j considered and of the field waveform.
By solving all the balance equations for the various compartments and the transport equations used in our model, we can evaluate the resulting net change in the Ca++ flux at the onset of the em exposure, i.e., after the initial em transient has ended but before large changes in the membrane voltage and in the internal free Ca++ and Mg++ concentrations occur (see Appendix). We conclude that the net Ca++ influx $\phi_m - A_m$ is given by:

$$(25)\qquad \phi_m - A_m \cong \phi_{m,o}(c\Delta y_\mu - \Delta y_e)$$

where $\phi_{m,o}$ is the steady value common to ϕ_m and A_m prior to the onset of the em exposure, c is a suitable constant (which is always positive irrespective of the parameter values used in the model) and $\Delta y_\mu = \Delta(K_\mu^-/K_{\mu,o}^-)$, $\Delta y_e = \Delta(K_e^-/K_{e,o}^-)$ are the relative changes in the rate constants corresponding to the channel-mouth escaping process and to the pump-release process. The result does not depend on the field induced change in K_E^-.

We have proved that an electric field can modulate the cell Ca++ flux through the proposed mechanism. A net Ca++ influx occurs if the field-induced relative change in the dissociation rate constant which models the outer mouth-escaping process is dominant with respect to the pump-release process ($c\Delta y_\mu > \Delta y_e$). A net efflux occurs in the opposite case ($c\Delta y_\mu < \Delta y_e$). These findings may account for the contradictory findings obtained by different researchers because of the different signal waveforms and types of cells used in their experiments. As an application of these results, we in the next section we further speculate about their implications on lymphocyte reactivation by lectins (PHA) [1].

LYMPHOCYTE REACTIVATION

The application of eq.(6) to the case of lymphocyte reactivation by means of PHA has been extensively discussed in [1]. The ratio G between the number of receptor-encounters linked by PHA and the number of spontaneous encounters in the absence of PHA has been evaluated in [1]:

$$(26) \quad G=1+(4/3)\pi R_A^3 C_L \exp(-yK_o^- t_D)$$

where C_L is the PHA concentration, and K_o^- is the inverse of the time t_A necessary for a lectin and encounter to move away, travelling a mean square distance equal to R_A, in the absence of em exposure. The dissociation rate constant K_o^- has been labelled r_-^o in [1].

The underlying working hypothesis is that the initial reactivation event is an increase in the influx of Ca++ ions across the cell membrane. The time interval necessary for an elementary reactivation event to occur is t_D, so that only the receptor couples which remain close for a longer time than t_D are assumed to be effective in promoting the Ca++ influx.

The relationship in eq.(26) is qualitatively plotted versus C_L in the second quadrant (clockwise) of Fig.12, for zero field (y=1) and for low-frequency electric exposure (y>1). A saturation effect at intermediate PHA doses is shown in Fig.12, whereas high PHA doses cause the formation of large PHA-receptor clusters, which are less effective in inducing a Ca++ influx. The net influx of Ca++ ions across the cell membrane is shown in the third quadrant of Fig.12, for controls and cells under em exposure. In this example, we assume that the em field per se increases the net Ca++ influx, in agreement with the experimental results reported by Grattarola et al. ("Electromagnetic Field Effects on Phytohemagglutinin (PHA) Induced Lymphocyte Reactivation", in this book). Assuming the Ca++ influx to be responsible for lymphocyte reactivation, the number of mitotic events should be a monotonically increasing function of the Ca++ net influx (fourth quadrant). The first quadrant of Fig.12 shows the plots of the resulting number of mitotic events versus C_L, with and without em exposure.

The final result (first quadrant of Fig.12) is quite interesting, irrespective of being qualitative. It allows us to predict that at very low doses of lectin, the effect of em exposure will be synergistic with PHA. At low doses, the opposite effect will occur, whereas at intermediate and high doses the em exposure will enhance the mitogenic action of PHA again.

CONCLUSIONS

The drift force exerted by an ac electric field on the movement of a charged ligand can affect the rate at which the ligand binds to the cell membrane. The corresponding dissociation constant assumes a value which is always larger than the value at zero exposure. Such a value may become a multivalued function of

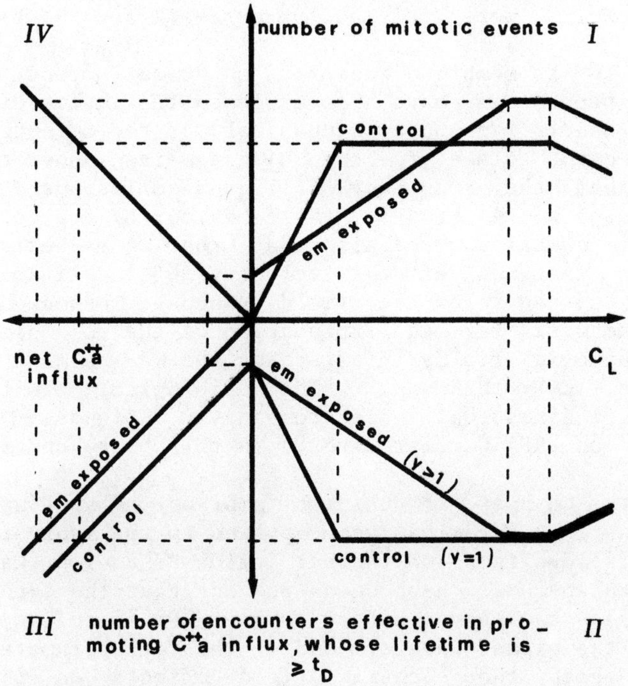

Figure 12 - Qualitative relationship between lymphocyte reactivation (number of mitotic events) versus PHA concentration (C_L) under low frequency low amplitude em exposure. The em exposure affects the number of encounters formed by those PHA receptors which are effective in promoting Ca^{++} influx and whose lifetime is longer than t_D (minimum time interval necessary to trigger an elementary reactivation event). The em exposure affects also the net Ca^{++} influx through the PHA-encounter aggregates and the Ca^{++} pumps.

the electric field parameters so that an unstable behaviour is likely to occur. For each type of ligand, a critical field intensity can be defined, above which strong field effects can be predicted. This threshold electric field is roughly proportional to the product of the effective local diffusivity and viscosity experienced by the ligand in proximity to its binding site. If the molecular arrangement of the binding site is such that both low effective diffusivity and low effective viscosity are achieved, the critical field can be less than a few mV/cm.

Such results were previously derived for the binding of a lectin (PHA) to its membrane receptors on human lymphocytes [1]. In the present paper, we have also explored the implications of the magnetic components (dc and ac sinusoidal) of the Lorentz force on the ligand motion close to the binding site, under the working hypothesis that the frequency of ligand collisions due to water molecules is low. A study of the behaviour of the corresponding Langevin-like equations reveals that the ac magnetic field may induce a dc response of the velocity of the ligand, which is drifted by the ac electric field. Such a behaviour offers a possible framework for the explanation of the role played by low intensity magnetic fields in affecting the Ca++ efflux from brain tissue (see C.F.Blackmann, "The Biological Influences of Low-Frequency Sinusoidal Electromagnetic Signals Alone and Superimposed on RF Carrier Waves", in this book) and in inducing morphological changes in chicken embryos [16]. The approach developed in this paper has been applied to the evaluation of the electrical field effects on the net influx of an ion (Ca++) across the cell membrane. It has been proved that the exposure to an electric field can give either a net Ca++ influx or efflux, depending on the signal waveform and on the cell parameters. Taking into account both the electric field effects on the receptor encounters and the Ca++ influx across the lymphocytes membrane, we are able to anticipate, in a qualitative fashion, the extent of lymphocyte reactivation versus PHA concentration. An example of such prediction is given which seems to be in agreement with the data reported by Grattarola et al. (see "Electromagnetic Field Effects on Phytohemagglutinin (PHA) Induced Lymphocyte Reactivation", in this book).

APPENDIX

The cell-surface model considered in this paper in order to evaluate the effects of the onset of the electric exposure on the net Ca++ flux across the cell membrane is depicted in Fig. 11. The order of magnitude of the cell surface area is 150μm . On the external side, the negative binding sites, S_E, are due to the membrane glycoproteins, which are embedded in the membrane and protrude, their ionized glycocalices into the external medium ($S_\varepsilon \cong 2\ 10^7$ per cell). The phospholipid heads pointing at the cell outside bear negligible charges. This compartment of negatively-charged binding sites attracts positive counterions. Application of polyelectrolyte theory (see M.Parodi et al., in this book) allows us to conclude that, at physiological levels, divalent cations (mainly Ca++) are the most effective in neutralizing the negative sites. On this basis, we introduce a compartment σ_e of free Ca++ ions (Mg++ ions are neglected) which can bind to the negative sites of the glycoproteins via the dissociation and association rate constants, K_ε^- and $K_\varepsilon^+ = a_\varepsilon K_\varepsilon^-$ (see eq.(3)). The compartment of Ca++ ions bound to the glycoproteins is labelled σ_ε. We further assume that about 28% of the membrane phospholipid heads pointing at the cell inside are negatively ionized. We introduce a compartment of free divalent ions, mainly due to Mg++ ions. Some of these ions bind to the inner negative sites. In any case, we assume that the concentration of Mg++ ions remains constant at the onset of the electric exposure. Irrespective of the fact that number of free (σ_i) and bound (σ_I) Ca++ ions is much smaller than the the number of Mg++ ions, the contribution of Ca++ ions is included in the model because it is linked to the external Ca++ compartments via the passive Ca++ channels (ϕ_m) and the active ATPase pumps (A_m). We label S_I the compartment of negative phospholipid heads which can be neutralized by Ca++ ions through a binding process characterized by the dissociation and association rate constants, K_I^- and $K_I^+ = a_I K_I^-$.

The free Ca++ ions, σ_e and σ_i, are allowed to randomly hit the external and internal mouths of the passive channels, which can accomodate S_μ and S_η ions, respectively. The Ca++ populations inside the external and internal mouths are labelled σ_μ and σ_η, respectively. The free Ca++ ions leave or enter the mouths at a rate which depends on the rate constants K_μ^- (K_η^-) and $K_\mu^+ = a_\mu K_\mu^-$ ($K_\eta^+ = a_\eta K_\eta^-$).

The transport mechanism of Ca++ ions through the passive channels can be modelled according to Eyring's theory, thereby obtaining $\phi_m = \phi_m(v_m, \sigma_\mu, \sigma_\eta)$, where v_m is the voltage drop across the channel. Because of the assumption used in Fig.11, it is :

(A1) $\partial\phi_m/\partial v_m < 0 \quad ; \quad \partial\phi_m/\partial\sigma_\mu > 0 \quad ; \quad \partial\phi_m/\partial\sigma_\eta < 0$

for small variations close to the physiological steady state.

The pump-efflux term A_m is given by the relationship in eq. (23):

(A2) $A_m = a_m K_e^- \sigma_i / (\sigma_i + a_i)$

where K_e^- ($K_e^+ = a_e K_e^-$) is the release (association) rate constant of free Ca++ ions on the outer side of a pump, and K_i^- ($K_i^+ = a_i K_i^-$) is the release (association) rate constant of free Ca++ ions on the inner side of a pump. This expression gives the closed-form dependence of A_m on K_E^-, and is in agreement with [15]. In the absence of em exposure, we label each variable by the subfix "o". In this case (physiological steady state), it is $\phi_{m,o}=A_{m,o}$, i.e., the net Ca++ flux is zero.

At the onset of the em exposure, it is reasonable to assume $v_m \cong v_{m,o}$, because the voltage drop across the membrane is also controlled by all the other ions close to the cell surface. Accordingly, the overall charges in the two space charge regions in Fig.11 should remain constant. Then:

(A3) $2(\sigma_e+\sigma_\mu+\sigma_E)-S_E=\text{const.}$

(A4) $2(\sigma_i+\sigma_\eta+\sigma_I)-S_I=\text{const.}$

Finally, the net influx of Ca++ ions from the outside is labelled ϕ_e, and the net influx of Ca++ ions into the cytoplasm is labelled ϕ_i.

The balance equations for each compartment are:

(A5) $d\sigma_E/dt=K_E^-[a_E\sigma_e(S_E-\sigma_E)-\sigma_E]$

(A6) $d\sigma_I/dt=K_I^-[a_I\sigma_i(S_I-\sigma_I)-\sigma_I]$

(A7) $d\sigma_\mu/dt=-\phi_m+K_\mu^-[a_\mu\sigma_e(S_\mu-\sigma_\mu)-\sigma_\mu]$

(A8) $d\sigma_\eta/dt=\phi_m+K_\eta^-[a_\eta\sigma_i(S_\eta-\sigma_\eta)-\sigma_\eta]$

(A9) $d(\sigma_e+\sigma_\mu+\sigma_E)/dt=\phi_e+A_m-\phi_m$

(A10) $d(\sigma_i+\sigma_\eta+\sigma_I)/dt=-\phi_i-A_m+\phi_m$

Due to the neutrality conditions in (A3) and (A4), eqs.(A9) and (A10) become

(A11) $\phi_m-A_m=\phi_e=\phi_i$

We can evaluate the resulting net change in the Ca++-flux region (ϕ_m-A_m) at the onset of the electric exposure, i.e., after the initial electrochemical transient is over, but before large changes in v_m and in the concentration of the other ions may occur.

The frequency content of the externally induced periodic ac electric field is low enough to affect the random walk of the Ca++ ions only in the outer space charge region, thus affecting the rate constants of the binding process to the glycoprotein sites (K_E^-), of the channel-mouth escaping process (K_μ^-) and of the pump release process (K_e^-). In other words, it is $K_I^-=K_{I,o}^-$, $K_\eta^-=K_{\eta,o}^-$, $K_i^-=K_{i,o}^-$.

The theory developed in the first section allows us to conclude that, if K_j^- is the value of any of the rate constants (j=E,μ,e),

$K^-_{j,o}$ is the value in the absence of exposure, and $y_j = K_j^-/K^-_{j,o}$, it is $\Delta y = y - 1 > 0$, irrespective of the process considered ($j=E,\mu,e$) and of the signal waveform.

According to our definition for the onset of the electric exposure, we can evaluate the changes induced by Δy_j ($j=E,\mu,e$) in the variables which appear in eqs.(A2,A5,A6,A7,A8,A11) by letting d /dt≅0 in eqs.(A5,A6,A7,A8).
After some cumbersome computations, it is possible to prove that, for small variations in the steady state conditions, it is:

$$\phi_m - A_m \cong \phi_{m,o}(c\Delta y_\mu - \Delta y_e)$$

where c is a suitable constant which is always positive irrespective of the parameter values used in the model, and $\Delta y_\mu = \Delta(K^-_\mu/K^-_{\mu,o})$ and $\Delta y_e = \Delta(K^-_e/K^-_{e,o})$ are the relative changes in the rate constants corresponding to the mouth-escaping process and to the pump-release process. This result does not depend on the field-induced change in $K_{\bar{E}}$. It has been obtained using the inequalities in (A1) and observing that it is always $K^-_\eta \sigma_{\eta,o} > \phi_m$.

REFERENCES

1. A.Chiabrera, M.Grattarola and R.Viviani. Interaction between electromagnetic fields and cells: Microelectrophoretic effect of ligands and surface receptors. Bioelectromagnetics, 5:173 (1984).
2. F.Caratozzolo, A.Chiabrera, G.Giannetti, M.Grattarola, M.Parodi and A.Raveane. Perturbation of calcium flux across the cell membrane induced by low-frequency em exposure. Proceedings URSI XXIst General Assembly, Florence, Italy,23 (1984).
3. A.Chiabrera and G.A.Rodan. The effects of electromagnetic fields on receptor-ligand interaction: A theoretical analysis. J. Bioelectricity, submitted, (1985).
4. G.I.Bell. Models for the specific adhesion of cells to cells. Science, 200:618 (1978).
5. S.M.Bawin and W.R.Adey. Sensitivity of calcium binding in cerebral tissue to weak environmental electric fields oscillating at low frequency. Proc.Natl.Acad.Sci. U.S.A., 73:1999 (1976).
6. C.F.Blackmann, S.G.Benane, L.S.Kinney, W.T.Joines and D.E.House. Effects of ELF fields on calcium-ion efflux from brain tissue in Vitro. Radiat.Res., 92:510 (1982).
7. G.Colacicco and A.A.Pilla. Chemical, physical and biological correlations in the Ca-uptake by embryonal chick tibia in vitro. Bioelectrochem. and Bioenergetics, 10:119 (1983).
8. R.Dixey and G.Rein. H-noradrenaline release potentiated in a clonal nerve cell line by low intensity pulsed magnetic fields. Nature, 296:253 (1982).
9. A.Chiabrera, M.Hinsenkamp, A.A.Pilla, J.Ryaby, D.Ponta,

A.Belmont, F.Beltrame, M.Grattarola, and C.Nicolini. Cytofluorometry of electromagnetically controlled cell dedifferentiation. J.Histochem.Cytochem., 27:375 (1979).
10. A.Chiabrera, R.Viviani, G.Parodi, G.Vernazza, M.Hinsenkamp, A.A.Pilla, J.Ryaby, F.Beltrame, M.Grattarola and C.Nicolini. Automated cytometry of electrochemically exposed frog erytrocytes. Cytometry, 1:42 (1980).
11. M.Grattarola, R.Viviani and A.Chiabrera. Modelling of the perturbation induced by low-frequency electromagnetic fields on the membrane receptors of stimulated human lymphocytes. I. Influence of the fields on the system free energy. Studia Biophysica, 91:117, (1982).
12. A.Chiabrera, M.Grattarola, R.Viviani and C.Braccini. Modelling of the perturbation induced by low-frequency electromagnetic fields on the membrane receptors of stimulated human lymphocytes. II. Influence of the field on mean lifetimes of the aggregation process. Studia Biophysica, 91:125, (1982).
13. R.Heinrich, M.Gaestel and R.Glaser. The electric potential profile across the erythrocyte membrane. J.Theor.Biol., 96:211 (1982).
14. H.Eyring. The theory of absolute reaction rates. Trans.Faraday Soc., 34:41 (1938).
15. H.G.Ferreira and V.L.Lew. Use of ionophore A23187 to measure cytoplasmic Ca buffering and activation of the Ca pump by internal Ca. Nature, 259:47 (1976).
16. A.Ubeda, J.Leal, M.A.Trillo, M.A.Jimenez and J.M.R. Delgado. Pulse shape of magnetic fields influence chick embryogenesis. J.Anat., 137:513 (1983).

CYCLOTRON RESONANCE IN MEMBRANE TRANSPORT

A.R. Liboff

Department of Physics
Oakland University
Rochester, Michigan 48063

INTRODUCTION

Recently, Blackman et al.[1] reported an interesting set of experiments in which the efflux of Ca^{2+} from chick brain is enhanced for certain combinations of (a) monochromatic, low-frequency, linearly polarized magnetic fields and (b) static magnetic fields of the order of the local Earth's field. In some respects these results seem to imply a nuclear magnetic resonance effect, but order-of-magnitude calculations show that the frequencies involved are much too low.

By contrast, we feel these observations are highly suggestive of cyclotron resonance phenomenon. In the following, we shall explore this idea first reviewing the basic theory of cyclotron resonance, second, showing that this type of resonance is a distinct possibility in biological systems in geomagnetic fields, and finally, proposing a membrane channel model that helps to explain the existing experimental observations and provides a pathway for designing new experiments to test for cyclotron resonance in cell membranes.

CYCLOTRON RESONANCE FOR FREE PARTICLES

When a charged particle enters a uniform, static magnetic field $\vec{B}$, it experiences a (Lorentz) force at right angles to both the particle's velocity $\vec{v}$ and the field $\vec{B}$:

$$\vec{F} = q\vec{v} \times \vec{B} \qquad (1)$$

where q is the charge of the particle (Figure 1). Despite this force, the magnetic field does not transfer energy to the particle, as long

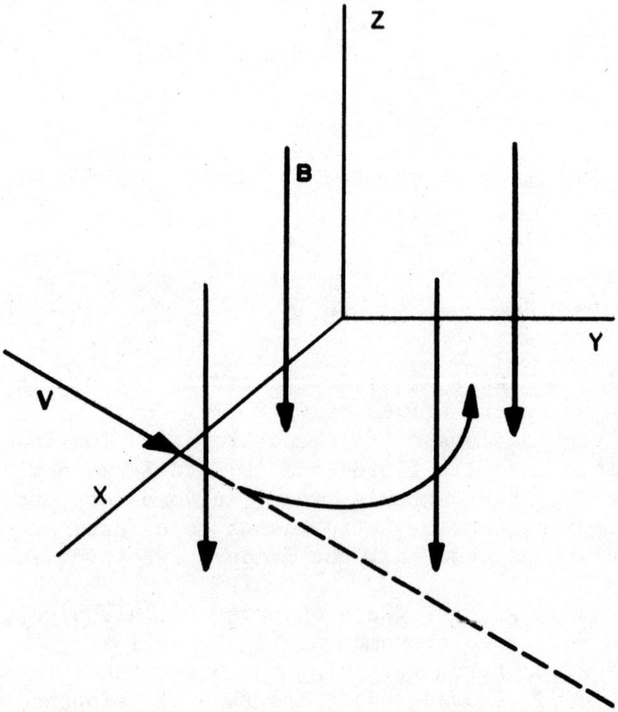

Figure 1 - Charged particle in magnetic field. A charged particle entering a uniform magnetic field will experience a force at right angles to its velocity and to the field. This force disappears when $\vec{v}$ and $\vec{B}$ are parallel.

as $\vec{B}$ is constant in time. When $\vec{v}$ and $\vec{B}$ are parallel, the force $\vec{F}$ vanishes. However, for all other orientations of the velocity, the charged particle will be forced to execute a curved path. Figure 2 shows two such paths, a circular one when $\vec{v}$ is originally at right angles to $\vec{B}$, and a helix when $\vec{v}$ is at some intermediate angle to the magnetic fields, greater than 0°, but less than 90°. For either path, the circle or the helix, one can calculate the forces acting on the charged particle (Figure 3). Balancing the centripetal force F_C against the Lorentz force F_L,

$$\frac{mv^2}{r} = qvB \ . \tag{2}$$

If the particle executes ν revolutions per unit time, we can write the velocity as

$$v = 2\pi r \cdot \nu \tag{3}$$

Substitution of this expression into Equation (1) yields the familiar condition for cyclotron resonance,

$$\nu = \frac{1}{2\pi}\frac{q}{m} B, \tag{4}$$

relating the gyrofrequency ν to only two parameters, the charge to mass ratio q/m and the static magnetic field B. Although this condition is in principle independent of other dynamical variables, boundary conditions normally impose limits on v and r.

RESONANCE CONDITIONS IN GEOMAGNETIC FIELD

It is rather instructive to probe the cyclotron resonance condition (Eq. 4) for those values of B corresponding to the geomagnetic field. The Earth's total field ranges from 0.25 gauss at the geomagnetic equator to 0.70 at the magnetic south pole (25-70 μT). In Figure 4 we plot the cyclotron resonance frequency against the charge-to-mass ratio, for different values of the magnetic field. Consider, in particular, those gyrofrequencies corresponding to B = 0.5 gauss, the Earth's field value that is often measured in the mid-latitudes. It is a remarkable fact that most of the singly and doubly charged ions of biological interest (Table I) have corresponding gyrofrequencies in the ELF range 10-100 Hz. We see, therefore, that the opportunity for cyclotron resonance effects in living systems in the geomagnetic field appears to be indeed favorable, especially at ELF frequencies.

The implication surrounding such frequencies being associated with different ionic species is that externally applied electromagnetic signals at frequencies close to a given resonance may couple to the corresponding ionic species in such a way as to selectively transfer energy to these ions and thus, indirectly, to the metabolic activities in which they are involved.

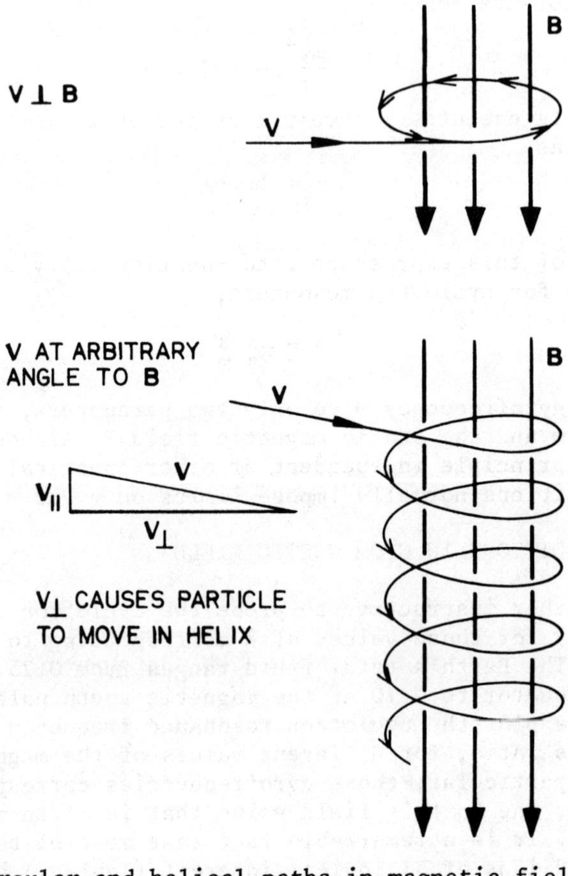

Figure 2 - Circular and helical paths in magnetic field. The angle between particle velocity and field direction determines type of path (in absence of any other field), a circle when $\vec{v}$ is normal to $\vec{B}$, a helix otherwise.

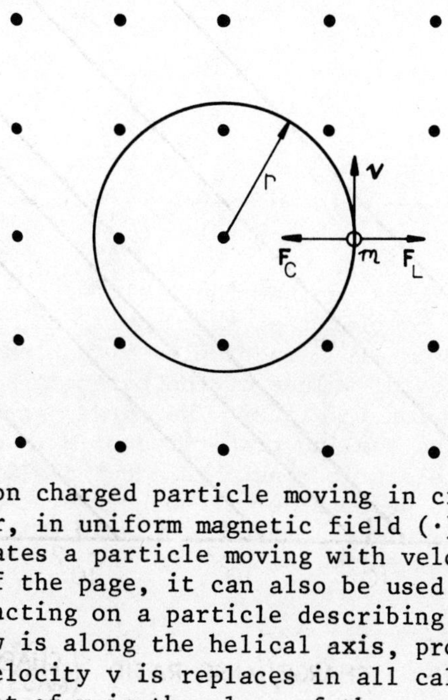

Figure 3 - Forces on charged particle moving in circular path of radius r, in uniform magnetic field (·). Although this illustrates a particle moving with velocity $\vec{v}$ in the plane of the page, it can also be used to describe the forces acting on a particle describing a helix, where the view is along the helical axis, providing that the total velocity v is replaces in all calculations by the component of v in the plane of the page, v_1.

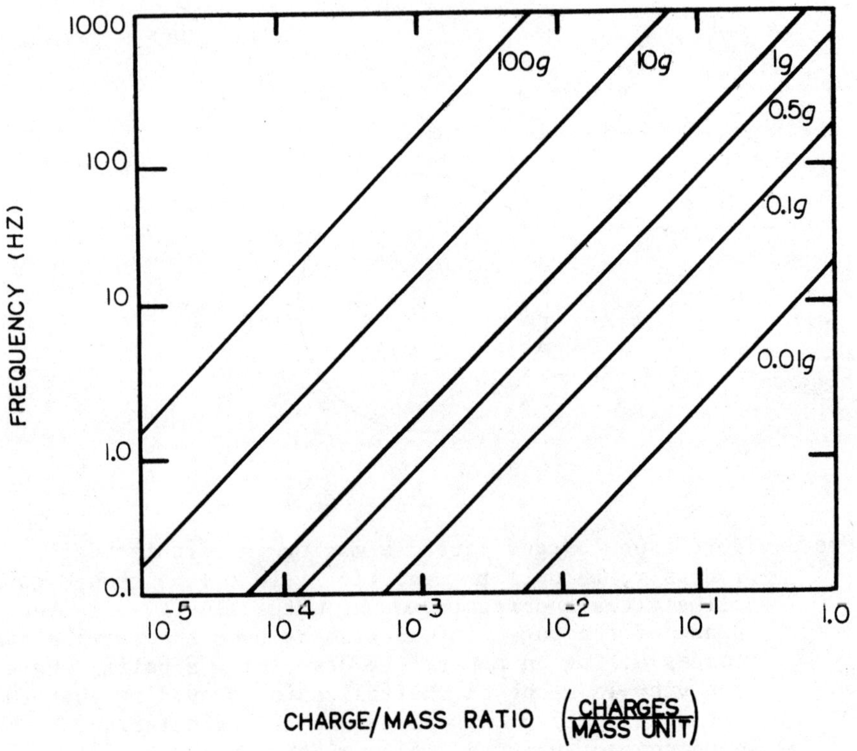

Figure 4 - Cyclotron resonance frequency plotted against the charge-to-mass ratio, for different values of the static field.

TABLE I

Calculated Cyclotron Resonances for selected free (unhydrated) ions of biological interest, assuming a geomagnetic field of 0.5 gauss.

IONIC SPECIES	q/m (in units of e/u)	CYCLOTRON RESONANCE FREQUENCY @ 0.5 g
H^+	0.99	760 Hz
Li^+	0.14	110
Mg^{2+}	0.082	61.5
Ca^{2+}	0.050	38.7
Na^+	0.043	33.3
Fe^{2+}	0.036	27.9
Zn^{2+}	0.031	23.4
K^+	0.026	19.6
Sr^{2+}	0.023	17.3
Rb^+	0.012	9.0

LIMITATIONS ON CELLULAR RESONANCE

Before examining the conditions governing such energy transfer, let us list some of the limitations on cellular cyclotron resonance mechanisms.

(a) Ionic Hydration

The examples listed in Table I are for "naked" ions, free of any attached water. However, such unattached ions in the cellular environment may be somewhat rare. Addition of one or more water molecules to any of the species listed in Table I will have the effect of decreasing the charge-to-mass ratio, thereby also decreasing the gyrofrequency for a given magnetic field. For example, the resonant frequency of the singly hydrated Na^+ ion is reduced, at 0.5 gauss, from 33.3 Hz to 18.6 Hz. It is nevertheless conceivable that free ions may play key roles in certain compartments of the cell, particularly those involving charge transport in the nuclear, mitochondrial and plasma membranes.

(b) Ionic Scattering

In order to maintain the circular or helical paths shown in Figure 2, the charged particles in question must have mean free paths $\gtrsim 2\pi r$. The author's brother looked at this problem nearly twenty years ago[2] and came to the conclusion that sustained cyclotron resonance by ions in cells is highly unlikely, except in extreme fields ($\gtrsim 10^{13}$ gauss); this is determined by the ratio of the gyrofrequency to the collision frequency. More recently Polk[3] reexamined this problem showing that thermal noise fluctuations prevent coherent currents from forming in cells, except along the perimeters of large (> 100 μm) cells, or again, in very large fields. We argue that such noise limitations may not be applicable to ionic transport through channels. The ion selectivity of membrane channels tends to reduce the scattering by other ionic species, enzymes, proteins, etc. in a given channel compared to that within the cytoplasm, leaving mainly the effects due to Coulomb repulsion by similar ions within the channel and lattice vibrations along the walls of the channel. In short, the permeability of ions through membranes is greater than would be expected if thermal scattering played the same role within the membrane as it plays in the cytoplasm or in the extracellular environment.

(c) Ion Energy

There are obvious constraints on the cyclotron radius r in Equation (3), ranging from a limit of 200 nm if the ion is moving transverse to the membrane, up to perhaps 50 μm if the ion moves parallel to the membrane. The kinetic energy of a given ion, in terms of its cyclotron frequency, is

$$\frac{mv^2}{2} = q\pi r^2 B\nu \ . \tag{5}$$

If we use the values for K^+ ions in Table I, we find that these two extremes in radius both correspond to energies far below kT:

$$E_{\perp} = 2.0 \times 10^{-35} \text{ joules} = 1.25 \times 10^{-16} \text{ eV}$$

$$E_{\parallel} = 1.25 \times 10^{-30} \text{ joules} = 7.8 \times 10^{-12} \text{ eV},$$

again indicating the instrinsic difficulty in maintaining a cyclotron resonance orbit in the presence of thermal fluctuations. Indeed the radius of the path of the same ion at an energy comparable to the thermal background would be nearly three meters!

RESONANCE IN HELICAL, CONSTRAINED CHANNELS

These problems are obviated to some extent by (1) restricting the cellular regions in which cyclotron resonance may possibly occur to membranes and (2) considering only constrained, pre-existing paths through such membranes.

Such a model is consistent with helical pathways through membrane-bound proceins. One example of a likely helical channel is found in the structure of the light-driven proton pump, bacteriorhodopsin,[4] in which the polypeptide chain spans the membrane as seven twisted α-helices. Still another is the acetylcholine receptor, a complex of five transmembrane subunits for which a helical structure has been proposed[5]. It is also possible that a helical channel can be performed as an interstitial gap between adjacent lengths of the protein chain, external to a section of α-helix, using it as a geometric template (and perhaps as an energetic one, as well). We can readily show that a charged particle constrained to such a path will enjoy the same cyclotron resonance condition that is evidenced by the particle when free. Figure 5 shows such a helical path. The coordinates of an ion describing this path are

$$\begin{aligned} x &= a\cos\theta \\ y &= a\sin\theta \\ z &= \frac{p\theta}{2\pi} \end{aligned} \tag{6}$$

where a is the helix radius, as viewed along its axis, Θ is the angle made by the radius vector in the xy-plane, and p is the pitch of the helix. The velocity components are

$$\begin{aligned} \dot{x} &= v_x = a\,\dot{\theta}\sin\theta \\ \dot{y} &= v_y = a\,\dot{\theta}\cos\theta \\ \dot{z} &= v_z = \frac{p\dot{\theta}}{2\pi} \ . \end{aligned} \tag{7}$$

When a magnetic field B is applied along the z-axis, this leads to the equations of motion:

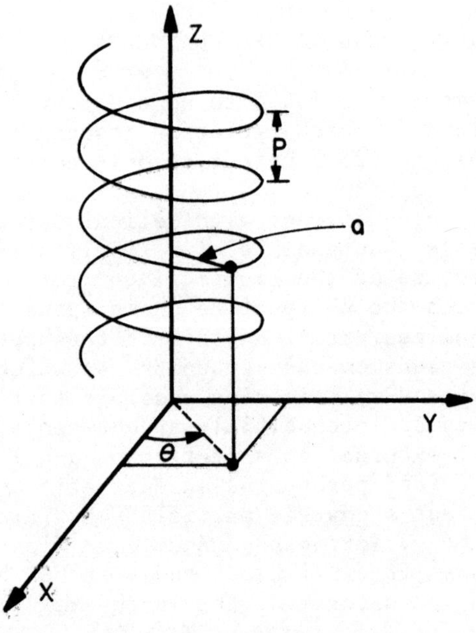

Figure 5 - Helical parameters. The helix backbone is along the z-axis, the pitch is p, the radius a, and θ the angle in the xy-plane used to describe any point on the helix.

$$\ddot{\theta}\sin\theta + \dot{\theta}^2\cos\theta = -\frac{q}{m}B\dot{\theta}\cos\theta$$

$$\ddot{\theta}\cos\theta - \dot{\theta}^2\sin\theta = -\frac{q}{m}B\dot{\theta}\sin\theta \tag{8}$$

$$\ddot{\theta} = 0\ .$$

For the special case $\dot{\Theta} = \omega$ = constant these equations reduce to

$$\omega^2\cos\theta = -\frac{q}{m}\omega B\cos\theta$$

$$-\omega^2\sin\theta = -\frac{q}{m}\omega B\sin\theta\ , \tag{9}$$

or simply

$$\omega = \frac{qB}{m} \tag{10}$$

which, because $\omega = 2\pi\nu$, is identical to Equation (4). The principal difference between free and constrained orbits lies in the fact that a free particle will increase its radius according to Equation (5) as the particle energy increases, in order to maintain the resonance condition (4).

ENERGY TRANSFER TO MEMBRANE FROM PERIODIC FIELD

We suggest that because of pre-existing helical channels and the likelihood of establishing cyclotron resonance for ions transported through these channels under geomagnetic fields, there is the distinct possibility that superposed oscillating magnetic or electric fields may act to transfer kinetic energy to channel ions when the frequency of the time-varying field is at or close to the gyrofrequency ν. Because the passage of these ions is necessarily single-file, the final energy level attained by a corresponding membrane electrogenic pump is in part limited by the drift velocity v_z of the ions in the direction of the axis of the helix. One can thus consider the possibility that the resonant energy transfer to the ions from a local time-varying field will act to increase the total velocity and therefore the ionic drift velocity through the membrane, thereby providing a more rapid utilization of such pumps. In particular, in the Na^+/Ca^{2+} or K^+/Ca^{2+} transmembrane exchange mechanisms are enhanced, a larger net transport of Ca^{2+} will result.

In Blackman's experiment, a larger Ca^{2+} efflux was observed for a combination of Earth's field and applied frequency of 0.38 gauss and 15 Hz, respectively. When these numbers were both doubled, to 0.76 and 30 Hz, the combination again gave rise to enhanced Ca^{2+} efflux. If we substitute these two sets of values for ν and B into Equation (4), the charge-to-mass ratio that results is 0.26 (in ratios of electron charge e to atomic mass units u). This is precisely the charge-to-mass ratio for singly ionized potassium (see Table I). The cyclotron resonance model that we propose is therefore consistent with channelized potassium ions in resonance at 15 and at 30 Hz in Blackman's experiment, in both cases extracting energy directly from the applied, time-varying field, resulting in extra Ca^{2+} efflux.

DISCUSSION

One critical question surrounding this explanation of Blackman's data is to ask to what extent geomagnetic cyclotron resonance can be used to explain the results obtained by many other recent observers linking low-frequency magnetic fields with varied biological responses. There are two classes of such reports, those involving responses occurring only at certain frequency windows and those occurring over a wide frequency passband. Presumably, the former experiments, when subected to a further analysis, may possibly be explained as frequency-selective conditions (on either $\vec{E}(t)$[6] or $\vec{B}(t)$[7]) corresponding to the specific level of the local geomagnetic field. It is more difficult to speculate as to how cyclotron resonance in cells can lead to responses that are frequency-independent. One clue may reside in the observation that many of the wide-band responses seem to fall into a category involving the nucleus--lengthening[8] or shortening[9] the mitotic cycle, inducing embryonic malformations[10], enhancing mRNA transcription[11], DNA synthesis[12], and tumorigenic activity[13,14] -- whereas the original Ca^{2+}-efflux reports clearly reflect changes at the membrane level.

Let us differentiate between the interaction site, where the electromagnetic field initially couples to the cell, on the one hand, and the subsequent region of expression, on the other. We can define the region of expression as that location in the system about which certain narrowly defined measurements are taken that reflect one set of physiological changes in the cell resulting from the original interaction. It should be clear that one physical interaction with a cell can result in many such sets of changes, and, further, that the location of such regions is a function of the experiment--one designed to measure changes in mRNA involves the nuclear region whereas an experiment designed around ionic efflux relates to the plasma membrane, or perhaps to the cytoplasm. Although this all appears quite obvious, the implications are rather interesting, and perhaps not altogether appreciated by the biological community. Consider Figure 6, showing an initial interaction followed by a cascade of responses in different parts of the cell. Despite the lack of specific details, it is reasonable to assume that the most immediate response will be found closest to the interaction site and that the time at which subsequent regions of expression become measurable will be successively delayed, due to factors such as messenger mobilization, cell phase, diffusion constants, etc.. Thus, in the example shown, measurements made on the plasma membrane would be affected more rapidly than those made in the nucleus or in subsequent protein construction. Furthermore, it is likely that the strength of responses in regions remote from the interaction site (e.g., 4, 5 and 6 in Figure 6) will be diffused by repair mechanisms,

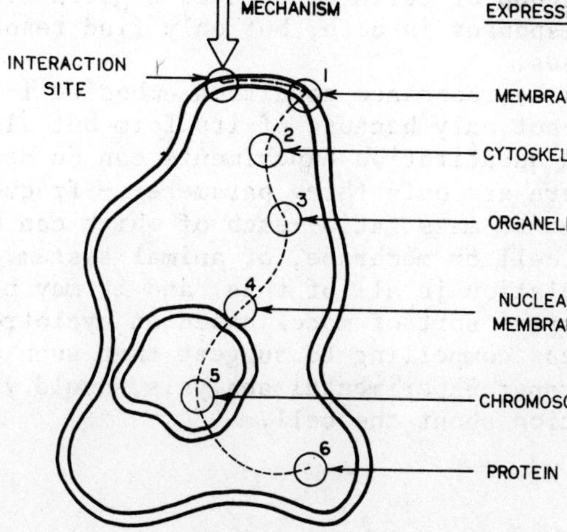

Figure 6 - Cascade of cell responses. Whatever the interaction site in a cell, there will be more than one subsequent region of expression. In this schematic example the initial interaction is at the membrane; a cascade of six events follow as shown. Different measurement techniques can be constructed for each of these regions, corresponding to the respective expressions.

differences in location within a given region, and cell phase. What emerges from all of this is a sequence in which the biological response remote from the initiating site may not only be delayed but also diffused. Figure 7 shows a group of three such responses corresponding to regions 1, 3 and 5 in Figure 1. It becomes clear that even if sharply timed biological resonances are occurring at a given frequency in one part of the cell, it does not follow that the response will be as sharp in other compartments of the cell. Another way of expressing this is that, to a first approximation, we may expect that the Q of successively removed regions of expression will be proportionately smaller; in Figure 7, $Q_1 > Q_3 > Q_5$. Indeed, it is conceivable that Q may be so small that the aplitudes of resonant responses are far less than the intrinsic noise-level of the experiment . It follows that one may perform two different types of measurements on a group of cells exposed to a given frequency, observe biological responses in both, but only find resonance effects in one of these cases.

The beauty of the resonance condition embodied in Equation (4) is its simplicity--not only because of its form but also because of the ease with which quantitative experiments can be designed around this condition. There are only three parameters--frequency, field strength, and charge-to-mass ratio--each of which can be separately varied for a given cell or membrane, or animal system. Of course, there is some speculation in all of this, and it may be premature to go beyond the barest sort of model based on cyclotron resonance, but it is nonetheless compelling to suggest that such a resonance mechanism, under proper experimental analysis, could yield immense amounts of information about the cell.

SUMMARY

We have shown that the cyclotron resonance condition that is obtained for biological ions in the Earth's magnetic field corresponds mainly to gyrofrequencies in the ELF region, 10-100 Hz. This is true for free ions or for ions moving in constrained helical pathways. In particular, it is shown that Blackman's data, in part, corresponds to cyclotron resonance for K^+ ions, implying that what he observed may have been K^+/Ca^{2+} membrane exchange.

There are at least three important limitations on cellular cyclotron resonance, (a) the presumption that the resonant ions are non-hydrated is unusual in the cellular environment, (b) the frequency of scattering events at thermal levels is theoretically larger than the calculated gyrofrequencies and should therefore swamp out any coherent resonant phenomena, and (c) the kinetic energies of the resonant ions appear to be much smaller than the thermal energies. These limitations rule out any cytoplasmic or extracellular possibilities for the site of cyclotron resonance, probly restricting consideration solely to membrane channels.

Therefore a membrane mechanism at the molecular level is suggested in which ions move in helical, ion-selective channels,

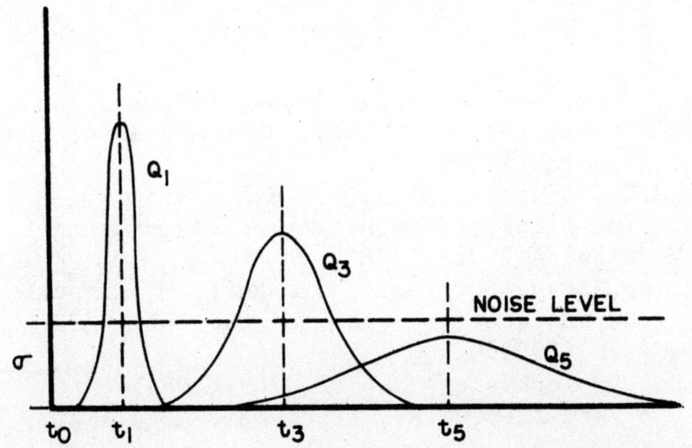

Figure 7 - Resonant response (vertical axis) in different regions. Regions of cell close to the interaction site exhibit sharper responses (Q_1) than those removed from the site (Q_3, Q_5). If the noise level of the measurement associated with a particular region of expression is appreciable, then it may swamp the resonant response (Q_5), leading to an apparent lack of frequency dependence.

enjoying specific resonant conditions determined by the local geomagnetic field, fully capable of extracting energy from properly tuned periodic fields. One aspect of this model is its analytical testability, the cyclotron resonance frequency depending only on two independent parameters.

It is conceivable that the cyclotron resonance interaction may also account for biological responses other than those exhibiting frequency windows. This will depend on the Q of the resonant response in each compartment of the cell.

In any event, if cyclotron resonance does indeed occur at geomagnetic levels in biological systems, there is great potential, through corollary experiments, for a more detailed understanding of the cell.

REFERENCES

1. Blackman, C.F., Benane, S.G., House, D.E. and Rabinowitz, J.R. The Bioelectromagnetics Society, abstract Sa-4, 6th Annual Meeting, Atlanta (1984)
2. Liboff, R.L., Biophysical J., 5, 845 (1965)
3. Polk, C., The Bioelectromagnetic Society, abstract P-36, 6th Annual Meeting, Atlanta (1984)
4. Alberts, B., Bray, D., Lewis, J., Raff, M., Roberts, K. and Watson, J.D., Molecular Biology of the Cell, 272, Garland Publ. Co., New York (1983)
5. Guy, H.R., Biolophysical J., 45, 249 (1984)
6. Bawin, S.M. and Adey, W.R., Proc. Nat. Acad. Sci., 73, 1999-2003 (1976)
7. Blackman, C.F., NATO Adv. Res. Workshop: "Interactions Between Electromagnetic Fields and Cells", Sept. 18-28, 1984, Erice, Sicily
8. Marron, M.T., Goodman, E.M. and Greenebaum, B., Nature, 254, 66-67 (1975)
9. AArholt, E., Flinn, E.A. and Smith, C.W., Phys. Med. Biol., 26, 613-621 (1981)
10. Delgado, J.M.R., Leal, J., Monteagudo, J.L. and Gracia, M.G., J. Anat., 134 (3), 533-551 (1982)
11. Goodman, R., Bassett, C.A.L. and Hendersen, A.S., Science, 220, 1283 (1983)
12. Liboff, A.R., Williams, T., Jr., Strong, D.M. and Wistar, R. Jr., Science, 223, 818-820 (1984)
13. Winters, W.D. and Phillips, J.L., The Bioelectromagnetic Society, abstract A2-6, 6th Annual Meeting, Atlanta (1984)
14. Liboff, A.R. and Kaplow, J.M., The Bioelectromagnetic Society, abstract A4-2, 6th Annual Meeting, Atlanta (1984)

CELL POLYELECTROLYTES

Fumio Oosawa

Department of Biophysical Engineering
Osaka University, Toyonaka, Osaka 560
Japan

INTRODUCTION

Biopolymers such as proteins and nucleic acids are polyelectrolytes, and biological systems such as muscle, membranes and cytoplasm may be regarded as organized polyelectrolytes. Some of the properties of living cells and their constituents can be understood from this standpoint.

The strong electric interaction between highly charged macromolecules (polyions) and their counter ions is a source of the characteristic properties of polyelectrolytes. Counter ions are bound around polyions, but retain their mobility. Their behaviors are affected by the electric field, so that the organization of polyelectrolytes is sensitive to the applied field both in vitro and in vivo.

COUNTER ION DISTRIBUTION

Let us consider a solution of linear polyions with their counter ions. When the charge of the polyion is increased, the number of counter ions increases. At a certain critical value of the charge density of the polyion, counter ions begin to be condensed in the vicinity of the polyion. With further increase of the charge density, the number of condensed counter ions increases to cancel the increase of the charge density. This "counter ion condensation" is due to the logarithmic electric potential from the linear polyion.[1] The condensed counter ions are not made free by dilution of the polyion, because the difference in energy and entropy between bound and free counter ions both increase with dilution in proportion to the logarithm of the free volume per polyion.

The critical value of the charge density of the polyion for condensation is determined by the energy-entropy balance of counter ions, which is given by

$$n^* e_o^2/\varepsilon l k T = 1 \qquad (1)$$

where e_o is the electronic charge, k the Boltzmann constant, T the absolute temperature, ε the dielectric constant of the solvent, l the length of the polyion; n^* is the critical value of the number of charges on the polyion and n^*/l is the critical charge density which corresponds to one electronic charge per about 7 A along the polyion at room temperature in water. When the number of charges n exceeds this critical value n^*, excess counter ions $(n - n^*)$ are condensed on the polyion and n^* counter ions are distributed outside the polyion.

The distribution of counter ions outside the polyion is not uniform, but their concentration is high near the polyion. The concentration of counter ions in the periphery of the free volume of the polyion where the average electric field vanishes can be estimated by analysing the geometric character of the Poisson-Boltzmann equation for the present system. Above the critical charge density of the polyion, the concentration of counter ions in the periphery is about half of the averaged concentration outside. Therefore, half of counter ions outside may be conventionally defined to be loosely bound to the polyion, and another half to be free. Thus, counter ions can be classified into three groups; (I) condensed counter ions, (II) loosely bound counter ions and (III) free counter ions, as illustrated in Fig. 1.

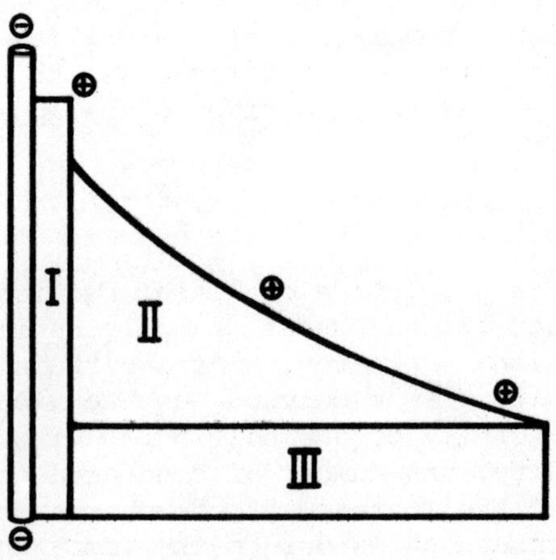

Fig. 1. Illustration of distribution of counter ions around a polyion.[1]

Introducing a non-dimensional quantity Q

$$Q = n e_o^2 / \varepsilon kT\, l \qquad (2)$$

to express the charge density, the fractions of counter ions in these three groups, β_I, β_{II} and β_{III} are given by [1]

	I	II	III
$Q \lesssim 1$	0	$Q/2$	$1 - Q/2$
$Q \gtrsim 1$	$1 - 1/Q$	$1/2Q$	$1/2Q$

The number of counter ions in three groups, $n\beta_I$, $n\beta_{II}$ and $n\beta_{III}$ increases with the increase of the charge density as shown in Fig. 2. Below the critical charge density, no counter ions are condensed, and loosely bound counter ions increases gradually. At the critical charge density, the number of loosely bound counter ions becomes nearly equal to the number of free counter ions. Above the critical charge density, only the condensed counter ions increase, and loosely bound and free counter ions are kept constant.

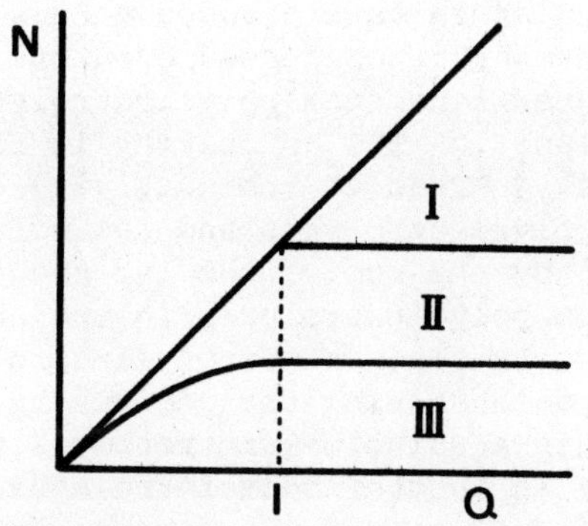

Fig. 2. The numbers of counter ions in three groups, condensed (I), loosely bound (II) and free (III) as functions of the charge density Q of a polyion.[1]

The critical charge density for condensation depends on the valency of counter ions. It is inversely proportional to the valency. When two kinds of counter ions having different valencies are mixed, counter ions of higher valency are preferrentially condensed. With increasing charge density, counter ions of higher valency are first condensed; counter ions of lower valency are not condensed until those of higher valency are all condensed, as shown in Fig. 3. The difference in the valency of counter ions is extremely amplified in polyelectrolytes.

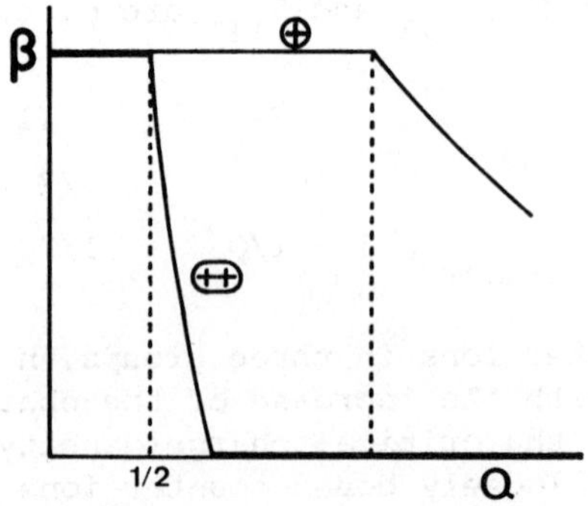

Fig. 3. Fractions of condensed counter ions, divalent and monovalent, in a mixture of two kinds of counter ions at a constant ratio as functions of the charge density of a polyion.[1]

EFFECTS OF SALTS

When low molecular salts are added to a solution of polyelectrolytes, the number of condensed counter ions is not much influenced, if counter ions from polyelectrolytes and from salts are of the same valency. The additivity is found in the activity of counter ions.[1] That is, the activity of counter ions in the mixture of polyelectrolytes and low molecular salts is approximately given by the sum of the independent contributions of counter ions from polyelectrolytes in the absence of salts and those from salts in the absence of polyelectrolytes. Fig. 4 gives experimental data showing the additivity of activity of counter ions (Na^+) in a solution of tropomyosin, a rodlike protein of about 400 A in length, containing NaCl.[2] Tropomyosin is

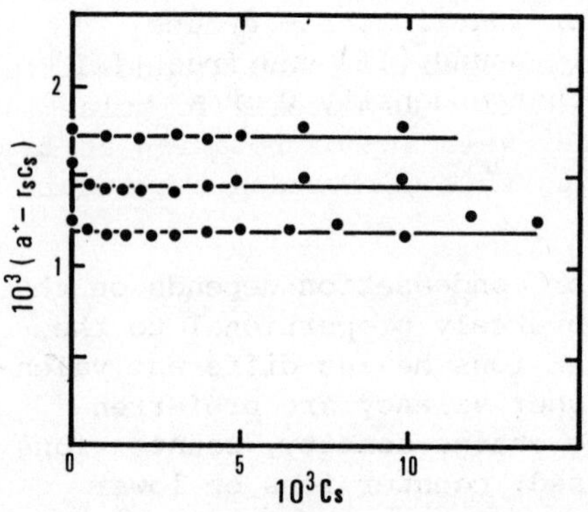

Fig. 4. The activity of counter ions (Na^+) in a solution of tropomyosin containing various concentrations of NaCl (c_s) at three different values of pH. The activity (a^+) is given by the sum of the contribution of salts ($\gamma_s c_s$) and the contribution of counter ions of tromyosin without salts, which depends on pH.[2]

highly charged and forms a long polymer, so that its property is well described as a polyelectrolyte. This "additivity" can be theoretically derived on the basis of the same geometrical characteristics of the electric field around the polyion, as remarked in the derivation of the counter ion condensation.

The additivity does not necessarily mean that low molecular salts have no effects on the state of the polyion. According to the additivity, the chemical potential of salts, u_s, is expressed in the following form;[1]

$$u_s = kT \ln(\gamma_p c_p + \gamma_s c_s)(\gamma_s' c_s') \quad (4)$$

where counter ions from polyelectrolytes and salts were assumed to be of the same species. c_p is the concentration of charges of polyions or counter ions from polyelectrolytes, γ_p the activity coefficient of those counter ions in the absence of salts, c_s is the concentration of counter ions from salts and γ_s the activity coefficient of those counter ions in the absence of polyelectrolytes; c_s' and γ_s' are the concentration and the activity coefficient of co-ions. The coefficient γ_p corresponds to β_{III} in Eq. (3). Therefore, above the critical charge density of the polyion the total activity of counter ions which appears in the logarithmic term of Eq. (4) is approximatewritten as[1]

$$\gamma_p c_p + \gamma_s c_s = c_p/2Q + c_s \quad (5)$$

The right hand side is nearly equivalent to the quantity $(1/2)\,\kappa^2$ where κ is often defined as a Debye-Huckell parameter in polyelectrolytes with salts.

The chemical potential of salts depends on the concentration of polyions as shown in Eqs. (4) with (5). Therefore, the chemical potential of polyions must depend on the concentration of salts. By applying the Gibbs-Duhem relation between the derivatives of chemical potentials of different components, it is found that the chemical potential of polyions u_p changes with the concentration of salts approximately in the following way;[1]

$$u_p(c_s) = \text{const.} + n\,kT\,(\gamma_p/\gamma_s)\ln(\gamma_p c_p + \gamma_s c_s) \quad (6)$$

At low concentrations of salts where the concentration of counter ions from salts is much smaller than the concentration of free counter ions from polyelectrolytes, the effect of salts is negligible. The effect begins to appear when these concentrations become comparable. At high concentrations of salts, the chemical potential of polyions increases in proportion to the logarithm of the concentration of salts.

Polyions of higher charge density have smaller values of the coefficient γ_p. According to Eq. (6), the increase of their chemical potential by salts must be smaller. Therefore, polyions of higher charge density are made relatively stable by the addition of salts, as shown in Fig. 5. In the case of flexible linear polyions, large extension of polyions results in increase of the coefficient γ_p. Polyions having smaller extension are made more stable by salts. In fact, the logarithm of the extension of flexible polyions was found to decrease in proportion to the term $\ln(\gamma_p c_p + \gamma_s c_s)$ or $\ln \kappa$.[1]

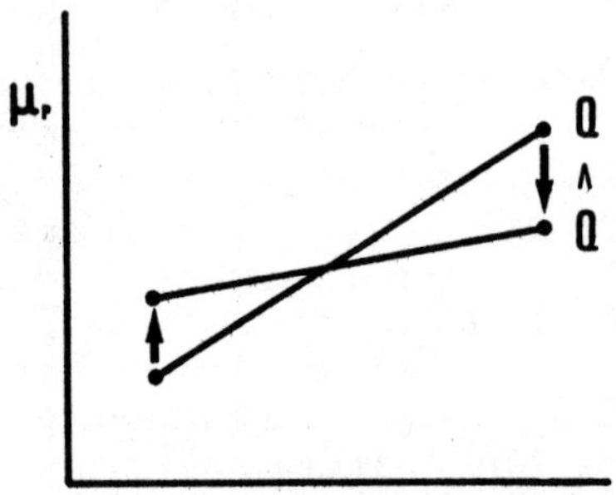

Fig. 5. The relation between the chemical potential u_p of polyion and the logarithm of parameter κ or the salt concentration at different Q's.[1]

The logarithmic dependence of the chemical potential of polyions on the concentration of salts means that an e-fold increase of the concentration of salts continues to have a similar additional effect on polyions even at high concentrations of salts. Saturation does not occur in the effect of salts. Eq. (6) is a result of simple combination of two characteristic properties of polyelectrolytes, "counter ion condensation" and "additivity". The long tail of the effect of salts at high concentrations is the third characteristic property of polyelectrolytes.[1]

POLYIONS AND COUNTER IONS IN THE ELECTRIC FIELD

Solutions of polyelectrolytes have very large dielectric constants. This is mainly due to polarization of counter ions condensed on polyions.[1] In the absence of the electric field, spontaneous fluctuation of distribution of condensed counter ions along the polyion produces a fluctuating dipole. The polarizability of polyions with counter ions in the electric field is proportional to the average of the square of this dipole. It is proportional to the product of the number of condensed counter ions, $n - n^*$, and the square of the length of the polyion, l^2. For polyions of a given charge density, the number of condensed counter ions is proportional to the length l. If the dielectric constants were compared in solutions of polyions of the same kind having different lengths at a given weight concentration, the dielectric constant is expected to be proportional to the square of the length l. This was confirmed in experiments, as shown in Fig. 6.[3]

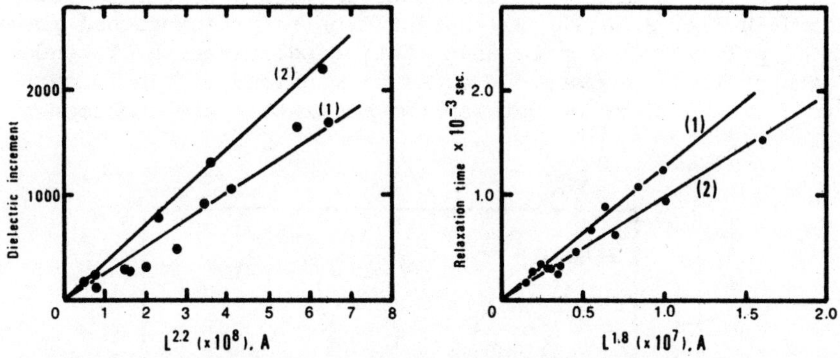

Fig. 6. The dielectric constant and the relaxation time of polarization of solutions of DNA having different lengths. Both are nearly proportional to the square of the length.[3]

The relaxation of polarization of condensed counter ions occurs by their diffusion along the polyion. Then, the relaxation time must be proportional to the square of the length l. Actually, the main dispersion of dielectric constant was found at the frequency which is inversely proportional to l^2. The observed value of the frequency, however, was much higher than that estimated by assuming free diffusion of counter ions along the polyion. The movement of condensed counter ions is not as free as counter ions outside.

Because of the large polarizability of condensed counter ions along the polyion, the electric field produces a torque to orient the polyion in parallel to the field. An optical birefringence due to such orientation of polyions is observable under the field. Very recently, the birefringence under an alternating electric field has been measured in solutions of polyelectrolytes at low concentrations.[4] In the field of frequency w, the birefringence is composed of two components, a component of frequency 2w and a DCcomponent (an average level), as shown in Fig. 7.

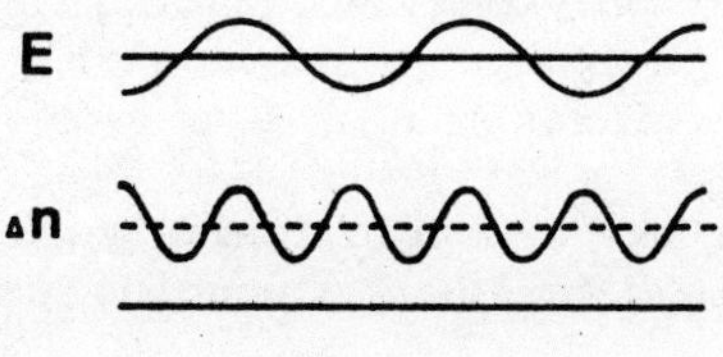

Fig. 7. Illustration of the applied AC field and the response of the birefringence of polyelectrolytes.[4]

The 2w component is due to orientational movements of polyions following the field. When the frequency is increased, this component vanishes first, where the DC component remains almost constant. The DC component begins to decrease at the same frequency as the dielectric dispersion. In polyions of about 1 micron in length, the 2w component decreases to half at about 10^2 Hz and the DC component decreases to half at about 10^3 Hz, as shown in Fig. 8. The relaxation time of orientational movements of polyions is longer than that of polarization of condensed counter ions in about one order of magnitude. In longer polyions, the difference between two relaxation times becomes larger.

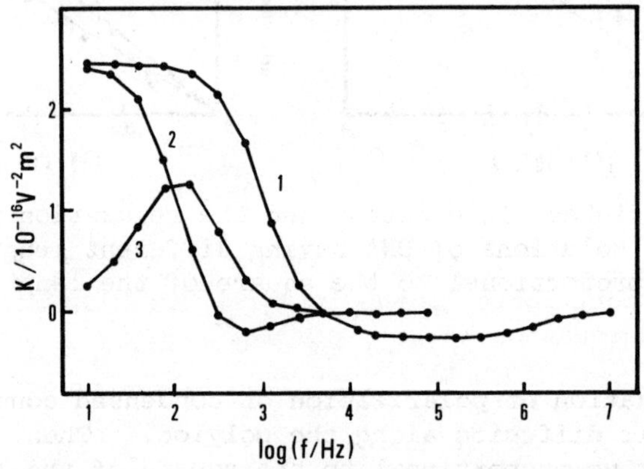

Fig. 8. The electric birefringence of a solution of polystyrene sulfonate under an AC field of various frequencies, f. The amplitude of the DC component (1) and two components, real and imaginary (2) and (3) of the 2w response.[4]

When the frequency of the applied field is further increased, the reversal of the DC component of birefringence takes place, as shown in Fig. 8. This means that at high frequencies the polyion or its parts are oriented in a direction perpendicular to the field. When the polarization of counter ions along the polyion can not follow the field of high frequencies, the polarization perpendicular to the polyion becomes predominant. The movements of condensed counter ions perpendicular to the polyion has a short relaxation time. The reversed birefringence decreases at a very high frequency where the dielectric constant shows the second dispersion. This frequency is independent of the length of the polyion. Instead, it increases with increasing concentration of polyions, suggesting contributions of counter ions outside the polyion, particularly the loosely bound counter ions to the relaxation.

These behaviors of polyions under the electric field, parallel orientation of polyions to the field at low frequencies and perpendicular orientation at high frequencies, have been observed in both synthetic polyelectrolytes such as polystyrene sulfonate and native polyelectrolytes such as DNA.[4] All rodlike or linear polyions show similar responses to the field at low concentrations (Fig. 9).

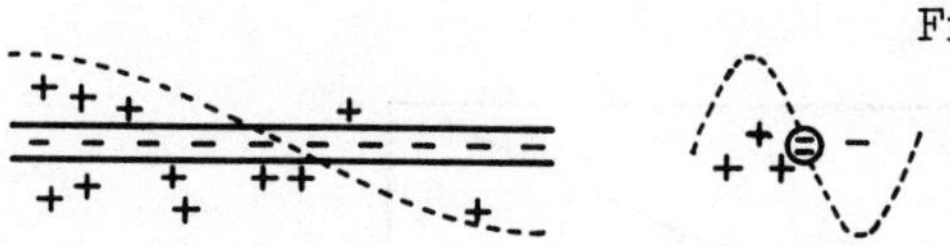

Fig. 9. Polarization of counter ions around a polyion in AC fields of low frequency and high frequency.

The addition of low molecular salts to solutions of polyelectrolytes greatly decreases both the dielectric constant and the birefringence at low frequencies. The increasing concentration of counter ions and co-ions outside the polyion moves to cancel the polarization of condensed counter ions along the polyion. In the absence of salts, the concentration of counter ions outside the polyion is much less than that of condensed counter ions. Therefore, the relaxation of polarization along the polyion occurs mostly by the movement of condensed counter ions. However, with increasing concentration of salts, ions outside the polyion make more contributions to the relaxation.

EFFECTS OF THE FIELD ON ORIENTED POLYIONS

Long rodlike or linear polyions can undergo free rotational movements only at very low concentrations. At higher concentrations, they make anisotropic domains by spontaneous parallel orientation even in the absence of the field. Each polyion in the domain has counter ions condensed on it. In addition, many counter ions are condensed in the assembly of parallel polyions. The concentration of counter ions in the assembly is higher than that outside the assembly. Under this condition, the electric field of low frequencies often induces the orientation of polyions perpendicular to the field.

A typical example was found in the case of F-actin, a helical polymer of globular protein, G-actin. Upon the addition of salts, G-actin polymerizes into F-actin. In the beginning of polymerization where a small number of short F-actins are formed, the electric field produces a positive birefringence. Short F-actins are oriented in parallel to the field. When a large number of long F-actins are formed, however, the birefringence changes to negative, showing that long F-actins are oriented perpendicular to the field.[5] (Fig. 10).

The effect of the electric field on F-actin oriented by flow was also examined.[5] A flow of low shear rates produces a large birefringence due to almost complete orientation of F-actin in parallel to the flow. When an electric field is applied in parallel to the flow, the birefringence decreases; that is, the orientation of F-actin is disturbed very much. On the other hand, when the field is perpendicular to the flow, the birefringence does not change much. These experiments indicate that the orientation of F-actin perpendicular to the field is more stable.

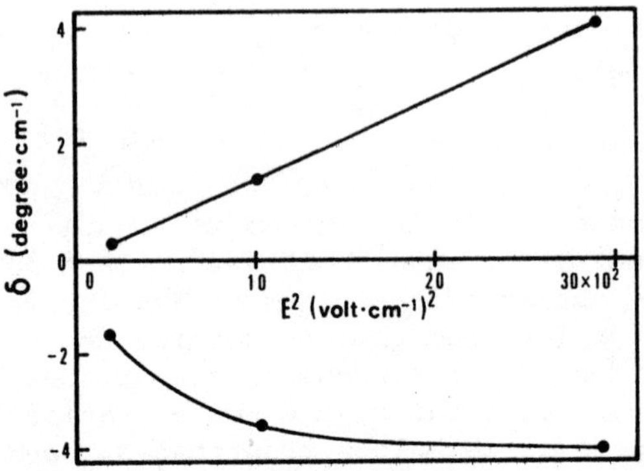

Fig. 10. The electric birefringence of solutions of F-actin at low concentrations or in the initial stage of polymerization (upper) and at high concentrations or in the later stage of polymerization (lower).[5]

Now, what does happen when an electric field is applied to F-actin in the cell? Experiments were made on a muscle fiber in which myosin filaments and F-actin filaments are aligned in parallel, interdigitating each other.[6] After removing the cell membrane of a fiber immersed in oil, an electric field was applied in parallel to the fiber. First, a large change was observed in the intensity of light diffracted by the fiber put in a laser beam. Then, under an optical microscope, a change of the birefringence was found. It came from I-bands in the striated muscle fiber. It means that the orientation of F-actin filaments was disturbed by the field. The birefringence decreased in one side of each z-line and increased in the other side. Since F-actin filaments in opposite sides of z-line have opposite structural polarities, the different responses in the birefringence in two sides suggest that F-actin filaments in the fiber have a permanent dipole, in addition to the induced dipole due to counter ions. In a solution, F-actin filaments have random polarities. Even when they are oriented in parallel, some of them may be antiparallel in terms of polarity. In the muscle fiber, F-actin filaments are arranged with the same polarity in each side of z-line. Therefore, the responses of F-actin filaments may be different between in the solution and in the fiber.

In an alternating field of frequency w, the total birefringence of the fiber changed with a frequency 2w. The 2w component decreases with increasing frequency. From the dispersion frequency of this component, the relaxation time of the rotational or bending movements of F-actin filaments was estimated. It was found to depend on the physiological state of the fiber, as shown in Fig. 11. In the absence of calcium ions, the fiber is in the resting state. Upon the addition of calcium ions, it is transferred to the active state. If the fiber was extremely stretched until myosin and F-actin filaments were inhibited to interact, the fiber does not contract even after the addition of calcium ions. In these two states, the response of the fiber to the alternating field was compared. The frequency where the 2w component decreases was very much larger in the absence of calcium ions than in the presence. In a certain range of the frequency, the birefringence of the fiber in the resting state changes following the field, whereas that in the active state does not. This means that F-actin filaments in the two states have different rotational or bending mobilities.

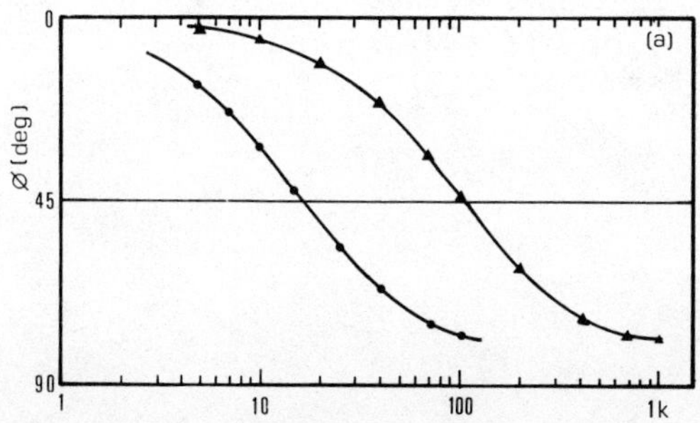

Fig. 11. The response of I-band or F-actin filaments in a muscle fiber to the AC field of various frequencies.[6] In this experiment, the measurements were made on the intensity of diffraction lines from the fiber, instead of the birefringence. The ordinate is the phase delay of the response; in the absence of calcium ions (right) and the presence (left).

Actually, F-actin filaments in the fiber bind regulatory proteins named tropomyosin and troponin. Calcium ions are bound to troponin, and this binding induces some change in the structure of F-actin which makes possible the interaction with myosin to generate a force for contraction of the fiber. The increase of the rotational or bending mobilities of F-actin may be important in the active interaction with myosin.

In living cells we often find parallel arrangements of protein polymers such as F-actin, microtubules and intermediate filaments. These polar structures are considered to have essential roles in the morphogenesis and motility of the cell. Recently, several experiments have shown that the electric field, even a weak field, has remarkable effects on these structures. The above experiment on F-actin filaments in the muscle fiber is one of them. It was also reported that in a culture cell, F-actin filaments are aligned perpendicular to the field applied from outside.[7] The response is slow but clear. Among various possible interactions between the electric field and the cell, the interaction between the field and cell polyelectrolytes to change the orientation of polyions may be a most effective one.

REFERENCES

1) F. Oosawa; Polyelectrolytes, Marcel Dekker, N.Y. 1971
2) S. Iida and N. Imai; J. Phys. Chem., 73, 75 (1969)
3) S. Takashima; J. Phys. Chem., 70, 1372 (1966), Biopolymers, 5, 899 (1967)
4) N. Ookubo, Y. Hirai, K. Ito and S. Hayakawa; Rep. Prog. Polym. Phys. Jap., 27, 475 (1984)
5) S. Kobayasi, H. Asai and F. Oosawa; Biochi. Biophys. Acta, 88, 528 (1964); S. Kobayasi; Biochim.Biophys. Acta, 88, 541 (1964)
6) S. Yoshino, Y. Umazume, R. Natori, S. Fujime and S. Chiba; Biophys. Chem., 8, 317 (1978)
7) P. Luther, B. Peng and J. Lin; Nature, 303, 61 (1983)

CONTROL OF CONFORMATION AND MOVEMENTS OF CELL POLYELECTROLYTES

Fumio Oosawa

Department of Biophysical Engineering
Osaka University, Toyonaka, Osaka 560
Japan

INTRODUCTION

The states of cell polyelectrolytes such as nucleic acids, proteins and their assemblies are controlled by the environmental condition. The structure of the complex of DNA and proteins in cell nuclei changes depending on the ionic condition. The assembly of protein filaments in the cell is influenced by the electric field. There are many examples of the control of conformation and movements of cell polyelectrolytes in vitro and in vivo. The control is made by 1) the ionic condition, 2) the electric current, 3) the electric and magnetic field and 4) the mechanical force. An electric current may be sometimes equivalent to the exchange of ions and an electric field may produce a mechanical force on polyions. The control appears in different levels of structures; 1) polyelectrolyte molecules, 2) their polymers or complexes, 3) assemblies of polymers or complexes and 4) the higher order structures of those assemblies. In this paper, some experimental results on the control of conformation and movements of protein polymers are described.

METHODS

In this study, a new optical microscopic technique has been employed for direct observation of single filaments of protein polymers such as F-actin, microtubules, and bacterial flagella. They are helical or tubular polymers of globular protein molecules. Their diameters are very much smaller than the wave length of visible light, so that it seemd to be very difficult to get their images under an optical microscope. However, it was found recently that they are made visible under a microscope by intense illumination in a dark field. Microtubules and bacterial flagella gave scattered light strong enough for direct observation. They appear as bright filaments under the microscope. If they are much longer than the wave length, their shapes can be defined and their movements can be followed.

Another method to make visible very thin filaments is to apply a fluorescence microscope after labelling the filaments with a fluorescent dye. F-actin is much thinner than microtubules or flagella, so that it is not visible even with dark field illumination. However, when labelled with a dye, single filaments of F-actin are clear observable under a fluorescence microscope. If the light for excitation of the dye is strong enough, the number of dye molecules on the filaments can be small. This method is very convenient if the labelling has no influence on the property of protein filaments.

F-ACTIN AS A SEMIFLEXIBLE FILAMENT

F-actin is a helical polymer of globular protein monomers called G-actin. The polymerization process consists of nucleation and growth, and F-actin is formed only above a certain critical concentration of G-actin, which is determined by the environmental condition.[1] With increasing salt concentration, the critical concentration decreases, and at the physiological salt concentration, most of actin are in the state of F-actin.

Phalloidin-rhodamine is specifically bound to F-actin without any influence on the physiological function of F-actin.[2,3] Under a fluorescence microscope, we can observe single F-actin filaments undergoing translational, rotational and bending Brownian movements, as shown in Fig. 1. Bending movements indicate that F-actin is not a rigid rod but a semiflexible filament. According

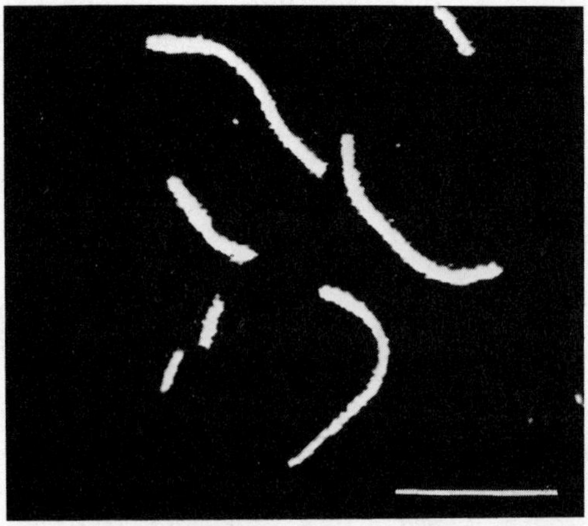

Fig. 1. Single filaments of F-actin labelled with phalloidin-rhodamine under a fluorescence microscope.
bar : 5 um.
by T. Yanagida.

to the statistical analysis of the relation between the contour length and the end-to-end distance of moving filaments, the elastic modulus for bending or the flexural rigidity is about 4 10^{-17} dyn cm^2. The relaxation time of the bending movements is a few seconds for F-actin of 10 um in length. These values are in good agreement with the values previously estimated by the quasielastic light scattering and other methods.[4] Such flexibility arises from the possible deformation of the intermonomer bond and the intramonomer conformation in F-actin.

The free bending movements of F-actin are observable at very low concentrations of F-actin around 1 ug/ml. If the concentration becomes high, around 1 mg/ml, the bending of long F-actin filaments is very limited. For observation of movements of single F-actin filaments at a high concentration, a small number of F-actin filaments labelled with FITC by chemical linking were mixed with a large number of unlabelled F-actin.[5] Then, labelled F-actin filaments showed "reptile" movements. They moved in a longitudinal direction along the filaments but not in a transverse direction. The longitudinal movements were bidirectional and the average speed was about 0.3 um/s for F-actin of 10 um in length.

CONTROL OF FLEXIBILITY OF F-ACTIN

The flexibility of F-actin is controlled depending on the physiolgical state of the cell. In a muscle fiber, F-actin filaments and myosin filaments are aligned in parallel, interdigitating each other. F-actin filaments bind tropomyosin and troponin, and the interaction of F-actin with myosin to produce a sliding force for muscle contraction is initiated by the binding of calcium ions to troponin on F-actin. Therefore, some conformational change is expected to be induced in F-actin by binding of calcium ions to troponin. Actually, the flexibility of the complex of F-actin with tropomyosin and troponin in solution was found to be larger in the presence of calcium ions than in the absence. The average amplitude and the relaxation time of bending movements of the complex are increased by addition of calcium ions at a micromolar concentration. In other words, troponin without calcium ions has an inhibitory effect on the bending movements of F-actin and calcium ions release this inhibition. (Table 1.)

Table 1. The flexural rigidity of the F-actin-tropomyosin-troponin complex with and without calcium ions and heavy-meromyosin (HMM).

	(10^{-17} dyn cm^2)
Complex – Ca^{2+}	7.4
+ Ca^{2+}	5.2
+ HMM – Ca^{2+}	8.3
+ HMM + Ca^{2+}	4.6

Such flexibility increase of the complex occurs in vivo. In the resting state of a muscle fiber, F-actin filaments with tropomyosin and troponin are rigid and upon activation of the fiber, calcium ions released from sarcoplasmic reticulum make the filaments flexible. The flexibility of F-actin seems to be required for generation of the sliding force by interaction with myosin coupled with splitting of ATP.[6]

When myosin is added to F-actin filaments in the flexible state, the bending movements are amplified and slowed down. The binding of myosin (or myosin fragments - heavy meromyosin) makes F-actin more flexible. However, in the case of the complex of F-actin with tropomyosin and troponin in the rigid state, myosin has just an opposite effect. As shown in Table 1, the difference in the flexibility of the complex between in the absence and the presence of calcium ions is enlarged by myosin.[2]

What does happen when F-actin filaments interact with myosin in the presence of ATP? As shown in Fig. 2, the amplitude of bending movements increases and the average period of bending decreases.[2] In the thermal bending movement, if the flexibility increased, the amplitude and the period both must increase. The increase of amplitude and the simultaneous decrease of period indicate that myosin with ATP induces "active" bending movements of F-actin, consuming the free energy of ATP. Such active movements are not induced by myosin with ATP when the flexibility of F-actin is suppressed by troponin in the absence of calcium ions.

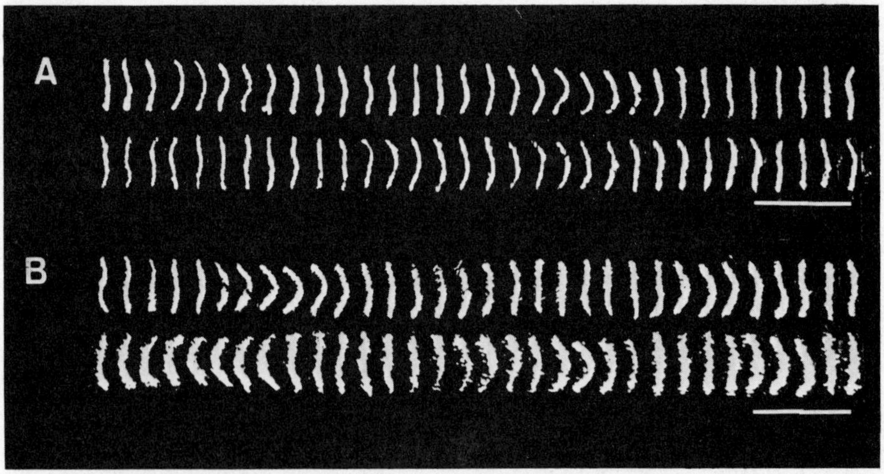

Fig. 2. Sequential photographs of single filaments of the complex of F-actin with tropomyosin and troponin interacting with heavy meromyosin in the presence of ATP. A) in the absence of calcium ions (resting) and B) in the presence (active). Length: 10 um; bar: for A) 0.75 sec; for B) 0.50 sec.[2]

Activation of bending movements of F-actin by myosin with ATP occurs at low concentrations of F-actin. At high concentrations where F-actin undergoes reptile movements, myosin with ATP has a different effect on F-actin.[5] When myosin filaments are added with ATP, F-actin filaments show occasionally very fast unidirectional slidinglike movements, of which the speed attains several um/s. Apparently, the restriction of bending movements in F-actin makes possible the transformation of active bending to active sliding.

These experiments suggest that living cells have appropriate control mechanisms of both the conformation of single filaments of F-actin and the state of assembly of those filaments to exhibit the physiological function.

POLYMORPHIC TRANSITION OF FLAGELLA

Bacterial flagella are tubular polymers composed of globular protein monomers, flagellin.[1,7] The tubes have certain degrees of bent and twist, and assume helical shapes. If all monomers in tubular polymers are equivalent, the tubes must be straight. The helical shapes are stable only when monomers in different conformations are arranged regularly in the tubes.[7] Flagella can take several different helical shapes which have discreet values of pitch and diameter. These shapes can be formed by different combinations of monomers in two conformations.[7,8] Fig. 3 shows flagella under an optical microscope with dark field illumination.

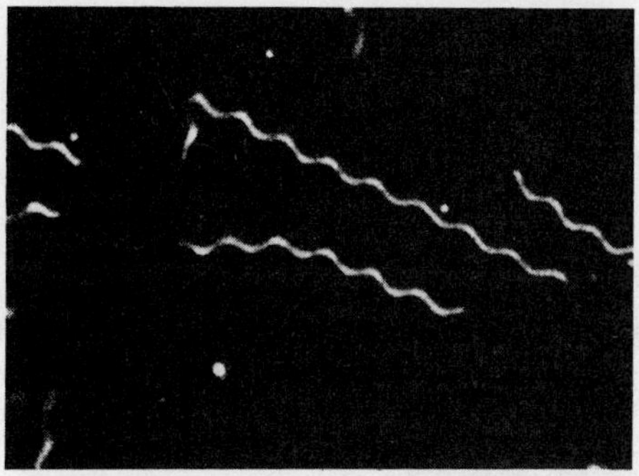

Fig. 3. single filaments of bacterial flagella under a optical microscope with dark field illumination (x2,300). by H. Hotani.

The transition of flagella between two helical shapes can be induced by changing the environmental condition; the salt concentration, pH, temperature and organic solvents.[9-11] The transition occurs reversibly in a narrow range of the environmental condition. In Fig. 4, the transition was caused by changing temperature. By direct observation of flagella under a microscope, it was confirmed that each flagellum changes its shape discontinuously during a change in the environmental condition.[10] The propagation of transformation of the shape along the filament can be followed. In the intermediate condition, flagella of two different shapes coexist. The analysis of the transition by pH or the salt concentration shows that very small changes in the number of protons or salt ions bound to flagellin monomers are associated with the transition.

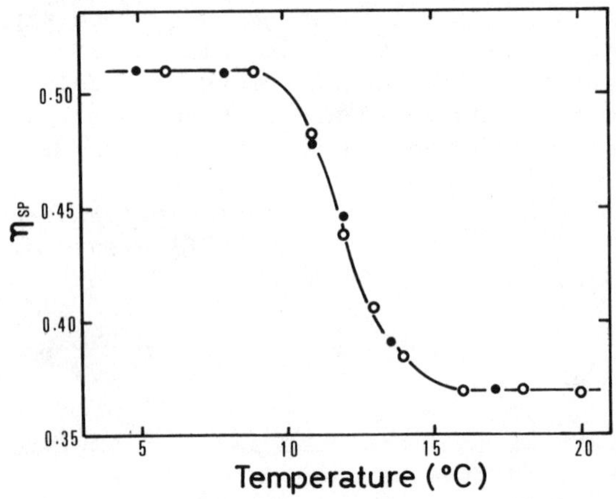

Fig. 3. The thermally induced transition of flagella. The ordinate is the viscosity of the solution during temperature rise (o) and drop (•). According to optical observation, flagella assume a left handed normal shape at high temperatures and a right handed semi-coiled shape at low temperatures.[10]

The transition is also induced by changing the concentration of organic solvents.[11] In Fig. 5, the concentration of ethylene glycol was raised from 0 % to 40 % in a solution under an optical microscope. In flagella, the end of which was fixed on a glass slide, the transition from the normal shape to the coiled shape started at the other end and proceeded along the filament.

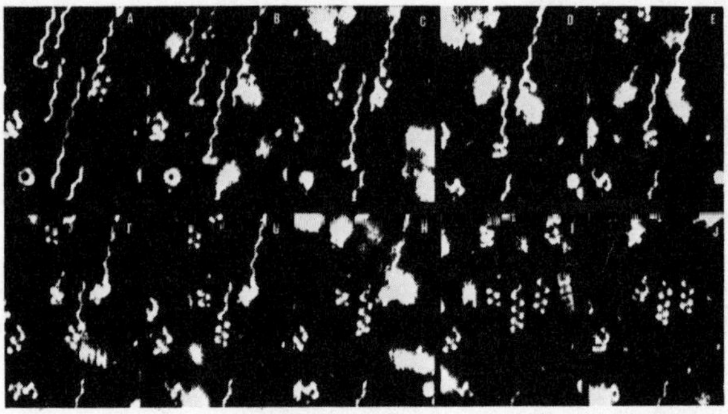

Fig. 4. Sequential photographs of flagella undergoing the transition from the normal shape to the coiled shape in ethylene glycol; about 30 seconds from upper left to lower right: (x 1,000).[11] by H. Hotani.

A mechanical force can also induce a transition of the helical shape of flagella.[12] If one end of a flagellum is fixed and a flow is given to the medium, a mechanical force acts on the flagellum to twist and stretch. Sometimes, a transition takes place. Even the handedness of the helix changes and the transition propagates along the filament repeatedly.

Bacteria swim by rotating helical flagella by a rotary motor at the base of each flagellum. During swimming, the helical shape is maintained to work as a screw. Therefore, the transition of the helical shape of flagella observed in vitro has no physiological significance, although it happens in swimming bacteria.

DISCUSSION

Two examples were briefly described on the control of conformation and movements of protein polymers. In the case of F-actin filaments, their thermal and active movements are conrolled by the environmental condition, and in the case of bacterial flagella, their overall shapes show discontinuous transitions depending on the condition. Besides the ionic condition and the mechanical force, the electric field also has influences on the conformation and movements of protein polymers. The polymers have polar structures. They have permanent electric dipoles along the filaments, in addition to the induced dipoles due to counter ions. Actually, as described in the other paper,* an electric field applied to a muscle fiber changes the orientation of F-actin filaments in the fiber, and this response of F-actin to the field is different between two states of the fiber, the resting state and the activated state. This arises from the difference in the flexibility of F-actin between two states. Very recently, a magnetic field was found to have a remarkable effect on the orientation of F-actin filaments.[13] Thus, the control of conformation and movements of protein polymers and the state of their assemblies by the electric and magnetic fields is a very interesting subject to be extensively studied in future.

REFERENCES

1) F. Oosawa and S. Asakura; Thermodynamics of the Polymerization of Proteins, Acad. Press. 1975
2) T. Yanagida, M. Nakase, K. Nishiyama and F. Oosawa; Nature, 307, 58 (1984)
3) E. Prochniewicz-Nakayama, T. Yanagida and F. Oosawa; J. Cell Biol., 97, 1663 (183)
4) S. Fujime; J. Phys. Soc. Jap.; 29, 751 (1970)
5) H. Honda and S. Asakura; reported at the annual meeting of Biophysical society of Japan, 1983 and 1984.
6) F. Oosawa; Macromolecular Assembly of Actin; Muscle and Nonmuscle Motility, Acad. Press. 1983
7) S. Asakura; Adv. Biophys., 1, 99 (1970)
8) C. Calladine; J. Mol. Biol., 118, 457 (1978)
9) R. Kamiya and S. Asakura; J. Mol. Biol., 106, 167 (1976)
10) E. Hasegawa, R. Kamiya and S. Asakura; J. Mol. Biol., 160, 609 (1982)
11) H. Hotani; Biosystems, 12, 325 (1980)
12) H. Hotani; J. Mol. Biol., 156, 791 (1982)
13) J. Torbet and M. Dickens; FEBS. Lett. 173, 403 (1984)

* : F. Oosawa; the other paper in this book.

ION INTERACTION WITH DNA: RESULTS BASED ON POLYELECTROLYTES THEORY

M. Parodi, A. Chiabrera, G. Giannetti, and M. Grattarola

Biophysical and Electronic Engineering Department
Via All'Opera Pia 11 A
16145 Genova Italy

INTRODUCTION

A large body of experimental data suggests that low frequency weak electromagnetic fields can affect cell functions. In the case of quiescent cells (e.g., frog erythrocytes), nuclear enlargement[1] and chromatin decondensation[2,3] are among the most documented early events (which later on will result into RNA synthesis), induced by electromagnetic exposure.

Considerations based on conductive and dielectric properties of cellular structures show that a direct action of low frequency electromagnetic fields on chromatin is extremely unlikely to occur. On the other hand, they seem to be able (through mechanisms which are still not well characterized) to affect Ca++ concentrations near the cell surface.

In the following we will assume, as a working hypothesis, that these ionic (e.g. Ca++) modifications do result, after unknown intermediate steps, in a variation in the concentration of Mg++ ions inside the nucleus, which causes structural changes into DNA-chromatin.

We shall attempt to offer an explanation of such structural changes within the framework of polyelectrolyte theory, which will be applied, as a first order approach, on simple DNA macromolecules in a solution of monovalent and/or divalent cations.

For the benefit of the reader, before entering the physico-mathematical section of the paper, we will shortly justify the two main statements of the previously quoted working hypothesis, which implies:

a)the existence of a coupling between Ca++ and Mg++ inside a cell;

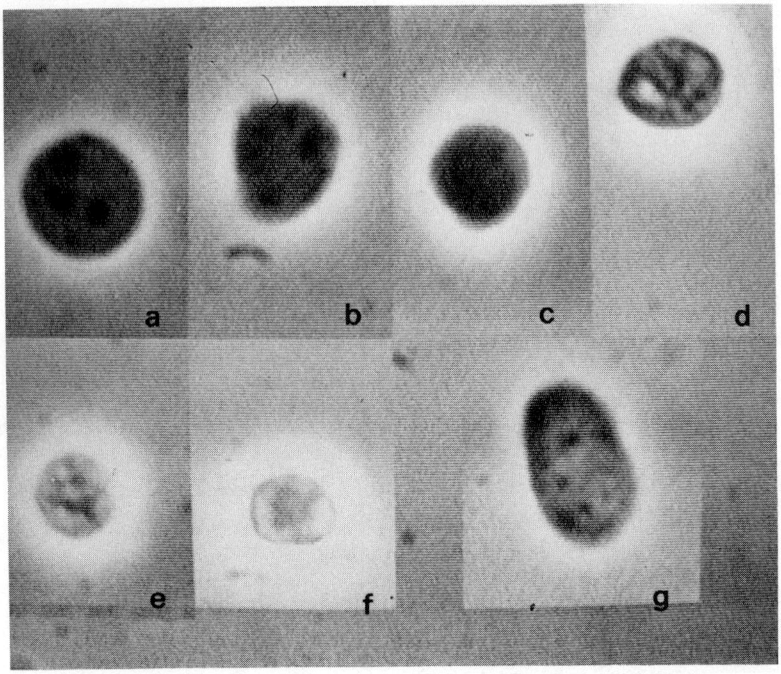

Fig.1 Collage of phase contrast images of nuclei, directly taken from one the ACTA system displayes. From the top left corner to the right bottom corner: 0. mM MgCl; 0.3 mM MgCl; 0.8 mM MgCl; 1. mM MgCl; 2.5 mM MgCl; 30. mM MgCl; 60. mM MgCl

b)the effectiveness of cations (and especially divalent ones)in affecting DNA-chromatin.
With reference to point a), an exchange between free Ca++ and Mg++ inside the cytoplasm has been recently proposed as a result of cell treatment with the ionophore A21378[4]. Moreover, the processes of microtubules assembly-disassembly are known to be modulated in an opposite way by Ca++ and Mg++[5]. Finally, Ca++ and Mg++ are known to compete in the functioning of the ATPase pump[6]. An extensive model, connecting electromagnetic fields, Ca++ ions ("cytoplasmic messanger") and Mg++ ions ("nuclear messanger")can be found in[7].
The effectiveness of cations in affecting DNA-chromatin structure is well documented at different levels of DNA-chromatin organization.
I) A, B type to Z type transition in DNA fragments can be induced by monovalent, divalent and trivalent cations, whose critical

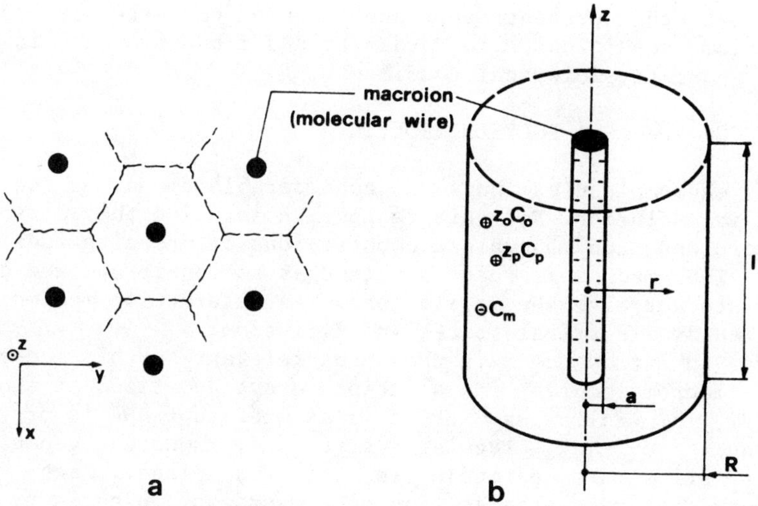

Fig.2 a) Ideal schematization of the DNA solution. Circles represent the transverse section of the macroion.
b) Structure of each elementary cell of the solution represented in a)

amount sharply decreases as a function of their valency [8] (1M, 100mM, 0.02mM ranges respectively).
II) Monovalent cations (less effectively) and divalent cations (more effectively)have been shown to affect the spatial organization of reconstructed nucleosomes and higher superstructures [9] (50mM, 0.3mM ranges respectively).
III) Cations do affect the overall structure of intact nuclei, as reported by many authors. A detailed study of the condensation-decondensation process induced by Mg++ on cell nuclei already equilibrated in presence of larger amount of KCl has been performed by our group [10,11]. In Fig. 1, an example of the resulting morphological effects is shown. The main aim of this paper is to show that such an effect can be predicted, at least qualitatively, by means of polyelectrolyte theory.
More specifically, we will show that the theory does predict the quoted relationship between cations valence and their effectiveness. Obviously, we do not exclude that other factors, not directly related to the electrical properties of DNA-chromatin, can be relevant in order to explain structural modifications induced in cell nuclei by ions. Finally, we are

aware that the present approach should be extended to non-equilibrium conditions, in order to describe chromatin inside a nucleus under physiological conditions.

MODELING OF DNA AS A POLYELECTROLYTE

DNA macromolecules have a considerable amount of ionizable groups in solution. For this reason, in solution they dissociate into macroions and monovalent counter-ions of opposite (positive) charge. The macroions can be thought as linear charged chains which interact with the mobile ions. The interaction mechanism is controlled by electrical forces and diffusion.

In order to study the main phenomena relevant to this mechanism, the DNA macromolecules in solution are schematized as shown in figs. 2a, b. Following[12], it is assumed that DNA is organized according to a regular structure, whose geometrical characteristics are periodic in the x-y plane. Each circle represents the transverse section of a macroion, which is regarded as a charged rod (DNA double strand) indefinitely extending along the z-direction. Each dashed contour delimits the free volume around each macroion. This volume depends on the polyelectrolyte concentration, and contains both the counter-ions and the mobile ions coming from the dissociated salt molecules. We reserve the denomination "counter ions" to the mobile ions arising from the dissociation of the DNA macromolecules. Each elementary region of such an ideal structure must be electrically neutral. Inside each elementary volume, the electric field exhibits an almost circular symmetry around the rod, so that any one of these volumes can be ideally represented by the cylindrical structure shown in Fig.2b.

The linear DNA double strand corresponds to the inner cylindrical rod where a uniform negative charge density $-q\lambda$ per unit of length is assumed (q is the absolute value of the electron charge). This is obviously an approximation, which rests on the fact that the DNA macroion can be thought as an array of unitary electron charges. The periodical distance d of these charges is half the DNA base-pair spacing (3.4 Å for the B type), so that the parameter value is 0.59 $Å^{-1}$.

The cylindrical domain $a \leqslant r \leqslant R$ models the free volume around the macroion, whose radius a is about 10 Å. The radius R depends on the DNA concentration (πR^2 is the area of any polygonal surface in Fig.2a).

We consider the following three kinds of mobile ions:

- positive counter-ions: volume concentration C_o, valence z_o;
- positive salt ions: volume concentration C_p, valence z_p;
- negative salt ions: volume concentration C_m, monovalent.

The negative salt ions are assumed monovalent in order to simulate the most common situations.

At thermodynamic equilibrium, the electric potential φ and the concentrations C_o , C_p , C_m are assumed to fulfil the

Poisson-Boltzmann equation:

$$\nabla^2\varphi=-q[z_o C_o+z_p C_p-C_m]/\mathcal{E} \quad (1)$$

where

$$C_o=C_{oa}\exp[-qz_o\varphi/KT] \quad (2)$$

$$C_p=C_{pa}\exp[-qz_p\varphi/KT] \quad (3)$$

$$C_m=C_{ma}\exp[q\varphi/KT] \quad (4)$$

K being the Boltzmann constant, T the absolute temperature, and $\mathcal{E}$ he solvent dielectric permittivity. Assuming $\varphi(a)=0$, the constants C_{oa}, C_{pa}, C_{ma} have the physical meaning of volume concentrations close to the rod surface for the three kinds of ions. The electric potential φ depends on the r-coordinate only, as well as C_o, C_p, C_m.
We are aware that eq. 1 can be correctly used at the macroscopic level only, so that it can be argued that its application on the angstrom scale is questionable. Nevertheless, we keep this approach because it is consistent with our very simplified model and with our purpose to avoid any cumbersome microscopic statistical analysis in searching for simple phenomenological guidelines for the ions interaction mechanism. Its main features should not be lost in applying the Poisson-Boltzmann equation under so extreme conditions[12]. The boundary conditions for the problem can easily be found. First of all, the electric field on the rod is given by:

$$-\left.\frac{d\varphi}{dr}\right|_{r=a}=-q\lambda/2\pi a\mathcal{E} \quad (5)$$

Then, due to the overall charge neutrality of the elementary cell, we must have

$$-\left.\frac{d\varphi}{dr}\right|_{r=R}=0 \quad (6)$$

An independent constraint is imposed on the quantity of mobile ions dissociated from the salt molecules. Let v be the generic cylindrical volume corresponding to a tract of length ℓ along the z coordinate, and let qn_p be the assigned amount of charge for both the positive and the negative ions inside v. Then, we have:

$$\int_v z_p C_p\,dv=n_p \quad (7)$$

$$\int_v C_m\,dv=n_p \quad (8)$$

A similar constraint could be imposed on the counter-ions, since their amount of charge in v must be equal to $q\lambda\ell$. However, this condition is not independent of the boundary condition (6) and the

constraints (7) and (8). We do not consider any binding sites on the rod surface, where some ions could adsorb.
The model depicted in Figs.2a, b and eqs. (1-8) completely define the problem, which can be summarized as follows: to find the electric potential φ and the unknown concentrations C_{oa}, C_{pa}, C_{ma} which fulfil both the Poisson-Boltzmann equation and the integral constraints (7), (8), the boundary conditions being $\varphi(a)=0$ and eqs. (5), (6). This problem cannot be solved in closed form nor by direct numerical integration. In looking for an approximate solution, the main difficulties arise from the nonlinear dependence on φ and on the aforesaid unknown concentrations, which affects both the Poisson-Boltzmann-equation and the integral constraints. We stress that such integral constraints make rather cumbersome any approximate method based on the discretization of the differential equation. The variational formulation outlined in Appendix permits one to handle the problem in a rather direct way. It is shown that the solution φ of the problem makes stationary a functional which represents the electrostatic energy of the structure. Then, approximate solutions for φ can be found looking for this condition when a proper basis of "trial functions" for φ is assumed.
In the next section, an average variable which summarizes the features of the solution for our present purposes is considered. Furthermore, it is worth noting that other kinds of information can be obtained once such an approximate solution is known. For instance, the partial derivative of the electrostatic energy with respect to the λ^{-1} term gives the intensity of the electrostatic repulsive force along the z-direction. The evaluation of this quantity plays a fundamental role in the study of the properties of the polyelectrolyte, because it balances the mechanical forces due to the chemical bonds and it determines the equilibrium conformation of the macromolecule.

RESULTS

The above model has been used in order to simulate the stationary distribution of the various kinds of ions of DNA and salt solutions at T=300° K. Our purpose is to offer extensive results of ion-DNA interaction at different concentrations of monovalent and divalent ions. In order to summarize the main features of such ion distributions under the various physical conditions, an average variable has been introduced for each kind of ions, that is:

$$r = \frac{1}{a} \frac{\int_v C_i \, r \, dv}{\int_v C_i \, dv} \qquad (i=o,\ p,\ m\ )$$

where v is the volume surrounding the macroion ($a \leq r \leq R$). The above quantity plays the role of the "center of mass" of the i-th charge

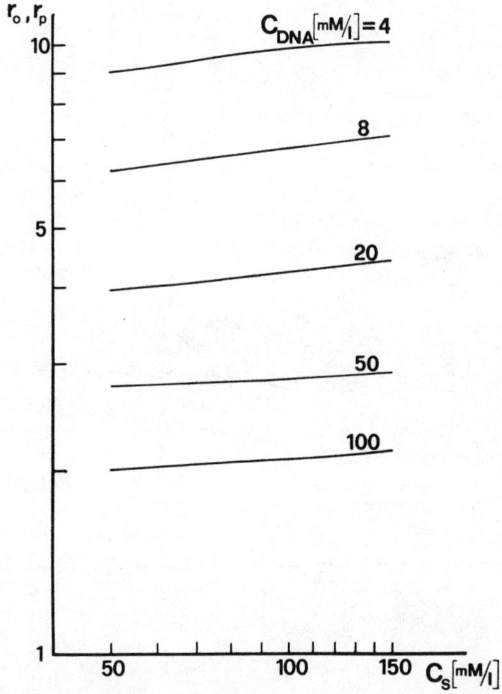

Fig.3 Plot of the normalized centers of mass r_o , r_p of the counter ions and of the positive salt ions vs. the salt molar concentration C_S . Both kinds of ions are monovalent. The various curves deal with different values of the DNA molar concentration C_{DNA}.

cloud inside v normalized with respect to the radius a of the DNA macroion. We recall that, in the case of an impulsive charge distribution at r=a, r_i attains its minimum (unit) value. The behaviour of r_o and r_p has been studied vs. the molar concentration C_s of the salt for various different values of molar DNA concentration C_{DNA} and for the cases of monovalent and divalent salt in solution. The molar concentrations C_{DNA} and C_S are obviously related to the previously introduced parameters, that is:

$$C_{DNA} = \frac{10^{27}\lambda}{\pi R^2 \, 2 N_A}; \qquad C_S = \frac{n_p}{z_p \cdot \ell} \, \frac{10^{27}}{\pi R^2 N_A} \qquad \text{(moles/l)}$$

where N_A is the Avogadro number, R is given in Å, $2/\lambda$ (in Å) is the

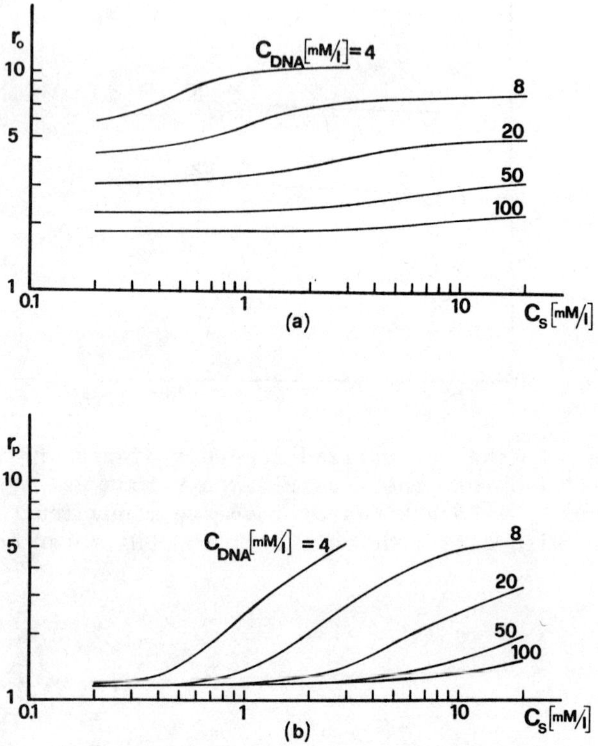

Fig.4a, b Plot of the normalized centers of mass r_o , r_p of the counter ions and of the positive salt ions vs. the salt molar concentration C_S . The positive salt ions (curves r_p) are divalent, while the counter ions are monovalent. The various curves deal with different values of the DNA molar concentration C_{DNA} .

base-pair spacing, n_P /z_P and $\lambda\ell$ are the quantities of positive ions and of counter ions in the volume v.
The range of variation for C_{DNA} is 4-100 mM/l, while C_S ranges from 50 to 150 mM/l and from 0.2 to 10 mM/l for the case of a monovalent and of a divalent salt respectively. These ranges have been chosen in order to include the most common physiological situations. Fig. 3 shows the behaviour of both r_o and r_P vs. C_S for the case of a monovalent salt (we recall that the counter-ions are supposed monovalent, so that there is no difference between the two kinds of ions in this case). For any given parametric value of C_{DNA} (that is, for any given value of R) the two centers of mass increase with C_S . The comparison between the various parametric curves shows that, for any given value of C_S, the centers of mass decrease as C_{DNA} increases. This is not surprising, since the higher is C_{DNA}, the lower is the outward radius R of v: the ion cloud is correspondingly forced to become closer and closer to the macroion.
Figs.4a and b show the behaviour of r_o and r_P vs. C_S for the case of a divalent salt. For any given value of C_{DNA}, the curve r_P pertinent to the center of mass of the positive divalent ion cloud is lower than the corresponding one for the (monovalent) counter ions, r_o . This fact points out the competitive mechanism between ions of different species: divalent ions practically replace monovalent counter ions near to the DNA macroion, thanks to the stronger electric attractive force. This causes the electric field close to DNA to drop rapidly to very low values. As a consequence, the counter ions distributions become practically independent on C_S when this is higher than few mM/l (see fig.4a).
A last consideration can be done on the set of r_P curves (fig.4b). For any given value of C_{DNA}, the center of mass r_P of the divalent ions is initially constant with respect to C_S , and very close to the limit unitary value of an "impulsive" distribution. This points out the fact that the DNA macroion initially behaves like a sink with respect to the (increasing) quantity of divalent ions. When the concentration gradient becomes too large, a fraction of these ions cannot be retained close to the macroion, and the r_P curves begin to increase with C_S . The threshold value of C_S after which this phenomenon occurs depends on the DNA concentration (i.e., on R) for reasons similar to those discussed with reference to fig.3.

CONCLUSIONS

The comparison between the case of a monovalent and of a divalent salt in the DNA solution should have put into evidence that relatively small quantities of divalent ions strongly influence the electric field distribution inside the free-volume surrounding the DNA macroion. Our working hypothesis is that this fact is sufficient to cause structural changes into DNA-chromatin. Such an hypothesis is in agreement with the experimental observations, so that we consider the above described polyelectrolyte approach a good starting point for the interpretation of the very complex phenomena shown in fig.1. The evaluation of the corresponding electrostatic energy offers the correct framework for the study of the equilibrium DNA conformation.

APPENDIX

The solution φ of the problem makes stationary the functional:

$$I=\int_{v}(\nabla\varphi\cdot\nabla\varphi)\,dv \qquad (A1)$$

when all the previously-stated constraints are fulfilled.
Let us consider the first-order variation δI of I due to a variation from φ to $\varphi+\delta\varphi$. We have:

$$\delta I=2\int_{v}[\nabla\varphi\cdot\nabla(\delta\varphi)]\,dv \qquad (A2)$$

Taking into account the Green relationships, the above expression becomes:

$$\delta I=2\cdot\int_{Sv}\delta\varphi\frac{d\varphi}{dn}dS-2\int_{v}\delta\varphi\nabla^2\varphi\,dv \qquad (A3)$$

where S_v is the boundary surface of v, that is, the cylinders r=a and r=R, while $d\varphi/dn$ is the normal derivative to S_v, that is, $\pm d\varphi/dr$. Now, φ can be assigned for r=a, so that the corresponding $\delta\varphi=0$. Moreover, eq. (6) must hold for r=R, where it gives $d\varphi/dn=0$. Consequently, the surface integral is zero. We observe that the second integral, thanks to eqs. (1-4), can be written as

$$2\int_{v}\delta\varphi\frac{q}{\varepsilon}\left[z_o C_{oa}\exp(-qz_o\varphi/KT)+z_p C_{pa}\exp(-qz_p\varphi/KT)-C_{ma}\exp(q\varphi/KT)\right]dv$$

so that it can be divided into three integrals, which are proportional to the first-order variations of the resulting overall charges corresponding to each kind of ions inside v. On the basis of the previous section, all the aforesaid overall charges are assigned (see, for instance, eqs. (7) and (8)), so that their variations must be zero. As a consequence, also the volume integral of eq. (A3) is zero, and we obtain:

$$\delta I=0 \qquad (A4)$$

which completes the proof.

REFERENCES

1. A. A. Pilla: "Electrochemical information transfer at living cell membranes" Ann. NY Acad. Sci., 238, pp.149-157, 1974

2. A. Chiabrera, M. Hindsenkamp, A. A. Pilla, J. Ryaby, D. Ponta, A. Belmont, F. Beltrame M. Grattarola, C. Nicolini: "Cytofluorimetry of electromagnetically controlled cell dedifferentiation " J. Histochemistry and Cytochemistry, 27: 375-381, 1979

3. A. Chiabrera, R. Viviani, G. Parodi, G. Vernazza, M. Hinsenkamp, AA. Pilla, J. Ryaby, F. Beltrame, M. Grattarola, C. Nicolini: "Automated cytometry of electrochemically exposed frog erytrocytes " Cytometry, 1, pp.42-48, 1980.

4. A. Cittadini, A. Scarpa: "Intracellular Mg++ homeostasis of Ehrlich ascites tumor cells " Archives of Biochemistry and Biophysics, 227, pp.202-209, 1983.

5. T. Haga, T. Abe, M. Kurokawa: "Polymerization and depolymerization of micro tubules in vitro as studied by flow birefringence" FEBS letters, 39, pp. 291-295, 1974.

6. E. Carafoli, A. Scarpa Ed.: "Transport ATPases", Ann. NY Acad. Sci., vol. 402, 1982

7. A. Chiabrera, M. Grattarola, M. Parodi in: "Modeling and analysis in Biomedicine" C. Nicolini Ed., Word Scientific Pub., pp.163-184, (1984).

8. M. Behe, G. Felsenfeld: "Effects of methylation on a synthetic polynucleotide: the B-Z transition in poly(dG-mdC)", Proc. Natl. Acad. Sci. USA, 78, pp.1619-1623, 1981.

9. F. Thoma, T. Koller, A. Klug: "Involvement of histone H1 in the organization of the nucleosome and of the salt-dependent superstructures of chromatin ", J. Cell Biol., 83, pp.403-426, 1979.

10. A. Chiabrera, G. Giannetti, M. Grattarola, M. Parodi, P. Carlo, R. Finollo: The role of ions in modifying chromatin structure In: "Electric and electromagnetic stimulation of bone growth", AG. Karger ed, Basilea, 1984.

11. M. Grattarola, P. Carlo, G. Giannetti, R. Finollo, R. Viviani, A. Chiabrera: "Laser flow measurements of scattering and fluorescence from cell nuclei in the presence of increasing Mg++ concentrations", Biophysical J., in press, 1984.

12. R. Fuoss, A. Katchalsky, S. Lifson: "The potential of an infinite rod-like molecule and the distribution of counter-ions", Proc. Natl. Acad. Sci., 37, pp.579-589, 1951

PHYSICO-CHEMICAL MODELS

OF VOLTAGE-DEPENDENT GATING IN MEMBRANES

Gerhard Schwarz

Department of Biophysical Chemistry
Biocenter of the University
CH 4056 Basel, Switzerland

INTRODUCTION

Electric signals play a very important role in biological information transfer, particularly that of the nervous system.[1] They are usually generated and then propagated on an excitable membrane in form of an action potential, i.e. a short-time significant depolarization of the membrane. The electric potential across the membrane exhibits a transient increase above its resting value inducing substantial changes of the strong electric field in the interior domain. These appear to exert crucial effects in the course of the molecular events of the excitation process, namely the opening and closing of ion-conducting channels which is called "gating" for short. Although the detailed mechanism is yet unknown, some relevant physico-chemical aspects can nevertheless safely be pointed out.

FUNDAMENTALS

Phenomenology

Excitation of biological membranes has been studied experimentally to a very large extent, preferentially with the squid axon. It turned out that the effect arises from considerable variations of the ionic conductance of the membrane, caused by changes of the membrane potential and coming off in about a millisecond. A voltage jump of 100 mV may result in a practically complete transition between two extreme levels. These apparently correspond to gating states with the existing channels being all closed or all open, respectively.

There are sodium specific channels, exhibiting a fast activation phase followed by spontaneous inactivation, and also separate channels for potassium ions which undergo activation only. The appropriate dependence on voltage and time is empirically described by the Hodgkin-Huxley equations.[2] These leave us in the dark, however, as far as the molecular mechanism of the voltage dependent gating is concerned.

More recently, it became feasible to block the ion currents through the channels so that displacment currents could be measured which are presumably be associated with the actual gating process.[3] Such a "gating current" presents more direct experimental evidence of the voltage-sensitive steps of gating than the comparatively indirect signal provided by the ionic current.

Structural Basis

Possible gating mechanisms naturally must be discussed with due regard to the inherent features of biological membranes. These are comprised in the now widely accepted "fluid mosaic model".[4] It implies that functional entities such as the "gating particles" are made of integral protein molecules which are firmly incorporated in a lipid bilayer matrix by means of hydrophobic interactions with the hydrocarbon moiety. Protein structures forming channels naturally must span the entire membrane. This calls for hydrophilic regions located towards the aqueous surroundings. In a fairly simple way a channel may be constructed as a circular arrangement of aggregated protomers (i.e. identical subunits) leaving a hole in the center.

Anyway, the molecular nature of native channels needs be elucidated in the course of exploring the gating mechanism. Regarding the voltage-dependent sodium channel some progress in its isolation and characterization has already been achieved.[5] It was shown to include at least a big protein molecule (of about 320 kd molecular weight) with the main constituent being a comparatively large polypeptide chain (of ca. 270 kd).

Electro-mechanical Effects

In addition to the issue of structure, one must raise the question how the field actually brings about the gating function. It has been suggested that this may be caused by a field induced rotation of an electric dipole linked up to the channel. This would require a dipole moment of some hundred Debye units to be turned around over practically 180 degrees. Permanent dipoles of that magnitude are quite reasonable for a protein molecule of moderate size. On the other hand, however, the proposed transverse rotation must be practically ruled out on energetic grounds.[6]

We note that rotation of comparable particles takes around 10^{-7}s in water corresponding to at least 10^{-5}s in a membrane-like medium (on account of the very much enhanced viscosity). Since the observed gating times are of the order of 10^{-3}s we then arrive at an apparent (specific) activation energy of less than 20 kJ/mol which is in fact highly unreasonable. The necessary implication that hydrophobic parts of the protein come across hydrophilic regions of the membrane (and vice versa) calls for a significantly greater expenditure of free energy (- as known from studies of partition equilibria even a small hydrophobic group, e.g. ethyl or propyl, already needs about 10 kJ/mol in order to be transferred into hydrophilic surroundings). The dipolar energy gained in the applied electric field (of $\lesssim$ 10 kJ/mol) is not nearly sufficient to fill the obvious energy gap. Accordingly the transverse rotary motion of an integral protein would become too slow by many orders of magnitude. Analogous argumentation applies to another electro-mechanical field effect, namely the field induced translatory motion of a gating particle.

POSSIBLE MODEL APPROACHES

A Basic "Chemical" Model

Many processes in molecular biology are apparently controlled by a certain conformational transition of a biopolymer constituent. Therefore it seems quite obvious that gating may simply result from appropriate conversions of the channel structure. Such steps are not necessarily subject to a highly restricted rate as pointed out for purely mechanical motion. They could well be fast enough to allow gating within the millisecond range. On the other hand, a sufficient voltage-dependence is readily explained in terms of a change of dipole moment upon the structural transformation under consideration. From a fundamental point of view even a basic model involving only two conformation states can provide a satisfactory picture of channel activation, being quantitatively consistent with experimental data.[6]

Let us assume an integral protein, P, which can exist in two functional states, either P_1 (inactive) or P_2 (active). We refer to our previous detailed discussion as far as the field effect on $P_1 \rightleftharpoons P_2$ is concerned.[7] In order to apply the results to pertinent data of practical work the electric field strength, E (inside the membrane), has to be expressed by means of the membrane potential, V (which is defined as the electric potential in the cell interior relative to the exterior). As is easily seen we have

$$V = E \cdot d + \Delta\psi \qquad (1)$$

where E has to be taken positive (negative) when directed towards the cell exterior (interior) and d is the thickness of the high resistance core of the membrane. The $\Delta\psi$ stands for the appropriate difference of the potential steps possibly generated at the inner and outer interfaces between the membrane and the aqueous phases. Recalling the eqs.(25a,b) of our previous article[7] it eventually turns out that under equilibrium conditions the fraction of P in the "active" state becomes

$$\Theta = \exp\{(V-V_m)/V_s\} / [1 + \exp\{(V-V_m)/V_s\}] \qquad (2a)$$

with the two characteristic parameters

$$V_s = (k_B T)/(m_{12}/d) \quad , \quad V_m = \Delta\psi - V_s \ln K_o \qquad (2b,c)$$

These quantities represent the mid-point (V_m) and measure the sharpness (V_s) of the transition when plotted as Θ versus V. Increasing V would activate more and more P provided $m_{12} > 0$ (i.e., P_2 has a dipole moment greater than that of P_1, taken in the direction of the normal towards the cell exterior). For

$$V \gtrsim V_m + 4V_s \quad \text{or} \quad V \lesssim V_m - 4V_s$$

almost all P are active or inactive, respectively. We note that in this model V_m can assume any value depending on $\Delta\psi$ and particularly on the equilibrium constant K_o which is determined by the zero-field thermodynamic properties of the conformational transition.

Any system of molecular structure is naturally associated with a specific distribution of atomic and electronic charges. Thus a change of structure must give rise to a displacement current, I. It is derived from the time-dependence of the overall dipole moment M. In the present case we have

$$M = M_1 + N\, m_{12}\, \Theta \qquad (3)$$

(M_1 refers to the all-P_1 state, N is the total number of P). For the transition under consideration this gives

$$I = (1/d)(dM/dt) = (m_{12}/d)N(d\Theta/dt) \qquad (4)$$

The time-dependence of Θ is described by the eq.(2b) in our previous article.[7] We apply it to the usual experimental condition involving an equilibrium established at $\Theta = \bar{\Theta}^o$ which is suddenly perturbed (at t=0) by a voltage jump ($\Theta \rightarrow \bar{\Theta}^o + \delta\bar{\Theta} = \bar{\Theta}$). One then finds that

$$d\Theta/dt = (\delta\bar{\Theta}/\tau)\exp(-t/\tau) \tag{5}$$

where we have introduced the relaxation time

$$\tau = 1/(k_{12}+k_{21}) = (1/k_m)(1-\bar{\Theta})^{\alpha}\,\bar{\Theta}^{(1-\alpha)} \tag{6}$$

(k_m is equal to $k_{12} = k_{21}$ at $V = V_m$, $\alpha = m_1^*/m_{12}$). With an inital V small enough to have $\bar{\Theta}^o = 0$ the displacement current evidently becomes

$$I = (m_{12}/d)N(\bar{\Theta}/\tau)\exp(-t/\tau) \tag{7}$$

By integration the corresponding transfer of charge is calculated as

$$Q = \int_0^{\infty} I dt = (m_{12}/d)N\cdot\bar{\Theta} \tag{8}$$

provided the perturbing voltage is maintained long enough ($t \gtrsim 4\tau$).

In early gating current measurements with squid axon only a single exponential time function was observed (apparently reflecting the sodium activation step because it has the smallest τ-value so that the amplitude of I would be rather large). Let us therefore tentatively describe the experimental results in terms of our basic model process. First we note that indeed the registered values of Q asymptotically approach an upper limit, Q_∞, at higher V. According to (8) this means

$$Q_\infty = (m_{12}/d)N \quad , \quad Q/Q_\infty = \bar{\Theta} \tag{9a,b}$$

Furthermore it follows from (2a) that

$$\ln\left\{Q/(Q_\infty-Q)\right\} = (V-V_m)/V_s \tag{10}$$

Actually this linear relation with respect to V fits very well the available data. With those reported by Meves[8] for instance we obtain $Q_\infty = 2.3\cdot10^{-4}$ C/m^2, V_m = -19 mV, V_s = 14.5 mV. Because of (2b) and (9a) one then gets

$$m_{12}/d = 8\ \text{D/Å} \quad , \quad N = 860/(\mu m)^2$$

which looks quite reasonable in view of the known properties of the voltage-dependent sodium channel. In addition, the relaxation times comply with (6) where $1/k_m = 1.4$ ms and $\alpha = 0.44$.

More Complex Models

Actual gating mechanisms are undoubtedly composed of more than just one reaction step. For this reason we must be prepared to treat the theoretical problems in terms of a greater number of conformational and possibly other "chemical" steps. This does not present fundamental difficulties, though the calculations may naturally become more involved. Nevertheless the general approach is quite versatile and very promising because of its potential in exploiting the methods of chemical thermodynamics and kinetics. In particular, the well-known effects on gating exerted by all kinds of chemical agents can be ascribed to different binding affinities of the various functional states of channel structure.

In order to demonstrate a more sophisticated approach let us briefly examine some additional experimental evidence (a detailed discussion has been given elsewhere[9]). Soon after the original gating current measurements it turned out that

(i) after switching off a voltage pulse the charge transfer, Q_{off}, is of different magnitude than Q_{on} depending on the pulse duration[10,11]

(ii) the apparent Q_∞ depends on temperature[12]

(iii) another component of current occurs which is comparatively slow[11]

All these facts are incompatible with a single-step model. They can, however, readily be understood on the basis of a model comprising transitions between three conformations P_1, P_2, and P_3. We may propose the scheme

$$P_1 \underset{}{\overset{\text{(fast)}}{\rightleftharpoons}} P_2 \overset{\text{(slow)}}{\rightleftharpoons} P_3 \tag{11}$$

This generally implies a displacement current which consists of a fast and a slow component (with the relaxation times τ_1 and τ_2, respectively), in other words

$$I = (Q_1/\tau_1)\exp(-t/\tau_1)+(Q_2/\tau_2)\exp(-t/\tau_2) \tag{12}$$

The Q_1, Q_2 can be calculated in terms of the relevant chemical equilibrium constants, dipole moment differences and total number of P. The relaxation time τ_1 refers to $P_1 \rightleftharpoons P_2$ with $P_2 \rightleftharpoons P_3$ being frozen whereas τ_2 applies to the latter step on the condition that the former one is equilibrated.

If $Q_2 \lesssim Q_1$, the magnitude of the slow current component may become too small to be picked up by the measuring instrument. For a pulse duration time τ_p the theory then yields

$$Q_{on} = Q_1(1-\exp[-\tau_p/\tau_1]) \tag{13a}$$

$$Q_{off} = Q_{on} - Q_c(1-\exp[-\tau_p/\tau_2]) \tag{13b}$$

Q_1 is of course the charge transferred in the fast step, $P_1 \rightleftharpoons P_2$, with the amount of P_3 remaining constant. During the pulse the slow process results in changes of P_3 corresponding to a final charge transfer of Q_c. When switching off the voltage pulse the amount of charge transferred back in the fast current thus becomes different from Q_{on}. Furthermore we note that Q_∞, i.e. the apparent upper bound of Q_{on}, is proportional to N_1+N_2 (i.e. the number of P in states P_1 and P_2). The equilibrium distribution of P over all three states must, however, generally be a function of temperature. This explains the temperature-dependence of Q_∞.

Simple Model Systems

A number of chemically well characterized low molecular weight substances form voltage-dependent ion-conducting pores in artificial and natural membranes.[13] The most prominent one among them is alamethicin, an antibiotic linear peptide of 20 amino acids. It brings about a substantial increase of the membrane conductance, g, which can be described by the relation[14]

$$g = g_o\, c_A^n \exp(V/V_s) \tag{14}$$

where g_o depends on the kind and amount of salt added; c_A is the total peptide concentration (the parameters are $n \approx 10$, $V_s \approx 4$ mV). The voltage must have a positive sign on the cis-side of the membrane (i.e. that aqueous compartment to which the peptide was added). In contrast to channels in nerve membranes no asymptotic level of g occurs at accessible values of V.

There is good reason to believe that pores are formed by aggregation of a certain number of monomeric peptides which span the membrane ("barrel stave" model). In single-pore measurements various conductance states corresponding to pores of different size could be identified.[15]

It is so far not clear how the electric field actually promotes the formation of a pore. Anyway it appears reasonable to assume that each pore is finally formed by aggregation of n "active" monomers, A*. The amount of these A* should be a very small fraction of the total peptide so that the concentration of A* could be written as $c_* = \gamma \cdot c_A$ with a "partition coefficient" γ. According to mass action the concentration of pores becomes

$$c_p = K_a\, c_*^n = K_a\, c_A^n\, \gamma^n \tag{15}$$

(K_a being the equilibrium constant of the aggregation). Most simply we may propose A* to contribute a dipole moment m_* to the overall dipole of the membrane (directed towards the trans-side) whereas the bulk of A has no effect. This implies

$$\gamma = \gamma_o \exp\left\{(m_*/k_B T)\cdot E\right\} \tag{16}$$

since γ can be formally treated as the equilibrium constant of $A \rightleftharpoons A^*$. The aggregation step would probably not cause a change of dipole moment. Then K_a remains independent of the field. Upon inserting (16) in (15) we can immediately derive (14) if we only set

$$V_s = k_B T/(n m_*/d) \tag{17}$$

With $V_s \approx 4$ mV, $n \approx 10$ and $d \approx 30$ Å this implies $m_* \approx 90$ D which seems quite possible. In fact, monomer alamethicin molecules in a lipophilic solvent such as octanol have been found to be elongated particles carrying a dipole moment of about 75 D parallel to the long axis (of 35 Å).[16] It must be emphasized, however, that the real molecular nature of the A* cannot be safely specified at the present. There is still a controversial discussion concerning the field-sensitive step. Further experimental data about the relevant physico-chemical properties of the system are apparently needed in order to clarify the situation.

REFERENCES

1. e.g. S.W. Kuffler and J.G. Nicholls, "From Neuron to Brain", Sinauer Associates, Sunderland MA (1977).
2. A.L. Hodgkin and A.F. Huxley, A quantitative description of membrane current and its application to conduction and excitation in nerve, J. Physiol. (London) 117: 500 (1952).
3. W. Ulbricht, Ionic channels and gating currents in excitable membranes, Annual Rev. Biophys. Bioeng. 6: 7 (1977).

4. S.J. Singer and G.L. Nicolson, The fluid mosaic model of the structure of cell membranes, Science 175: 720 (1972).
5. R.P. Hartshone and W.A. Catteral, Purification of the saxitoxin receptor of the sodium channel from rat brain, Proc. Nat. Acad. Sci. USA 78: 4620 (1981).
6. G. Schwarz, On the physico-chemical basis of voltage-dependent molecular gating mechanisms in biological membranes. J. Membrane Biol. 43: 127 (1978).
7. G. Schwarz, Electric field effects on rate processes and on cooperative transitions of macromolecular structures, in these Proceedings.
8. H. Meves, The effect of holding potential on the asymmetry currents in squid giant axons, J. Physiol. (London), 243: 847 (1974).
9. G. Schwarz, Quantitative analysis of activation and inactivation of asymmetry currents in biological membranes, based on a conformational transition model, J. Membrane Biol. 43: 149 (1978).
10. H. Meves and W. Vogel, Inactivation of the asymmetrical displacement current in giant axons of Loligo forbesi, J. Physiol. (London) 267: 377 (1977).
11. C.M. Armstrong and F. Bezanilla, Inactivation of the sodium channel. II Gating current experiments, J. Gen. Physiol. 70: 567 (1977).
12. J.E. Kimura and H. Meves, Effect of temperature on the asymmetrical charge movement in squid giant axons, J. Physiol. (London) 271: 14 (1977)
13. R. Latorre and O. Alvarez, Voltage-dependent channels in planar lipid bilayer membranes, Physiol. Rev. 61: 77 (1981).
14. M. Eisenberg, J.E. Hall and C.A. Mead, The nature of the voltage-dependent conductance induced by alamethicin in black lipid membranes, J. Membrane Biol. 14: 143 (1973).
15. G. Boheim and H.A. Kolb, Analysis of the multi-pore system of alamethicin in a lipid membrane, J. Membrane Biol. 38: 99 (1978).
16. G. Schwarz and P. Savko, Structural and dipolar properties of the voltage-dependent pore former alamethicin in octanol/dioxane, Biophys. J. 39: 211 (1982).

LINEARITY, ADMITTANCE AND ION CONDUCTION IN NERVE MEMBRANE

Harvey M. Fishman

Department of Physiology and Biophysics
University of Texas Medical Branch
Galveston, Texas 77550 USA

INTRODUCTION

Two important matters in a discussion of the physiological effects of external electromagnetic fields on nerve are: (1) the field strength that produces effects and (2) the changes that are induced in membrane conduction. In both (1) and (2) stimulus and membrane properties are involved. For example, consider the relationship between threshold amplitude and frequency of a sinusoidal current applied to a uniform length of squid axon. As Guttman and Hackmeister (1971) found (Fig. 1), a resonance between stimulus frequency and membrane response is indicated with an optimum frequency for stimulation of 69Hz at 10°C. Both the optimum frequency and the sharpness of the curve increase with increasing temperature. The dependence of threshold on stimulus frequency was also found to be very sensitive to the external calcium concentration. How can the threshold dependence on stimulus frequency be explained? The answer can be obtained from the membrane stimulus-response behavior in the subthreshold (linear) domain of applied signal amplitudes and, in particular, from determinations of the membrane impedance and its dependence on ion conduction. Furthermore, there is the additional problem that the field variation across a 50A membrane is totally unknown. The local field strength at important sites of structures constituting ion channels through the membrane could vary considerably from the assumption which is made here, namely, that the control of the voltage drop across a membrane produces a constant field within the membrane.

EARLY IMPEDANCE MEASUREMENTS

The first demonstration that spike generation in nerve involves

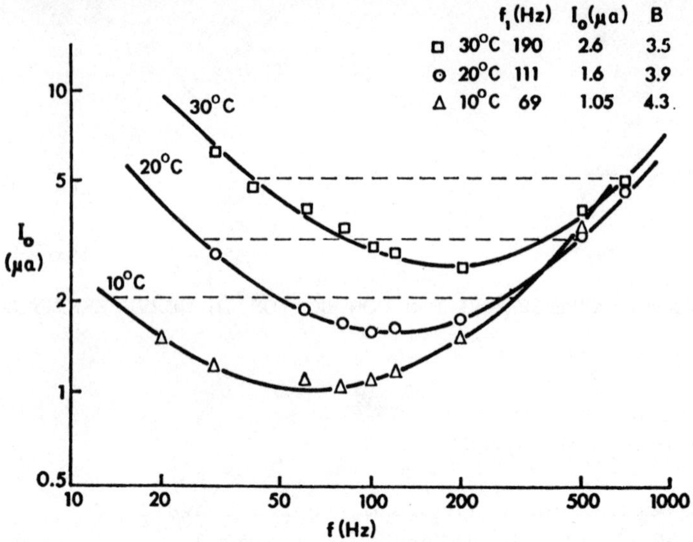

Fig. 1. Threshold current (I_o) as a function of the frequency (f) of a sinusoidal stimulus applied as a current source to a "space-clamped" squid axon at three temperatures. f_1 is the optimum frequency of stimulation and B the width between points $\sqrt{2}\ I_o$ on each curve. (From Guttman and Hackmeister, 1971, by permission of J. Gen. Physiol.).

changes in membrane conductance was made by measurement of the transverse impedance change of a squid axon during impulse propagation (Cole and Curtis, 1939). These results were followed by the observation of an anomalous inductive reactance in the membrane impedance locus of a resting squid axon (Cole and Baker, 1941), which was extracted from measurements between two external longitudinal electrodes (Fig. 2). Although the anomalous inductive reactance seemed odd at the time, it eventually was explained as the frequency-domain manifestation of a time and voltage dependent membrane ion conductance (Cole, 1947; Hodgkin and Huxley, 1952; Chandler et al., 1962) (Fig. 3). However, the actual demonstration that K and Na conductances and their kinetics determine the low frequency impedance of squid axon was not made until almost 40 years after the Cole and Baker measurements. The amplitude range for a linear response to a perturbation in membrane current or voltage was also not quantified during this period.

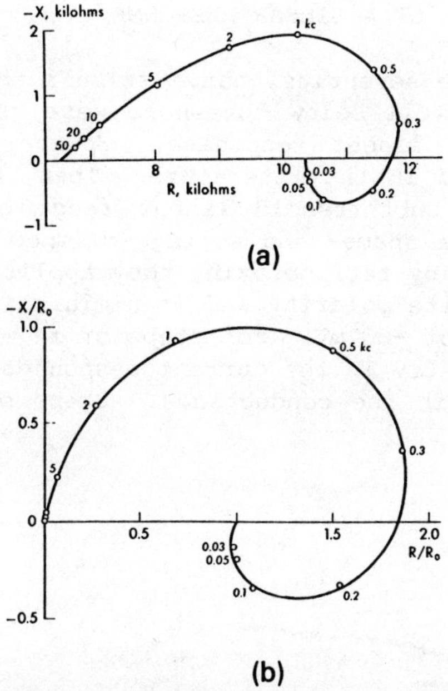

Fig. 2. (a) Longitudinal impedance of a squid axon, measured by Cole and Baker (1941), between two distant electrodes. (b) The membrane impedance after removal of the cable impedance in (a). After Cole (1972), by permission of University of California Press.

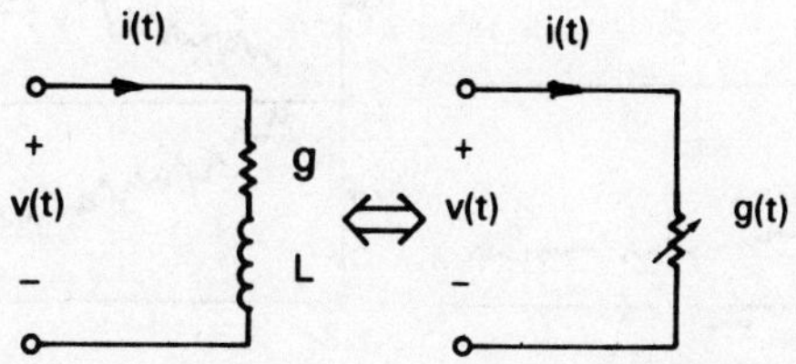

Fig. 3. The explanation of the Cole and Baker inductive-like reactance in the impedance locus of Fig. 2 is due to the temporal variation of membrane conductance which is indistinguishable from the response of a gL circuit.

THE AMPLITUDE RANGE OF A LINEAR RESPONSE

Until the late seventies, nerve stimuli that produced membrane potential changes well below threshold were generally assumed to produce reasonably linear responses. However, no quantitative information existed in the literature. Then, Moore et al. (1980) characterized the subthreshold linear range by three different measurements. In a space- and voltage-clamped squid axon, with Na conduction blocked by tetrodotoxin, they applied step voltage clamps of equal and opposite polarity and of diminishing amplitudes about a holding potential of -60 mV. For steps of ±4 mV (±8 kV/cm) (Fig. 4) a pronounced asymmetry in the current responses was observed indicating nonlinearity in the conductance. Steps of ±1 mV (±2 kV/cm)

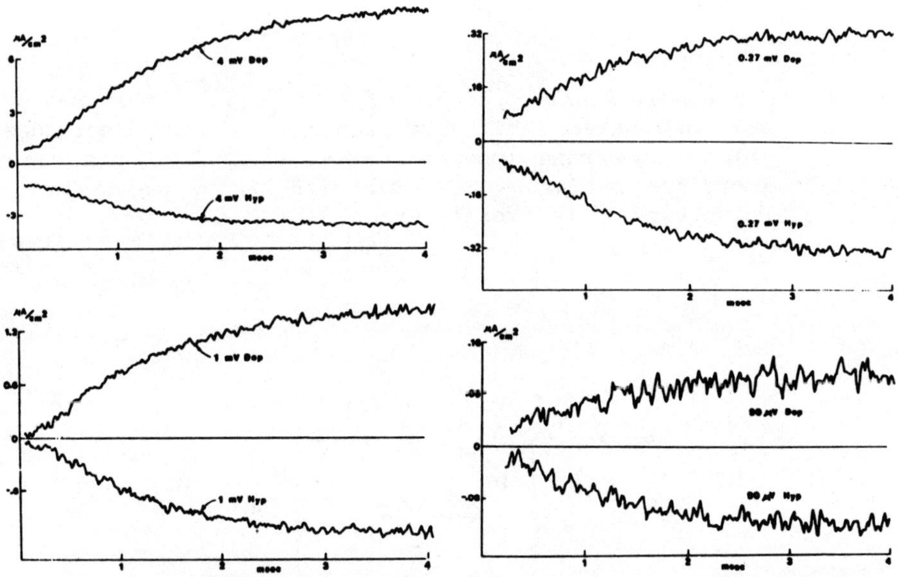

Fig. 4. Relaxation of K conductance of a squid axon in response to equal amplitude but opposite polarity step changes of membrane potential. The response to the 4-mV (8kV/cm) steps were averaged 16 times and the 1-mV (2kV/cm) responses, 128 times. The 0.27-(0.54kV/cm) and 0.09-(180 V/cm) mV responses were averaged 1024 times. The asymmetry in the responses develops with time but disappears at the smallest amplitude approaching spontaneous fluctuations. (From Moore et al., 1980, by permission of J. Membr. Biol.).

produced much less, but still measurable, asymmetry down to steps of ±90 μV (±180 V/cm), which are comparable to spontaneous fluctuations, where the asymmetry was imperceptible. In the second set of measurements, the membrane was voltage clamped to a 10 Hz sinusoid and nonlinearity in the response was measured by spectrum analysis of the amplitude of harmonics generated in the response as the amplitude of the applied sinusoid was increased. Fig. 5 shows that for applied fields up to 3 kV/cm the nonlinearity is due to the generation of second harmonic and is quite small (less than 10% (-20dB) of the response at the fundamental frequency). Electric fields above 3 kV/cm elicit third and fourth harmonics in addition to the second, and the amplitude of the harmonics becomes significant with respect to the amplitude of the fundamental, thus

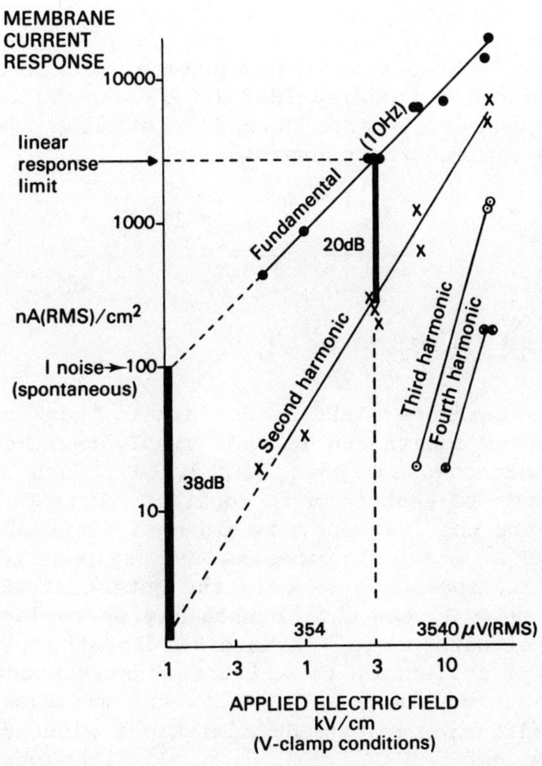

Fig. 5. Harmonic analysis of a squid axon membrane current in response to a sinusoidal (10 Hz) voltage clamp of various amplitudes to explore the range of a linear response. Applied electric field strength assumes constant field throughout membrane and a membrane thickness of 50A. (After Moore et al., 1980, by permission of J. Membr. Biol.).

producing substantial nonlinearity at applied fields of about 10 kV/cm (where the firing of an action potential would normally occur). For fundamental frequencies higher than 10Hz less nonlinearity was apparent at the same applied field. This latter result can also be seen in the temporal response to small step clamps in Fig. 4 in which the asymmetry develops with time and is most pronounced at long times after the step voltage change. DeFelice et al. (1981) confirmed the essential character of harmonic generation and the development of nonlinearity under conditions in which Na as well as K conduction was operative in the squid axon. FitzHugh (1983) provided a comparison with calculations of harmonic generation based on the nonlinearities in the Hodgkin-Huxley (1952) formulation. FitzHugh obtained the interesting result that the slope of the amplitude versus frequency relation for any given harmonic response to a sinusoidal voltage, in a double log plot, is equal to the order of the harmonic.

MEMBRANE ADMITTANCE

Linearization of the nonlinear equation for current flow in a squid axon by Hodgkin and Huxley (HH) (1952) was followed by derivation of the membrane admittance (Chandler et al., 1962; Mauro et al., 1970) which yielded the expression:

$$Y(j\omega) = j\omega C + g_\infty + \frac{g_n}{1+j\omega\tau_n} + \frac{g_m}{1+j\omega\tau_m} + \frac{g_h}{1+j\omega\tau_h} \qquad (1)$$

and

$$g_\infty = g_L + n_\infty^4 \bar{g}_k + m_\infty^3 h_\infty \bar{g}_{Na} ,$$

where each of the parameters are as defined in Hodgkin and Huxley (1952). The complex admittance is most simply realized as a circuit with 7 parallel branches as shown in Fig. 6a. Each branch of the circuit corresponds to each term in equation (1). The terms of the form $g(1+j\omega\tau)^{-1}$ are the frequency-domain equivalent of the time-domain relaxation of the n, m and h processes and are modeled simply as gL branches. However, because $g_m < 0$ in the potential range $V_K < V_K < V_{Na}$ and τ_m must be > 0, the $g_m L_m$ branch may be replaced by a $g_m C_m$ branch in parallel with a $-|g_m|$ branch (Mauro et al., 1970) as in Fig. 6b. Finally, collection of all the shunt conductances into a single equivalent conductance g'_∞ yields the circuit in Fig. 6c, which is the realization of the HH admittance with all conducting systems operative.

Equation (1) has been used to calculate membrane admittance and to compare the calculations with actual admittance measurements under a variety of conditions (Fishman et al., 1977a; Fishman et al., 1977b; Poussart et al., 1977; Fishman et al., 1979; Fishman et al., 1981). First, under normal conditions with all ionic conductances operative, a qualitative comparison between data in a squid

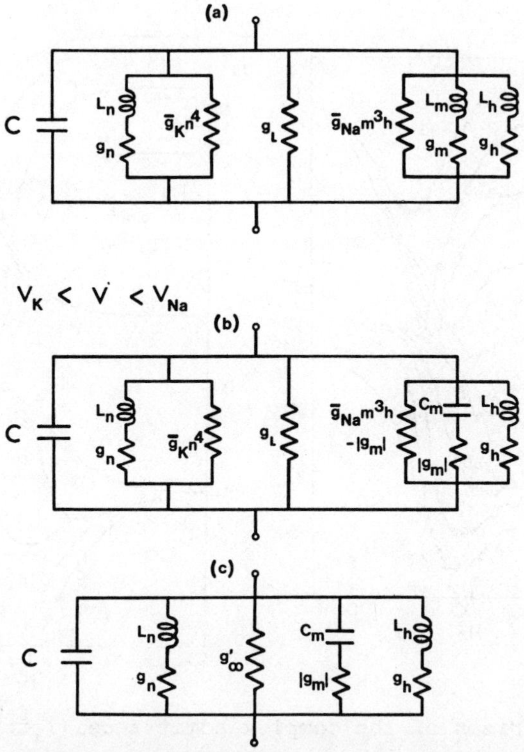

Fig. 6. Electrical circuit realization of squid axon membrane admittance (equation 1) from the linearized Hodgkin-Huxley equations. See text for details. (From Fishman et al., 1977b, by permission of J. Membr. Biol.).

axon and those computed from the HH admittance show some similarities and some differences (Fig. 7). At hyperpolarized potentials (< -60mV) the membrane shows RC behavior and with depolarizations antiresonances occur at increasing frequencies. These features are also in the computational model, although the model shows an extremely sharp antiresonance at -40mV and humps in magnitude and phase at -30 which are not observed experimentally. An explanation of these discrepancies is deferred until discussion of the Na conduction system.

In order to simplify interpretation of the admittance it is more convenient to study the admittance of a membrane with only one of the ion conduction systems operative at a time. The K conduction system has first-order kinetics in the HH description, but it yields

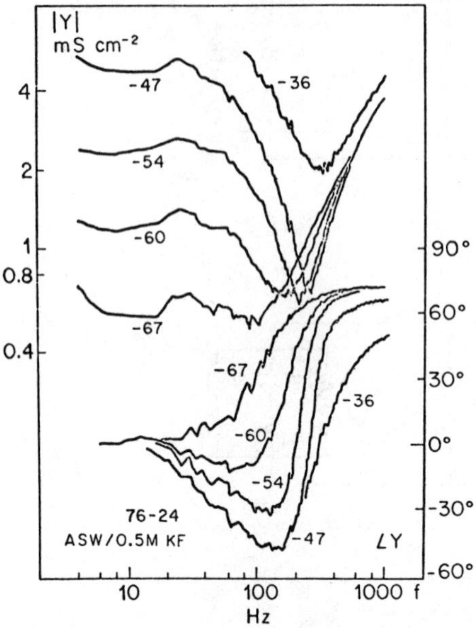

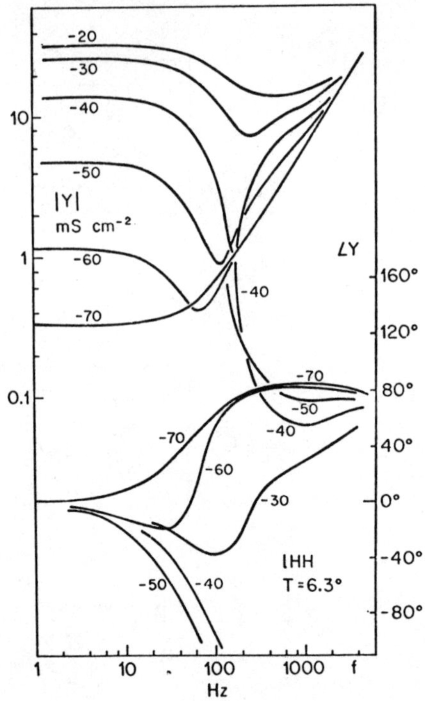

Fig. 7. Comparison of the complex admittance (left) measured in a squid axon with all ion conductances operative with (right) the calculated admittance from equation (1). (After Poussart et al., 1977, by permission of Ann. NY Acad. Sci.).

an antiresonant admittance when shunted by membrane capacitance. This is demonstrated in Fig. 8 in which the admittance of the HH K conductance system alone and in parallel with an ideal membrane capacitance is plotted for displacements of membrane potential from a fixed holding value. By experiment, the antiresonance in the admittance (Fig. 9) was shown, as Cole (1947) suggested, to be due to the "anomalous reactance" of the K conductance interacting with membrane capacitance. During acquisition of the data of Fig. 9, Na conduction was suppressed with tetrodotoxin while K conduction remained operative by internal perfusion with 0.5M K. The admittance shows an antiresonance. Upon suppression of K conduction by changing the internal perfusate to 0.3 Cs, the antiresonance disappears at all membrane potentials. A complete set of admittance functions at 5mV increments from -59 to 16mV with the K conductance the dominant membrane conduction system is shown in Fig. 10. By comparison with plots of the computational admittance under

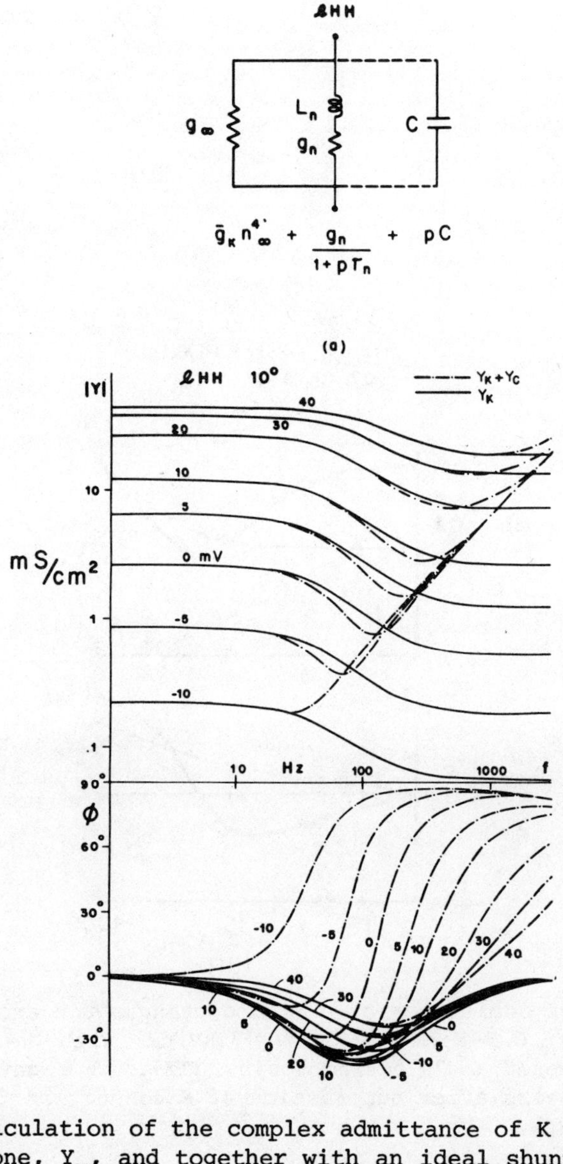

Fig. 8. Calculation of the complex admittance of K conduction alone, Y_K, and together with an ideal shunt capacitance, $Y_K + Y_C$. Magnitude, $|Y|$, and phase angle, ϕ of admittance were calculated assuming a specific resistance of $7\Omega\cdot cm^2$ in series with the K-related terms in equation (1). (After Fishman et al., 1979, by permission of J. Membr. Biol.).

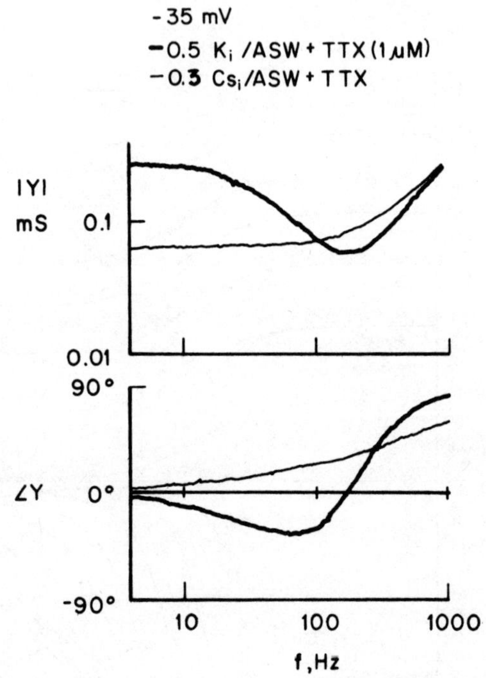

Fig. 9. Complex admittance of a K conducting squid axon (thick curves, 0.5 K_i) at -35 mV (70kV/cm) with Na conduction suppressed with tetrodotoxin (TTX). The antiresonance disappears after suppression of K conduction (thin curves, 0.3 Cs_i).

the same conditions, it is clear that there is good qualitative agreement between the linearized HH description for K conductance and the squid axon admittance.

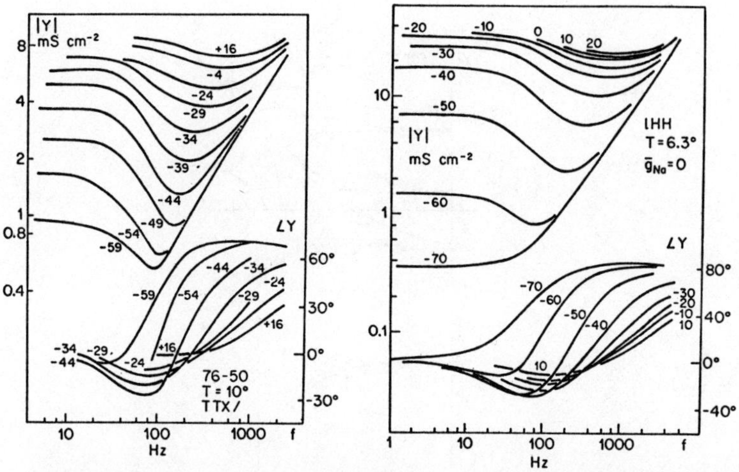

Fig. 10. Comparison of the admittance (left) measured in a K conducting squid axon over a 70mV range with (right) the computed admittance from equation (1) with $\bar{g}_{Na}=0$. (After Poussart et al., 1977, by permission of Ann. NY Acad. Sci.).

The Na conduction system, even when linearized, presents much more complicated admittance functions, in conjunction with membrane capacitance, than the K conduction system. It has second order kinetics and it contributes a steady-state negative conductance. As indicated in Fig. 11 the computed admittance of Na conduction with and without membrane capacitance shows several features. A 180° phase angle at low frequencies, which is associated with the steady state negative conductance ($|g_{Na}| < 0$), is important for understanding how the Na system can interact with the K conduction system. Pronounced humps and antiresonant phenomena are also possible. In experiments in which axons are internally perfused with Cs solutions, K conduction is eliminated and Na conduction is the dominant remaining system (Chandler and Meves, 1970). Under such conditions, the phase angle of membrane admittance at a potential of -30mV is nearly 180° (Fig. 12) as predicted from the calculations on the HH model (Fig. 11). Upon suppression of Na conduction by

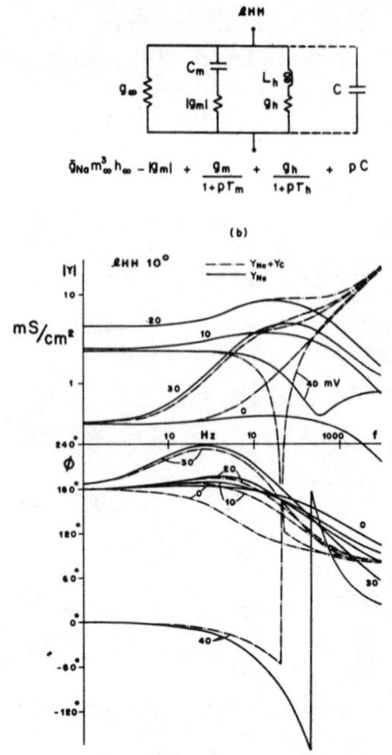

Fig. 11. Calculation of the complex admittance of Na conduction alone, Y_{Na}, and together with an ideal shunt capacitance, $Y_{Na}+Y_C$. A specific resistance of $7\Omega\cdot cm^2$ was assumed in series with the Na-related terms in equation (1). (After Fishman et al., 1979, by permission of J. Membr. Biol.).

addition of 1μM tetrodotoxin to the external sea water, the phase function changes to one that does not exceed 90° (Fig. 12). Thus, the Na conducting system produces the 180° phase shift. The 180° phase means that the real part of the complex admittance is negative.

Comparison of membrane admittance for a Na conducting membrane over a 40-50mV potential range with computations from the HH admittance (Fig. 13) shows good qualitative agreement. Humps in the magnitude function and approaches to 180° in the phase at low frequencies occur in both cases. However, the computational admittance contains an antiresonance at -20mV whereas the measured admittance does not. The reason for this disagreement and the disagreement between computation and data at -40mV in Fig. 7 lies in the neutralization of positive conductances by the negative conductance contributed by the Na system. Whenever the real part of the admittance is negative, an antiresonance is not observed. However, if there is sufficient residual leakage or contribution from K conduction to neutralize the negative Na conductance, such that the real part is positive and large, an antiresonance will occur. These

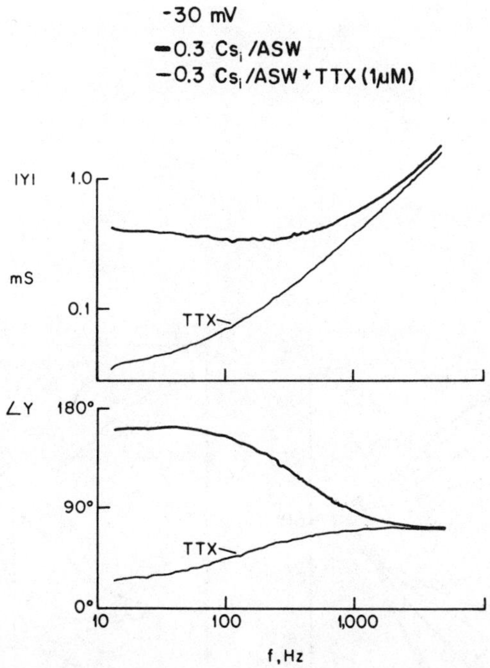

Fig. 12. Complex admittance of a Na conducting squid axon at -30mV (60kV/cm) without K conduction (0.3 Cs_i). The 180° phase shift at low frequency indicates that the real part of the admittance is negative (steady-state negative conductance). The negative real part disappears (phase function <90°) after suppression of Na conduction with TTX. (After Fishman et al., 1983, by permission of Biophys. J.).

effects are demonstrated in Fig. 14, where a Na conducting axon and the computational axon are both shunted by a positive conductance to produce an antiresonant admittance.

Resonant impedances due to the time variation of membrane ion conductances have also been measured in skeletal muscle (Moore, 1983) and in snail neurons (Miyamoto and Fishman, 1983). In the muscle and neuron impedances both K and Na conductances play a role similar to those described here for the squid axon. Furthermore, in the snail neuron Ca conductance also contributes significantly to the neuronal impedance.

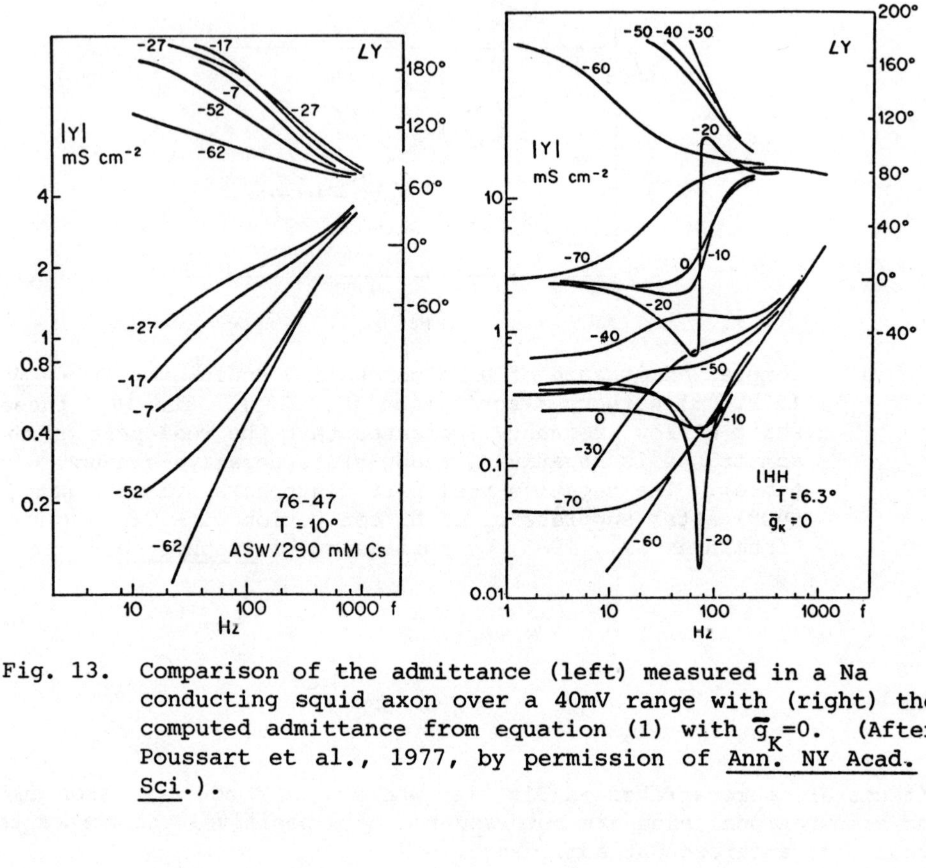

Fig. 13. Comparison of the admittance (left) measured in a Na conducting squid axon over a 40mV range with (right) the computed admittance from equation (1) with $\bar{g}_K=0$. (After Poussart et al., 1977, by permission of Ann. NY Acad. Sci.).

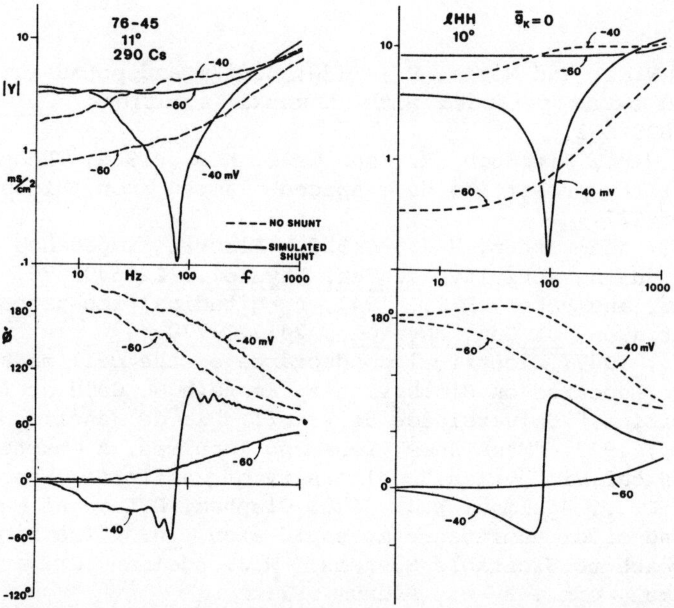

Fig. 14. Comparison of the complex admittance (left) measured in a squid axon with (right) the calculated admittance from equation (1) with parameters that take into account the experimental conditions. No antiresonance occurs at -40 or -60mV (dashed curves); however, after simulation of a positive shunt conductance to make the net shunt conductance positive (ϕ < 90°) experimentally and by increasing g_L in eq. (1), an antiresonance appears at -40mV (solid curves) in both the measured and computed admittance. (From Fishman et al., 1979, by permission of J. Membr. Biol.).

CONCLUSIONS

Nonlinearity becomes significant in axons at field strengths of about 3 kV/cm. This level is about a third of the threshold value. The frequency of stimulation also is a factor in determining threshold because axon membrane impedances resonate at about 100 Hz due to the reactance-like behavior of the time- and voltage-dependent ionic conductances. The membrane impedance at frequencies below 1 kHz is affected by all conditions that alter membrane ion conductances because conduction in ion channels dominates membrane impedance at low frequencies.

I thank Ms. Sarah Adams for typing this manuscript. This work was supported by NIH grants NS11764 and NS13778.

REFERENCES

Chandler, W.K., and Meves, H., 1970, Sodium and potassium currents in squid axons perfused with fluoride solutions, J. Physiol., 211:623-652.

Chandler, W.K., FitzHugh, R., and Cole, K.S., 1962, Theoretical stability properties of a spaced-clamped axon, Biophys. J., 2:105-127.

Cole, K.S., and Curtis, H.J., 1939, Electric impedance of the squid axon during activity, J. Gen. Physiol., 22:649-670.

Cole, K.S., and Baker, R.F., 1941, Longitudinal impedance of the squid giant axon, J. Gen. Physiol., 24:771-778.

Cole, K.S., 1947, Electrical conductance of the cell membrane, in: "Four Lectures on Biophysics by Kenneth S. Cole at Instituto de Biofisica," Universidade de Brasil, Rio de Janeiro.

Cole, K.S., 1972, "Membranes, Ions and Impulses, A Chapter of Classical Biophysics," University of California Press, Berkeley.

DeFelice, L.J., Adelman, W.J., Jr., Clapham, D.E., and Mauro, A., 1981, Second-order admittance in squid axon, in: "The Biophysical Approach to Excitable Systems," W.J. Adelman, Jr. and D.E. Goldman, eds., 37-63, Plenum, NY.

Fishman, H.M., Moore, L.E., and Poussart, D.J.M., 1977a, Asymmetry currents and admittance in squid axons, Biophys. J., 19:177-183.

Fishman, H.M., Poussart, D.J.M., Moore, L.E., and Siebenga, E., 1977b, K^+ conduction description from the low frequency impedance and admittance of squid axon, J. Membr. Biol., 32:255-290.

Fishman, H.M., Poussart, D., and Moore, L.E., 1979, Complex admittance of Na^+ conduction in squid axon, J. Membr. Biol., 50:43-63.

Fishman, H.M., Moore, L.E., and Poussart, D., 1981, Squid axon K conduction: Admittance and noise during short- versus long-duration step clamps, in: "The Biophysical Approach to Excitable Systems," W.J. Adelman, Jr. and D.E. Goldman, eds., 65-95, Plenum, NY.

Fishman, H.M., Leuchtag, H.R., and Poussart, D., 1983, Fluctuation and linear analysis of Na current kinetics in squid axon, Biophys. J. 43:293-307.

FitzHugh, R., 1983, Sinusoidal voltage clamp of the Hodgkin-Huxley model, Biophys. J., 42:11-16.

Guttman, R., and Hackmeister, L., 1971, Effect of calcium, temperature and polarizing currents upon alternating current excitation of space-clamped squid axons, J. Gen. Physiol., 58:304-321.

Hodgkin, A.L., and Huxley, A.F., 1952, A quantitative description of membrane current and its application to conduction and excitation in nerve, J. Physiol., 117:500-544.

Mauro, A., Conti, F., Dodge, F., and Schor, R., 1970, Subthreshold behavior and phenomenological impedance of squid giant axon, J. Gen. Physiol., 55:496-523.

Miyamoto, S. and Fishman, H.M., 1983, Ion conductance effects in the low frequency impedance of snail neuron, Biophys. J., 41:58a.

Moore, L.E., 1983, Effects of voltage-dependent ion-conduction processes on the complex admittance of single skeletal muscle fibers, in: "Structure and Function in Excitable Cells," D.C. Chang, I. Tasaki, W.J. Adelman, Jr., and H.R. Leuchtag, eds., 175-196, Plenum, NY.

Moore, L.E., Fishman, H.M., and Poussart, D.J.M., 1980, Small-signal analysis of K^+ conduction in squid axons, J. Membr. Biol. 54:157-164.

Poussart, D.J.M., Moore, L.E., and Fishman, H.M., 1977, Ion movements and kinetics in squid axon, I. Complex admittance, Ann. NY Acad. Sci., 303:355-379.

NONLINEAR IONIC FLUXES IN EMF EXPOSED CELLS

Lucia Marconi, Giovanna Morgavi, Sandro Ridella
and Claudia Rolando

Istituto per i Circuiti Elettronici C.N.R.
Via All'Opera Pia 11 - 16145 Genova, Italy

INTRODUCTION

During the Erice School on "Modelling and analysis in biomedicine" in 1982 we presented a paper about the interaction between an electrical field and a biological system (Drago et al. 1982 and 1984): all the elements were modelled by linear equations. This simplification was necessary ought to the complexity of the nonlinear analysis and to the fact that the nonlinear analysis itself needs the results of the linear computations to be carried on as we will show subsequently. The reason of these studies is in understanding the origin of possible "non thermal" phenomena in the interaction between non ionizing radiations and living systems (Lerner 1984). The problem has been carried on experimentally by many research groups in the world (Adey 1981, Becker et al. 1982, Beltrame et al. 1980), but few papers have been published on the mechanisms of the interactions (Taylor et al. 1979). Moreover, often these mechanisms are described qualitatively since it is not possible to specify numerically all the parameters in the model. The main result of the linear analysis in Drago et al. (1982 and 1984) was to discover that with the present knowledge, a cell behaves like an electrical filter: that means that many transfer functions between a bioelectrochemical variable of the system and the externally applied electrical field show a non monotonic behaviour in the frequency range from 0 to 10 MHz. It must be stressed that this result has been obtained without adjusting any system parameter. At the same time we are well aware that this is not a proof of the existence of "nonthermal" effects, since the amplitude of the signals are not involved and the time behaviour of the variables is periodic which means that in the mean no modification occours. We decided then to carry on as a second step the nonlinear analysis in order evaluate the values of the rectified ionic fluxes and possible related biological effects.

THE MODEL

In figure 1 the model of the cell exposed to an electric field is shown.

The model we used has the following features:

1) The external medium and the cytoplasm are modelled by complex dielectric constants ε_1, and ε_3 respectively:

$$\varepsilon_{1,3} = \varepsilon_{\infty 1,3} + \frac{\varepsilon_{0 1,3} - \varepsilon_{\infty 1,3}}{1 + J\, f/f_{0 1,3}} + \frac{\sigma}{J 2\pi f \varepsilon_v} \qquad (1)$$

where ε_∞ is the high frequency dielectric constant; ε_0 is the low frequency dielectric constant; f_0 is the dipolar relaxation frequency; σ is the ions conductivity, f is the frequency (Hz) and $\varepsilon_v = 8.854 \times 10^{-12}$ (F/m).

2) The cell membrane is described by the well established Hodgkin and Huxley equations (Hodgkin et al. 1952).

3) The cell radius is R=6.5μm.

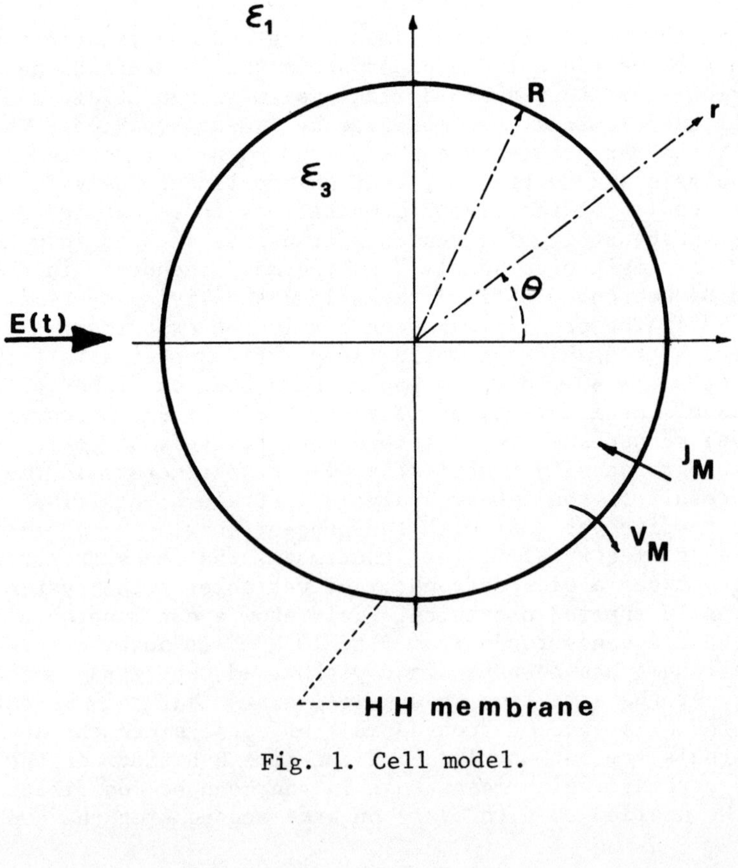

Fig. 1. Cell model.

This model looks like the one presented by Drago et al.(1982 and 1984): the strong simplifications in this paper are justified by our interest in low frequency range of the electromagnetic spectrum since Drago et al. (1982 and 1984) showed that, in this range, the most important effects occur at the cell membrane level. The values of the complex dielectric constants are in the range of the measured ones by Drago et al. (1982 and 1984) while the choice of HH membrane lets to have non linear equations describing separately Na, K and Cl fluxes. This was not available with other membrane models (Drago et al. 1982 and 1984). In order to test the electrochemical transfer hypothesis probably an analysis of the Ca and Mg fluxes should be better (Beltrame et al. 1980): this was not done ought to the non availability of quantitative data.

The HH equations were slightly modified since Na, K and leakage currents at steady state, without any externally applied field, are not zero. Then, following Marchesi et al. (1982) and Casaleggio et al. (1984), we introduced to current generators (one for Na and the other for K) which can be considered to represent sodium and potassium pumps. The equivalent circuit is drawn in figure 2.

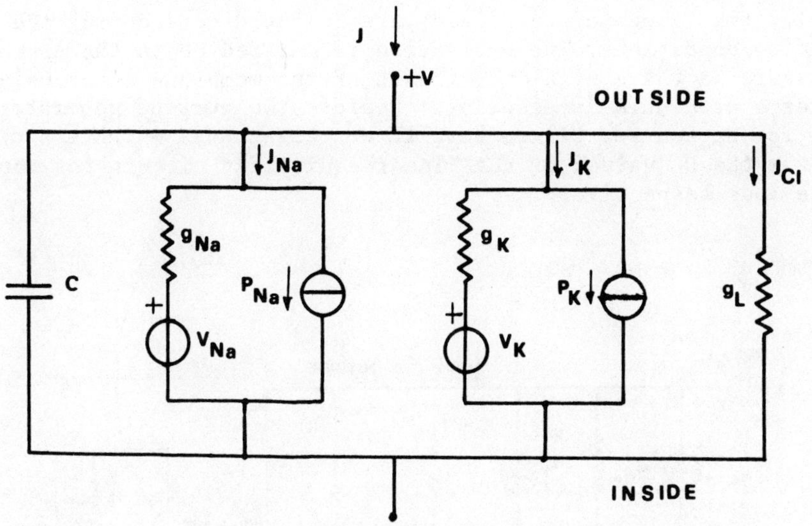

Fig. 2. Hodgkin and Huxley modified membrane model.

This is a very simplified model and we are conscious of its limitations expecially from the biological point of view. Our purpose was to use available data to analyze quantitatively the hypotheses of electrochemical information transfer through living cell membrane (Pilla 1974). Modification of the model are certainly necessary when one is interested in discussing a particular experiment.

RECTIFIED FLUXES COMPUTATION

The linear analysis of the model has been presented by Morgavi et al. (1983) and by Casaleggio et al.(1984). A non monotonic behaviour of the electrochemical variables versus frequency was obtained confirming previous results (Drago et al.1982 and 1984).

The value of the membrane voltage obtained through the linear analysis is used to carry on the evaluation of the rectified ionic fluxes, in the hypothesis that the amplitude of the externally applied electrical field is low: that happens in many therapeutical applications and in many experiments (Chiabrera et al.1979).

Then the procedure suggested by Volterra (Bussgang et al.1974) may be used: it consists in evaluating all quantities at all positions in linear situation and in inserting ideal current or voltage generators in points where non linearity are located. Their values are computed approximating the true value of the generator's controlling variables with the linear value previously obtained. Finally the values of the fields are computed considering the new boundary conditions. The derivation is carried on in the Appendix. In figure 3a the equivalent circuit of the membrane after using Volterra procedure is described: besides two current generators (one for Na and for Cl) we have three resistances which are obtained from the DC values of the linear equivalent circuit for the three ions respectively.

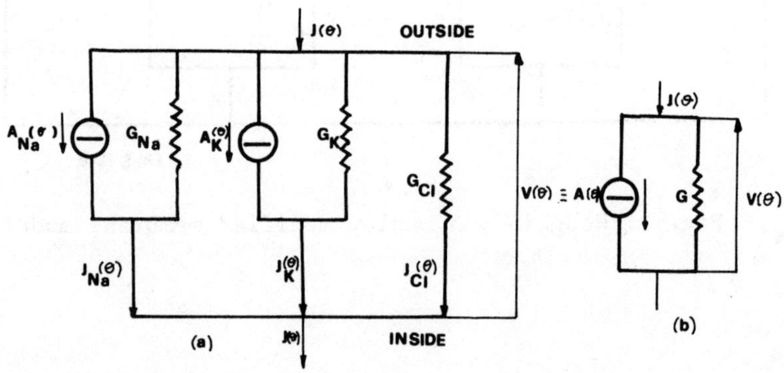

Fig. 3. a. The equivalent circuit of the membrane after using the Volterra procedure.
b. Norton equivalent of the circuit a.

By considering a Norton equivalent (fig. 3b) we obtain

$$A(\theta) = A_K(\theta) + A_{Na}(\theta) \tag{2}$$

$$G = G_{Na} + G_K + G_{Cl} \tag{3}$$

and

$$A(\theta) = A_0 * \cos^2\theta = A_0 * (2/3\ P_2 + P_0/3) \tag{4}$$

where $P_0=1$ and $P2= (3\cos^2\theta - 1)/2$ are Legendre functions.

The Laplace equations outside and inside the cell may be solved:

$$\phi_{int} = K_0 * P_0 + K_2 * (r/R)^2 * P_2 \tag{5}$$

$$\phi_{ext} = H_0 * P_0 * R/r + H_2 * (R/r)^3 * P_2 \tag{6}$$

where ϕ_{int} and ϕ_{ext} are the potentials inside and outside the cell respectively. The current densities at the cell surface are:

$$J_{int}(\theta) = \sigma_3 \frac{\partial \phi_{int}}{\partial r} = \frac{\sigma_3 * K_2 * 2 * P_2}{R} \tag{7}$$

$$J_{ext}(\theta) = \sigma_1 \frac{\partial \phi_{ext}}{\partial r} = -\frac{\sigma_1}{R} * (H_0 * P_0 + 3 * H_2 * P_2) \tag{8}$$

The current density into the membrane is given by:

$$J(\theta) = [\phi_{ext}(R,\theta) - \phi_{int}(R,\theta)] * G + A_0 * (2/3 * P_2 + P_0/3) \tag{9}$$

the continuity condition for the current imposes that:

$$J_{int}(\theta) = J_{ext}(\theta) = J(\theta) \tag{10}$$

and from these equations the unknown constants may be obtained:

$$K_0 = A_0 / (3 * G) \tag{11}$$

$$H_0 = 0 \tag{12}$$

$$K_2 = \frac{2 * \sigma_1 * R * A_0}{6 * \sigma_1 * \sigma_3 + R * G * (3 * \sigma_1 + 2 * \sigma_3)} \tag{13}$$

$$H_2 = -\frac{4 * \sigma_3 * R * A_0}{3 * [6 * \sigma_1 * \sigma_3 + R * G * (3 * \sigma_1 + 2 * \sigma_3)]} \tag{14}$$

It is possible now to compute the rectified quantities everywhere and especially on the membrane: in particular it is possible to evaluate the ions exchange through the cell surface and the pattern of the current outside and inside the cell. It is easy to see from (10) and (11) that there is no current with a space dependence P_0. From (9) one obtains that the rectified ionic currents originate a DC voltage on the membrane which is independent from the angle ϑ whose value is:

$$V_{P_0} = -A_0/(3*G) \quad (15)$$

From Fig. 3 we have computed the total rectified fluxes of K, Na and Cl entering the cell through the membrane surface.

$$I_i = \frac{\oint_{\text{membrane surface}} J_i(\vartheta) * ds}{Z_i * e} \quad (16)$$

$$I = \sum_i I_i \quad (17)$$

where: i=Na, K, Cl; e is the ion charge ($1.6*10^{-19}$ C); Z_i=1 for i=Na, K and Z_i=-1 for i=Cl.

The wave shape of the applied electric field is a pulse train (with period T) with zero mean (fig. 4):

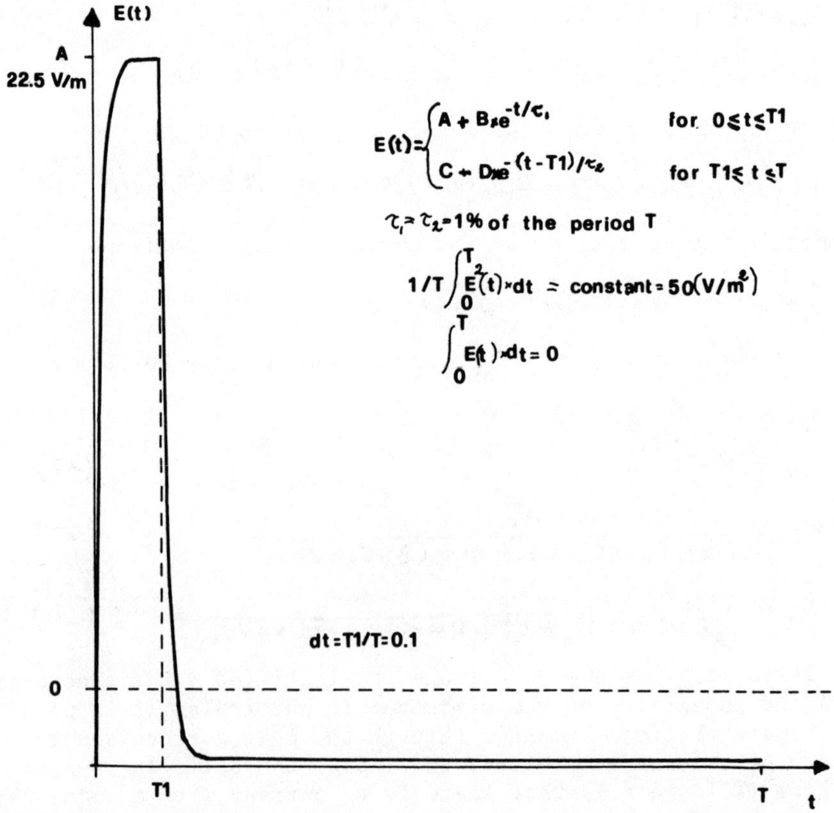

Fig. 4. Pulse train of the applied electric field.

$$E(t) = A + B * \exp(-t/\tau_1) \qquad 0 \leq t \leq T1$$

$$E(t) = C + D * \exp(-(t-T1)/\tau_2) \quad T1 \leq t \leq T \tag{18}$$

$$\tau_1 = \tau_2 = 1\% \text{ of the period } T$$

$$dt = T1/T$$

Signals are considered with constant power: the reference is a sine wave with peak value of 10 V/m:

$$1/T * \int_0^T E^2(t) * dt = 50\ (V/m)^2 \tag{19}$$

The membrane voltage, which is used in Appendix, was computed using the Fourier series of the applied electric field and the linear transfer function between V and E which was obtained by Casaleggio et al. (1984).

In fig. 5 one can see the frequency behaviour of the current generators of the circuit of fig. 3a and 3b for a signal with E=22.5 V/m, dt=0.1.

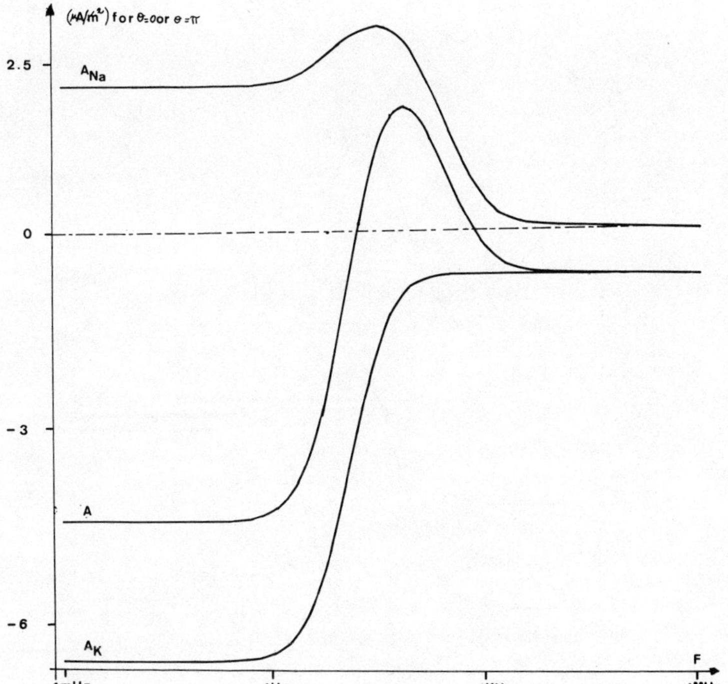

Fig. 5. DC component of the Na (A_{Na}), the K (A_K) currents for E=22.5 V/m and dt=0.1.

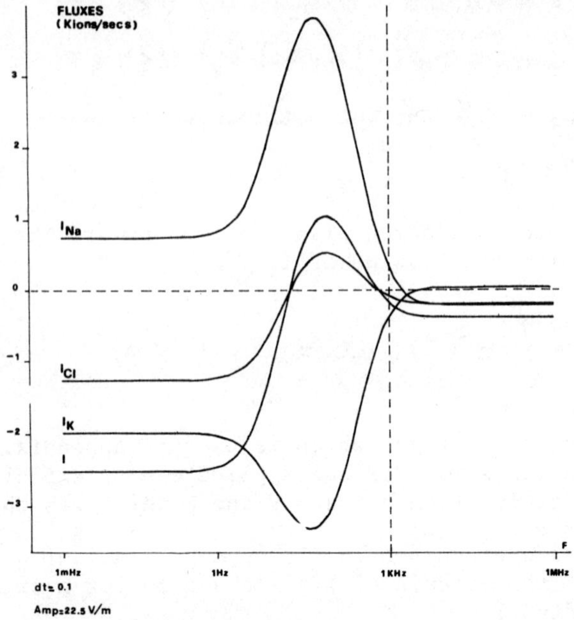

Fig. 6. Total rectified fluxes against the frequency for E=22.5 V/m and dt=0.1

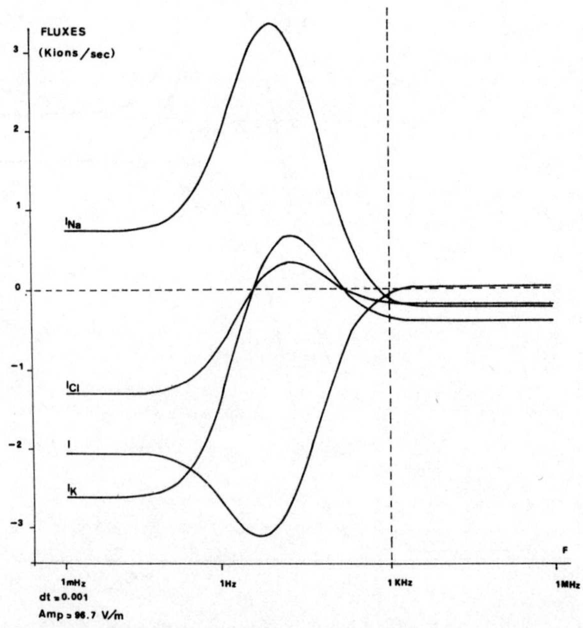

Fig. 7. Total rectified fluxes against the frequency for E=96.7 V/m and dt=0.001

In fig. 6 and 7 two examples of the total rectified fluxes against the frequency are shown for (E=22.5 V/m, dt=0.1) and for (E=96.7 V/m, dt=0.001). These fluxes are strongly frequency dependent: they are proportional to the square of the amplitude of the applied field (Appendix).

If a plot of the total current density outside the membrane is made we obtain the results of figure 8.

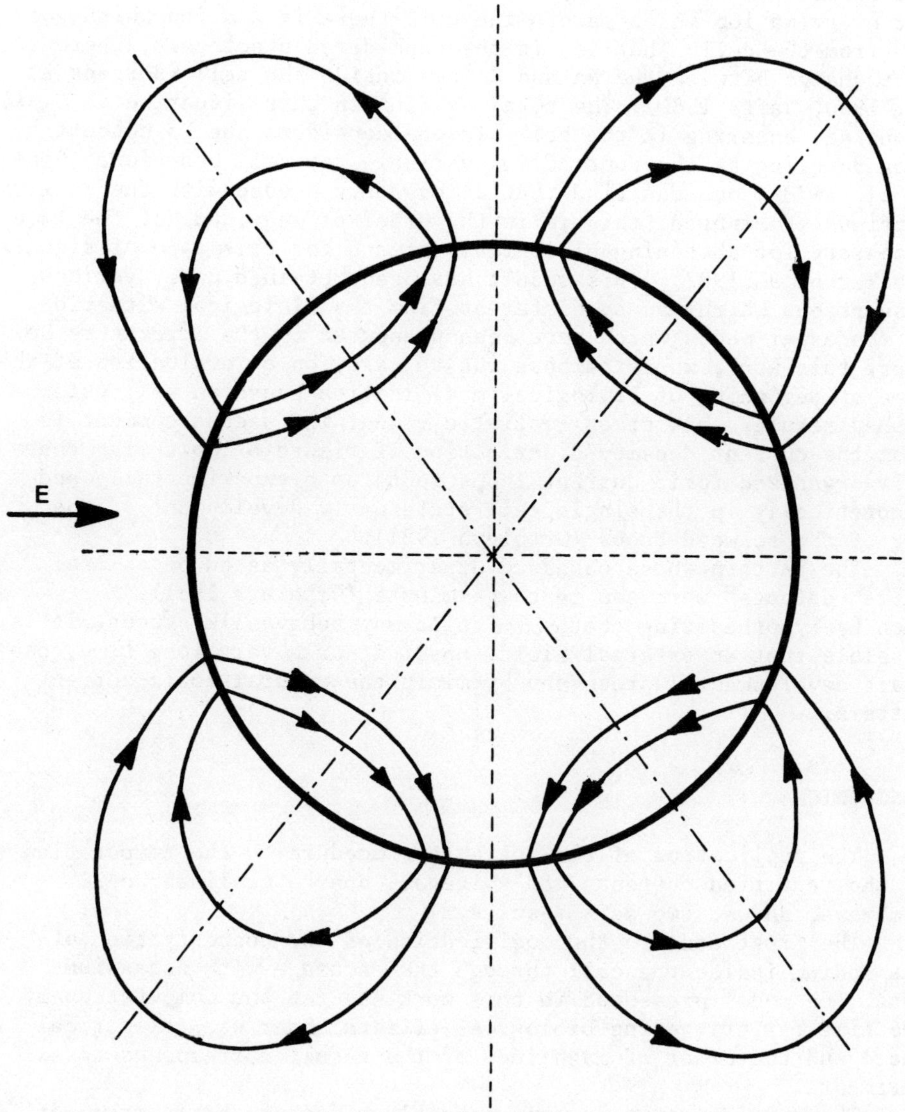

Fig. 8. Total current density distribution outside and inside the cell.

DISCUSSION

Two main comments follow from the previous results. If one considers the ions concentration inside the cell as the cell controlling biological variable, it is evident that they are modified by the rectified K, Na and Cl fluxes. Particularly modifications of internal Na concentration has been found to modify strongly the behaviour of a cell. Let us suppose that a cell has an internal Na concentration of 0.01M which corresponds to $7*10^9$ ions/cell and let us consider an externally applied signal E=22.5 V/m,dt=0.1:if one considers the frequency f=100 Hz at which I=0, one obtains that, for every Na ion which enters the cell there is a K ion which goes out from the cell. That is, in the considered hypotheses, there is an exchange between the Na and K ions inside the cell (Borgens et al. 1980, Jaffe 1980). The total Na flux in this situation is $I_{Na}=3.4$ Kions/sec entering in the cell. In one considers the Na concentration doubling as the controlling variable for cell behaviour (Cone et al. 1973), one can find that 24 days are needed with the Na flux previously computed: this is in the order of magnitude of the time necessary for obtaining biological effects for this type of signals (Becker et al.1982). This result has been obtained under various assumptions which may be different from the biological situations of the experiments which have been presented in the literature.However, this work, whose purpose was to carry on an evaluation of the time of beginning of biological modification based on well established results, has given realistic value. The second comment is that the current density distribution of figure 8 looks like the self organized ionic current loops found both experimentally and theoretically in the single cell state early development of the egg of the seaweed Fucus (Ortoleva 1981).

The pattern whose observed experimentally as an occasional "birth defects" were two roots germinate (Ortoleva 1981).
Then by hypothesizing that other cells may behave like Fucus, it is possible that an external field, applied for a very long time, can start developmental processes by mimic the natural ionic current pattern.

CONCLUSION

The application of the Volterra procedure to the computation of the rectified currents and voltages, due to nonlinear cell membrane, showed two main results.

The first concern the modification of the concentration of the sodium inside the cell through the exchange with potassium ions. The model presented in this work has let the computation of the time for triggering biological effects under weak electrical field and the order of magnitude of the result corresponds to experiments.

The second result is that possible effect on early stage of

cell development are suggested by the pattern of ionic fluxes.

APPENDIX

We developed the HH equations in Taylor series at the third order (a small signal voltage V(t) is applied when the membrane is in a quiescent state):

$$\alpha_a = \alpha_{a0} + \alpha_{a1} * V + \alpha_{a2} * V^2 \qquad (A.1)$$

$$\beta_a = \beta_{a0} + \beta_{a1} * V + \beta_{a2} * V^2 \qquad (A.2)$$

where:

$$\alpha_{a0} = \alpha_a \Big|_{V=0} \qquad (A.3) \qquad \beta_{a0} = \beta_a \Big|_{V=0} \qquad (A.4)$$

$$\alpha_{a1} = \frac{d\alpha_a}{dV}\Big|_{V=0} \qquad (A.5) \qquad \beta_{a1} = \frac{d\beta_a}{dV}\Big|_{V=0} \qquad (A.6)$$

$$\alpha_{a2} = \frac{d^2\alpha_a}{dV^2}\Big|_{V=0} \qquad (A.7) \qquad \beta_{a2} = \frac{d^2\beta}{dV^2}\Big|_{V=0} \qquad (A.8)$$

$$\frac{da}{dt} = \alpha_a * (1 - a) - \beta_a * a \qquad (A.9)$$

$$a = a_0 + a_L + a_{NL} \qquad (A.10)$$

where a=n,m,h and the symbols O,L,NL refer respectively to the steady state, to the linear component and to the nonlinear component and:

$$J_{Na} = \bar{g}_{Na} * m^3 * h * (V - V_{Na}) - P_{Na} \qquad (A.11)$$

$$J_K = \bar{g}_K * n^4 * (V - V_K) - P_K \qquad (A.12)$$

By substituting (A.1), (A.2), (A.3), (A.4), (A.5), (A.6), (A.7), (A.8), (A.9), (A.10) in (A.11) and (A.12) we obtained:

$$J_{Na} = J_{Na0} + J_{NaL} + J_{NaNL} \qquad (A.13)$$

$$J_K = J_{K0} + J_{KL} + J_{KNL} \qquad (A.14)$$

where:

$$J_{Na0} = -\bar{g}_{Na} * V_{Na} * m_0^3 * h_0 - P_{Na} = 0 \qquad (A.15)$$

$$J_{NaL} = \bar{g}_{Na} * (m_0^3 * h_0 * V - V_{Na} * (m_0^3 * h_L + + 3 m_0^2 * h_0 * m_L)) \qquad (A.16)$$

$$J_{NaNL} = \bar{g}_{Na} * (V * (m_o^3 * h_L + 3 * m_o^2 * h_o * m_L) + - V_{Na} * (m_o^3 * h_{NL} + 3 * m_o^2 * m_L * h_L + + 3 * m_o * h_o * m_L^2 + 3 * m_o^2 * h_o * m_{NL})) \quad (A.17)$$

and:

$$J_{Ko} = - \bar{g}_K * V_K * m_o^4 - P_K = 0 \quad (A.18)$$

$$J_{KL} = \bar{g}_K * (V * m_o^4 - 4 * V_K * m_o^3 * m_L) \quad (A.19)$$

$$J_{KNL} = \bar{g}_K * (4 * m_o^3 * V * m_L - 6 * V_K * m_o^2 * m_L^2 + - 4 * V_K * m_o^3 * m_{NL}) \quad (A.20)$$

then we developed the signals V(t) in Fourier series $(\omega_i = \omega * i, -\infty < i < \infty)$

$$V(t) = \sum_i V_i * \exp(J\omega_i t) \quad (A.21)$$

and we found the current linear terms as:

$$J_{NaL} = \sum_i Y^L_{Na,i} * V_i * \exp(J\omega_i t) \quad (A.22)$$

$$J_{KL} = \sum_i Y^L_{K,i} * V_i * \exp(J\omega_i t) \quad (A.23)$$

where the linear model $Y^L_{Na,i}$ and $Y^L_{K,i}$ can be found from Casaleggio et al. (1984).
Then we solved the HH system for non linear currents neglecting all terms depending from the amplitude of the applied signal with order greater than 2.

Due to this approximation we obtained linear differential equations in nonlinear currents controlled by functions depending from linear terms. As a consequence, the nonlinear terms are described by the equations:

$$J_{NaNL} = \sum_i \sum_K Y^{NL}_{Na,i,K} * V_i * V_K * \exp(J\omega_i t) \exp(J\omega_K t) \quad (A.24)$$

$$J_{KNL} = \sum_i \sum_K Y_{K,i,K} * V_i * V_K * \exp(J\omega_i t) * \exp(J\omega_K t) \quad (A.25)$$

where $Y^{NL}_{Na,i,K}$ and $Y^{NL}_{K,i,K}$ come from the following relations:

$$H_{1a}(\omega_i) = \frac{\alpha_{a1} * (1 - a_o) - \beta_{a1} * a_o}{J\omega_i + \alpha_{ao} + \beta_{ao}} \quad (A.26)$$

$$H_{2a}(\omega_i,\omega_k) = \frac{(-\alpha_{a1}-\beta_{a1})*(H_{1a}(\omega_i)+H_{1a}(\omega_k))+(\alpha_{a2}*(1-a_0)-\beta_{a2}*a_0)*2}{2*(J\omega_i+J\omega_k+\alpha_{a0}+\beta_{a0})} \quad (A.27)$$

$$a_L = \sum_i H_{1a}(\omega_i)*V_i*\exp(J\omega_i t) \quad (A.28)$$

$$a_{NL} = \sum_i \sum_k H_{2a}(\omega_i,\omega_k)*V_i*V_k*\exp(J\omega_i t)* *\exp(J\omega_k t) \quad (A.29)$$

where a=n,m,h.

By substituting the relations (A.28) and (A.29) in (A.17), (A.20), (A.24) and (A.25) one can find:

$$Y^{NL}_{Na,i,k} = \bar{g}_{Na}*\{m_0^3*H_{1h}(\omega_i)+3*m_0^2*h_0*H_{1m}(\omega_i)+ - V_{Na}*[m_0^3*H_{2h}(\omega_i,\omega_k)+3*m_0^2*H_{1m}(\omega_i)* *H_{1h}(\omega_k)+3*m_0*h_0*H_{1m}(\omega_i)*H_{1m}(\omega_k)+ +3*m_0^2*h_0*H_{2m}(\omega_i,\omega_k)]\} \quad (A.30)$$

and:

$$Y^{NL}_{K,i,k} = \bar{g}_K*(4*m_0^3*H_{1m}(\omega_i)-6*V_K*m_0^2*H_{1m}(\omega_i)* *H_{1m}(\omega_k)-4*V_K*m_0^3*H_{2m}(\omega_i,\omega_k)) \quad (A.31)$$

From (A.24) and (A.25) one can obtain easily the DC component of the Na (A_{Na}) and K (A_K) currents (fig. 5). Since $V=V_0(t)\cos\vartheta$ from linear analysis (Casaleggio et al. 1984), one obtains that the current generators A_{Na} and A_K are proportional to $\cos^2\vartheta$.

REFERENCES

Adey, W. R., 1981, Tissue interactions with non ionizing electromagnetics fields, Physiological R., 61:435.

Becker, R. O., and Marino, A.A., 1982, "Electromagnetism and Life", State of University of New York Press, Albany.

Beltrame, F., Chiabrera, A., Grattarola, M., Guerrini, P., Parodi, G., Ponta, D., Vernazza, G., and Viviani, R., 1980, Electromagnetic control of cell function, Alta Frequenza, 49:101.

Borgens, R. B., Jaffe, L. F., and Cohen, M. J., 1980, Large and persistent electrical currents enter the transected lamprey spinal cord, Proc. Natl. Acad. Sci. USA, 77:1209.

Bussgang, J. J., Ehrman, L., and Graham, J. W., 1974, Analysis of nonlinear systems with multiple inputs, Proc. of IEEE, 62:1088.

Casaleggio, A., Marconi, L., Morgavi, G., Ridella, S., and Rolando, C., 1984, Rectified ionic fluxes in an electrically stimulated cell, Bioelectromagnetics Society Meeting, July 15-19, 79.

Casaleggio, A., Marconi, L., Morgavi, G., Ridella, S., and Rolando, C., 1984, Evaluation of ionic fluxes in a cell with nonlinear membrane, stimulated by an electric field, J. of Bioelectricity, 3 (1 & 2): 305.

Chiabrera, A., Hinsenkamp, M., Pilla, A. A., and Nicolini, C., 1979, Electromagnetic induction of electrochemical information at cell surface: application to chromatin structure modification, in: "Chromatin Structure and Function", C. Nicolini, ed., Plenum Press, New York.

Cone, C. D. Jr., and Tonglier, M. Jr., 1973, Contact inhibition of division involvment of the electrical transmembrane potential J. Cell Physiol., 82:373.

Drago, G.P., Marchesi, M., Ridella, S., and Parodi, M., Erice 1982, Electrical modelling of cells, School on "Modelling and analysis in Biomedicine". Published by World Scientific Publishing Co, Singapore.

Drago, G.P., Marchesi, M., and Ridella, S., 1984, The frequency dependence of an analytical model of an electrically stimulated biological structure, Bioelectromagnetics, 5:47.

Hodgkin, A. L., and Huxley, A.F., 1952, A quantitative description of membrane current and its application to conduction and excitation in nerve, J. Physiol., 117:500.

Jaffe, L. F., 1980, "Calcium Explosions as Triggers of Development", New York Academy of Sciences, 86.

Lerner, E. J., 1984, Biological effects of electromagnetic fields, IEEE Spectrum, May: 57.

Marchesi, M., and Parodi, M., 1982, Some theoretical results on the behaviour of Na and K currents in excitable membranes under voltage waveforms of various shape and frequency, Bioelectrochemistry Bioenergetics, 9:603.

Morgavi, G., and Ridella, S., 1983, Interaction of electric fields and ionic currents in an excitable membrane: a nonlinear analysis, Bioelectromagnetics Society Meeting, June 12-16,15.

Ortoleva, P. J., 1981, Biological effects of non ionizing radiation, in: "ACS Symposium Series", K.H. Illinger, ed., Washington D.C.

Pilla, A. A., 1974, Mechanisms of electrochemical phenomena in tissue repair and growth, Bioelectrochemistry Bioenergetic, 1:227.

Taylor, L. S., and Cheung, A.Y., 1979, The mechanisms of microwave biological effects, University of Maryland, College Park, Maryland, Workshop May 14-16.

EM-FIELD INDUCED FORCE EFFECTS

Herman P. Schwan

Max-Planck-Institut für Biophysik
Frankfurt/Main, West-Germany and
Dept. of Biomedical Engineering*
University of Pennsylvania
Philadelphia, PA, USA

INTRODUCTION

It is well known that DC-fields exert forces on charged particles. Electrophoresis as well as microelectrophoresis have become established tools in biological laboratories. However, it is less well known that alternating EM-fields can just as well exert significant forces on particles. Particles whose properties differ from those of the surrounding medium experience separation of charges at their boundaries as they are exposed to alternating fields and "induced" dipoles are created. Particles which by their chemical nature possess distributed charges have a "permanent" dipole or a distribution of such dipoles if the distribution of charges of opposite sign is not the same. Both particles with permanent or induced dipole characteristics respond to alternating fields. The theory of permanent dipole responses has been originally developed by Debye and refined by many others. It predicts relaxational effects in the RF-range. However, dipolar properties of biological molecules contribute little to the tissue properties at ELF- and RF-frequencies. The dipolar responses of proteins and other biological macromolcules have been investigated extensively. It appears that these responses are for fields of interest in this context "weak", i.e. their response to the EM-field is well below the thermal threshold and biological effects are not expected. However, the induced dipole type of response of cellular particles deserves consideration in this context since usually field induced forces increase with the size of the particle and are, therefore, more significant at the cellular level than at the molecular one.

---

* Mailing address

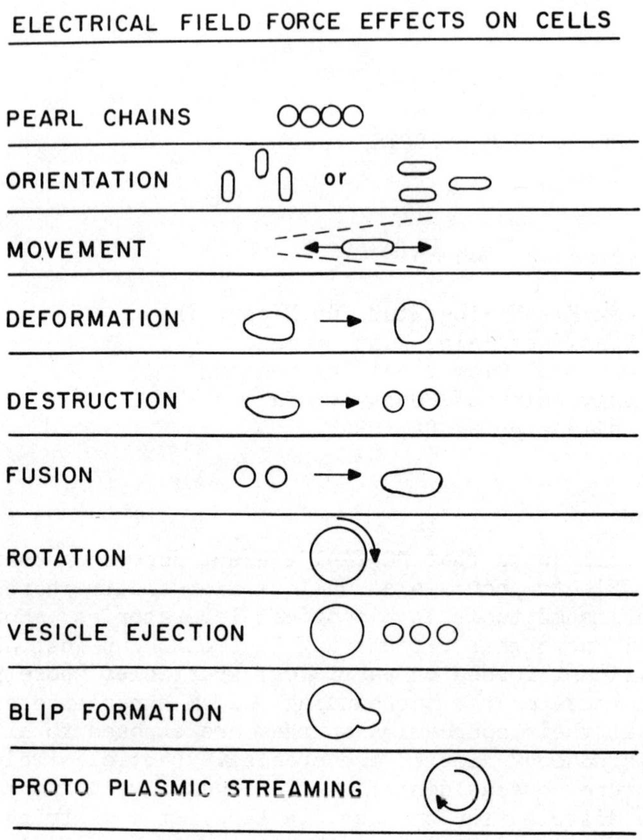

Fig. 1. Schematic indicating various cellular effects which can be induced by alternating fields external to the cell. See text.

The simplest example of the forces which are generated by electrical charges is the attraction of the two plates of a capacitor. The force is given by

$$F\ (\mathrm{dyn/cm^2}) = 10^7\ QE = 10^7\ Cd\ E^2 \qquad (1)$$

Here Q is the charge in Amp-sec, C is the capacity in Farad, E the field strength (V/cm) and d the plate distance (cm). We note that the force is proportional to the square of the field strength. Hence there exists also an effective average force if an AC potential is applied. If a particle is transported into the field between the capacitor plates, work has to be performed to accomplish the transport from the free field space into the field. Corresponding forces have been known for some time as ponderomotoric forces, but only considered for DC-fields. We shall discuss these forces for biological media and alternating fields.

BIOLOGICAL MANIFESTATIONS

Figure 1 presents a summary of various biological effects so far observed. Particles align in the direction of the field and form "pearl chains". Nonspherical particles experience orientation with their long axis either pointing in the field direction or perpendicular to it depending on frquency. In an inhomogeneous field, particles move either towards or from the region of higher field strength. This effect has been named dielectrophoresis by Pohl [1]. Cellular particles may deform under the influence of the alternating field. In extreme cases they may experience destruction. Under other circumstances cellular particles fuse together so that large cells with several nuclei are formed. Cells have also been observed to rotate. Cytoplasmic streaming can be affected. The cell wall may develop extrusion (blip formation) or eject small vesicles, suggesting novel technology to form such vesicles.

Initial studies of these effects concentrated on the pearl chain formation effect and then dielectrophoresis. The other phenomena noted above are more recent origin and the manipulation of cells for all sorts of purposes is now a rapidly growing field of great interest to many. Pohl's book on dielectrophoresis [1] and Zimmermann's review articles [2] are excellent sources of information. The earlier work on pearl chain formation has been summarized by Schwan and Sher [3] and by Schwan [4].

An example of the variety of effects which may be achieved is indicated in figure 2. A cell, originally of spherical shape, is distorted to an elongated shape. If the field is high enough this elongation may proceed to the point where the cell breaks into two parts. This can be achieved by use of alternating field whose frequency and magnitude are chosen to enhance the elongation effect. The effect may be reversed by first using an appropriate field to join the two particles using the pearl chain formation process. When pearl chains form, appropriate pulses can then be chosen to cause dielectric destruction of the membrane. The pulse duration must be sufficiently long, i.e. longer than the time constant of the β-dispersion, to assure that the applied pulse delivers its potentials to the cell membrane and not to the cell interior. Membrane breakdown will take place at the cellular points adjacent to each other since current density is there largest as a result of dipole interaction.

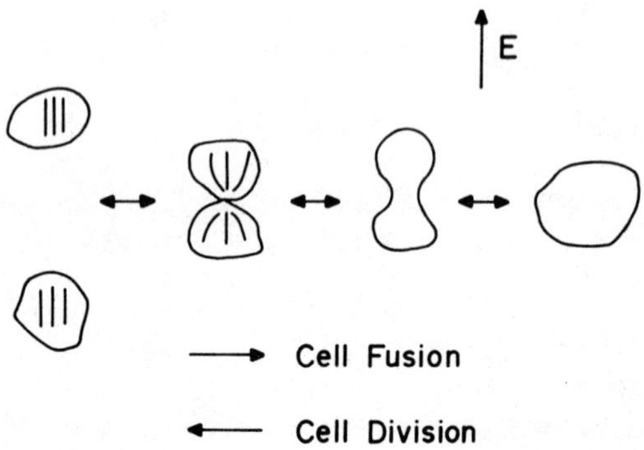

Fig. 2. Example of a reversible chain of events evoked by alternating fields. See text.

The fusion of cells using appropriate combinations of AC-fields and pulses has been of particular interest in recent years since it permits easy gene transfer. How to optimize the fusion of nuclei inside intact cells remains to be investigated. It is possible to suggest optimal conditions for joining nuclei. The frequency range above the cellular β-dispersion must be chosen to affect the joining of nuclei using the pearl chain principle since at lower frequencies the cell interior is effectively shielded by the outer membrane[5]. And the frequency content of the pulse which may be briefly applied to break down nuclear membranes should not be so high as to decrease the nuclear membrane potential significantly.

Some of the above indicated phenomena are hypothetical while others such as fusion, orientation, pearl chain formation and dielectrophoresis have been realized.

AMOEBA RESPONSES

Friend[6,7] has investigated the responses of amoebae to alternating fields in detail. Since this work provides a particularly suitable demonstration of the many different responses possible we shall quote from it. Friend choose the fresh water amoeba Chaos Chaos since its large size is convenient (nearly 0.5 mm) and its dilute culture fluid largely eliminates electrode polarization problems and undue heating. Table 1 summarizes the phenomena observed at various frequenices and field strength levels. Some of these effects had been observed before, others resemble effects of DC fields such as the well known crawling towards the cathode. We cite from Friend's detailed report the following condensed section.

Table 1. FIELD EFFECTS ON AMOEBA CHAOS (FRIEND[6,7])

PERPENDICULAR ELONGATION (10 V/cm)
PARALLEL ELONGATION
CYTOPLASMIC WITHDRAWAL
MEMBRANE DAMAGE
BLEB FORMATION
MICROBLEB FORMATION

"Chaos is surrounded by a membrane which has the ability for immediate self-repair and is highly extensible. It is fairly impermeable for most ions, except potassium. The cytoplasma exists in two phases, the plasmasol and the plasmagel, with the plasmagel forming a semi rigid hollow structure through which the plasmasol moves. The amoeba moves by means of advancing pseudopodia. These are essentially tubes of plasmagel through which the plasmasol flows to the ends of the advancing tips. It is converted back into plasmagel along the sides of the pseudopodia near the tips. In other regions of the amoeba, principally in withdrawing pseudopodia, plasmagel is reconverted into plasmasol and flows toward the center of the cell to supply plasmasol for the advancing pseudopodia.

The second kind of motion that Chaos, and many other cells, can carry out is a local invagination of the membrane-surface-layer-structure. Such invaginations are used to capture food (phagocytosis) or to take in other materials (pinocytosis) found in the environment. The invaginated region forms a spheroid or vesicle which detaches from the surface and floats in the cytoplasm. These vesicles usually redivide into thousands of smaller ones. The opposite process also occurs. Many tiny vesicles can come together to form one larger vesicle, which then moves to the surface and expels its contents. Water is expelled, for example, by contractile vacuoles to control osmotic balance. In fact, nearly the whole physiological status of Chaos is regulated by such vesicular inputs and outputs. The interior of a Chaos is filled with an enormous number of these vesicles of all sizes, along with many other inclusions, and appears to be highly compartmented".

Friend observed the following effects: "The most unusual effect in Chaos is the elongation of the cell perpendicular to the field at quite low field strengths of about 10 V/cm or less at certain frequencies. The effect is most evident between 100 Hz and 1 kHz. Three successive photographs of a single Chaos undergoing perpendicular elongation over a period of about one minute in a field of about 15 V/cm and 316 Hz are shown in Fig. 3.

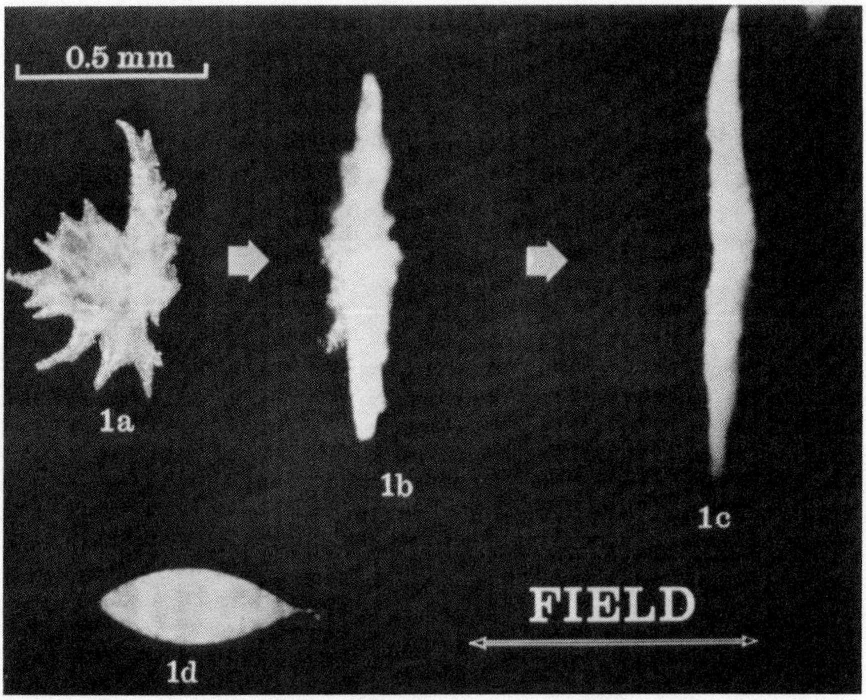

Fig. 3. Perpendicular and parallel extension of Chaos by the field. In a, b and c the same Chaos is shown without field, then partially extended perpendicular to a 316 Hz field of about 15 V/cm, and finally fully extended. The last photograph in fig.3d shows a Chaos extended parallel to a field of about 1 MHz and 250 V/cm. From Friend, ref.6.

In the middle range of frequencies from 10 Hz to 10 kHz, one of the first effects noticed in a freely crawling amoeba is a slight drift or withdrawal of the cytoplasm toward the center parallel to the field at only a few V/cm. If the field is not increased further the amoeba accommodates and continues to move freely. As the field is gradually increased, the parallel pseudopodia are withdrawn and the perpendicular ones are extended until finally the amoeba can no longer crawl freely, but only in the perpendicular direction.

A little above the threshold field level for withdrawal, but before the appearance of the other effects, sudden internal motions of the cytoplasm may occur, as if an internal membrane or restraint had broken. Also, the clear, very thin, hyaline layer just beneath the membrane tends to disappear, especially at higher frequencies.

The most important phenomena to occur at higher field levels are disturbances of the membrane. Often these occur as the sudden appearance of a clear bleb or blister on a side of the amoeba facing the electrodes. The bleb may vary in size from 25 microns to a size covering a large part of the cell. These blebs either detach themselves or break down and the membrane usually repairs itself rapidly.

As the voltage increases more blebs appear and eventually the membrane disintegrates. It is possible with great care to raise the field level to a point where just one small bleb appears and pinches off, then a second bleb appears and pinches off at the same place, etc. A whole string of tiny clear spheres moving away from the cell may be produced.

At the lowest frequencies (1 to 10 Hz) the amoeba may appear nearly normal up to the level of membrane disturbance. Often one or more of the pseudopodia parallel to the field will pinch off, before membrane damage occurs, into one or more small intact balls of apparently normal cytoplasm surrounded by a membrane. It is possible to confuse this process with bleb formation. At times a small "hole" will open in the side of the amoeba and cytoplasmic material will "pump" in and out in response to very slow fluctuations of the field.

As the frequency increases the field level required to cause membrane damage increases rapidly. Eventually a field strength level is reached beyond 100 kHz which causes the amoeba to undergo thermal injury and die before a standard bleb appears. Just below these higher frequencies at 50 kHz to 100 kHz, the cell assumes an elliptical shape of low eccentricity perpendicular to the field and blebs pop out simultaneously on both sides during membrane breakdown. Thin filamentous stream of material shoot out very rapidly from the broken blebs.

At the highest frequencies (0.5 MHz to 10 MHz) little effect on the amoeba is observed until a level of perhaps 25 to 50 V/cm or more is reached. The amoeba then tends to round up and becomes quite rigid in appearance. No obvious perpendicularity can be observed. The cell does not seem to be permanently harmed until a level just below the level of thermal injury is reached. Often just below or at the level of thermal injury the amoeba will begin to pull out parallel rather than perpendicular to the field in a lenticular shape. Suddenly material will begin pouring out of the ends toward the electrodes and a violent circulation develops in the surrounding medium. The amoeba never recovers once this process begins to happen, even if no thermal injury is evident.

Other, anomalous, membrane disturbances occur in the highest frequency region even at field levels below those at which membrane disturbances occur at lower frequencies. These anomalous disturbances can be characterized as <u>microblebs</u>. They are tiny, clear, elongated blebs much smaller than the blebs at lower frequencies, and appear on the surfaces of the membranes facing the electrodes. They cannot be produced as easily or as consistently as blebs at lower frequencies. Sometimes chain-like structures will form on the membrane. Occasionally one of the microblebs will develop into a full size bleb, then break down, and cytoplasm may even be lost. Short pulses seem to have anomalous effects also".
A summary of all effects observed by Friend is presented in Fig. 4.

So much from Friend's observations. Not all phenomena observed by him are of the same origin. Some are clearly related to membrane induced potentials. The threshold field strength required to elicitate bleb formation for example increases sharply above 1 kHz. This may be the characteristic frequency of the β-dispersion predicted for the large amoeba. Hence the membrane induced potential drop rapidly as the cytoplasmic resistance becomes comparable and larger than the membrane reactance above that frequency[5]. Furthermore, for an average radius of 0.3 mm and a threshold field E_o=20 V/cm below 1 kHz the membrane potential is estimated to be V_m=1.5 E_o R=1 Volt. This value is close to values which have been reported by many observers to cause reversible membrane breakdown.

The small change in the solid threshold curve near 50 Hz is not necessarily statistically significant. It could well relate to the α-dispersion effects discussed elsewhere in this volume[5].

Other phenomena such as the orientation with or perpendicular to the direction of the field are most probably field induced force effects to be discussed later. We have pointed out before that threshold field strength values for field induced force effects and biologically significant membrane potentials are often comparable[8]. This makes it sometimes difficult to readily identify the mechanism responsible for the observed effect.

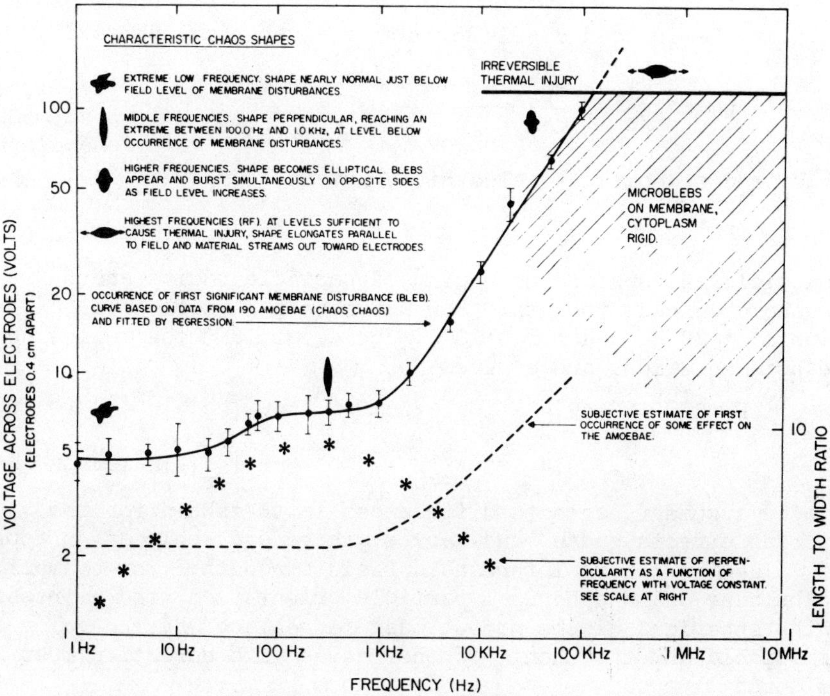

Fig. 4. A summary indicating the frequency dependance of the phenomena observed. The solid curve shows the frequency dependence of the onset of membrane damage as indicated by bleb formation. Also indicated are estimates of the frequency dpendence of perpendicular extension and parallel cytoplasmic withdrawal. Medium is standard cultural fluid. Vertical bars represent 1 S.D. of measurements on 10 amoeba at each frequency and of uniform size (about 0.2×0.4×0.2 mm). From Friend, ref. 6.

THRESHOLD CONSIDERATIONS

A semiquantitative argument can be advanced to predict the threshold field strength needed to observe above quoted effects. We argue that the Langevin criterion for permanent dipoles may be applicable also for induced ones and propose an equation for the threshold field strength E_{th}:

$$\mu \cdot E_{th} = KT \tag{2}$$

Here μ is the induced dipole moment of a spherical particle

$$\mu = \varepsilon_a' R^3 \frac{\varepsilon_p^* - \varepsilon_a^*}{\varepsilon_p^* + 2\varepsilon_a^*} E_{th} \tag{3}$$

with ε_p and ε_a complex dielectric constants

$$\varepsilon^* = \varepsilon' - j\varepsilon'' \tag{4}$$

of particle and suspending medium, R particle radius and E_{th} the "threshold" field strength. For water or a physiological external medium $\varepsilon' = 78\ E_o$ and for $4 \cdot 10^{-21}$ Joules at 18°C the following threshold equations are derived

$$E_{th}R^{1.5} = 2.4 \cdot 10^{-5}\ \text{V/cm}^{-0.5} \quad \text{for } |\varepsilon_p^*| > |\varepsilon_a^*| \tag{5}$$

$$= 3.4 \cdot 10^{-5}\ \text{V/cm}^{-0.5} \quad \text{for } |\varepsilon_p^*| < |\varepsilon_a^*| \tag{6}$$

The argument neglects differences in threshold for the different effects quoted and serves, therefore, only as an approximate value for the real threshold field strength. For example, for the case of nonspherical particle orientation, the expression for the spherical dipole moment must be replaced by the value for an ellipsoid and the change of this value with direction must be considered.

Equation (2) can be justified for the case of DC-fields and real dielectric constants, i.e. negligible losses. Then μE is the potential enegry assumed by the particle as it is moved into the field E. Only if this energy is larger than KT can one anticipate significant interaction phenomena which involve dipole interactions with the field or other particles. For alternating fields and complex dielectric constants, i.e. lossy medium and particle, the meaning of potential energy becomes obscure as pointed out by Sauer[9] and threshold field calculations based on the force which a particle experiences as compared to brownian forces are appropriate. This force equation was derived by Sauer[10] as follows:

$$F = -\pi\varepsilon_a' R^3\ \text{Re}\left(\frac{\varepsilon_p^* - \varepsilon_a^*}{\varepsilon_p^* + 2\varepsilon_a^*}\right) \nabla|E|^2 \tag{7}$$

Table 2. Experimental Determination of the Product $E_{th}R^{1.5}$($V/cm^{1/2}$)

Effect	Particles	Author	$E_{th}R^{1.5}$($V/am^{1/2}$)
Pearl chain formation	Silicone	Sher and Schwan	6.6×10^{-5}
"	"	Sher at al.	8.6
"	"	Takashima	11.0
"	Erythrocytes, Silicone, <u>E. coli</u>	Schwan	8.0
Orientation	<u>E. coli</u>	Sher and Schwan	3.2
		Average	8.5×10^{-5}

For detailed references, see H.P.Schwan (1981) Dielectric properties of biological tissues and biophysical mechanisms of electromagnetic-field interaction. In: ACS Symposium Series No. 157, Biological Effects of Nonionizing Radiation (ed. Illinger), p. 109, and H.P. Schwan (1982) Biophysics of the interaction of electromagnetic energy with cells and membranes. In: Advances in Biological Effects and Dosimetry of Low Energy Electromagnetic Fields, Plenum Press - London.

PEARL CHAIN FORMATION

There are many qualitative reports in the literature on the phenomenon of pearl chain formation. Early manifestations of pearl chain formation were reported by Muth who studied colloidal particles. Krasny-Ergen suggested that dipole-dipole interactions are responsible for the effect and developed a theory assuming biological cells to be highly conductive compared to the suspending medium. To our knowledge, the only quantitative efforts to determine E_{th} were undertaken by Schwan and his colleagues[3,4]. These extensive investigations included the total frequency range from ELF- to microwave frequencies and a variety of different particles and cells. Some of these efforts are summarized in table 2. Average values for the quantity $E_{th}R^{1.5}$ are given in $V/cm^{0.5}$ for each of the investigations undertaken at different times. The overall average value of 8.5 compares fairly well with the simple predictions in equation (5,6).

The data summarized in table 2 are taken from the frequency range where E_{th} is nearly independent of frequency as shown in fugure 5. This independence is anticipated from equations 5, 6 and 7 since the dielectric constant of the medium ε_a' does not change throughout the ELF- and RF-range if it is an electrolyte. However, sometimes a noticeable decline of E_{th} at lower frequencies is noted

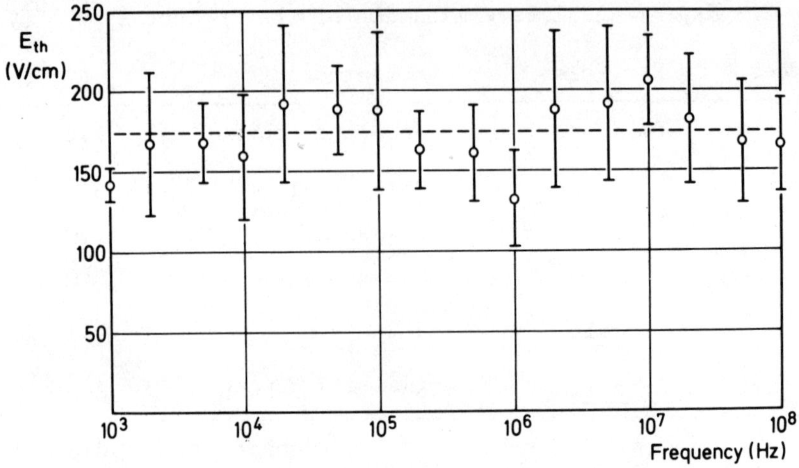

Fig. 5. Threshold field strength for pearl chain formation as function of frequency for polysterene spheres of 1.17 μm diameter. Bars indicate one standard deviation. From Sher, ref.20.

as demonstrated in figure 6. It may reflect the α-dispersion discussed elsewhere in this volume[5]. We conclude from these results:

1. The available theory, while at best only approximate, appears to be applicable;
2. The effective dielectric constants of cells are very large compared to those of silicone, at least at RF frequencies. The results shown in table 2 demonstrate therefore that the particle dielectric constant is not very important as anticipated from equations 5, 6 and 7.
3. The proportionality of E_{th}^2 and $1/R^3$ is confirmed;
4. Threshold values for sizes typical of biological cells (10 to 100 μm) are anticipated near 0.1 to 5 V/cm. Other effects at 0.1 V/cm can therefore not be excluded;
5. The marked decline of E_{th} at lowered ELF-frequencies for charged particles and cells emphasizes the conclusions 4 and suggest even lower threshold values in the ELF-range.

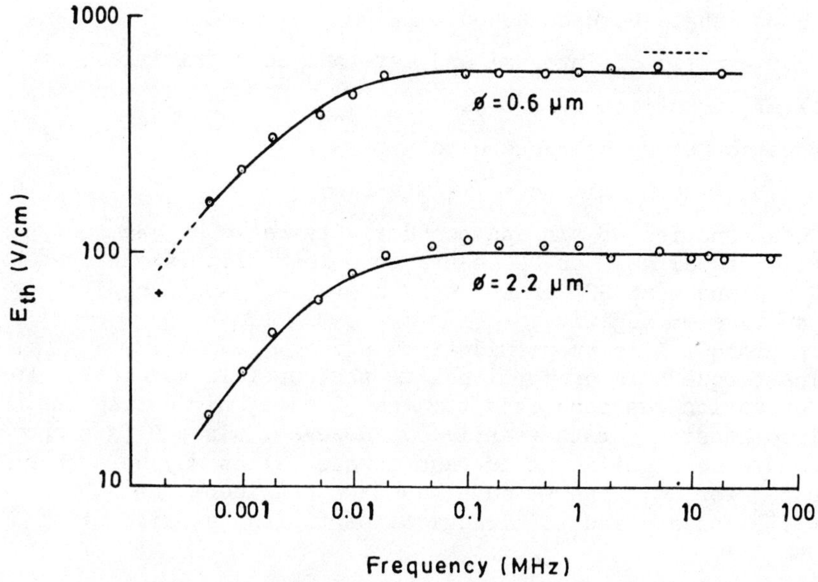

Fig. 6. Threshold field strength for pearl chain formation as function of frequency for silicon particles of 0.6 and 2.2 μm diameter. From Sher, ref.20 and Takashima and Schwan, ref.19.

ORIENTATION

Fuerth[11] calculated the torque which an ellipsoidal particle experiences in an alternating field. He also carried out experiments which supported his theory. Equation 8 gives Fuerth's expression for the torque D

$$D = \frac{2\varepsilon_a E^2 \sin 2\delta}{\{x \sim (4/\mu - 1)\}\{y + (4/\mu - 1)\}} VF(a/b) \qquad (8)$$

The symbols have the following meaning:

- a, b — axes of ellipsoid (a axis of rotation)
- V — particle volume
- F — a function of a/b alone
- ε_a — dielectric constant of suspending medium
- E — external field strength

δ angle between a and E

μ ratio of internal and external conductivities

X, Y functions of a/b alone

The equation 8 may be reduced to

$$D = \varepsilon_a E^2 \sin 2\delta \cdot V \cdot G(a/b,\mu) \qquad (9)$$

with G a function of a/b and conductivity ratio μ. We recognize:

1. For δ equal to 0 or 90° a stable orientation is achieved with D = 0 and with 90° or 0° being unstable, depending if G is positive or negative.
2. The torque D is proportional to E^2.
3. The torque D is proportional to the particle volume V. Fuerth's derivation was conducted for the DC case, neglecting the dielectric constant of the particle. However, since D is proportional to E^2 significant average torque values may be also anticipated for AC. The AC case has been considered by Schwarz et al.[12] and applied to biological particles by Saito et al.[13].

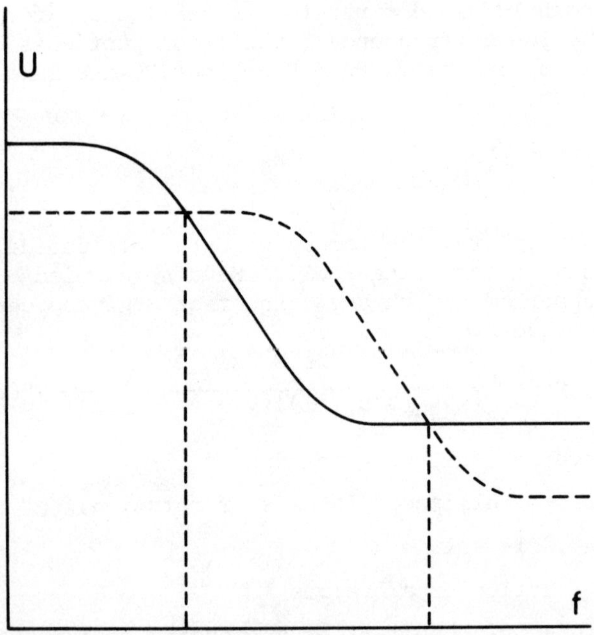

Fig. 7. Electric potential energy of an ellipsoid as function of frequency for various directions. Intersects indicate "turn-over-frequencies". The frequency dependence is only anticipated for dispersive or dissipative properties such that the ratio of dielectric constant to conductivity is different for particle and medium.

The mechanism responsible for orientation may be based on the principle of minimal potential energy. Clearly, an orientation will be achieved for which the G quantity in equation (9) suggests a stable orientation provided that the threshold to overcome thermal agitation is exceeded. The frequency dependence of the potential energy of an ellipsoid U was studied theoretically by Schwarz et[12] and is summarized in figure 7. The potential energy U is frequency dependent and, in the simplest case, simply changes from one to another level as capacitative currents surpass conductive ones at sufficiently high frequencies. For the case of frequency independent properties of the medium and particle, a relaxational behavior such as indicated in the figure is anticipated for U.

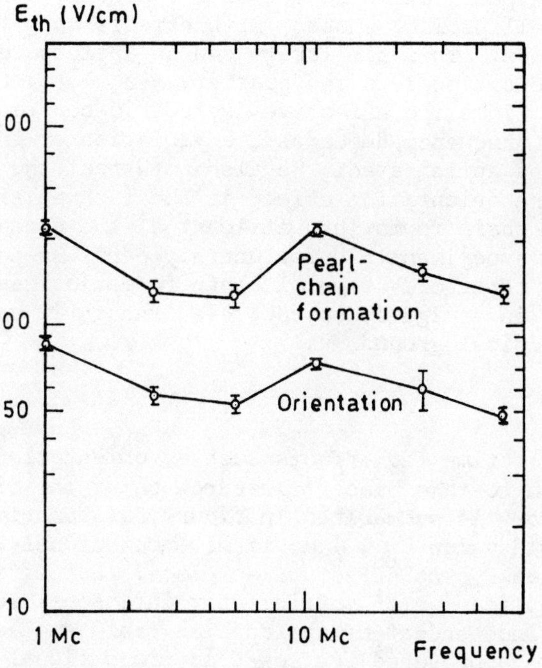

Fig. 8. Comparison of threshold field strength E_{th} for orientation and for pearl chain formation of E.Coli in water. Bars indicate range of three measurements. From Sher, ref. 20.

However, the magnitude of the U-values depends also on the orientation of the particle with respect to the field. Frequently the curves which pertain to different orientation cut each other as indicated. Stable orientations are achieved for that orientation which provides minimal U-values. Thus the particle reorients at the "turn-over" frequencies f_1 and f_2. This probably explains the observation of Teixeira-Pinto et al.[14] that unicellular organisms move in the field direction and perpendicular to it depending on frequency. Indeed, the theoretical prediction of Saito et al.[13] that turn-over frequencies occur in the upper RF-range confirms the observations by Teixeira-Pinto et al.

The theoretical formulations of Schwarz et al. and Saito et al. are based on the potential energy expression developed by Schwarz[15]. They may be therefore subject to the criticism stated by Sauer indicated above. However, the principles outlined above are very likely to be correct. They are already suggested by Fuerth's early work summarized above.

In figure 8 we present threshold field strength values for orientation and pearl chain formation for E.Coli, a bacterium of approximately ellipsoidal shape. Both effects show the same frequency dependence. This similarity can be anticipated since both effects are proportional to the quantity $(\varepsilon_p^* - \varepsilon_a^*)/(\varepsilon_p^* + 2\varepsilon_a^*)$ defined before. Since the effective dielectric constant of bacteria ε_p^* is strongly frequency dependent, a variation of this ratio with frequency can be anticipated. We also note that the threshold field strength for the orientation effect is about three times lower than that for pearl chain formation, at least at the concentration level chosen in this experiment. Unfortunately particle concentration effects on the threshold of pearl chain formation has not been systematically investigated so far, even though it can be anticipated on theoretical grounds[16].

TIME CONSTANT

It requires time for effects such as orientation or pearl chain formation to take place. Research about the time constant in our laboratory is summarized in figure 9. For fields in excess of the threshold value E_{th}, T declines proportional with $1/E^2$. For fields less than E_{th} no effect takes place. But effects previously achieved by application of sufficiently intense fields disappear at a rate almost independent of the applied field smaller than E_{th}. The value of T_o approached for low E is proportional to R^3 as indicated in figure 10. Typical T_o values for particle sizes in the μm range are of the order of ten seconds and T is about 0.1 seconds for $E \sim 10\ E_{th}$. A theory for the time constant of pearl chain formation has been outlined by Saito and Schwan[17] and experimental studies by Sher et al.[18] and Takashima and Schwan[19] support it. One important consequence of the behavior indicated in figure 9 is that pulsed fields can be no more effective than continuous fields of the same average power to achieve pearl chains and, presumably, orientation and other field effects. Optimal pulsing conditions are achieved if the time between succeeding pulses is no longer than T_o. If it is longer, pulsed fields will be less effective than continuous ones. We conclude, therefore, that field effects cannot be optimized by using pulsed fields in order to reduce undesirable heating.

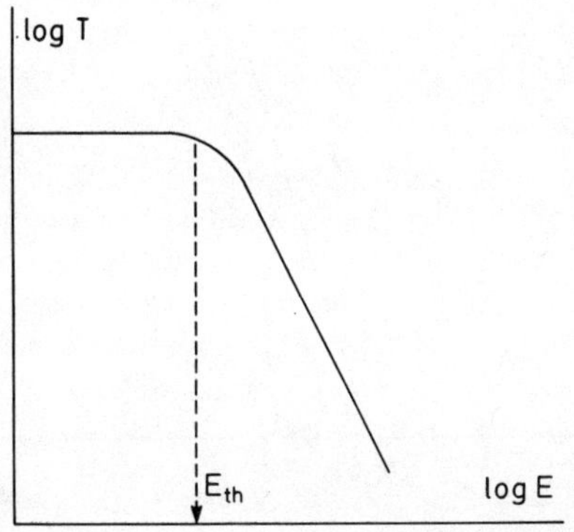

Fig. 9. Time constant of pearl chain formation or reversal for fields E greater or less than the threshold field strength E_{th} as function of field strength.

ROTATION

The phenomenon of cells spinning if exposed to an alternating field is a particularly interesting one. From symmetry considerations one would conclude that no torque can be imposed on the cell. However, it was noted that cellular spinning appears to demand some disturbance of external field such as created by neighboring cells. In discussions with Zimmermann it was postulated that a rotating field component might be involved and we suggested to test the idea by exposing cells to rotating fields. This is achieved

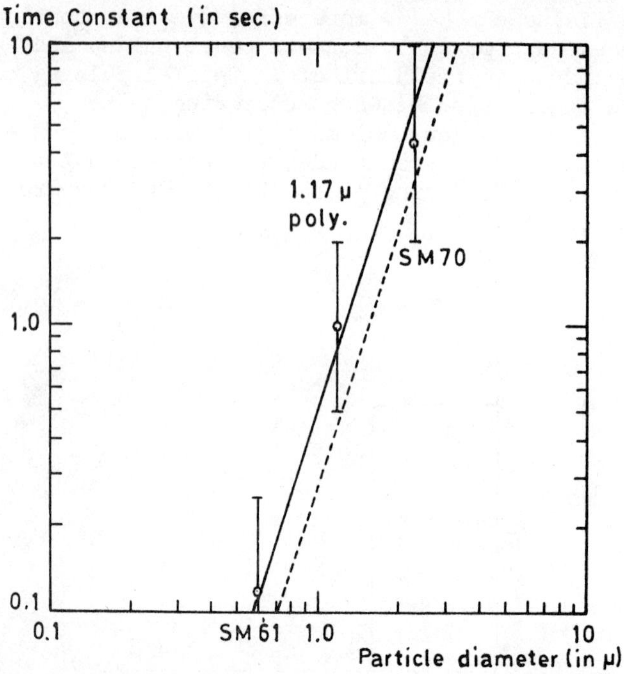

Fig. 10. Time constant vs. particle diameter for pearl chain formation. Points are geometric mean values of the range of estimates indicated by bars. The solid line fits the data and has a slope of 3. The dotted line is theoretically predicted for E=0 by Sher. From Sher, ref. 20.

by superpositioning two fields at right angles with a phase difference of 90°. This is now an established technique and Zimmermann as well as Pohl have studied the phenomena. Zimmermann derived an equation which relates the spinning rate to cellular parameters such as membrane capacitance and obtained values comparable to those obtained from dielectric studies. Cellular rotation rates peak at frequencies which appear to be identical to that of the β-dispersion.

REFERENCES

1. H.A. Pohl, "Dielectrophoresis", Cambridge University Press, London (1978).

2. U. Zimmermann, Electric field-mediated fusion and related electrical phenomena, Biochimica et Biophysica Acta, 694:227-277 (1982).
3. H.P. Schwan and L.D. Sher, Alternating-current field-induced forces and their biological implications, J. Electrochem. Soc., 116:170 (1969).
4. H.P. Schwan, Nonthermal cellular effects of electromagnetic fields: AC-field induced ponderomotoric forces, Br. J. Cancer 45, Suppl., V:220-224 (1982).
5. H.P. Schwan, Electric properties of cells and tissues, this volume.
6. A.W. Friend, Jr., Low frequencies, motile cells, measurements and models. Research report Project MF 12.524.015-0001B No. 4, Naval Medical Research Institute, Bethesda, Maryland 20014.
7. A.W.Friend, E.D.Finch and H.P.Schwan, Low frequency electric field induced changes in the shape and motility of amoebae. Science 187, 357-359 (1974).
8. H.P.Schwan, Biophysical principles of the interaction of ELF-fields with living matter, II.Coupling considerations and Forces, Plenum Press, in press.
9. F.A.Sauer, Private communication (1981).
10. F.A.Sauer, Forces on suspended particles in the electromagnetic field, in: "Coherent Excitations in Biological Systems", H.Froehlich and F.Kremer, eds., Springer-Verlag, Berlin - Heidelberg (1983).
11. R. Fuerth, The absolute determination of dielectric constants with the ellipsoid method, ZS Physik, 44:256 (1927).
12. G. Schwarz, M. Saito and H.P, Schwan, On the orientation of non-spherical particles in an alternating electrical field, J. Chem. Physics, 43:3562-3569 (1965).
13. M. Saito, G. Schwarz and H.P. Schwan, Response of non-spherical biological particles to alternating electric field, Biophys.J., 6:313 (1966).
14. A.A. Teixeira-Pinto, L.L. Nejelski, Jr., J.L. Cutler and J.H. Heller, Exp. Cell. Res., 20:548 (1960).
15. G. Schwarz, J. Chem. Phys., 39:2387 (1963).
16. F.A. Sauer, this volume.
17. M. Saito and H.P. Schwan, The time constants of pearl chain formation, in: "Biological Effects of Microwave Radiation", Plenum Press, New York, 1:85-97 (1961).
18. L.D. Sher, E. Kresch and H.P. Schwan, On the possibility of nonthermal biological effects of pulsed electromagnetic radiation, Biophys. J., 10:970-979 (1970).
19. S. Takashima and H.P. Schwan, Alignment of microscopic particles in electrical fields and its biological implications. Biophys. J., in press.
20. L.D. Sher, Mechanical effects of AC-fields on particles dispersed in a liquid: Biological implications, Ph.D. thesis, University of Pennsylvania, 1961.

THE EFFECTS OF PULSED RF FIELDS ON THE SHAPE AND VOLUME OF NORMAL AND SICKLED ERYTHROCYTES

Shiro Takashima, Stephen Chang and Toshio Asakura

Department of Bioengineering D3, University of Pennsylvania and Children's Hospital, Philadelphia, Pa. 19104, U.S.A.

INTRODUCTION

It is now known that chemical agents can be used to reverse the sickling of erythrocytes containing Hb S (SS-erythrocytes) (1,2). The most effective way to inhibit sickling or reverse the sickling is to increase water influx through the membrane and reduce the effective concentration of intracellular Hb S (3,4,5,6). It is well-known that the length of delay time prior to gelation is inversely proportional to the 30th power of Hb S concentration (7). Thus, a slight decrease in the concentration of Hb S would prolong the delay time significantly and, thereby, delays or inhibits sickling.

In search of a non-invasive and non-chemical means to desickle SS erythrocytes, we found that pulsed radio frequency (RF) fields are effective in reversing the sickling process at moderate field intensities of 3-5 kV/cm (8). The most likely mechanism of the observed shape change by RF fields may be due to the formation of pores in the membrane and subsequent influx of ions and water thus causing the dilution of Hb S in the erythrocyte. Application of a field of 3 kV/cm, for example, to a spherical cell having a diameter of 4μm would create a potential of 1.5-1.8 volts across the membrane (9,10,11). These potentials are large enough to cause partial or local degradation of the erythrocyte membrane leading to pore formation. Thus, the field-induced shape change of sickled

SS-erythrocytes may be primarily due to the interaction of applied fields with the membrane rather than with intracellular Hb S gels.

The work reported in this paper is an extension of the previous investigation (8) by emplying a variety of experimental conditions. Of particular importance is finding a set of conditions under which we can achieve desickling of SS-erythrocytes with minimum damages to the membrane. Previously, we observed that the reversibility of field-induced shape change was poor. In other words, once sickled erythrocytes were transformed into spherocytes by electrical fields, the process of shape change could not be arrested even by the removal of the electrical field. In general, under these conditions, the shape change led to hemolysis unless the field was turned off at an early stage of desickling. It is therefore important to find an optimum condition for achieving reversible shape change of sickled SS-erythrocytes. In the present work, short pulse width of 20-50 μs were used for most of the experiments. Also the use of physiological media such as Hank's solution was found to enhance this tendency. Further refinement of the technique would achieve reversible desickling of SS-erythrocytes.

Experimental Procedure

The electronic system used for these experiments is shown in Fig.1. a) Hewlett-Packard TM500 RF and pulse generators. This system generates pulsed as well as continuous RF fields. b) ENI

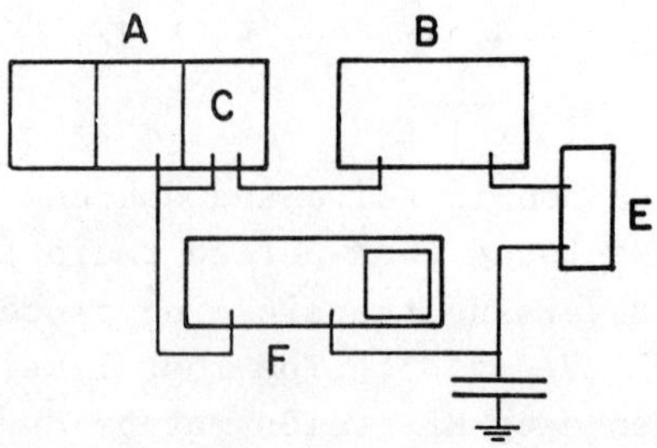

Figure 1 Electronic system used for exposure of erythrocytes See text for explanation.

power amplifier 1140LA with a maximum output of 2 kW with the frequency response of 10 KHz and 500 KHz. The range of this instrument can be extended to 1 MHz with a reduced output. c) Hewlett-

Packard oscilloscope 1207B. d) General Radio attenuator model 350B. e) Sample holder and electrodes. f) Leitz microscope and Panasonic TV monitor system. Fig.2 illustrates microelectrodes consisting of two Pt-Ir wires with a diameter of 50 μm. The distance between two electrodes is 0.3-0.5 mm. A small amount of sickled SS-erythrocytes suspended either in 0.9% NaCl or Hank's

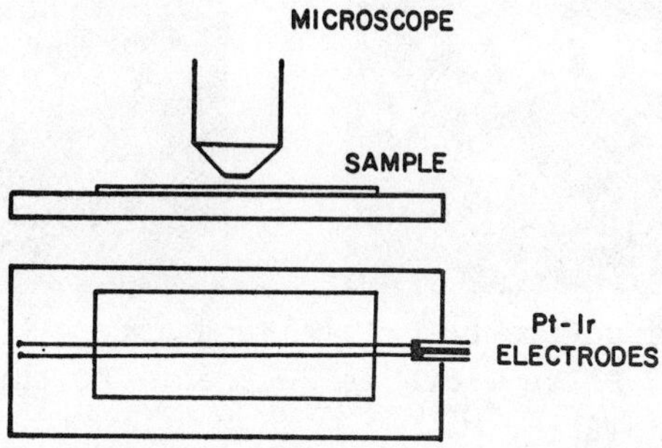

Figure 2 Electrode assembly used for RF exposure of SS-erythrocytes. See text for detail.

solution was placed in the chamber and the voltage of applied field was gradually increased using an attenuator. The observation of shape change was continued for 10-15 minutes at each voltage step. The threshold was defined as the field intensity at which the onset of shape change becomes visible. Usually, a few of most extensively sickled erythrocytes were singled out on the screen and their shape change with increasing field intensities was followed.

The pulse width was varied between 0.02 msec and 10 msec with a constant interval of 1 second. In addition, the carrier frequency was varied from 10 KHz to 1 KHz. However, care was taken to select the lowest frequency for each pulse width in such a way to have at least one complete cycle in the pulse. For example, the lowest frequency for 0.02 msec pulse was 50 KHz, etc. As mentioned before, Hank's solution as well as NaCl solution were used in the present work. The pH of the suspending solution was 7.0-7.2 throughout this work and the temperature of ambient air was maintained at 35-40°C using an incandenscent lamp near the sample holder.

Results and Discussion

Fig. 3 shows sickled SS-erythrocytes before and after exposure to pulsed RF fields. The photograph a) shows typical sickled SS-erythrocytes and b) shows the same erythrocytes after the shape change has progressed to near completion. These pictures were obtained using a short pulse width of 20 μsec with an interval of

(a)

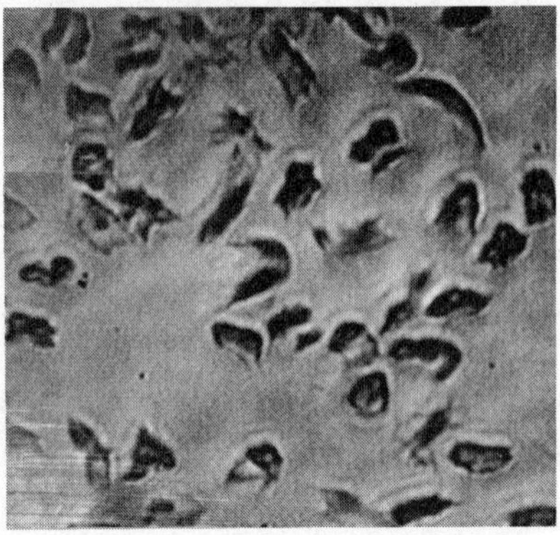

(b)

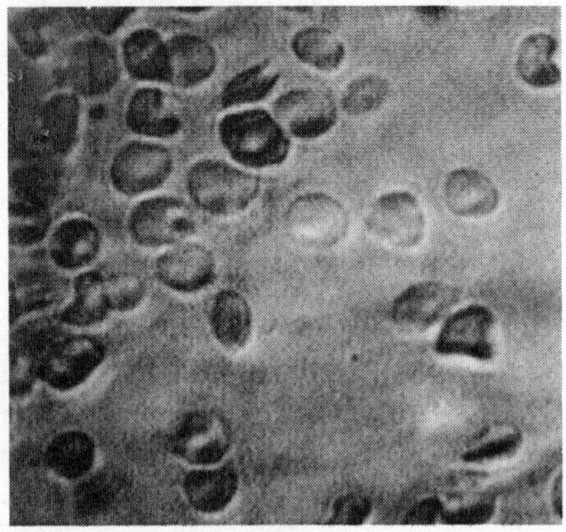

Figure 3 SS-erythrocytes before (a) and after (b) desickling by pulsed RF fields. Pulse width is 20 μsec. Erythrocytes were suspended in Hank's solution.

1 sec.(duty cycle = 0.00002). In our preliminary experiments (8), a pulse width of 5 msec was used (0.9% NaCl as suspending medium). With long pulse widths, however, the shape change of erythrocytes often led to hemolysis. It was stated by Zimmermann (12,13) that in general, reversible shape change of biological cells can be achieved more readily using high amplitude short pulses rather than low amplitude long pulses. We observed that the combination of short pulses and the use of Hank's solution effectively prevented extensive hemolysis of erythrocytes as shown in Fig.3.

Determination of thresholds

Having observed qualitatively that short pulses seem to induce the shape change of SS-erythrocytes without serious damage to the membrane, we investigated quantitatively the dependence of threshold E_{th} on pulse duration. Fig.4 shows the threshold intensity with various pulse widths using 0.9% NaCl solution (curve 1) and Hank's solution (curve 2) as the suspending media. As shown, the increase in threshold with decreasing pulse duration was very slow until the pulse duration reached a value of 0.1 msec where the threshold showed a sudden increase. On the other hand, as the pulse width increases above 1 msec, the threshold seemed to approach a value of between 2 and 3 kV/cm asymptotically. This means that a further increase in pulse width will not cause a further decrease in the value of the threshold, in other words, E_{th} becomes nearly

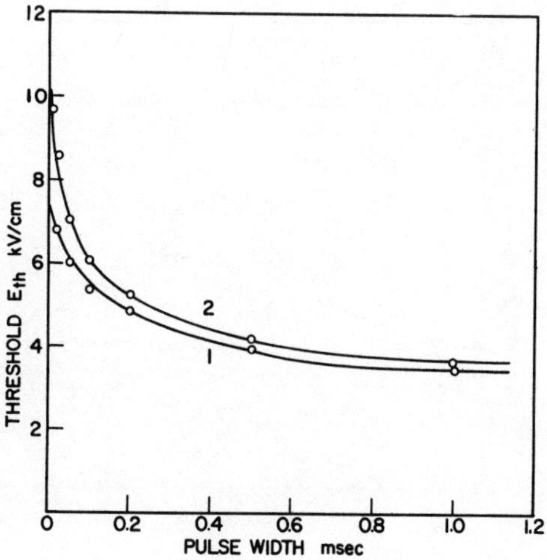

Figure 4 The dependence of threshold on pulse widths. Curve 1 was obtained in 0.9% NaCl solution and curve 2 in Hank's solution as suspending medium. Pulse width is expressed in msec and threshold in kV/cm.

independent of pulse duration. Note that the threshold is higher in Hank's solution than in NaCl solution, an indication that erythrocyte membrane is more stable in Hank's solution than in the latter.

In general, the threshold of shape change depends on the frequency of applied fields and an increase in the frequency causes an increase in threshold. The frequency dependence of threshold in Hank's solution is shown in Fig.5 between 10 KHz and 1 MHz for various pulse durations. The value of threshold gradually increases and reaches a plateau at about 300-500 KHz. The broken curve shown in this figure was calculated using the equation derived by Schwan (14) for spherical particles which are surrounded by a thin shell.

$$V = \frac{1.5\ RE}{\sqrt{1 + (\omega T)^2}} \qquad (1)$$

where R is the radius of the sphere, E is field intensity, ω is angular frequency ($2\pi f$) and T is time constant. We calculated the threshold, using this equation, assuming $R = 4\mu m$ and $E = 4 kV/cm$. The time constant T was assumed to be about 4×10^{-7} sec. Obviously, the agreement between calculated and observed threshold values is not satisfactory, particularly at high frequencies.

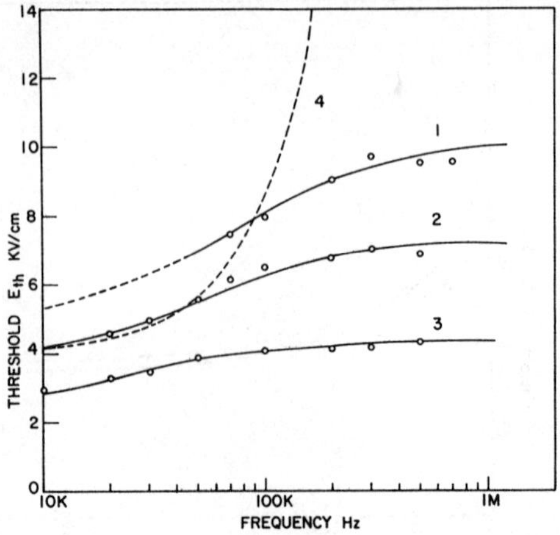

Figure 5 The dependence of threshold values on carrier frequency. The pulse widths are 20, 50 and 100 μsec for curves 1, 2 and 3. Curve 4 was calculated using Eq. 1. See text for detail.

While the theory predicts that the threshold continues to increase indefinitely with increasing frequencies, the observed values level off at high frequencies. This observation predicts that there may be an additional mechanism causing the desickling of SS-erythrocytes.

An interesting observation is that erythrocytes are more stable mechanically when suspended in Hank's solution than in NaCl solution. Under this condition, the field-induced shape change of SS-erythrocytes does not necessarily lead to hemolysis. There may be a number of factors which affect the stability of biological membranes, i.e., temperature, pH and surfactants. However, there are numerous observations which attest to the importance of divalent ions for the stability of biological membranes (15). Thus, it is our current interest to investigate, in the near future, the effect of divalent ions on the threshold of field-induced shape change of biological cells using chelating agents such as EDTA or EGTA.

Membrane Potential Induced by Applied Fields

It was postulated by previous investigators that the cause of shape change of biological cells by applied fields is the field induced potential across the membrane. For example, Kinosita and Tsong (9) estimated the potential induced by a field of 5 kV/cm across erythrocyte membrane would be about 1.5 volts. A potential of this magnitude is sufficient to perforate the membrane and cause the formation of pores.

The calculation of field induced potentials using various shapes such as sphere, spheroid and cylinder have been performed by various investigators (10,14,16). For example, the field induced potential for spherical cells is given by (14):

$$V = 1.5\ RE \quad (2)$$

and that for a long cylinder with a field across it is given by (17):

$$V = 1.0\ RE \quad (3)$$

Since neither normal red blood cells nor sickled erythrocytes are spherical or infinitely long, the calculation of field-induced potential must be generalized using a more realistic model for sickled erythrocytes. In this paper, we will discuss two cases. 1) Prolate elliposids with a thin shell and the field along the major axis and 2) the same ellipsoid with the field perpendicular with the major axis. The model used for the calculation is shown in Fig.6. The major and minor axes are a,b and c. For the sake of simplicity it was assumed that b = c. The thickness of the shell

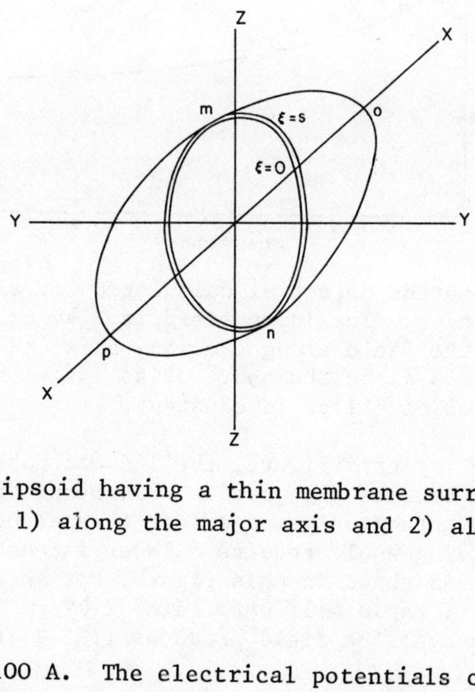

Figure 6 An ellipsoid having a thin membrane surrounding it. A field is applied 1) along the major axis and 2) along the minor axis.

is assumed to be 100 A. The electrical potentials can be computed for the interior, the shell and the exterior separately by solving Laplace's equations. The detail of the calculation for ellipsoidal coordinates is discussed by Stratton (18) and it will not be repeated. The membrane potential induced by applied fields E is obtained by the following equation.

$$\Delta V = \phi_i(\xi=0) - \phi_a(\xi=s) \quad (4)$$

The field-induced potential depends on the coordinates. For example, the potential must be zero at sites m and n(Fig.6). On the other hand, the maximum potential will be induced at sites o and p if the field is parallel to the major axis. Fig.7 shows the dependence of induced potentials on the axial ratio calculated at the apex of prolate ellipsoids. These calculations are performed using ellipsoids of various axial ratios with, however, the same

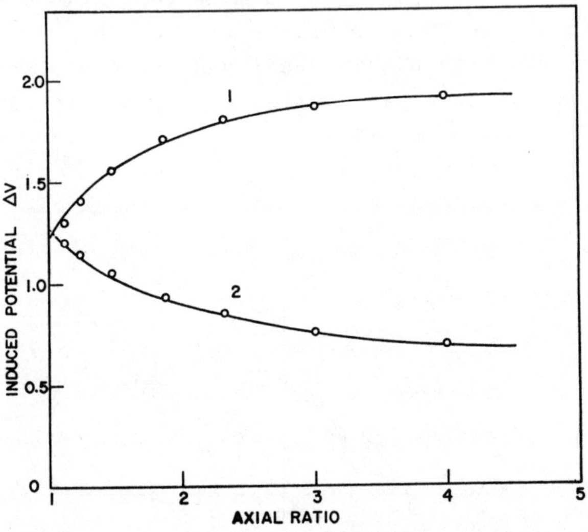

Figure 7 The membrane potential calculated for ellipsoids having various axial ratios. The input field is 2 kV/cm. Curve 1 was calculated with the field along the long axis and curve 2 was obtained with a field along the short axis. Axial ratio was varied keeping the volume of ellipsoid constant.

volume. As shown by this figure, the induced potential increases as the ellipsoid becomes elongated. This means that as the erythrocytes sickle extensively and orient themselves along the field vector, the rounding would require a lower intensity than for those less elongated. As shown in this figure, for an elongated ellipsoid with an axial ratio 4:1, a field of 2 kV/cm induces a potential of 1.9 V whereas the same field produces 1.55 V if the axial ratio is 2:1.

In the above, we discussed the induced potential when the field is parallel to the major axis "a" of the ellipsoid. In the following, a discussion for the case where the field is parallel to the minor axis "b" will be presented. As shown by curve 2 in Fig.7, the induced potential decreases with the elongation of ellipsoids. However, the dependence of induced potential on the axial ratio is less pronounced in this case than for a parallel field. Both curves converge to the same value when axial ratio is reduced to 1:1, the case of a sphere.

Usually, however, the cells are randomly oriented. In this case, the internal and external potentials ϕ_i and ϕ_a are given by the following equations.

$$\phi_i = -(E_x X + E_y Y + E_z Z) \tag{5}$$

$$\phi_a = -(E'_x Y + E'_y Y + E'_z Z) \tag{6}$$

where E_x, E_y, E_z and E'_x, E'_y, E'_z are induced fields along X, Y and Z directions. The third term in these equations can be ignored assuming that the ellipsoid is lying on the X-Y plane. Using these equations, we can calculate the induced potential for random distribution.

S.T. and S.C. are supported by ONR N00014-82-K-0321 and T.A. is supported by NIH HL-20750.

References

1. J. Dean and A.N. Schecter, The New England J. of Med., 14: 752-763 (1978).
2. I. M. Klotz, D. N. Haney and L. C. King, Science, 213:724-731 (1981).
3. R. B. Guy, R. K. Gavrilis and S. P. Rothenberg, Am. J. Med. Sci., 266:267-277 (1973).
4. J.P. Knochel, Arch. Intern. Med., 123:160-165 (1969).
5. T. Asakura, T. Ohnishi, K. Adachi, M. Ozguc, M. Hashimoto, M. Singer, M. O. russel and E. Schwartz, Proc. Nat'l. Acad. Sci., USA, 77:2955-2959 (1980).
6. R. M. Rosa, B. E. Bierer, T. Rogelio, J. S. Stoff, M. Mruskall, S. Robinson, H. F. Bunn and F. H. Epstein, N. Engl. J. Med. 303:1138-1143 (1980).
7. J. Hofrichter, P. D. Ross and W. A. Eaton, Proc. Nat'l. Acad. Sci., USA, 71:4864-4868 (1974).
8. S. Takashima and T. Asakura, Science, 220:411-413 (1983).
9. K. Kinosita and T. Y. Tsong, Biochim. Biophys. Acta, 471:227-242 (1977).
10. K. Kinosita and T. Y. Tsong, Proc. Nat'l. Acad. Sci., USA, 74:1923-1927 (1977).

11. K. Kinosita and T. Y. Tsong, Biochim. Biophys. Acta, 554:479-497 (1979).
12. U. Zimmermann, P. Scheurich, G. Pilwat and R. Benz, Angewandte Chemie, 20:325-344 (1981).
13. U. Zimmermann, Biochim. Biophys. Acta, 694:227-277 (1982).
14. H.P. Schwan, In: "Biological Effects and Dosimetry of Non-Ionizing Radiation. M. Grandolfo, S. M. Michaelson and A. Rindi (Eds.), Plenum Publishing Corp., New York, NY. (1982).
15. P. F. Baker, Progress Biophys. Mol. Biol., 24:177-222 (1972).
16. J. Berhardt and H. Pauly, Biophysik, 10:89-98 (1973).
17. J. A. Stratton, "Electromagnetic Theory", McGraw-Hill, New York, NY, (1941).

ELECTROMAGNETIC FIELD EFFECTS ON PHYTOHEMAGGLUTININ (PHA) INDUCED LYMPHOCYTE REACTIVATION

M. Grattarola ◊, A. Chiabrera ◊, G. Bonanno +,
R. Viviani ◊, and A. Raveane ◊

◊Biophysical and Electronic Engineering Department
Via all'Opera Pia 11a, 16145 Genova, Italy
+Pharmacology Institute, Via Capo Santa Chiara, Genova
Italy

INTRODUCTION

Weak electric currents, injected by means of electrodes in the culture medium or induced by means of time-varying magnetic fields, have been successfully applied to reactivate frog erythrocytes on a time scale of 24 h (Harrington and Becker, 1973; Becker and Murray, 1967; Pilla, 1974; Smith et al, 1978; Chiabrera et al, 1979a, 1980). Since 1977, our group has successfully attempted to monitor the reactivation process on single cells through the measurement of the fluorescence intensity of dye molecules intercalated in chromatin-DNA via laser flow microfluorimetry (Chiabrera et al, 1979a; Beltrame et al, 1981b) and by measuring densitometric and geometric parameters of Feulgen stained cell nuclei smeared on a slide via automated absorption image cytometry (Chiabrera et al, 1980).
The results of these analyses suggest a well-defined sequence of reactivation events at the nuclear level, i. e., the overall decondensation of nuclear chromatin, which becomes an active template for RNA synthesis, followed in some cases by DNA synthesis. In the absence on any em exposure, the reactivation process can be induced, to some extent, by changing the ionic composition of the medium (mainly Ca++ and Mg++) (Pilla, 1974; Chiabrera et al, 1978); besides, the process can be inhibited by La++ addition to the culture medium (Smith et al, 1978). Depending on the waveform used, em exposure can have a sinergistic or an antagonistic effect with respect to chemically induced reactivation (Chiabrera et al, 1979a, 1980).
The early chromatin decondensation process that always occurs in

these experiments strictly mimics the "activation process" described by Ringertz and Bolund (Ringertz and Bolund, 1969) for hen erythrocytes in a serum-free medium under different ionic conditions.

While the complete set of molecular steps involved in the reactivation phenomena is still unknown, nevertheless, the available data suggest the presence of at least two general features. Firstly, a relationship exists between the electrical perturbation, the redistribution of ions (very likely Ca++) at the cell surface, and their influx inside the cell, as already proposed also for other cellular systems, like brain tissue (Bawin et al, 1978; Tenforde, 1980), cartilage cells (Rodan et al, 1978) and embryonic chick tibia explants (Assailly et al, 1981). Secondly, a consequent global process of chromatin decondensation and nuclear volume increase occurs, as a first reactivation step, before the onset of RNA, protein, and DNA syntheses. Such a process seems to be quite general an event shown by many cell systems during transition from quiescence to proliferation, and it is referred, using the cell cycle terminology, as transition from G0 phase to G1 phase (Nicolini et al, 1977). Some experimental evidence, both in situ (Aaronson and Woo, 1981, Chiabrera et al, 1985) and in reconstitution experiments (Thoma et al, 1979), indirectly suggests the control of this process by nuclear ions (very likely Mg++). These two features (i. e., Ca++ modifications at the cell surface and Mg++ modifications inside the nucleus) have been included in a cell model (Chiabrera et al, 1978, 1979b, 1984b; Beltrame et al, 1980c) which accounts for the dependence of chromatin decondensation on the frequency content of the electromagnetic signal.

The available body of experimental data suggests that the primary target of the field is at the cell membrane level, while the process of chromatin decondensation seems to be a subsequent step, possibly electrochemically mediated by cations or by more complex biochemical events. Moreover, published data on synaptosomes (Lin-Liu and Adey, 1982) show that the em exposure causes the onset of physico-chemical phenomena (e. g., modifications to Ca++ fluxes) which involve membrane structures only. On the other hand, electromagnetic signals have been shown to inhibit also the in vitro response of bone cells to the parathyroid hormone, thus interfering with hormone-receptors interactions at the cell surface (Luben et al, 1982). All this evidence suggested that we should analyze the effect of em signals on human lymphocytes in vitro. In fact, human lymphocytes can be easily perturbed in vitro by external "messangers" (lectins) which interact with their membrane receptors via reversible binding processes, so that, on a purely speculative basis, an interference of the field with this process can be anticipated.

The addition of the lectin PHA to a suspension of mixed lymphocyte populations is able to fully reactivate the T lymphocyte population (in the presence of macrophages (Oppheneim et al, 1968)), bringing many of them to DNA duplication and to mitosis in 72 h. The mitogenic stimulus is initiated by the increase in the clustering of lymphocyte membrane receptors induced by lectin binding to receptors. It has been shown that the reactivation process started by lectins in human lymphocytes is slowed down if biopolyanions are added to the medium (Darzynkiewicz and Balazs, 1971), and that it can occur only if Ca++ is present in the culture medium (Whitney and Sutherland, 1972). Moreover, the stimulation by lectins can be mimicked if a Ca++ ionophore alone is used instead of lectins (Maino et al, 1974), and the amount of free Ca++ increases inside the cytoplasm a short time after the interaction with lectins(Tsien et al, 1982). In addition there are reports showing that one of the first events during the process of lymphocyte stimulation is a global phenomenon involving nuclear volume enlargment and chromatin decondensation (Rigler et al., 1969; Kushch et al, 1980; Abraham et al, 1980).

Few preliminary experimental results suggesting an antagonism between weak, low frequency electromagnetic fields and lectins induced stimulus were published by our group in 1980 -1982 (Beltrame et al 1980b, Grattarola et al, 1982). Later on other groups described the effectiveness of weak, low frequency em perturbations on the lectin induced stimulation of lymphocytes. Interestingly enough, both an antagonistic effect (Conti et al, 1983) and a synergistic effect (Hellman et al, 1983) has been reported. A stimulating effect on PHA treated leukemic cells has been also reported (Emilia et al, 1983).

A theoretical model focused on the interactions between lectins, receptors and electromagnetic fields has been proposed by our group in 1982 (Chiabrera et al, 1982), further developed (Grattarola et al, 1983, Chiabrera et al, 1984) and generalized to other ligands than lectins (i. e., Ca++ and pharatyroid hormon) in another paper of this volume (cfr. Chiabrera et al: "Electric and magnetic field effects on ligand binding to cell membrane "). In that paper it is anticipated that an antagonistic effect at low lectin doses (prevalence of the decrease in the mean life-time of small lectin-receptor aggregates) and a synergistic effect at high lectin doses (prevalence of the increase in the free Ca++ availability) may occur (Fig. 1). For extremely low doses (i. e., approaching the limit case of zero dose), a possible field- induced activation is also qualitatively predicted. The experiments reported here investigate the influence of weak, low frequency electromagnetic fields on human lymphocytes stimulated by sub-optimal doses of PHA.

MATERIALS AND METHODS

I. -Lymphocyte preparation

The separation of mononuclear cells from the heparinized

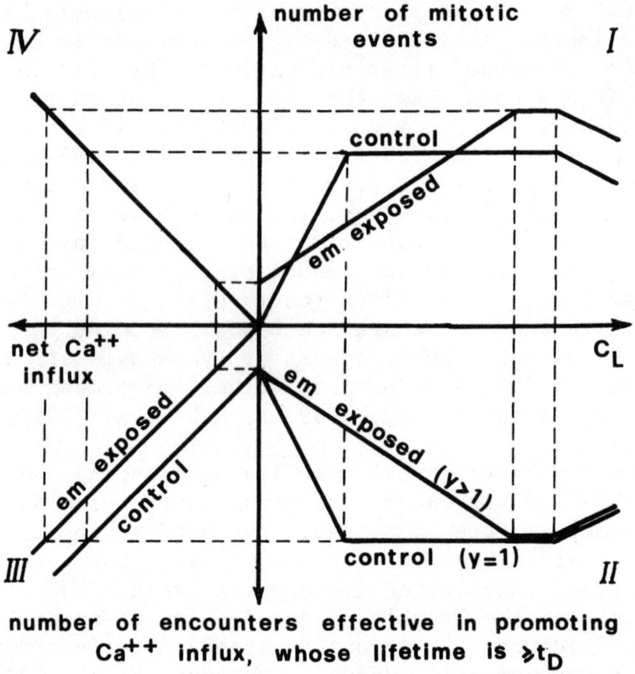

Figure 1 - Pictorial representation of the predictions of the model described in this book by Chiabrera et al. For more details, see the text and the paper by Chiabrera et al.: "Electric and magnetic field effects on ligand binding to cell membrane"

peripheral blood of one healthy volunteer (M. G.) was performed according to Boyum (Boyum, 1968). Peripheral blood leukocytes (PBL) were obtained by centrifugation of the samples on Histopaque 77 (Gibco, USA) gradients. Lymphocytes were pauperated of adhering cells by incubation, for 45 min. in humidified air, 5% CO2, in plastic flasks. The isolated lymphocytes were cultured in Leighton tubes at a cellular concentration of 10^6 cells/ml, in the presence of sub-optimal doses (2.5 - 10 μl/ml) of PHA (Gibco, USA) at pH = 7.6.

The cells were cultured in incubator (37° C, 5% CO_2) in RPMI 1640 complete medium after addition of 20% fetal calf serum (FCS), under one of the following experimental conditions:

a) Controls, without any treatment.
b) In the presence of phytohemagglutinin alone (PHA).
c) In the presence of em field alone (EMF)
d) In the presence of phytohemagglutinin and em field (PHA-EMF).

In one experiment, the cells were cultured in the presence of PHA and em exposure for 72 h, and then the em signal was switched off for further 24 h. By doing so, the cells were cultured for

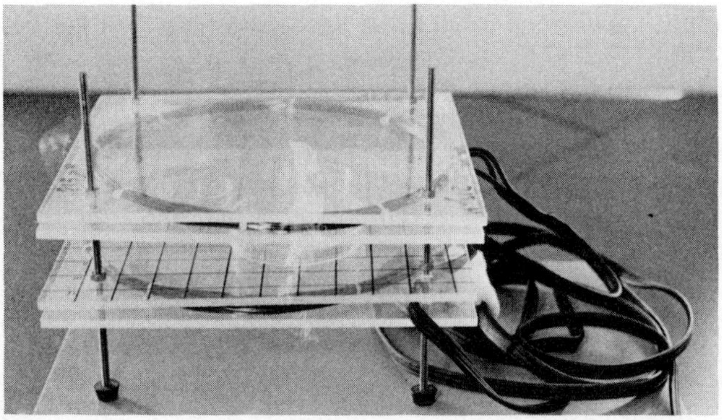

a)

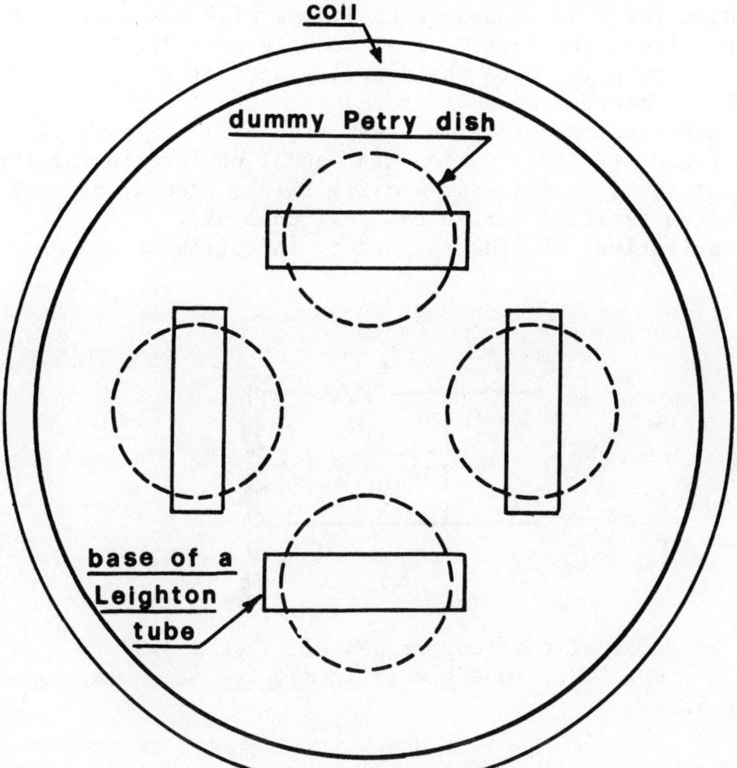

b)

Figure 2 - a) Exposure system. The Leighton tubes are placed in the middle plane. Dummy petri dishes are shown.

b) Top view of the exposure system. The four rectangles are the bases of the Leighton tubes. The four dotted circles indicate the positions of dummy Petri dishes.

other 24 h in the presence of PHA only, in order to test possible toxic effects that might have occured during the initial 72 h, while em exposure was active.
The lymphocytes were acridine orange stained in suspension and used for flow microfluorimetry (Beltrame et al, 1981b) or smeared on slides and Feulgen stained (Beltrame et al, 1982).
Cell viability was checked at the beginning and the end of each experiment by means of the trypan blue exclusion test in a counting chamber.

II. Em exposure

During em exposure (72 h), four Leighton tubes for each set of experiments are inserted in the middle plane between two coils (21 cm inner diameter, 30 loops each, 1.5 mm wire diameter, 6 cm far apart), in fixed concentric positions (Figure 2). The coils are connected in parallel with the signal generator (EBI, New Jersey, USA), which is external to the incubator.
The coil currents generate a resulting time-varying magnetic induction field B_o (t), which is about uniform in the central region, about 14 cm in diameter and 2cm thick, around the center of the coil system (spatial variations less than 5%).
The time derivative of the magnetic induction field has been

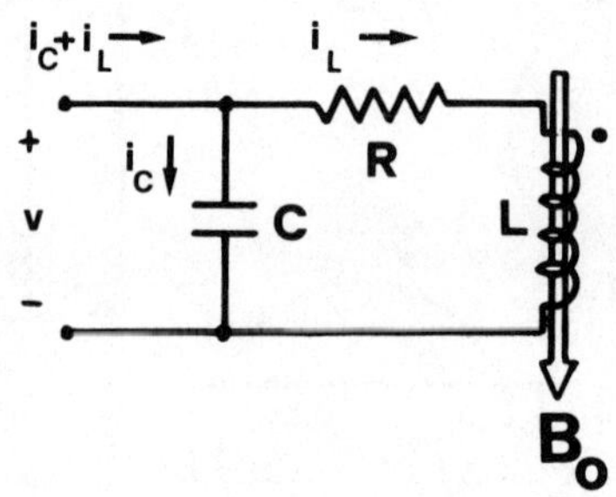

Figure 3 - a) Model of the loop probe: $v \cong .8\, i_L + 3.37\ 10^{-6}\ di_L/dt + 1.25\ 10^{-4}\ dB_o/dt$, where v is in mV, i_L is in mA, B_o is in gauss, t is in s.

The magnetic induction field B_o (t) is generated by two exposure coils and ranges between 0 and 20 gauss. the values L, R, C are obtained by fitting the model parameters to the impedance of the probe, measured by means of network analyzers, in the range 10 Hz to 10 MHz, letting B_o =0. The resonance frequency of the probe is about 3.8 MHz. In the frequency range of interest for the signal shown in b) and c), the effects of R, L, and C on the open-circuit voltage measurements of dB_o/dt are negligible.

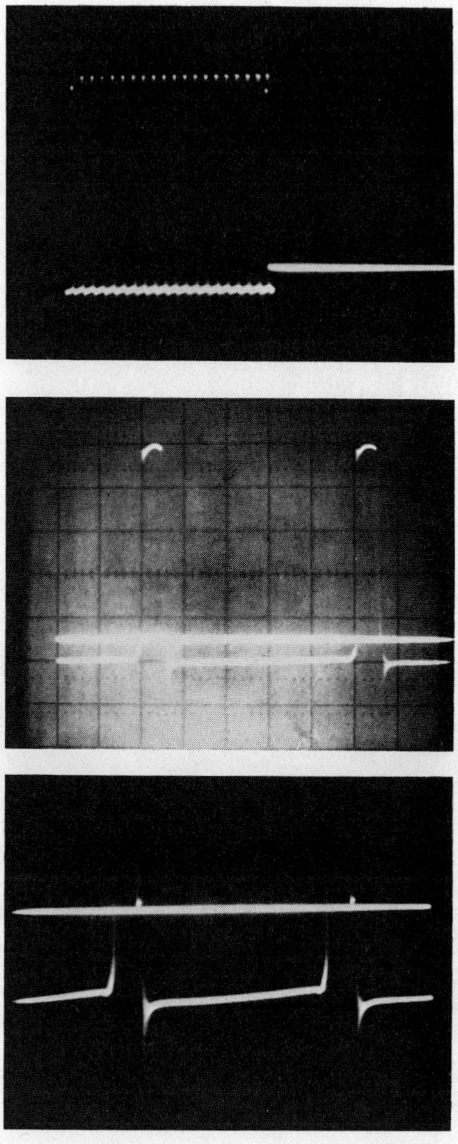

Figure 3 - b) Electric field (typical) induced at 2.5 cm from the exposure dish center; the top panel shows the burst of pulses (horizontal axis, 1ms/div; vertical axis, 20 mV/div.); the middle panel shows the upper portion of a single pulse within a burst (horizontal axis, 50 μs/div; vertical axis, 20mV/div.) and the bottom panel shows its lower portion (horizontal axis 50 μs/div.; vertical axis 5mV/div.)

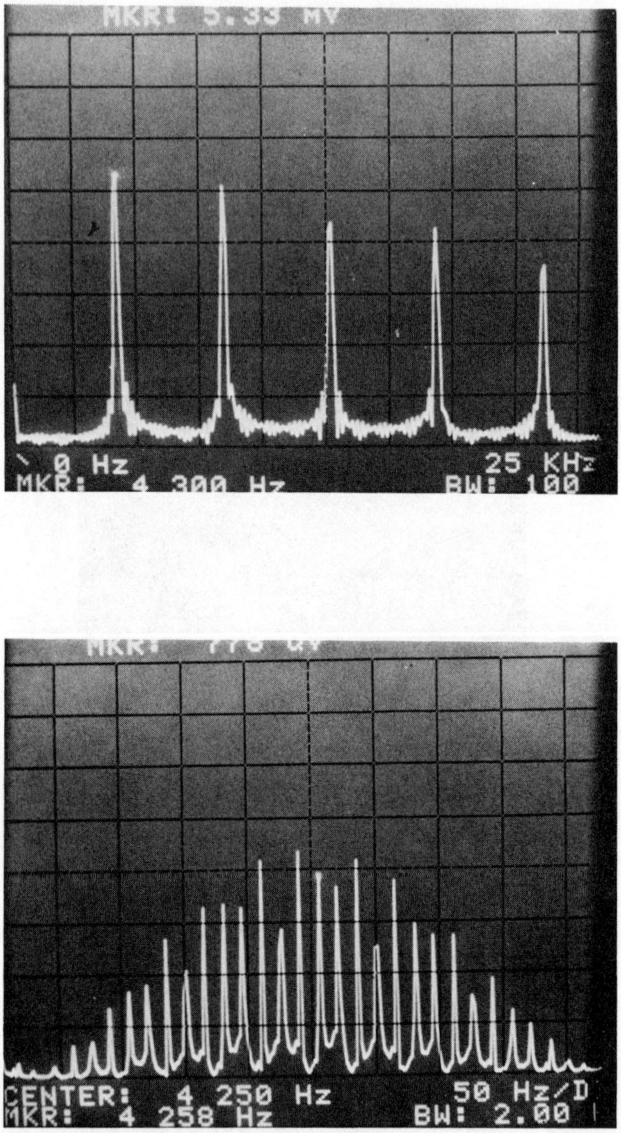

Figure 3 - c)Power spectrum of the periodic signal shown in b). The fundamental frequency is about 15.4 Hz; the top panel shows the overall spectrum from dc to 25 KHz. The bottom panel shows the spectrum from 4 KHz to 4.5 KHz. The relative amplitudes can be deduced from the marker (MKR in µV) values.

region, about 14 cm in diameter and 2cm thick, around the center of the coil system (spatial variations less than 5%).
The time derivative of the magnetic induction field has been measured by means of the open-circuit voltage of a small loop probe (about 0.5 cm in diameter), whose equivalent scheme and impedance are given in Figure 3a. The resistivity of the culture medium is high enough so that the contribution of the currents induced in the medium to the total magnetic field is negligible and not detectable by the loop probe. The magnetic field uniformity is not affected by the presence of the various culture vessel. For the user's convenience, the probe design and calibration (i. e. the number of loops) are such that the mV readout divided by 10 gives the value of the rotational component of the induced electric field in mV/cm at a radius of 2.5 cm from the center of a dish. In other words, the induced electric field intensity in a Petri dish 5 cm in diameter, lying in the region of the exposure system where the magnetic induction field B_o (t) is about uniform, increases linearly from a zero value, in the center, up to the value measured by the probe, on the edge of the dish.
The signal used is a 5 ms burst of pulses, which is repeated every 65 ms. The effects of em exposure on the culture medium have been checked using dummy Petri dishes, 5 cm in diameter, which are more easily accessible for temperature and field measurements. The magnetic induction field varies from 0 to a peak of about 20 Gauss. The estimated value of the average component of the magnetic induction field is about 1 Gauss. The temperature variation due to em exposure inside the culture medium is less than ± 0.1 C. The temperature changes have been monitored with a termocouple, which is designed and inserted in the exposure system in such a way as to minimize any eddy current induced by B_o (t), which otherwise would affect the termocouple's open-circuit voltage reading.
The induced electric field waveform used in the experiments is shown in Figure 3b. The overall accurancy of the vertical scale is 10%, whereas the time scale accurancy is 2%. Our estimate of the electric field intensity induced in the Leighton tubes (1.5 cm x 5.5 cm rectangular base) gives values no larger than the field shown in Figure 3b. Finally, Figure 3c shows the power spectrum of the signal, as given by a spectrum analyzer. The maximum occurs at 4.25 KHz, about corresponding to the repetition rate of the individual pulses within the burst. The order of magnitude of the peak frequency is coincident with the frequency value anticipated by our cell model, corresponding to the maximum sensitivity of changes in chromatin- DNA charges, due to an electric field induced externally to the cell (Chiabrera et al, 1978, 1979b, 1984b). The r. m. s. value E of the induced field is about 1.5 mV/cm.

III. -Staining and static absorption image cytometry

After fixation, a drop of cell suspension from each of the

Lighton tubes was used for smears. The smears were Feulgen stained under the following conditions: 25 minutes of 5 NHCL hydrolysis at 37° C and 2 hours of staining in the Schiff reagent (De Cosse and Aiello, 1966). The smears were theen analyzed in absorbance with the ACTA 500 static image analyzer at DIBE. The system, which has been described in more detail elsewere (Beltrame et al, 1980a), incorporates a Plumbicon TV camera for absorbance analysis (Bosh, Germany), interfaced via a buffer memory of 512 x 512 x 8 bits (Tesak, Italy) with an Hp 1000 minicomputer. The system also includes, as a module, an improved version of a TAS image analyzer (Leitz, Germany). Each nuclear image, as detected by the Plumbicon target, was digitalized and stored in the buffer memory, sent to the minicomputer, and transferred to a tape for further analysis. Four images were stored for each nucleus and averaged, in order to increase the signal-to noise ratio.

Later on, the images were software equalized for uneven illumination and non-uniformity of the tube target by using blank images (black and white) which had previously been stored during the acquisition process. Nuclear area (A), nuclear integrated optical density (D) (i. e., DNA amount), and available binding sites (S) (Beltrame et al, 1981a, 1982) were computed from the corrected images. The parameter S indicates the number of accessible chromatin sites for the intercalation of small molecule, like the fluorochrome AO. It can be shown (Beltrame et al, 1980) that, under some simplifying hypotheses, it is possible to evaluate S by means of absorption image cytometry instead of using flow microfluorometric cytometry. The corresponding histograms were obtained from at least 200 nuclei randomly chosen from each slide and are normalized to 100. The overall measurement standard errors are: D (integrated optical density) 1.5%; A (area) 0.6%; S (sites) 3%.

IV. Staining and flow microfluorometry

About 10^6 cells were recovered from each culture (each Leighton tube contains about 2 10^6 cells).

Prior to flow analysis, the cells were resuspended in PBS buffer up to a final concentration of 3 10^5 cells/ml, and the fluorescent dye, acridine orange (AO), was added up to a final concentration of about 2 10^{-5} M. Under these conditions, the green emission of AO, F_G , is expected to come out mainly from monomers of the dye intercalated between DNA base pairs (Nicolini et al, 1979, Beltrame et al, 1981b). The cell suspensions have been analyzed with an ICP flow system, a FMF flow system (Ortho Instruments, Massachusetts, USA)and with our FACS III flow system (Becton Dickinson, California, USA) equipped with a filter suited to detecting green emission (515-580 nm) at a rate of bout 1000 cells per second. The experiments have been repeated four times and the lymphocytes preparation has been performed in two different biological laboratories.

The histograms of the green emission distribution were obtained from at least 10^4 cells and subsequently analyzed by means of a HP 21MX minicomputer. All the histograms are renormalized to 10^4 . The microfluorometers have been calibrated with fluorescent beads (4.43 ± 0.3 μm in diameter, Polysciences, USA) and the worst CV is 7 %. This figure gives an estimate of the overall error in measuring F_G .

RESULTS

As already pointed out, the experiments described in this paper were mainly originated by theoretical considerations. These considerations relate the external em signal to the mean life-time of ion -receptor and ligand-receptor aggregates and suggest, as reported in detail in another paper of this book (Chiabrera et al "Electric and magnetic field effects... "), an antagonistic action of the electric field signal on the mitogenic activity of lectins for low doses of mitogens.
By increasing the doses, the model predicts that a decrease and even a reversal of the effect may occur. The experiments reported here were planned to cover only the low dose range. Moreover, all the samples analyzed originated from the same healty volunteer (M. G.). Therefore, no attempt has been made to test possible different responses of different subjecs or to utilize higher lectin doses.
First of all, we tested if any toxic effect was induced by the em exposure. The trypan blue exclusion test did not prove any detectable difference between cells in the four different treatment conditions, giving a viability percentage of 90% ± 5% in all cases with no apparent correlation to the treatment type.
Moreover, in one case the cells were further cultured for 24 h in the presence of PHA after 72 h in the presence of the same PHA dose and em exposure. Then, they were stained with acridine orange and analyzed by means of our FACS III. Figure 4 (left) shows a picture directly taken from the display of the microfluorometer. As can be easily seen, the cells are actively proliferating, many of them being in S-G2 phases. This result proves that they were not damaged by EMF treatment, and suggests that they were only delayed in entering the proliferative cycle. This point is better elucidated by Figure 4 (right), which shows a comparison between cells under the same experimental conditions as just described, after the first 72 h of treatment. The histogram with enhanced trace refers to PHA-stimulated cells, and the dimmer histogram refers to PHA-stimulated cells (same dose)in the presence of EMF.
Besides the evident difference between the fraction of cells in the S-G2 compartment, this picture shows an interesting feature regarding the G0-G1 transition (Abraham et al, 1980). In fact, while many PHA-stimulated cells are in the G1 compartment (that

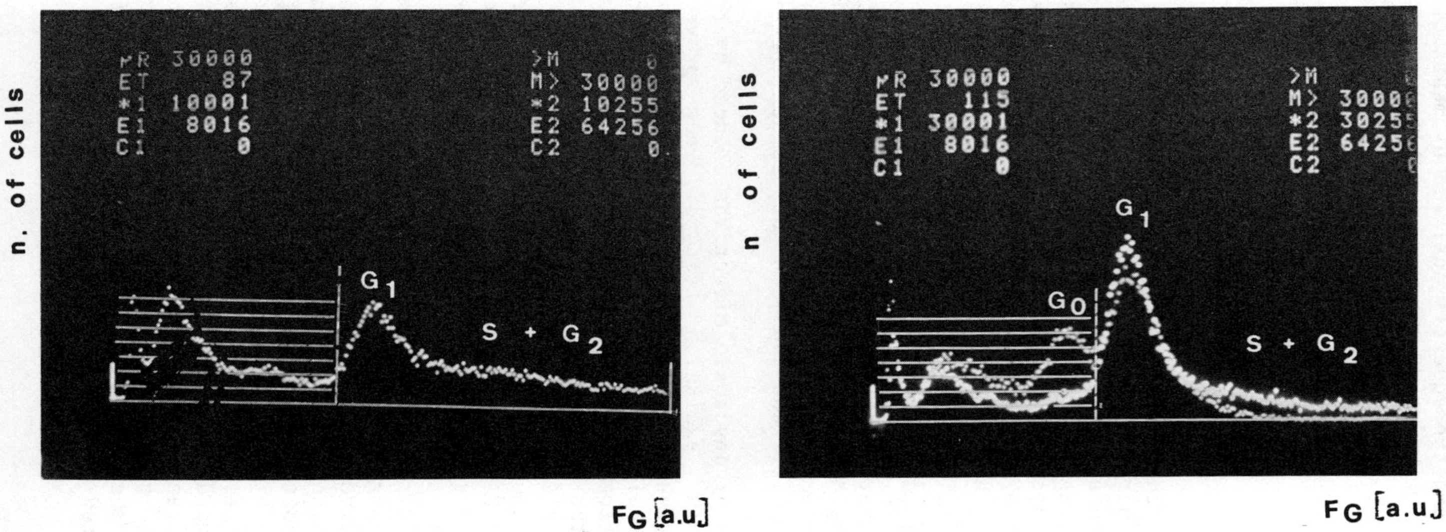

Figure 4 - The figure shows two pictures directly taken from the display of a FACS III system. The left histogram refers to the green fluorescence emission of a population of lymphocytes after 96 h (72 h + 24 h) of PHA stimulation, exposed to EMF during the first 72 h only.
On the right, the enhanced trace refers to the same lymphocyte population as above, stimulated by the same dose of PHA for 72 hours. The dimmer trace refers to lymphocytes stimulated for 72 hours by the same dose of PHA in the presence of EMF.
For the reader's convenience, an identification of cell compartments in terms of cell cycle terminology, (i. e., G0, G1, S and G2) is given. The vertical dotted line shows the location of the electronic thresholding used in Figure 5. For more details, refer to the text.

is, they are entering the cell cycle), many PHA-EMF treated cells are still in the G0 compartment (that is, they are still in a quiescent state).

A collection of results relevant to the 72 h treatment is given in Figure 5. Because of the electronic thresholding chosen, cells in the G0 compartment are not displayed in the histogram referring to flow data (first column)as we were mainly concerned with proliferating cells in the G1 and S-G2 compartments.

The histograms of the green emission from the four samples, corresponding to the experimental conditions a), b), c), d) of the Materials and Methods section, are shown in the first (left) column in Figure 5. The area of each histogram, originated by at least 10^4 cells, has been normalized to the same number of cells (10^4). The initial control showed no detectable differences as compared with

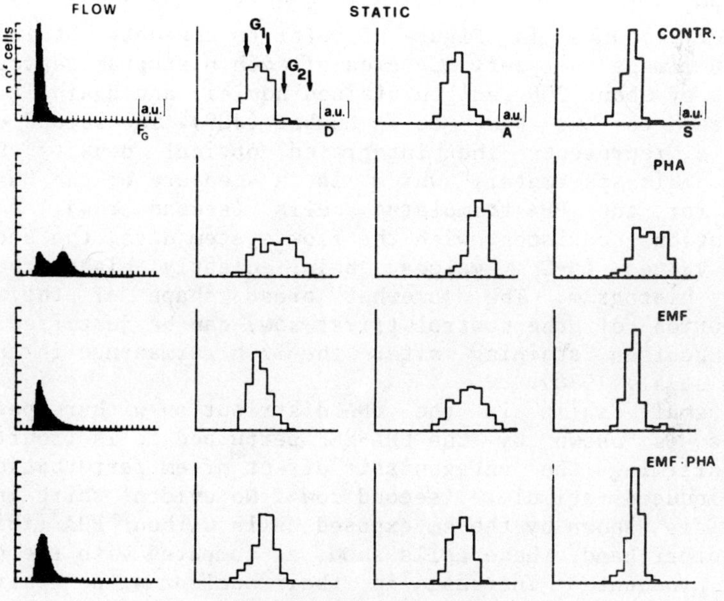

Figure 5 - First column (left): Histograms of the green fluorescence F_G emitted from the chromatin-DNA intercalated dye AO. Second column: Histograms of nuclear integrated optical density D (DNA amount) after Feulgen staining. Third column: Histograms of area A of the same Feulgen stained nuclei as in the second column. Fourth column: Available binding sites S of the cell nuclei. From top to bottom: Control cells; PHA stimulated cells; EMF exposed cells; PHA stimulated and EMF exposed cells.

Note: Arrows in the D histogram indicate the chosen 2C DNA window (first and third row, G1 window in Fig.6)and the chosen 4C window (second and fourth row, G2 window in Fig.6).

the 72 h control, and it is not presented in the Figure. The stimulating effect is clearly shown by the bimodal distribution of the histogram of the second row, as compared with the 72 h control (first row). This stimulating effect is strikingly reduced when em perturbation is added to PHA perturbation (fourth row). Similar results have been obtained also with the fluorochrome propidium iodide (data not shown). On the other hand, it is difficult to understand what happens with em perturbation alone (third row), where only a very small increase in the intensity of green emission is detectable, as compared with the control. Because of the nature of the intercalation process, this increase can be due to an increase in the number of binding sites caused by an increase in the DNA amount and/or by a conformational transition of the chromatin structure(Beltrame et al, 1981[b]) to a less condensed state. This point is clarified by the analysis of the data obtained by the entirely independent technique of static absorption image cytometry.

The other columns in Figure 5 refer to the data obtained by the ACTA 500 image analyzer. The area of each histogram represents the analysis of about 200 Feulgen stained nuclei, and again it has been normalized to the same cell number (100). The second column in Figure 5 represents the integrated optical density D of each sample. This parameter, which is a measure of the DNA amount, shows, for the PHA-stimulated cells (second row), a bimodal distribution consistent with the flow system data, the second peak of DNA values (S-G values)being clearly higher than in the control histogram. The somewhat broad shape of the unimodal distribution of the control (first row) can be justified in terms of the Feulgen staining after the 72 h permanence in the medium (Kushch et al, 1980).

Only a small shift in the DNA distribution with respect to the controls is shown by the PHA-EMF perturbed cells (fourth row), thus confirming the antagonistic effect of em perturbation on the lectin-induced stimulus (second row). No evident shift in the DNA content is shown by the em-exposed cells without PHA (third row). On the other hand, these cells show, as compared with the controls, a very pronounced increase in their nuclear areas (third row), which are comparable with those of stimulated cells. As a consequence they are characterized by a larger number of available sites S (last column). Figure 6 presents the values of A and S for nuclei of the four samples with, respectively, a 2C (first and third columns) and a 3-4c DNA content (second and fourth columns). If we focus our attention on the values of the areas, we can notice that, while very few cells in the em exposed samples (third row) are out of the G1 window, the areas displayed by their nuclei are comparable with those of the corresponding 2C-DNA nuclei of the PHA stimulated, cycling cells (second row). The PHA-EMF treated nuclei (fourth row) exhibit intermediate values of the parameters as compared with the PHA and EMF separated cases. The Feulgen staining and image analysis are time consuming procedures, so that only one

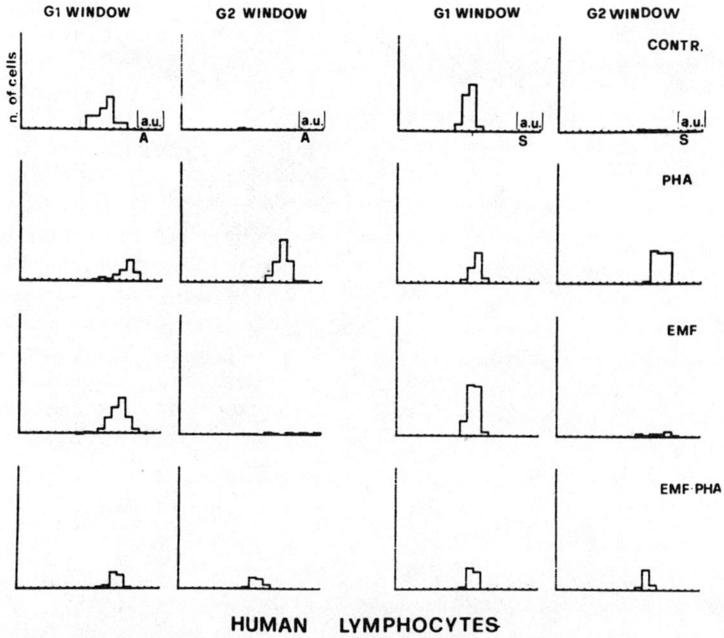

Figure 6 - First two columns (from left to right): Area histograms of Feulgen stained nuclei, with, respectively, a 2C DNA content (left, G1 window) and a 4C DNA content (right, G2 window). Last two columns: Available binding sites for the same two classes of Feulgen stained nuclei.
Note: For the choice of the two windows, refer to Figure 5.

experiment has so far been fully analyzed.
In order to offer the reader a qualitative picture of the extent of the chromatin decondensation process induced by em exposure alone, Figure 7 shows the digitalized images of two Feulgen stained nuclei with the same 2C amount of DNA and extreme area values.
As far as the microfluorimetric data are concerned, the histograms obtained in all the experiments performed so far, always exhibited features qualitatively in agreement with those of the first column of Figure 5. On the other hand, a high variability in the results was found, the inhibitory effect ranging from the one displayed in figure 5 to much weaker cases. Optimal or saturation doses of PHA were not utilized. At the same time, no experiments were done with extremely low PHA concentrations (i. e., 1 order of magnitude less than the optimal dose or lower).

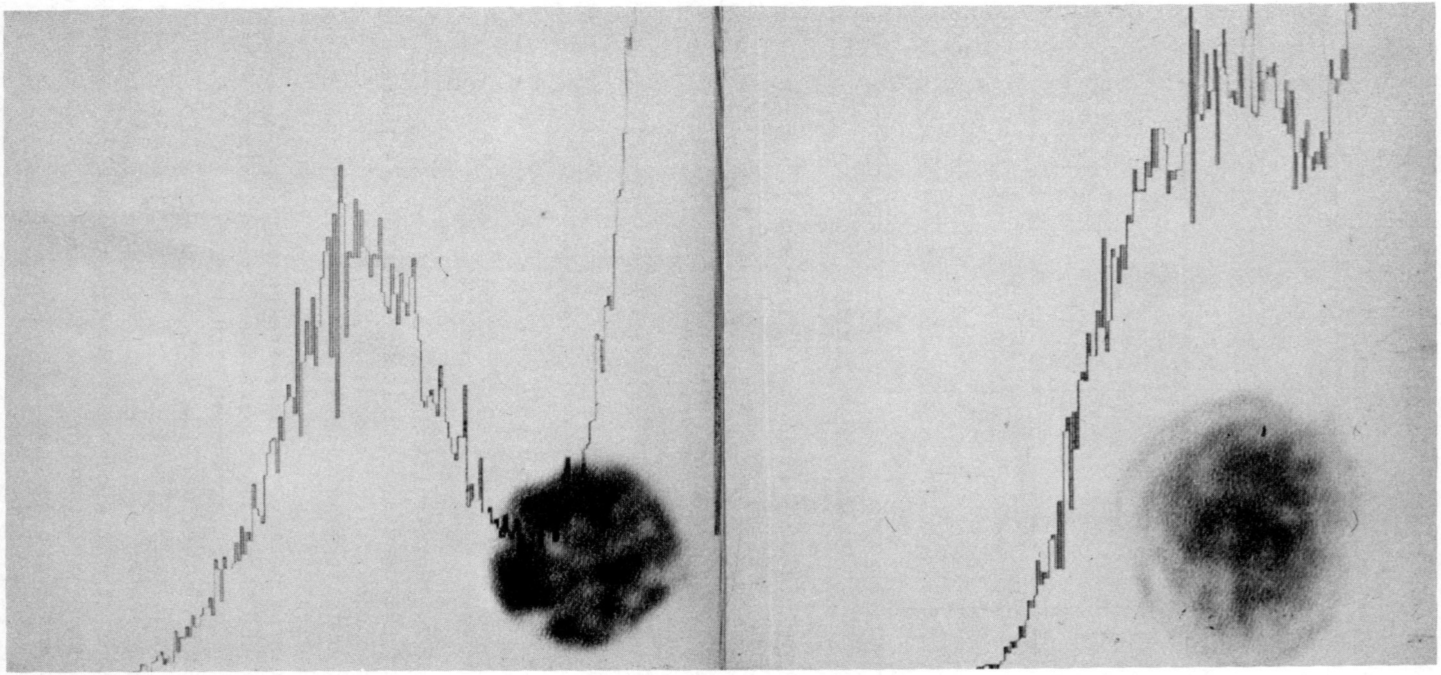

Figure 7 - Digitalized images of the Feulgen stained nuclei with the same 2C DNA amount and, respectively, a condensed conformation (quiescent cells, left) and a decondensed conformation (reactivated cell, right), after 72 h of em exposure alone. The superimposed histograms display the corresponding gray level distributions, from black to white. Both cells belong to the G1 window of Figure 5 (third row).

CONCLUSIONS

The analysis of the effect of weak em fields on cultured human lymphocytes, performed by the two independent and complementary techniques of automated flow microfluorometry and automated absorption image cytometry, allows us to draw the conclusion that the em signal, which has been shown to be able to reactivate frog erythrocytes, exerts a clear antagonistic effect on the mitogenic effect induced by sub-optimal doses of PHA on human lymphocytes (Figures 4, 5).

Under our experimental conditions, the em signal, per se, is able, after 72 hours of exposure, to trigger human lymphocytes to accomplish only the first step on the process of electrochemical reactivation, that is, to induce chromatin decondensation (Figure 6) without any (or with very little) DNA synthesis. These results are in agreement with the predictions of our model (see Chiabrera et al "Electric and magnetic field effects... ", in this book), which postulates both an increase in free Ca++ availability and a decrease in the mean life-time of small lectin receptor aggregates caused by the em field. A decondensation process also occurs in PHA stimulated cells with or without em exposure in a few hours.

Even though much more work is needed before attempting any comprehensive picture of the process of electrochemically induced cell reactivation, nevertheless, we regard our data on human lymphocytes as a new contribution in favour of the existence of (at least) two aspecific, physico-chemical mechanisms (one at the cellular surface level and the other at the chromatin level), which are involved in the electrochemical triggering of cells from quiescence to an active state. These mechanisms are not necessarily unique, and other more specific pathways can be followed by quiescent cells when activated by particular"mitogens".

ACKNOWLEDGEMENTS

This work was supported by the National Research Council (CNR) of Italy.

REFERENCES

Aaronson, RP and Woo, E, 1981, Organization in the cell nucleus: Divalent cations modulate the distribution of condensed and diffused chromatin. The Journal of Cell Biology 90: 181-186.

Abraham, S, Vonderheid, E., Zietz, S., Nicolini, C., 1980, Reversible (GO) and non readily reversible (Q) noncycling cells in human peripheral blood. Immunological structural and biological characterization. Cell Biophysics 2: 353-371.

Assailly, J., Monet, J-D., Goureau, Y., Christel, P. and Pilla, AA, 1981, Effect of weak inductively coupled pulsating currents on

calcium uptake in embryonic chick tibia explants. Bioelectrochem. and Bioeneregetics 8: 515 -521.
Bawin, SM, Sheppard, AR, Adey, WR, 1978, Possible mechanisms of weak electro magnetic field coupling in brain tissue. Bioelectrochem. and Bioenergetics 5: 67-76.
Becker, RO, Murray, D, 1967, A method for producing cellular dedifferentiation by means of very small electrical currents. Trans. NY Acad. Sci. 29: 606- 615.
Beltrame, F., Chiabrera, A., Grattarola, M., Guerrini, P., Parodi, G., Ponta, D., Vernazza, G., Viviani, R., 1980a, ACTA: Automated image analysis system for absorption, fluorescence and phase contrast studies of cell images. Proceedings 2nd Annual Conference of the IEEE Engineering in Medicine and Biology Society 58-60
Beltrame, F., Chiabrera, A., Gliozzi, A., Grattarola, M., Parodi, G., Vecchio, D., Ponta, D., Vernazza, G., Viviani, R., 1980b, Electromagnetic control of cell reactivation. Proceedings Symposium URSI Electromagnetic Waves and Biology 33-39.
Beltrame, F., Chiabrera, A., Grattarola, M., Guerrini, P., Parodi, G., Ponta, D. Vernazza, G., Viviani, R, 1980c, Electromagnetic control of cell functions, Alta Frequenza, XLIX, 2, 101-114
Beltrame, F., Chiabrera, A., Grattarola, M., Guerrini, P., Parodi, G., Ponta, D. Viviani, R., 1981a, Correspondence between flow microfluorometry and absorption image analyses of chromatin. Acta Phatologica Scandinavica, Suppl.274: 82-86.
Beltrame, F., Chiabrera, A., Grattarola, M., Guerrini, P., Parodi, G., Ponta, D. Vernazza, G., Viviani, R., 1981b, RE-evaluation of acridine orange stain for flow cytometry. I. Application to electrochemically induced chromatin reactivation. Bioelectrochem. and Bioenergetics 8: 387-404.
Beltrame, F., Chiabrera, A., Giannetti, G., Grattarola, M., Parodi, G., Ponta, D., Vernazza, G., Viviani, R., 1982, Re-evaluation of Feulgen stain for image cytometry. Application to electrochemically induced chromatin reactivation. Bioelectrochem. and Bioenergetics.9, 645-662.
Boyum, A., 1968, Separation of leukocytes from blood and bone marrow. Scand. J. Clin. Lab. Invest. 21, Suppl. 97: 1-12.
Chiabrera, A., Parodi, G., Ponta, D., Vernazza, G., Viviani, R.,, 1978, Bioelectrochemical effects of low frequency electric field. Presented at: Biological effects of electromagnetic waves URSI General Assembly, Helsinky, Finland.
Chiabrera, A., Hinsenkamp, M., Pilla AA., Ryaby, J., Ponta, D., Belmont, A., Beltrame, F. Grattarola, M., Nicolini, C., 1979a, Cytofluorometry of electromagnetically controlled cell dedifferentiation. J. Histochem. Cytochem. 27: 375-381.
Chiabrera, A., Hinsenkamp, M., Pilla A. A., Nicolini, C., 1979b, Electromagnetic induction of electrochemical information at cell surfaces: Application to chromatin structure modification. In Nicolini, C. (ed): Chromatin structure and function Part B: Plenum, pp.811-840.

Chiabrera, A., Viviani, R., Parodi, G., Vernazza, G., Hinsenkamp, M. Pilla, AA., Rybay J., Beltrame, F., Grattarola, M., Nicolini, C., 1980; Automated cytometry of electrochemically exposed frog eritrocytes. Cytometry 1: 42-48

Chiabrera, A., Grattarola, M., Viviani, R., Braccini, C., 1982, Modelling of the perturbation induced by low-frequency electromagnetic fields on the membrane receptors of stimulated human lymphocytes. II. Influence of the field on the mean lifetimes of the aggregation process. Studia Biophysica, 91, 125-131.

Chiabrera, A., Grattarola, M., Viviani, R., 1984, Interaction between electromagnetic fields and cells: Microelectrophoretic effect of ligands and surface receptors. Bioelectromagnetics, 5, .173-191.

Chiabrera, A., Grattarola, M. Parodi, M., 1984b, Modelling of electromagnetic processes in biological media under electromagnetic exposure, in"Modelling and Analysis in Biomedicine", C. Nicolini ed., Word Scient. Pub., 163-183.

Chiabrera, A., Giannetti, G., Grattarola, M., Parodi, M., Carlo, P. Finollo, R., in press, 1985: The role of ions in modifying chromatin structure. In: "Reconstruction surgery and traumatology. Vol.19. Electric and electromagnetic stimulation of bone growth", J. G. Rens Ed., Karger, Basel (Switzerland).

Conti, P., Gigante,, Cifone, M. G.,, Alesse, E., Ianni, G., Reale, M. and Angeletti, U. 1983 Reduced mitogenic stimulation of human lymphocytes by extremely low frequency electromagnetic fields FEBS Letters, 167: 156-160.

Darzynkiewicz, Z., Balazs, EA., 1971, Effect of connective tissue intercellular matrix on lymphocyte stimulation. Experimental Cell Research 76: 95-98

Decosse, JJ., Aiello,, N, 1966, Feulgen hydrolysis: Effect of acid and temperature J. Histochem. cytochem., 14, 601-606

Emilia, G., Cadossi, R., Torelli, G., Ferrari, S., Donelli, A., Zucchini, P: Low- frequency -low energy pulsing electromagnetic fields: Stimulating effect on human lymphocytes of chronic lymphocytic leukemia (CLL) Presented Third Annual Meeting Bioelectrical Repair and Growth Society, 2-5 Oct., 1983 S. Francisco, USA.

Grattarola, M., Viviani, R., Chiabrera, A 1982 Modelling of the perturbation induced by low-frequency electromagnetic fields on the membrane receptors of stimulated human lymphocytes. I. Influence of the field on the system's free energy. Studia Biophysica, 91, pp.117-124.

Grattarola, M., Caratozzolo, F., Viviani, R., Chiabrera, A., 1983, Modelling of the interaction between ligands and cell membrane receptors in the presence of exogenous electric fields"Int. Conf. Mathematics in Biology and Medicine, Bari, Italy, and Springler Verlag Notes in Biomathematics, in press.

Harrington, DB, Becker, RO, 1973, Electrical stimulation of RNA and protein synthesis in the frog erythrocyte. Experimental Cell Research 76: 95-98

Hellman, K. B., Hellman, A., Srickland, A. G., Fowler, A. K.: Effects of pulsed electromagnetic fields on cells of the immune system. Presented Third Annual Meeting Bioelectrical REpair and Growth Society, 2-5 Oct .1983, S. Francisco, (USA) 1983.

Kushch, AA., Niyazmatov, AA., Zelenin AV, 1980, The changes in the properties of chromatin from activated and non-activated lymphocytes at different stages of maturation under the action of salt solutions of different concentrations. Cell Differentiation 9: 291-304.

Lin-Liu, S., Adey, WR., 1982, Low frequency amplitude modulated microwave fields change calcium efflux rates from synaptosomes. Bioelectromagnetics 3: 309-322

Luben, RA, Cain, CD, Chen, MGY, Rosen, DM, Adey, WR, 1982, Effects of electromagnetic stimuli on bone and bone cells in vitro: Inhibition of responses to parathyroid hormone by low energy, low frequency fields. Proc. Natl. Acad. Sci. USA, 79: 4180-4184

Maino, VC, Green, NM, Crumpton, MJ 1974: The role of calcium ions in init trasformation of lymphocytes. Nature 251: 324-327.

Nicolini, C., Giaretti, W. De Saive, C., Kendall, F., 1977, The G0-G1 transition of WI 38 cells. Experimental Cell Research. 106: 119-125.

Nicolini, C., Belmont, A., Parodi, S. Lessin, S. Abraham, S., 1979, Mass action and acridine orange staining: static and flow cytofluorimetry. The Journal of Histochemistry and Cytochemistry 27: 102 - 113.

Oppenheim, JJ., Leventhal, BG, Hersh, E., 1968, The transformation of column -purified lymphocytes with nonspecific and specific antigenic stimuli. J. Immunol. 101: 262-270.

Pilla, AA, 1974, Electrochemical information transfer at living cell membrane. Ann. NY Acad. Sci. 238: 149-158.

Rigler, R., Killander, D., Bolund, L., Ringertz, NR., 1969, Cytochemical characterization of deoxyribonucleoprotein in individual cell nuclei. Experimental Cell Research.55: 215-224

Ringertz, NR., Bolund, L.,, 1969, Activation of hen erythrocyte deoxyribonucleo protein. Experimental Cell Research. 55: 205-214.

Rodan, G., Bourret, L., Norton, L., 1978, DNA synthesis in cartilge cells is stimulated by oscillating electric fields. Science, 199, 690-692.

Smith, SD, Thomas, CL, Frasch, SF., 1978, Lanthanum inhibition of electrically induced dedifferentiation in frog erythrocytes. Bioelectrochem. and Bioenergetics 5: 177-184.

Tenforde, TS., 1980, thermal aspects of electromagnetic field interactions with bound calcium ions at the nerve cell surface. J. Theor. Biol. 83: 517-521.

Thoma, F., Koller, T., Klug, A., 1979, Involvement of histone H1 in the organization of the nucleosome and of the salt-dependent superstructures of chromatin. The Journal of Cell Biology 83: 403-426.

Tsien, R., Pozzan, T., Rink, T, 1982, T-cell mitogens cause early changes in cytoplasmic free Ca++ and membrane potential in lymphocytes. Nature, 295, 68-70.

Whitney, R., Sutherland, R., 1978, Requirement for calcium ions in lymphocyte transformation stimulated by phytohemagglutinin. Cell Physiol.80, 329-233

ELECTROMAGNETIC MODULATION OF BIOLOGICAL PROCESSES: CONSIDERATION OF CELL-WAVEFORM INTERACTIONS

A.A. Pilla, R.E. Schmukler, J.J. Kaufman and G. Rein

Bioelectrochemistry Laboratory, Department of Orthopedic Surgery, Mount Sinai School of Medicine, New York, NY 10029 U.S.A.

INTRODUCTION

Inductively coupled pulsating current waveforms having specific waveform parameters have been utilized to affect a variety of cell and tissue systems. Examples are: Ca^{2+} uptake in isolated bone and cartilage tissue (1,2); increased DNA synthesis (3,4) and lysozyme activity (5) in isolated bone cells; lengthened membrane receptor aggregate lifetime in lymphocytes (6); increased neurotransmitter release in a clonal nerve cell line (7); stimulated transcriptional events in a dipteran genetic model (8); inhibited isolated bone cell response to parathyroid hormone (9); modulated rate of amphibian RBC response to trauma (with resultant DNA uncoiling) (10-12) or fracture healing in the rat (13,14) and of limb regeneration in the salamander (15); direct and synergistic (with cytotoxic and immunomodulating chemotherapy) effect on survival time of mice with B-16 melanoma, Ridgeway osteogenic sarcoma and spontaneous C3H mammary adenocarcinoma (16,17) and has been successfully employed clinically for non-unions and pseudarthroses of bone (18-20, 21-28).

It is important to note that the role of ions as transducers of information relevant to cell structure and function has gained widespread acceptance (29). Since induced current effects are believed to be regulated initially by ionic control mechanisms, the incorporation of these techniques will be beneficial for the development of mechanistic models of the action of these currents on cells. Examples of ionic control mechanisms include: growth factor activation of Na-K ATPase in fibroblasts (30,31), nerve growth factor effects regulated by Na-K ATPase (32,33), Ca^{2+} regulation via calmodulin of the cell cycle (34,35), differential Ca^{2+} requirements of neoplastic vs. non-neoplastic cells (36,37), and Ca^{2+} dependent adenylate cyclase activation in macrophages (38). This represents therefore a coupling mechanism which can be quantitatively analyzed. It is the purpose of this study to examine the interdependence of real time electrical responses of cells and the predictability of waveform configuration for biological effect. The biological assay system is the rate of macrophage cell adhesion on plastic culture dishes.

ELECTROCHEMICAL CELL SURFACE MODELS

The real time response of living cell membranes has been quantitated using impedance techniques. This is carried out by first using an electrochemical kinetic approach, which is described in detail elsewhere (39-47). For the purposes of this study, it is necessary to recall that the plasma membrane (and other organelle surfaces) exhibit electrical charge separation giving it capacitor-like properties. The structure of this capacitor is determined primarily by the interactions of water dipoles and protein and carbohydrate components of the membrane. In addition, this capacitor is leaky since transmembrane ion transport occurs. These general properties represent a class of electrochemical processes that characterize and govern the passage of current across the cell membrane. The following electrochemical steps have been modeled:

a) Dielectric charge storage due to the asymmetrical molecular organization (one-sidedness) of most membrane structures. This physical process can be represented by a simple capacitor.

b) Finite rate ion or dipole binding at specific molecular sites on the aqueous interfaces of the cell surfaces. The kinetics of this step can be represented by a series resistor/capacitor combination if uncoupled from membrane transport or by a parallel resistor/capacitor combination if part of the latter.

c) Transmembrane ion transport with finite entry (partitioning) kinetics. This two-step process can be represented by a resistor (for entry kinetics) in series with a parallel resistor/capacitor combination (for the transport portions).

Consideration of all of the above shows that current can cross the cell membrane via a complex, but readily discernible, set of electrochemical steps that are shown to be representative of the cell's real-time interaction with its charged environment. Impedance measurements that have been thus far carried out have allowed the relative magnitudes of the time constants (relaxation times) of these processes to be determined (43, 46-48). As expected, dielectric and specific adsorption time constants are somewhat smaller (1-100 μsec) than those involving membrane transport (1-100 msec). Of importance also is the observation from these and other (49-55) studies that the steady state (D.C.) current pathway (due to transport) exhibits a specific resistivity several orders of magnitude above that of the extracellular fluid.

EXPERIMENTAL IMPEDANCE DETERMINATION

The objective is to obtain electrical relaxation measurements over the anticipated wide frequency (time) range of the kinetics of electrochemical surface (membrane) process for isolated (non-epithelial) cells in physiologically meaningful conditions. To avoid membrane puncture with microelectrodes, previous techniques have utilized cell suspensions (56-60). The sensitivity of this approach is severely compromised, particularly in the frequency range of interest, due to the relatively high conductivity of isotonic salt suspending solution which shunts a majority of current around the cells. The methodology of the present study utilizes a filter composed of polycarbonate insulating material, containing well-defined and uniform cylindrical pores. Freshly isolated human red blood cells are washed three times in isotonic Ringers solution (0.15M NaCl, 4mM KCl, 1 mM $CaCl_2$, phosphate buffered to pH 7.4). The washed cells are placed in a vertical conductivity cell containing the insulating filter. The cells are forced tightly into the pores using a hydrostatic pressure head of 20 cm H_2O which is maintained during the impedance measurements. Blockage of at least 99% of the pores is ascertained by constant pressure flow measurement. The geometry of the HRBC in the pore is that of a half-dumbell as established using scanning and transmission electron microscopy (61,62). The pseudoepithelium thus created restricts the majority of the electrical leak pathway to the thin liquid film existing between the outer cell membrane and the inner wall of the filter pore. The dimensions of this film are determined essentially by the resultant force of the hydrostatic pressure head and the surface tension of the membrane/electrolyte/pore wall geometry. (Note: the HRBC is deformable, but the surface area is constant (63,64) at the pressure employed, which is within physiological range). In order to obtain a membrane segment likely to exhibit "lumped" (as opposed to dispersion or distributed) type behavior two filter pore characteristics were tested. The first had 4×10^5 (± 10%) pores/cm^2 of 1.85 ± 0.05 μm diameter and 13 ± 0.5 μm length. The ratio of exposed to embedded surface area was approximately 13 to 1. The second had 2×10^6 (± 10%) pores/cm^2 of 1.2 ± 0.05 μm diameter and 11 ± 0.5 μm length. In this case the surface area ratio was approximately 37 to 1. Measurements with both filters showed that the lumped parameter model derived from electrochemical kinetic theory (47) could validly be employed, as described below. A four-electrode technique is utilized to obtain impedance data with the filter pores occupied by cells. A current step of 10-50 μA is applied via two large surface area carbon electrodes. The voltage response is obtained with two large surface area porous tantalum electrodes having 1 mm in all outside dimensions, placed within 2 mm from the planes of the filter. Data is now acquired using a 2 channel digital oscilloscope (Biomation Model 4500) capable of a maximum sampling rate of 10 nsec per point, simultaneous for each channel. To resolve all transients of interest, data is initially obtained at 10 nsec per point for (usually) 120 points allowing sufficient baseline and early time information. Sampling is then automatically increased to 40 nsec per point for the remainder of the data record (1000 points total per channel). These time domain records are transferred to a Hewlett Packard 1000 computer using a specially constructed handshake interface program. This program also allows, via an appropriate menu selection, the creation of final time domain files to be used for the Laplace transformation routines. Generally 150-200 points are employed for analysis with the time spacing between points after the time base change (40 nsec portion of the record) completely user controlled.

The time domain results obtained are analyzed using real axis Laplace transformation. This is performed by applying the digital Laplace transformation algorithm previously described (65,66) to the input perturbation and output response time functions. This algorithm requires only that each function be described by a series of points, the minimum spacing of which determines the highest resolvable frequency. The time points need not be uniformly spaced and only enough points to satisfactorily describe the curves need be included. This factor results in a significant savings in both computation time and data storage, as compared to other identification techniques requiring uniform spacing between sample points (e.g., FFT and Z-transform). In addition, both input and output response functions can be of arbitrary shape, thus allowing all portions of waveforms, including finite rise time, to be used. The integral transform is carried out by assuming an exponential or trapezoidal function between two successive time points and summing the contributions of the closed form transform evaluated at those points.

The real-axis impedance function Z (σ) obtained using the above approach is utilized to evaluate the parameters of various electrochemical models. The models have been established such that impedance responses of various cell systems to changes in experimental conditions can be tested. A graphical method is used to get an estimate of the equivalent circuit elements. This method has been described in detail elsewhere (43,47,48). For the purposes of this study it is merely necessary to state that these models describe the form and the number of time constants by which the system may be described. For both pore configurations the HRBC could be easily described by three relaxation pathways two of which (RA, CA and RB, CB) have the form of specific adsorption kinetics (see previous section). The results are shown in Table I. Note that the value of CD (the dielectric membrane capacitance) compares well with the literature (50,55). These results were subjected to the F statistic which clearly shows ($p < .001$) that the system is better described by three time constants instead of two. Of importance to this study is the fact that the two specific adsorption-type time constants are very different (2 μsec vs. 300 μsec) thereby allowing induced current waveforms to be subjected to a model dependent frequency analysis.

FREQUENCY CHARACTERIZATIONS OF INDUCED CURRENT WAVEFORMS

In order to assess the relationship between waveform parameters for a given controlled geometry and cell response it is important to remember that the class of physical processes to which the weak pulsating current waveforms of the present study are designed to couple do not represent the overall metabolic activity of a cell. Rather, the step involved is a physicochemical one that is informational or catalytic in nature. Thus, these weak currents do not supply the energy for a particular biochemical process. In this context the analysis given here attempts to consider the real-time electrochemical response as providing a net change in cellular boundary conditions.

It has already been established that the geometric dependence of the dosage of inductively coupled current is predictable (67-69). Further it has been shown that both Laplace and Fourier transforms are useful to describe various frequency characterizations of induced waveforms (48,71). Here this anlaysis is extended to a description of the waveform via the real-axis Laplace transformation

Table I. Isolatable Impedance Parameters for HRBC Using an Electrochemical Kinetic Model

Membrane Capacitance ($\mu F/cm^2$)	Equivalent Adsorption Resistance ($ohm\text{-}cm^2$)	Equivalent Adsorption Capacitance ($\mu F/cm^2$)
	RA = 1.42 ± 0.14	CA = 1.2 ± 0.13
CD = 0.625 ± 0.012	RB = 20.5 ± 2.4	CB = 15.4 ± 1.4

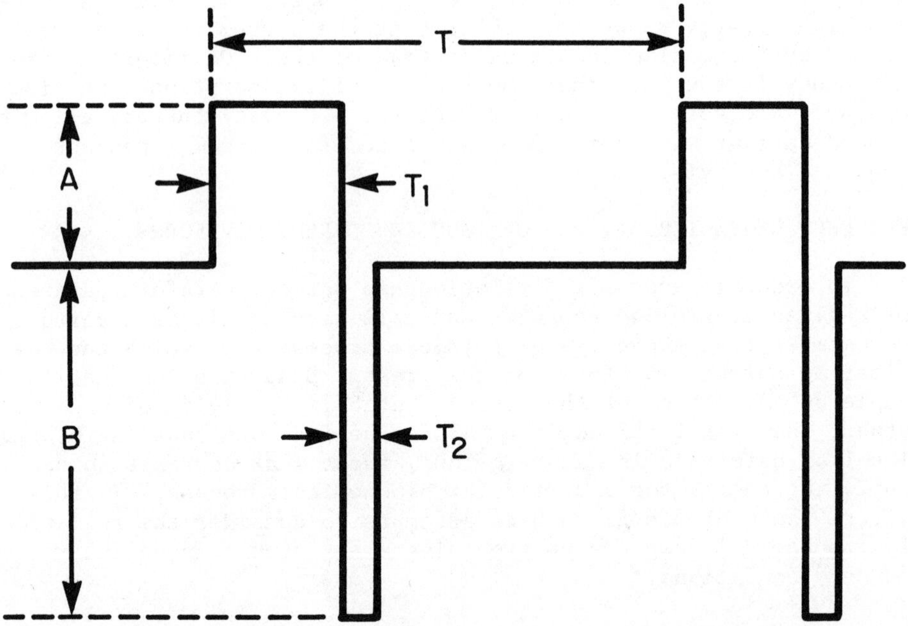

Figure 1. Schematic of the basic waveform of the induced electric field vector utilized in this study. Amplitude, defined as that of the main polarity (A) portion, is controlled by the coil driving voltage. The opposite polarity amplitude (B) is controlled using a zener diode of the appropriate voltage. Note that A x T1 = B x T2 (T1 and T2 are the respective polarity pulse durations) so that for a larger T2, the opposite polarity amplitude (B) is lower when A is constant. The pulse period, T, was maintained constant at 15 Hz. This waveform exists in tissue and represents that voltage which can drive current through the specific adsorption-like pathway. This induced electric field was generated using two identical (10 x 10 cm) air-gap coils wound with 50 turns of No. 14 copper magnet wire set for approximately Helmholtz (magnetically-aiding) behavior (i.e., the intercoil distance was 6.5 cm). The effective amplitude range (see Fig. 2) appeared to be that for which dB/dt ranged from 1-3.5 Gauss/μsec.

of the time domain power into a defined pathway. In this case the parameters for the pathway are derived from the above impedance study and assumed to be of the form of specific adsorption kinetics, i.e., a series RC equivalent circuit.

In order to test for the possible relationship between the kinetics of cell surface electrochemistry and induced current waveform the signal shown in Fig. 1 was employed. The main parameters examined were pulse width in both polarities (T1 and T2) for a given amplitude (A) and repetition rate (T). The real axis Laplace plane analysis for the spectrum of the power, P(σ), into a single specific adsorption-type time constant, TA, is given by:

$$P(\sigma) = \frac{A^2}{\sigma+1/TA}\left[1-e^{-(\sigma+1/TA)T1} + \frac{T1^2}{T2^2}\, e^{-\sigma T1}\left(1-e^{-(\sigma+1/TA)T2}\right)\right]$$

where σ is the real axis frequency designation. The variable of interest in this equation is TA which is obtained using RA x CA or RB x CB as evaluated from impedance measurements. A typical set of spectra is shown in Fig. 2, wherein the solid lines designate the boundaries within which the dose response curve for cell adhesion exists (75). These lines represent the limits for a signal having T1 = 200 μsec and T2 = 20 μsec (200/20, see Fig. 1). The other two lines give the spectra for signals having T1 = 100 μsec, T2 = 10 μsec (100/10 - dashed line) and T1 = 300 μsec, T2 = 30 μsec (300/30 - dotted line) respectively. It is important to note that the 100/10 signal used for Fig. 2 required only 1A, vs. 1.5A for the 200/20 signal (see Fig. 1 for definition of A), to achieve identical bio-effects. In contrast, the 300/30 signal used for Fig. 2 was at 1.5A (i.e., identical to the 200/20 signal) and was ineffective in the bioassay system. In fact, as will be shown in the next section, only those signals which have sufficient power to fall adequately within the frequency range enclosed by the dotted lines are biologically effective. These correlations exist only for TA = 1-5 μsec. There is no meaningful relationship when the system is presumed to have a longer time constant (e.g., RB x CB ≃ 300 μsec). Of note is the fact that the effective amplitude of the 100/10 signal was predicted on the basis of the spectra shown in Fig. 2.

ELECTROMAGNETIC MODULATION OF CELL ADHESION

In order to test the quantitative signal correlations proposed above for biological efficacy, an *in vitro* cell adhesion system was chosen. For a model system to accurately correlate different signal parameters with electrochemical cell surface phenomena it is important to choose a bioassay system which 1) involves direct interaction

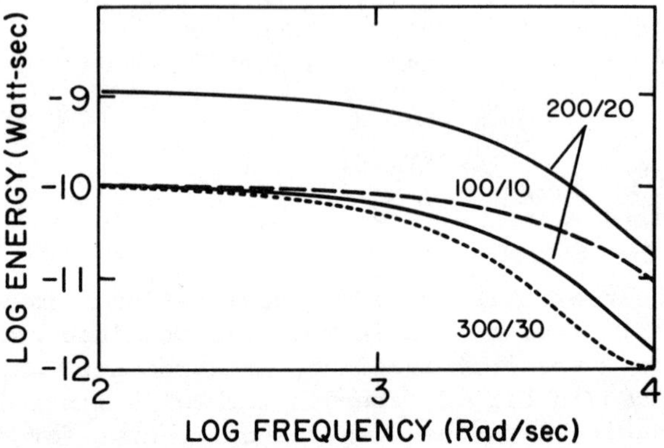

Figure 2. Real axis Laplace Transform of the power into a single time constant pathway using the driving voltage shown in Fig. 1. This is termed the spectrum of the power. These curves were obtained using a time constant of 3 μsec (see Table I for the values of specific adsorption-like relaxation times for the human red blood cell). The solid lines represent the experimentally determined amplitude boundaries for macrophage cell adhesion using the 200/20 waveform (see text). The dashed and dotted lines represent the 100/10 and 300/30 waveforms respectively. These were utilized respectively at the same, and 1.5 times, the amplitude A (see Fig. 1) as that of a marginally effective 200/20 signal (less than 50% of the clinical signal effect - see text). Only the 100/10 was maximally effective, whereas the 300/30 was ineffective. The region of this frequency spectrum extending from approximately 10^3 to 10^5 rad/sec appears to be that within which the effective amplitude range for this bioassay system should be set.

of induced currents with specific surface membrane events 2) represents an early response to such a stimuli and 3) is simple enough to avoid competing or sequential biochemical events which may also respond to this stimulus. Adhesion to the extracellular matrix (ECM) and cohesion to other cell membranes is an early time event which is the initial trigger for several universal biological phenomena including cell division, differentiation and migration (72). To avoid the subsequent biochemical events which are triggered by cell adhesion, the experimental protocol can be manipulated to ensure that only cell adhesion is actually being measured. This in turn assures the relative simplicity of the assay so there are a minimum of interfering biochemical reactions. Initial cell adhesion only involves ionic binding of two charged surfaces mediated by extracellular cations, notably calcium.

Although cell adhesion is a universal phenomenon, care must be taken in choosing a suitable cell type. Ideally a cell type should be chosen which utilizes adhesive and cohesive properties in the course of its normal biological function. Furthermore an experimental advantage of using cells which do not adhere too quickly or strongly is that the binding to the ECM will be slow enough to allow adequate induced current exposure. Finally, the cells should form a single-cell suspension with a minimum of cell clumping to ensure accurate cell counting. Macrophages which utilize cell adhesion as a basic mechanism in their phagocytotic function also satisfy these other criteria. Furthermore they are involved in many basic biological processes including the inflammatory response, tissue repair, bone morphogenesis and cell death phenomenon associated with cartilage blastemas (74). A virally transformed cell line, RAW 264.7, is available which contains several properties of normal macrophages and can be used to grow cells in large enough quantities for large scale studies and for biochemical analysis. Preliminary research from our laboratory indicates that macrophage rosetting, a surface membrane phenomenon involving binding of antibody to F_c receptors, decreases in response to current exposure (73).

Cell adhesion was measured as described elsewhere (75). Confluent culture dishes are treated with calcium and magnesium free Hanks buffered salt solution (HBSS) containing EDTA to liberate the cells which are then suspended in HBSS containing sodium bicarbonate, glucose, calcium and magnesium. The single cell suspension is then added to 35 mm tissue culture dishes placed in horizontally arranged Helmholtz coils or to control dishes located 4 feet from the experimental dishes in a warm room maintained at 37°C. After 15 min. the supernatant is removed and the free floating cells counted in a Coulter counter. The results are normalized with respect to the total number of cells added to the dish at the beginning of the experiment. The percent effect for a given signal is then calculated as the percent increase in the experimental values relative to the control values. In all cases the rate of adhesion was lowered. This value is then divided by the percent effect obtained for a standard signal which is a burst form of that shown in Fig. 1. The burst portion contains 21 pulses for a duration of 5 msec. The basic repetition rate is identical at 15 Hz. This reference signal is that most commonly employed in the clinical application for non-unions of bone (18,27). The effect of varying pulse widths on cell adhesion is shown in Table II wherein it can be seen that the signal designated 300/30 (see Fig. 2) is ineffective. Comparison of these results with the spectra shown in Fig. 2 reveals that the signal which has insufficient power within the indicated frequency range is indeed that designated 300/30.

The above study represents an initial attempt to provide a quantitative correlation between the real time electrical response of the cell and its biological response to induced currents. A strong correlation exists when the short (1-5 μsec) specific adsorption-type impedance pathway is incorporated in a model dependent frequency analysis. This is the first evidence, to the authors' knowledge, that the real time electrical response of the cell provides a useable quantity (a time constant) to predict the effectiveness of a particular induced current waveform configuration.

Table II. Effect of Induced Current Pulse Width on the Rate of Macrophage Cell Adhesion

Main Polarity Width T1 (sec)	Opposite Polarity Width T2 (sec)	Percent of Standard Clinical Signal
100	10	86 ± 19
200	20	91 ± 10
300	30	NS

Note: The 100/10 was at 1A vs. 1.5A for the other signals (see Fig. 1 for definition of A). All repetition rates were identical.

REFERENCES

1. J. Assailly, J.-D. Monet, Y. Goureau, P. Christel and A.A. Pilla, Bioelectrochemistry and Bioenergetics, 8, 515 (1981).
2. G. Colacicco and A.A. Pilla, Bioelectrochemistry and Bioenergetics, 10, 119 (1983); Calcif. Tiss. Int., 36, 167 (1984).

3. A. Shteyer, L.A. Norton, A.A. Pilla and G. Rodan, 5th International Symposium on Bioelectrochemistry, Weimar, East Germany, Sept. 1979.
4. A.R. Liboff, T. Williams, Jr., D.M. Strong and R. Wistar, Jr., Science, 223, 818 (1984).
5. L. Norton, Clin. Orthoped., 167, 280 (1982).
6. A. Chiabrera, M. Grattarola, R. Viviani, and C. Braccini, Studia Biophysica, 91, 125 (1982).
7. R. Dixey and G. Rein, Nature, 296, 253 (1982).
8. R. Goodman, C.A.L. Bassett and A.S. Henderson, Science, 220, 1283 (1983).
9. R.A. Luben, C.D. Cain, M. C-Y. Chen, D.M. Rosen and W.R. Adey, Proc. Natl. Acad. Sci. USA, 79, 4180 (1982).
10. A. Chiabrera, M. Hinsenkamp, A.A. Pilla, J. Ryaby, D. Ponta and C.Nicolini, J. Histochem. and Cytochem, 27, 375 (1979).
11. A. Chiabrera, M. Hinsenkamp, A.A. Pilla and C. Nicolini, in "Chromatin Structure and Function," C. Nicolini, ed., Plenum Press, New York, 1979, p. 811.
12. A. Chiabrera, R. Vivani, G. Parodi, G. Vernazza, M. Hinsenkamp, A.A. Pilla, J. Ryaby, F. Beltrame, M. Grattarola and C. Nicolini, Cytometry, 1, 42 (1980).
13. P. Christel, G. Cerf and A.A. Pilla, in "Mechanisms of Growth Control," R.O. Becker, ed., Charles C. Thomas Press, Springfield, IL, 1981, p. 237.
14. C.A.L. Bassett, M.G. Valdes and E. Hernandez, J.B.J.S., 64A, 888 (1982).
15. S.D. Smith and A.A. Pilla, in "Mechanisms of Growth Control," R.O. Becker, ed., Charles C. Thomas Press, Springfield, IL, 1981, p. 237.
16. L. Norton, A.A. Pilla, W. Regelson and L. Tansman, J. Electrochem. Soc., 127, 130C (1980).
17. L. Norton, L.S. Tansman, W. Regelson, S. Geller and A.A. Pilla, J. Electrochem. Soc., 129, 132C (1982).
18. C.A.L. Bassett, A.A. Pilla and R.J. Pawluk, Clin. Orthop., 124, 117 (1977).
19. C.A.L. Bassett, S. Mitchell, L. Norton and A.A. Pilla, Acta Orthop. Belg., 40, 706 (1978).
20. C.A.L. Bassett, S.N. Mitchell, L. Norton, N. Caulo and S.R. Gaston, in "Electrical Properties of Bone and Cartilage," C.T. Brighton, J. Black and S.R. Pollack, eds., Grune and Stratton, New York, 1979, p. 605.
21. C.A.L. Bassett, S.N. Mitchell, S.R. Gaston, J. Bone Joint Surg., 63A, 511 (1981).
22. L. Sedel, P. Christel, J. Duriez, R. Duriez, J. Evrard, C. Ficat, J. Cauchoix, J. Witovoet, Rev. Chir. Orthop., 67, 11 (1981).
23. M. Hinsenkamp, Rev. Med. Brux, 3, 19 (1982).
24. C.A.L. Bassett, N. Caulo and J. Kort, Clin. Orthop., 154, 136 (1981).

25. J.S. Kort, M.M. Schink, S.N. Mitchell and C.A.L. Bassett, Clin. Orthop., 165, 124 (1982).
26. J.C. Mulier and F. Spaas, Arch. Orthop. Traumat. Surg., 97, 293 (1980).
27. C.A.L. Bassett, S.N. Mitchell, M.M. Schink, J.B.J.S., 64A, 1214 (1982).
28. C.A.L. Bassett, S.N. Mitchell and S.R. Gaston, J.A.M.A., 247, 623 (1982).
29. Conference on "Ions, Cell Proliferation, and Cancer." Sponsored by the National Cancer Instutute, NIH, Lake PLacid, July, 1982.
30. J. Mendoza, N.M. Wigglesworth, P. Pohjanpelto and E. Rozengurt, J. Cell Physiol., 103, 17 (1980).
31. A. Lopez-Rivas, E. Adelberg and E. Rozengurt, Proc. Natl. Acad. Sci. (USA), 79, 6275 (1982).
32. J. Boonstra, P.T. Van der Sagg, W.H. Moolenaar and S.W. DeLaat, Exp. Cell Res., 131, 452 (1981).
33. J. Boonstra, S.D. Skaper and S.J. Varon, J. Cell. Physiol., 113, 28 (1982).
34. J.G. Chafouleas, W.E. Bolton, H. Hidaka, A.E. Boyd and A.R. Means, Cell, 28, 41 (1982).
35. J.F. Whitfield, A.L. Boynton, J.P. MacManus, R.H. Rixon, M. Sikorska, B. Tsong, P.R. Walker and S.H. Swierenga, Ann. N.Y. Acad. Sci., 339, 216 (1981).
36. A.L. Boynton, J.F. Whitfield, R.J. Isaacs and R.G. Trembley, Cancer Res., 37, 2657 (1977).
37. B. Hazleton and J. Tupper, J. Cell. Biol., 81, 538 (1979).
38. D. Gearsa, M. Seitz, W. Kramer, W. Grimm, G. Till and K. Resch, Exp. Cell Res., 118, 55 (1979).
39. A.A. Pilla, in Proceedings 7th Intersociety Energy Conversion and Engineering Conf., San Diego, 1972, Amer. Chem. Soc., Washington, D.C., p. 761.
40. A.A. Pilla, Ann. N.Y. Acad. Sci., 238, 149 (1974).
41. A.A. Pilla, Bioelectrochem. and Bioenergetics, 1, 227 (1974).
42. R.O. Becker and A.A. Pilla, in "Modern Aspects of Electrochemistry," J.O'M. Bockris, ed., Plenum Press, New York, vol. 10, 1975.
43. A.A. Pilla and G.S. Margules, J. Electrochem. Soc., 124, 1697 (1977).
44. A.A. Pilla, Adv. in Chem., A.C.S., 188, 126 (1980).
45. A.A. Pilla, in "Electrical Properties of Bone and Cartilage," C.T. Brighton, J. Black and S.R. Pollack, eds., Grune and Stratton, New York, 1979 p. 455
46. G.S. Margules, S.B. Doty and A.A. Pilla, Adv. in Chem., A.C.S., 108, 108 (1980).
47. R. Schmukler and A.A. Pilla, J. Electrochem. Soc., 127, 130C (1980); 129, 526 (1982).
48. A.A. Pilla, in "Bioelectrochemistry," H. Keyzer and F. Gutman eds., Plenum, New York, 1980, p. 353.

49. K.S. Cole and R.H. Cole, J. Chem. Phys., 9, 341 (1941).
50. H. Schwan and K.S. Cole, in "Medical Physics," The Year Book Medical Publishers, Inc., Chicago, vol. 3, 1960.
51. H. Schwan, Adv. Biol. Med. Phys., 5 (1957).
52. R. Plonsey, Bioelectric Phenomena, McGraw Hill, New York, 1969, p. 124.
53. L.R. Blinks, J. Gen. Physiol., 13, 361, 495 (1930).
54. C.Y. Kao, J. Gen Physiol., 40, 107 (1956).
55. K.S. Cole, Membranes, Ions and Impulses, Univ. of Calif. Press, Berkeley (1972).
56. H. Fricke, Phys. Rev., 24, 575 (1924).
57. J.L. Oncley, J. Am. Chem. Soc., 60, 1115 (1938).
58. K.S. Cole,J. Gen. Phys., 12, 29 (1928); 15, 641 (1932).
59. H.P. Schwan and E.L. Carstensen, Science, 125, 985 (1957).
60. H. Pauly and H.P. Schwan, Biophys. J., 6, 621 (1966).
61. R. Schmukler, E. Sc.D. Thesis, Columbia University (1981).
62. J.D. Brailsford, Blood Cells, 3, 25 (1977).
63. S. Chien, Science, 157, 827 (1967).
64. P.R. Zarda, S. Chien and R. Skalak, J. Biomechan., 10, 211 (1977).
65. A.A. Pilla, J. Electrochem. Soc., 117, 467 (1970).
66. A.A. Pilla, in "Electrochemistry: Calculations, Simulation and Instrumentation," J.S. Mattson, H.C. MacDonald and H.B. Mark, Jr., Editors, p. 139, Marcel-Dekker, New York (1972).
67. A.A. Pilla, P. Sechaud and B. McLeod, J. Biol. Phys., 11, 51 (1983).
68. B. McLeod, A.A. Pilla and M.W. Sampsel, Bioelectromagnetics, 4, 357 (1983).
69. B. McLeod and R.A. Parker, J. Bioelectricity, 3, 41 (1984).
70. B.F. Sisken, B. McLeod and A.A. Pilla, J. Bioelectricity, 3, 81 (1984).
71. A.A. Pilla and J.J. Kaufman, J. Bioelectricity, 3, 3 (1984).
72. See Ann. Rev. Biochem., 52, 761 (1983).
73. K. Fox, A.A. Pilla, K. Hellman and J. Ryaby, Transactions 4th Annual Bioelectrical Repair and Growth Society Meeting, Kyoto, Japan (1984).
74. D.O. Adams and T.A. Hamilton in Ann. Rev. Immunology, 2, 283 (1984).
75. G. Rein and A.A. Pilla, in Proceedings, 4th Annual Bioelectrical Repair and Growth Society Meeting, S. Inoue and H. Takakashi, eds., Nishimura Co., Niigata, Japan, 1985, in press.

AC FIELD EFFECTS OF AND BY LIVING CELLS

Herbert A. Pohl

Francis Bitter National Magnet Laboratory
Massachusetts Institute of Technology
Cambridge, MA 02139 U.S.A.

ABSTRACT

Electrical fields are important to the growth of living cells, and since many diseases, including cancer, are associated with cell growth, it is desirable to learn all one can about this aspect.

Several new and highly sensitive bioelectrical techniques successfully developed here are used to examine the natural ac electromagnetic emissions from living cells. The immediate focus of the research described here is to study the natural ac field as it is related to the various phases of the cellular growth cycle, and how it may vary with cell type. It is found to be ubiquitous, and to be present in all types of cells capable of reproducing.

A new technique, the formation of strikingly beautiful test patterns about living cells by various dielectric powders, reveals the site, type, and intensity of the ac fields. These patterns can be used to direct the more quantitative ensuing studies of the natural ac fields using either direct instrumental examination, cellular spin resonance, or micro-dielectrophoresis.

In light of the results, comments are made on the possible origin and role in cell life. It appears probable that the natural ac fields are either the signal of, or a feature of, certain critical cellular reproduction processes.

INTRODUCTION

The study of the natural ac electromagnetic oscillations of living cells has been made so a to learn their relation to cell division, and hence to the embryonic, somatic repair, regeneration, and tumor states. The origins, timing, and relations with other processes were sought. Their role in inter-cellular interactions and communications should be of significance.

Evidence demonstrates that cells undergoing mitosis act as tiny ac sources. The natural ac oscillations have been detected in bacterial, algal, fungal, annelidal, amphibian, avian, and mammalian cells by means which can leave no reasonable doubt as to their presence. It would thus appear that the natural ac oscillations are universal among living systems. They are most readily detected from dividing cells.

These natural ac fields appear to be of very short range (microns) and to be linked more to intra- rather than inter-cellular processes. A number of critical questions suggest themselves as ways to test the hypothesis that such natural ac oscillations are possibly a necessary accompaniment to cellular division.

To the extent that favorable answers are found to the questions posed, our knowledge of the origins, meaning, and control of these natural ac oscillations associated with cell growth could mean better insights into somatic repair and thus to improved diagnosis and treatment of disease. They may also have implications for ship bifouling and food preservation.

BACKGROUND INFORMATION

Occurrence of Natural AC Fields in and about Living Cells

Living cells are a source of natural ac electrical oscillations, as is shown by several laboratories using experimental approaches. These include (1) their microscopic action of dielectrophoresis upon tiny dielectric powder particles, (2) cellular spin resonance, (3) direct instrumental detection, and (4) the development of striking electric field patterns among detector powders near living cells. This section reviews and compares these recent results.

Active search for natural electrical oscillations and ac fields in living cells began about 15 years ago, following the initiative of Herbert Fröhlich (1968, 1980). He predicted that there would be oscillation of very high frequency (ca. 100 GHz), but left open the possibility that cooperative oscillations of a rather lower frequency could also occur. We shall focus in the present section upon low frequency electrical oscillations; for the evidence concerning the very high frequency range has been adduced and summarized by Fröhlich (1980). More recent evidence from laser Raman spectroscopy (Webb, 1980) indicates that metabolically active cells have responses to applied weak EM fields in the multigigaherz range.

Among the mechanisms which should be considered to be involved in natural ac oscillations are those involving the Fröhlich mechanism of dipolar oscillations driven by cellular metabolic energy, and mechanisms involving oscillating chemical reactions capable of producing charge waves. The latter will involve limit cycle operation as discussed by Fröhlich (1977), by Kaiser (1982), and by Albanese (1983). A model for oscillating chemical reactions occuring in a cell, and giving rise to charge waves was discussed by Pohl and Braden (1982).

In this model it is proposed that the natural intrinsic ac oscillating cellular dipoles stem from oscillating chemical reactions that couple spatially within the cell to mobile sets of ions so as to produce charge density waves. Based upon what is known of several well-studied oscillating reactions, (Noyes and Field, 1974; Noyes, 1977; Epstein et al., 1981), the cyclic course of the reaction runs between an ionic and a (branching) free radical phase. As the reaction develops, it develops in both temporal and spatial coordinates from the initiating site(s). It is therefore clear that if, during the ionic phase of the reaction cycle the outward speed of the positive ions does not exactly match that of the negatives, then a charge wave will develop. It is an example of a longitudinal electrical wave. Such charge waves, if uncoordinated, could begin at any time and at any place within the cell, and be incoherent. However, as these charge waves develop, they encounter structures within the cell which will tend to make them run in parallel. A collimation of the charge waves will follow. Since they interact electrically, a Coulombic correlation (in the sense of a Bose-Einstein cendensation) to form a coherent set of charge waves, one involving large regions of the cell in a cooperative oscillating dipole or multipole, will tend to be formed. Such a testable model suggests a variety of experimental studies as to the energy source, the magnitude, the control, and causes.

Oscillating reactions in biological systems are numerous and well-known. Excellent reviews are to be had on the subject (Treherne et al., 1979; Berridge and Rapp, 1979; Rapp, 1979). A number of well-known structures within cells are candidates for the means by which such cyclic charge waves might be channeled and made more or less parallel. These include the walls of the endoplasmic reticulum, the laminae of the cristae in mitochondria, the granae in chloroplasts, or the mitotic spindle apparatus.

The charge waves as proposed above might be traveling either geometrically parallel to, or at right angles to the laminar structures restricting them in the cell. Kinetic data on the presumed charge waves imply that the charge waves travel across rather than along the laminae, especially in the case of high frequency oscillations (Pohl and Braden, 1982). A theoretical analysis by Schmidt and Ortoleva (1979) suggests how oscillating reactions might be coupled to electrical charge waves.

For the natural cellular oscillations in the frequency range of 10 mZ to 1 MHz, it is likely that the ionic double layers at the many interfaces within the cell would play a role as a source of the set of mobile ions involved. Conversely, it is also possible that mechanical motion acting on biological objects produces electromagnetic waves having feedback into the natural ac oscillations of cells. We refer here to that mechanically-induced electromagnetic radiation (M.I.S.E.R. mechanism) discovered recently (Pohl, 1980a).

Another possible mechanism capable of producing resonant oscillations from cells is that which would be produced by local fluctuations in the charge density at the surface of the cell, which arise in turn from internal reaction oscillations. It will then affect the nearby charge distribution in the ionic double layer at the cell-medium interface surrounding the cell. This perturbation will then travel around the cell periphery and reverberate.

THE EXPERIMENTAL EVIDENCE

Several lines of experiments provide proof for the existence of natural ac electrical oscillations in living cells. Rowlands et al. (1982a, 1982b, 1983) showed that interactions of extraordinarily long range exist between living red blood cells in their plasma. These measurements imply that the interactions extend out to some 4 microns from the cell periphery, a fact that strongly suggests the presence of long range cooperative interactions of the type proposed by Fröhlich.

Several experimental techniques have been used by Pohl and co-workers, and by Mischel and Lamprecht to demonstrate the presence of natural electrical ac oscillations in the frequency range below 100 GHz. There are four such techniques. (1) DEP pattern formation, (2) Micro-dielectrophoresis (DEP), (3) cellular spin resonance (CSR), and (4) direct instrumental detection. In this section we briefly review the recent evidence from these.

Dielectrophoresis (DEP) is the motion of bulk or particulate matter induced by the action of nonuniform electric fields (Pohl, 1951, 1978). It can be used to explore for the presence of nonuniform fields about cells or other objects. In the present instance the motion of tiny polarizable particles near cells is used to help decide if ac fields are present near the cell surface. The motion will be towards the region of highest field intensity (positive dielectrophoresis) if the effective dielectric constant of the tiny polarizable particle exceeds that of the suspending medium. Contrarywise, the motion of the particles will be away from the region of greatest field intensity (repulsion, or negative dielectrophoresis) if they have a lower effective dielectric constant than that of the medium. Because the phenomenon occurs on a microscopic scale, it is referred to as micro-dielectrophoresis (micro-DEP).

A. DEP Field Patterns about Living Cells

The presence of the small ac electric fields produced by living cells is shown by the gentle dielectrophoretic force on tiny dielectric particles. The beauty of the field patterns is made evident as the living cell and the nearby tiny dielectric particles settle in a hanging drop.

That living cells produce static electric fields and even electrical pulses has been known for almost a century. The static field across the cell membrane of all living cells, as well as the involvement of field changes in nerve pulses is well known and has been intensely studied since the days of Galvani in 1791. The presence of sinusoidal ac fields in and about cells has only recently been established. Let us now consider a particularly simple and electrodeless technique for displaying the ac field distributions about living cells. Tiny dielectric particles of two types—those seeking field maxima or those avoiding them—are let settle freely about the cell in a sessile drop of aqueous medium of low conductivity (less than 100 micro-mho/cm). Patterns displaying the field distributions shortly develop that are reminiscent of those formed by magnetizable particles about a magnet (Rivera, et al., 1984).

The technique is simple and straightforward. A cell or two is suspended in a drop of high-resistivity medium (isoosmotic preferred) together with several hundred field-detector particles (average diameter 2 mm). The loaded drop hangs from a cover slip in the pocket of a hanging drop microscope slide and is examined with the aid of a microscope, a camera, and a TV (Vidicon) camera and tape recorder. The patterns develop in a matter of minutes.

The DEP field patterns of large algae such as <u>Netrium</u> or <u>Closterium</u> are stable for hours, for these cells are "at home" even in quite pure water.

The powders are prepared by grinding, then washing repeatedly with deionized water with centrifugation between washes. The final powder suspension is to be made up in 0.1% soluble potato starch (aqueous) as a stabilizer-dispersant. Sucrose or mannose (ca. 300 mM) is added as necessary to adjust the osomotic character.

The particles are usually of about 2 to 3 microns diameter, i.e., much smaller than that of the cells examined. The particles attracted to the high field region can be of barium titanate, strontium titanate, calcium zirconate, sodium niobate, or various ekaconjugated polymers, all of which have dielectric constants well in excess of 300. On the other hand, the particles repelled from the high field region can be of barium sulfate, silica, or alumina, all of which have dielectric constants less than about 12, and less than that of water, 78. Inspection of Plates I, II, III and IV and their captions give further detail. Note that dead cells are inactive, and also that, in accord with theory, the presence of evel low salt concentration (100 micromolar KCl) suffices to destroy the patterns about living cells, provided that it is <u>ac</u>

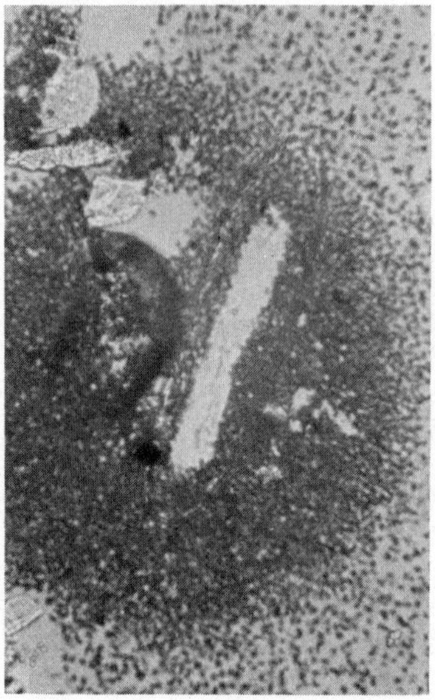

1a

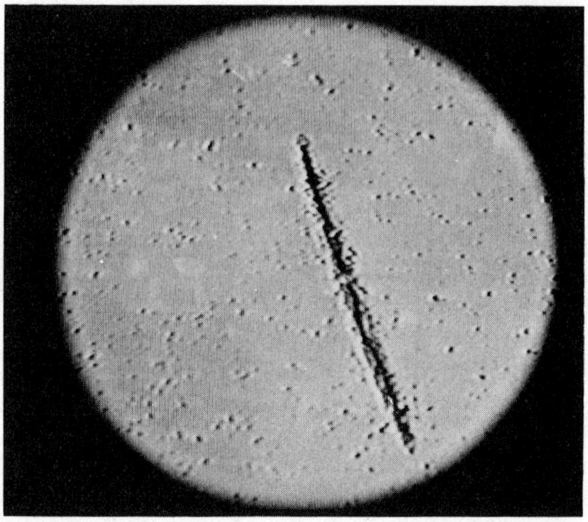

1b

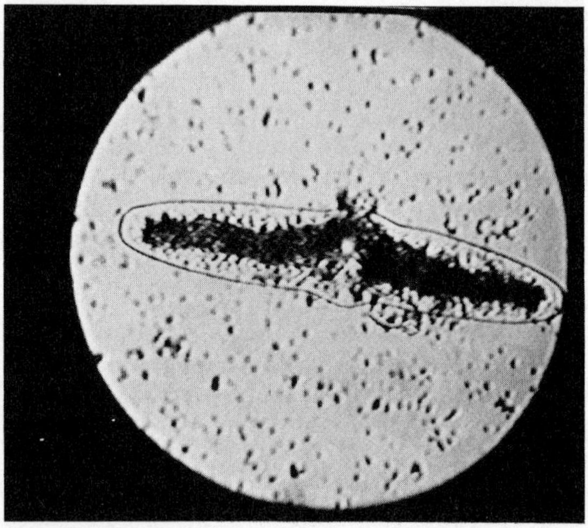

1c

PLATE I. Powder Patterns about Cells in Hanging Drops
INERTNESS

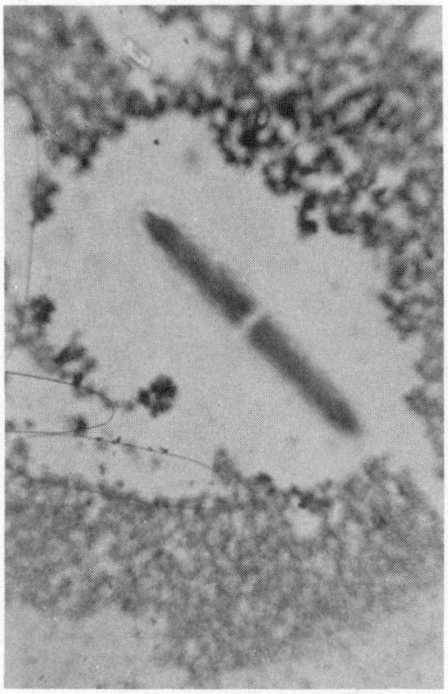

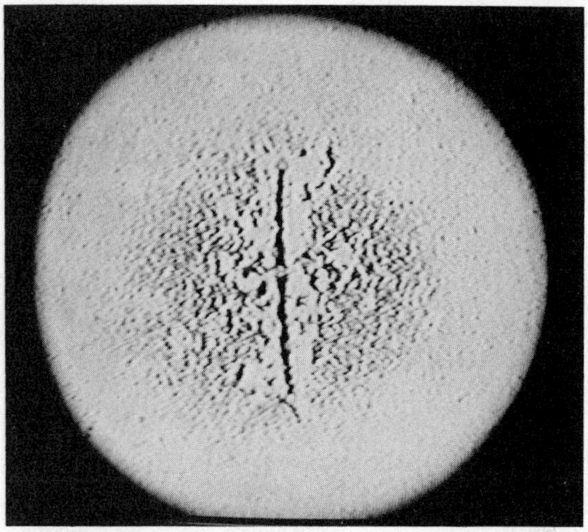

IIb

IIa

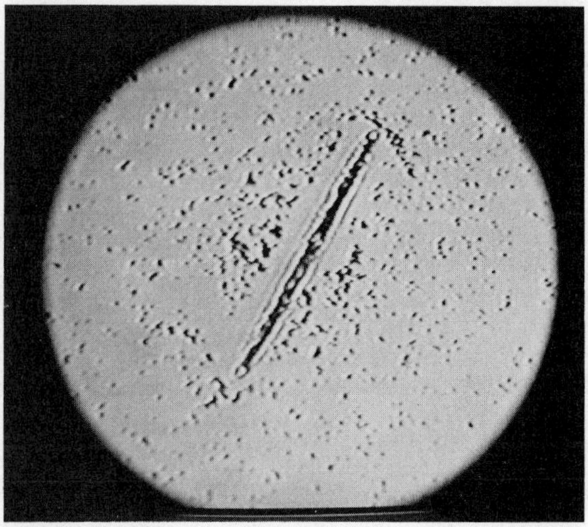

IIc

PLATE II. Powder Patterns about Cells in Sessile Drops
REPULSION

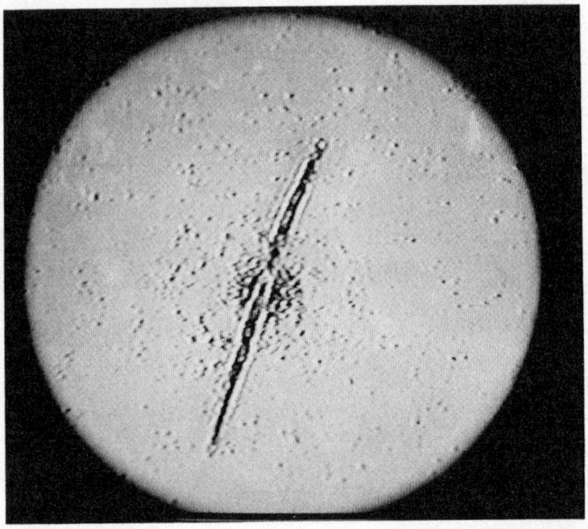

IIIa

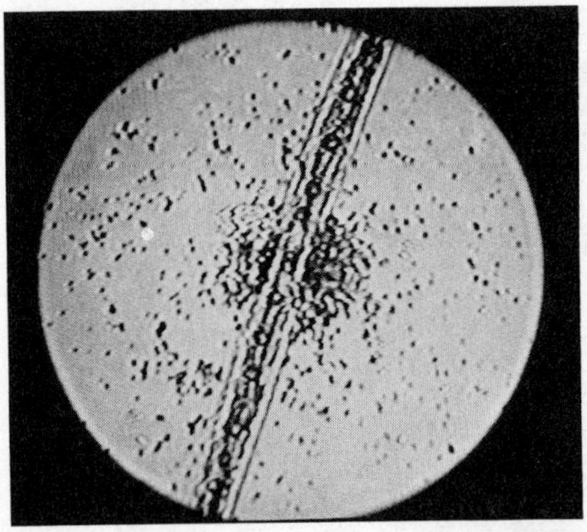

IIIb

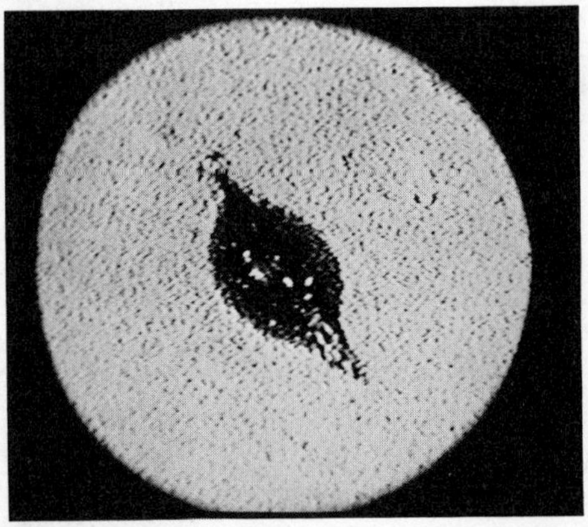

IIIc

PLATE III. Powder Patterns about Cells
in a Hanging Drop of Fresh Water

ATTRACTIONS, DIPOLAR PATTERNS WITH CLOSTERIUM ACEROSUM

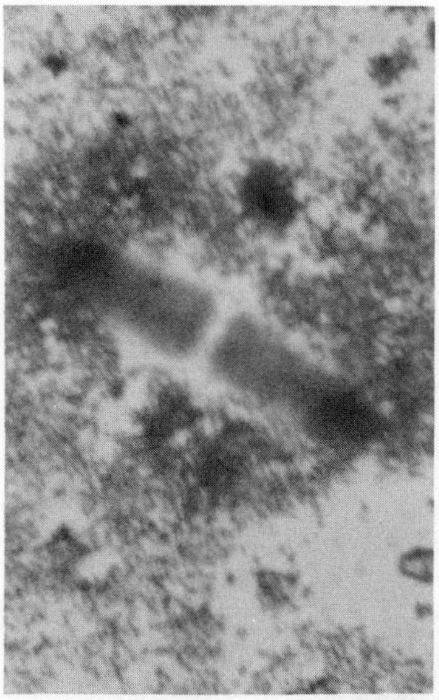

IVa

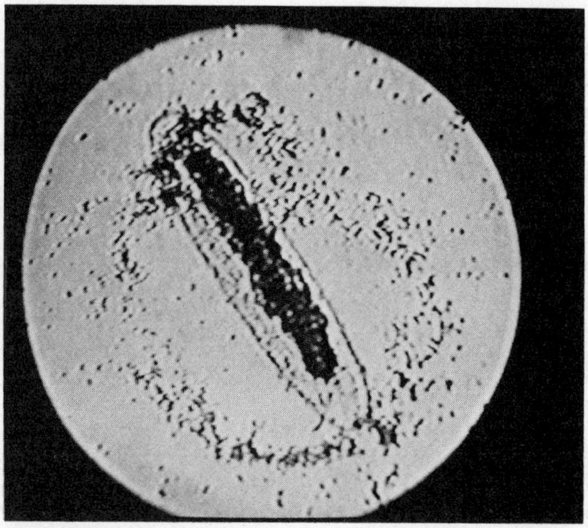

IVb

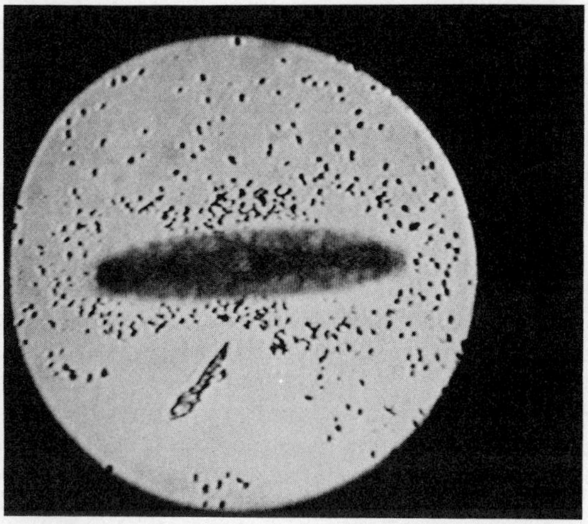

IVc

PLATE IV. Powder Patterns about Cells
in a Hanging Drop of Fresh Water

ATTRACTIONS, QUADRIPOLAR PATTERNS WITH NETRIUM DIGITUS

PLATE I CAPTION: I(a). Dead cell (Netrium digitus) surrounded by $BaTiO_3$ particles of ca. 2 micron diameter while in fresh (high resistivity, 200 kOhm cm) water. Note that the particles, although highly polar ones, do not form a particular pattern about the dead cell. The particles, in vigorous Brownian motion simply gather at the low point of the drop, surrounding the dead cell; I(b). Live cell (Closterium acerosum) in a weak salt solution (specific resistivity = 100 Ohm cm.) surrounded and covered by $BaTiO_3$ particles, and exhibiting electrical inertness. Here the salt raises the effective polarizability of the water, shunting out the electrical ac field put out by the cell. The cell appears "electrically dead" to the polar particles; I(c). Dead Netrium digitus amid the highly polarizable reduced form of $SrTiO_3$ particles of ca. 2 micron diameter. Note that the particles appear to form no particular pattern about the cell. The cell appears to the particles as "electrically dead."

PLATE II CAPTION: II(a). Live Netrium digitus in the presence of barium sulfate. Note the repulsion of the ca. 2 micron diameter particles that have a dielectric constant less than that of water, and would tend to be repelled by a nonuniform electric field. Note, too, that the circum-cellular field so indicated is quadripolar in character. Observe that the three nearby dead cells are quite ignored by the same type of particles strongly avoiding the live cell; II(b). Live Closterium acerosum cell amid tiny silica particles of ca. 2 micron diameter. Note the repulsion of the particles that have a dielectric constant less than that of water, indicating the presence of a (simple dipolar) ac field about the cell; II(c). Live Closterium acerosum cell amid tiny ca. 2 micron diameter particles of barium sulfate. Note the repulsion of the particles that have a dielectric constant less than that of water, indicating the presence of an ac field about the cell.

PLATE III CAPTION: III(a). Barium titanate (highly polarizable) particles of ca. 2 micron diameter form a pattern showing the dipolar character of the cullular ac field of this unicellular alga while in fresh water; III(b). Closeup of the ac field pattern formed by barium titanate particles near the alga in fresh water; III(c). The dipolar field pattern produced by the extremely polarizable reduced form of strontium titanate as the 2 micron diameter particles freely gather about an alga in fresh water.

PLATE IV CAPTION: IV(a). Quadripolar pattern of barium titanate particles about a live cell in fresh water; IV(b). Quadripolar pattern of reduced strontium titanate particles formed about a live algal cell in freshwater; IV(c). Quadripolar pattern of calcium zirconate particles formed about a live algal cell in fresh water. $BaTiO_3$, $SrTiO_3$, and $CaZrO_3$ all are highly polarizable particles having dielectric constants rather higher than that of the water. They are therefore attracted to the regions of greater ac field intensity.

and not static electric fields that are involved here. The large extent, ca 20 μm, of the fields confirms that they are ac fields for if they were dc fields acting from static ionic double layers, they would decay away in only several Debye lengths which would then extend only a few nanometers.

We conclude that the hanging drop DEP pattern technique is a simple and very demonstrative proof of the presence of natural oscillating fields about living cells. Moreover, the technique nicely pictures the type, site, and extent of the field intensities so that experiments such as direct instrumental assessment of such natural ac fields can be done with more exactness and confidence.

We have just received word (and photo) from colleagues in W. Germany that they duplicate and confirm our DEP pattern results on algae (Lamprecht, 1984, private communication).

B. Micro-Dielectrophoresis or Micro-DEP

Micro-DEP experiments can be done with simple apparatus, since no external fields, other than that of the cells themselves, are required. In brief, one puts a mixture of the cells and selected particles under the microscope and counts the fraction of particles associated with the cells. Static fields such as from ionic double layers are immediately suppressed to very short ranges, in the order of nanometers. Long range effects upon the particles, such as we are discussing here and observed to be in the order of micrometers, cannot be ascribed to the effects of static electric fields.

The presence of natural ac oscillating fields near cells is confirmed by studies of micro-DEP in several ways. The tests show that there is a preference for particles of high polarizability over those of low polarizability:

I. Shown by living but not by dead cells (Pohl, 1981b).

II. That is supressable by metabolic inhibitors such as would block ATPases (Pohl et al., 1981).

III. Which is maximal at or near mitosis, as observed in the yeast Saccharomyces cerevisiae, where the stages in the life cycle are conveniently determinable by observing the bud development (Pohl et al., 1981).

IV. Which is suppressable in accord with expectation from dielectric theory by simply increasing the conductivity of the medium, which in turn raises its effective dielectric constant to exceed that of the test powder and erasing, in effect, the DEP forces (Pohl, 1980a,b,c; 1981b; Pohl et al., 1981). The effective dielectric constant is taken to be the absolute value of the complex dielectric constant,

$$K = K' - jK'' = K' - js/(\varepsilon_o\omega).$$

$$K_{eff} = [(K')^2 + (K'')^2]^{\frac{1}{2}}$$

where K is the complex relative dielectric constant, j is the square root of minus one, s is the specific conductivity, K' is the in-phase and K" is the out-of-phase dielectric constant, ε_o is the permittivity of free space, and ω is the angular frequency of the electric field. We note from the above equation that the effective dielectric constant, and hence the DEP forces, involves the presence of all charges present ... those seen in dielectric constant measurements and those seen in conduction measurements.

V. Which can be observed for a variety of test powders of widely differing chemical natures (Pohl, 1981a,b, 1981b; Pohl et al., 1981; Roy et al., 1981).

VI. That is observable in a wide range of organisms, including bacteria, fungi, algae, as well as in avian blood cells, and mammalian fibroblasts of embryonic, normal, and tumor origins (Polh, 1980b, c, 1981b; Pohl et al., 1981).

VII. The micro-DEP results vary with the conductivity of the medium in a manner that yields an estimate of the minimum frequency of the natural ac fields (about 10 kHz), which agree with that deduced from CSR experiments (Pohl, 1980c, 1981a).

VIII. The behavior of the ferroelectric powder, $BaTiO_3$, in being attracted to cells, provides an estimate of the field strength provided by the cell at its surface (ca. 100 V/cm) which is consonant with that inferred from torque measurements in the CSR studies (Pohl, 1980c, 1981a).

The force exerted by the natural ac fields of cells upon external test objects such as the powder particles is of very short range, and to vary as the Z^{-7}, where Z is the distance between the centers of the cell and the test object.

The natural question arises, "Is the observed preference of living cells for polarizable test particles in the micro-DEP experiments possibly due to static fields?" Several facts strongly rebut this, and give evidence that the preference is an ac phenomenon. First, the evidence cited in points (IV) and (VII) above show that the presence of even slight conduction in the medium due, say, to 100 micromolar Nacl or KCl suffices to suppress the effect, and does so in accord with the theory for dielectrics in ac fields. Second, as is well known from the theory of polar media such as water, the static fields are damped out by distances of the order of magnitude of a Debye relaxation length. Because this is in the order of only about 100 Ångstroms or less in aqueous media, and the micro-DEP effects are observed to be active over distances many-fold larger (several microns), it is therefore quite unlikely that static field effects are dominant here. We conclude then that the evidence from micro-DEP experiments supports the existence of natural ac oscillations by cells, for the effect of the ac fields is detectable by this delicate but simple technique.

C. Cellular Spin Resonance, CSR.

Let us now turn to the evidence for natural ac oscillations as gotten from the study of cellular spin resonance (CSR). The application of ac fields to a suspension of cells can cause them to spin or rotate selectively. This provides a new and straightforward method for determining the dielectric properties of individual cells or even inert particles. The technique is particularly useful and transparent when rotating fields, such as those produced by three- or four-pole electrode systems, are used (Mischel, Voss, and Pohl, 1982; Arnold and Zimmermann, 1982a,b,; Pohl, 1983). The techniques and the theory are now well in hand, and future progress in this new area should be rapid. An especially interesting aspect of cellular spin in applied ac fields is that in which more or less parallel electrodes are used, and a sinusoidal field is applied.

A number of researchers have observed the spinning of cells while in such a simple two-pole ac field (Teixera-Pinto et al., 1960; Furedi and Valentine, 1962; Pohl and Crane, 1971; Pohl, 1978; Mischel and Lamprecht, 1980; Pohl 1981a; Pohl and Braden, 1982). The first account of the resonant spinning of living cells was that of Pohl and Crane (1971). They observed that the yeast *Saccharomyces cervisiae*, when subject to an ac electric field between two parallel wire electrodes, spun in a sharply responsive manner to the frequency of the applied field. Cells were seen to spin alone whether against an electrode or out in the suspension, and to spin collectively while in pearl-chain stacks. The resonant frequency, moreover, depended upon the age of the colony examined. The sharply resonant nature of the spinning response led to the use of the descriptive term, "cellular spin resonance" (CSR). Mischel and Lamprecht (1980) quantitatively studied the rate of spin of budding yeast cells. Vienken and Pilwat (1981) described the CSR of a number of species, including human erythrocytes and their ghosts as well as Friend cells, and mesophyll protoplasts of *Avena sativa*. The variation of the CSR spectrum with phase in the life cycle in yeast cells (*Saccharomyces cerevisiae*) has been studied by Rivera (1983).

There are two circumstances for the occurrence of CSR. On the one hand there is the spin of lone cells in simple harmonic ac fields (i.e., 4-pole), and while little affected by local field perturbations such as those due to rotating fields or delayed polarizations. On the other hand, there is much more general phenomenon of spin while under the influence of rotating ac fields.

The spinning of lone cells in two-pole ac fields was reported by a number of workers, including Pohl and Crane (1971), Chen (1973), Pohl and Braden (1982), and by Mischel and Lamprecht (1983). For several physical reasons, the observation of lone cells spinning in the mid region of a two-pole field must be of brief and occasional character, for the cells in that region usually move more or less towards an electrode or join with other cells to form pearl chains under the influence of DEP. In addition to the action of DEP in causing the lone cells to leave the mid-field region, there are other dispersive actions such as thermal upsets, ionic injection, etc. causing fluid streaming. Lone cells are, however, readily observed to spin freely and in a resonant way when against a mirror-smooth Pt electrode, as all investigators, but especially Mischel and Lamprecht (1983), have noted. The theory for the CSR of lone cells spinning in CSR in a two-pole ac field was discussed by Pohl (1981a, 1982). At the present writing, the only evident mechanisms clearly accounting for the CSR of lone live cells — i.e., where the frequency of the cellular spin rate is much less than that of the applied field, and where the spinning ceases upon the death of the cell — are mechanisms wherein the natural cellular dipolar oscillation exists and can interact with the applied field to provide a rotational torque.

Particles can also be made to spin by the presence of rotational electric fields. Here the rotating field induces a polarization which need not be instantaneous, but lags the field. The interaction of the delayed polarization (i.e., the dielectric loss) with rotating field then produces a torque capable of spinning the particle. The particle need not be a living cell, but might be a vesicle, a dead cell, or even an inanimate mineral particle. Rotational fields can be provided by the use of phased three- or four-pole electrode systems, for example. This method is useful for determining the dielectric properties as a relaxation spectrum of small particles (Pohl, 1983). Rotational fields capable of causing particle rotation can also arise as first pointed out by Schwan (1982), from the interaction of the particles with each other while present in an otherwise homogeneous field. Consider the case of cells, say, subject to an ac field provided by a two-pole field, but also very near another particle. The field due to the induced dipole of the second particle is phase-delayed compared to that of the applied field, and is therefore capable of providing a rotating field at the first cell, and thereby evoking a rotation. The spinning of cells by this mechanism is easily produced and observed (Holzapfel et al., 1982). The CSR of crowded cells in a two-pole ac field is usually due to this mechanism. The CSR of lone cells in such a two-pole ac field cannot, however, be due to such rotational field, and another mechanism such as that due to the interaction of natural cellular oscillating dipoles is needed.

The evidence that living cells naturally produce ac oscillating fields, as derived from CSR studies, supports the conclusions reached from the studies of micro-dielectrophoresis. The CSR evidence is not as firm as that for that obtained from micro-DEP studies, in view of the fact that there are several ways to interpret the CSR studies. One must be careful in assigning observable CSR as being due to natural cellular ac oscillations in view of the fact that rotating fields due to the presence of a nearby polarizable object, off axis from the lines of the applied external field, can evoke spinning. Careful observation to assure that the spinning lone cell is well away from other polarizable objects is required.

D. Direct Instrumental Detection

Direct instrumental observations of natural ac oscillations of cells have been made by several investigators. The presence of such oscillations in the frequency range of 50 to 500 Hz has been reported from studies of nerve cells (cf. Treherne et al., 1979). Oscillating enzyme reactions have been studied and are well recognized (Hess and Chance, 1978). Action potentials of macrophages derived from human monocytes were observed to oscillate at frequencies from 20 to about 80 Hz (McCann et al., 1983). The power spectra of mormyrid fish (*Pollimyrus isidori*) ranges from 1000 to about 50,000 Hz and is involved with the electric organs (Westby and Kirschbaum, 1982). The output is used by such "weakly electric" fish as the South American gymnotoid fish and the mormyriform fish of Africa for "electro-location" in their surroundings. It is difficult to imagine the origin of these radio frequency oscillations as being other than cellular.

Yano et al. (1981) have directly observed much slower oscillations (at a few Herz) of the syncytiotrophoblast of the human term placenta. Similar observations on murine fibroblasts were made by Okada et al. (1979). Very recent direct instrumental observations of electrical oscillations of high frequency have been reported by Smith (1983). His group used a Hewlett Packard 8553B Spectrum Analyzer and found the yeast, *Saccharomyces cerevisiae*, to emit oscillations in the neighborhood of 8 MHz. Rivera et al. (1983) also recently observed similar results with yeast. The alga *Netrium digitus* was observed to provide very weak signals in the order of microvolts at several frequency bands at about 7 and 33 kHz. Detection was done in a shielded enclosure with the aid of Pt electrodes externally placed in the medium at the ends of the cell and connected to a preamplifier mediating the signals to a wave analyzer (Hewlett Packard, Model 312A) (Fig. 1).

There are, there, four excellent lines of experimental evidence confirming the existence of natural cellular ac oscillations: that from micro-DEP, from CSR, from DEP patterns, and from direct instrumental observation. Much more needs to be done to establish the nature and range of the oscillations, and to understand their origins and use in the physiology of the cell.

OVERVIEW

We are presented therefore with the firm proof that cells produce natural ac fields. It is fitting then to ask where, how, and why such oscillations should arise in living cells. We have, in effect, two main experimental results at present. The first broad conclusion from the presently available data is that the cells of many types give evidence of the presence of natural oscillating electric fields. The micro-DEP studies and the DEP pattern studies in which fields about cells are detected by the

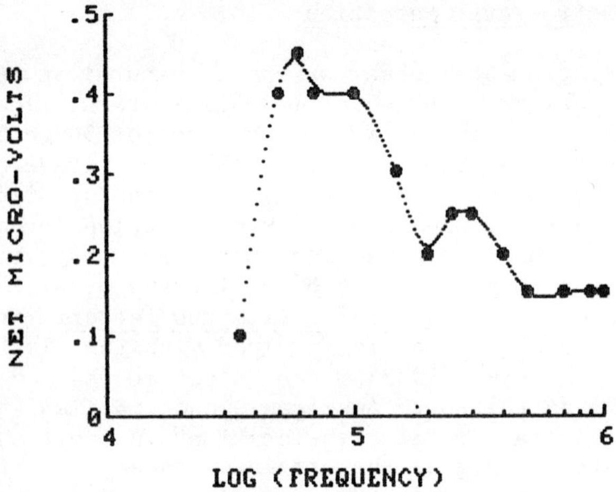

Figure 1 - RF signal from the alga, Netrium digitus.

way in which the cells favor the accumulation of highly polarizable particles over that of weakly polarizable ones, clearly show that the ac fields about the cells occur in a wide range of cellular species. This is evident in Table I.

The second broad conclusion about the presence of natural ac oscillations in cells again comes from micro-DEP studies. The n/p values indicating the strength of the preference of the cells for polarizable particles were observed to be maximal at or near mitosis. These results, obtained on the yeast Saccharomyces cerevisiae can be easily obtained because the phase in the life cycle is readily determined from the shape of the cell (Hartwell et al., 1974). The variations of the n/p values for the highly polarizable particles $BaTiO_3$, or $NaNbO_3$ show a peak at the period in the yeast life cycle when the small budding state is attained, e.g., when the cell is at or near mitosis.

QUESTIONS AND IMPLICATIONS

Given the evidence that cells do indeed produce natural ac oscillations; given the two principal results, i.e., that (1) the oscillations are "universal" among the species, and (2) that the oscillations are maximal at or near mitosis, it is interesting to speculate as to their meaning. Perhaps the first question to pose is: "Are the observed natural ac oscillations evidence of a process which is merely an unnecessary frill, or do they signal a necessary process in the life processes of cells?" The first principal result, the "universality" of the oscillations, indicates that the natural ac oscillations have been carried on for billions of years

Table I. The Preferences of Living Cells for Highly Polarizable Particles ... (as determined with micro-dielectrophoresis). (n = no. particles on a cell, p = no. of particles per unit volume).

Cell Type	Ratio, Polarizable to Non-Polarizable Particles; (n/p) polar
Bacteria, Bacillus cereus, synchronized	4
Yeast, Saccharomyces cerevisiae	2.0
Alga, Chlorella pyrenoidosa	2.2
Avian, Chicken red blood cell	2.0
Mammalian, Mouse "L" fibroblasts	2.0
Mammalian, Mouse Ascites tumor fibroblasts	2.2
Mammalian, Mouse, Fetal fibroblasts	2.0

throughout the evolutionary process. Nature is not inclined to be wasteful. With this in mind it is a fair presumption that indeed the ac oscillations signal phenomena which are essential. Secondly, it is also likely that the oscillatory phenomena must preceed or accompany the reproductive process, as evidenced by the second principal observation. Together, these suggest that there is an electrical (ac) aspect to the process of reproduction in the cells. The natural ac fields therefore disclose, tag, label, or reflect some process associated with cellular reproduction; probably in the S, G, or M phases. That the fields themselves are essential rather than serving as lables or other epiphenomena is not known at this time, and needs study.

In accord with this we then can make the provisional, postulate that: Electrical oscillations must accompany cellular reproduction, and without them, the reproductive process cannot proceed. This assumption can be tested and examined experimentally in a number of ways. The necessity for electrical oscillations to accompany the process of cellular reproduction may play a role, for example, in the well known process of "contact" or "density" inhibition of cell growth. Contact inhibition is often observable when cells are cultured in the laboratory, for their growth is seen to be checked as the cells become confluent and cover the growth medium. The known characteristics of contact inhibition offer a check on the electrical theory.

The ususal explanations given for such contact or density inhibition suggest that it occurs because of a shortage of food or, alternatively, because cells put out growth inhibitor factors. But if these were the only explanations, then growth would be expected to resume when fresh medium is flushed across the colony. It does not, however, which implies that there must be some other factor(s) at work, possibly through the gap junctions and not the outer medium. Let us consider an electrical one. If a cell must °oscillate° to reproduce, as suggested above, and the oscillation is an electrical one that can be damped out or suppressed by the presence of neighboring cells, but not necessarily by the presence of the normal cellular medium, then a simple mechanism can be suggested.

One such mechanism which suggests itself is the following: In the spectrum of electrical absorptions characteristic of any material, there are normally one or two sharp regions in the spectrum where strong energy absorption occurs. If a feeble electrical oscillator tries to oscillate in the medium at that particular frequency it will be strongly damped. It will only succeed in maintaining its output if it can be permitted to shift to a frequency at which the damping is so small that the energy drain is acceptable. With this picture in mind, let us view the cell as being normally equipped to leave the G_0 and to enter the growth (G, S, G_2, M) phase with an electrical oscillator system producing its charge waves at a particular frequency, f. But if the cell is amid other cells, and the strong damping frequency of this melange now coincides with the frequency f, the cell will then be unable to reproduce, for its necessary oscillations cannot now be maintained. According to this picture, in other words, it will be the density of cells surrounding the individual cell that will modify the electrical neighborhood, and control the ability of the cell to reproduce. Once the local cell density is sufficient to produce high electrical damping around the cell in question, the further oscillation of it at that frequency is repressed, and the reproductive process also repressed. Such a model can be tested in a number of ways, as by the deliberate injection of foreign material capable of electrically initiating the damping characteristics of massed cells, or by using electrical means to augment the oscillations at the presumed frequency, f, or by selected chemical agents.

Another test of the oscillation postulate is to account for the well known ability of cells to grow on certain occasions while in a high concentration of other cells. There are at least four such occasions in the life of mammalian cells, for example. These are: during embryonic growth, during normal body cell replacement, during wound healing, and during the growth of tumors. During these periods it will be necessary, in this picture, for the necessary oscillations to occur at a frequency somewhat removed from that at which the surrounding cell mass would damp out or suppress them. We suggest that this can be accomplished by a small shift in the biochemistry of oscillating reaction to a frequency either above or below that of the neighborhood's high loss frequency, and that this frequency shift is possibly under the control of a normally repressed gene. It is interesting in this connection to note that Ahuja and Anders (1977) reported on finding a gene (Tu gene) which had a repressor gene (R gene) judged to lie on a separate portion of the genetic materials, and that these genes were present in all cells capable of reproducing. They associated the control of cell growth with the activity of such a gene pair. It is interesting to ask if it is these genes which control the frequency f, and cause it to shift into the frequency range required for the successful operation of the reproductive cycle. Again, these ideas are testable in a variety of ways, and the testing of them may provide interesting insights into the mechanism of cellular control.

ACKNOWLEDGEMENTS — This research was sponsored by the Pohl Cancer Research Laboratory, Inc., to whom thanks are due. The Francis Bitter National Magnet Laboratory is supported by the National Science Foundation through its Division of Materials Research.

REFERENCES

Ahuja, M.R., and Anders, F., 1977, in "Recent Advances in Cancer Research," Vol. 1, Edited by R.C. Gallo, CRC Press, pp. 103-117.
Albanese, R.A., 1983, "Radiofrequency radiation and chemical reaction dynamics," International Conf. on "Nonuniform Electrodynamics in Biological Systems," Loma Linda, CA, Edited by W.R. Adey, June 1983.
Arnold, W.M., and Zimmermann, U., 1982a, Naturwissenchaften, 69:297.
Arnold, W.M., and Zimmermann, U., 1982b, Naturforsch., 37c:908.
Berliner, M.Z., 1978, Pflanzenphysiol., 88:341.
Berridge, M.J., and Rapp, P.E., 1979, J. Exp. Biol. 81:217.
Chen, C.S., 1973, "On the nature and origins of biological dielectrophoresis," Ph.D. Thesis, Oklahoma State University, Stillwater, OK 74078.
Chen, C.S., and Pohl, H.A., 1974, Trans. N.Y. Acad. Sci., 238:176.
Chou, K.C., and Chen, N.Y., 1977, Sci. Sin., 20:447.
Chou, K.C., Chen, N.Y., and Forsen, S., 1981, Chem. Scr., 18:126.
Chou, K.C., 1984, Biophys. J., 45:881.
Crane, J.S., and Pohl, H.A., 1977, J. Biol. Physics, 5:49.
Davydov, A.S., 1980, Sov. Phys. JETP, 51:397.
Del Guidice, E., Doglia, S., and Milani, M., 1982, Phys. Scripta, 26:232.
Dukhin, S.S., and Shilov, V.N., 1974, "Dielectric Phenomena and the Double Layer in Disperse Systems and Polyelectrolytes, J. Wiley and Sons, N.Y.
Epstein, I.R., Kustin, K., de Kepper, and Orban, M., 1981, J. Am. Chem. Soc., 103:2133
Fermi, E., Pasta, J., and Ulam, S., "Collected Works of Enrico Fermi," Vol. II., Univ. of Chicago Press, pp. 977-88.
Frohlich, H., 1968, Int. J. Quantum Chem., 2:641.
Frohlich, H., 1977, Neurosci. Res. Program Bull., 15:67.
Frohlich, H., 1980, Adv. Electron. Electron Physics, 53:85.
Frohlich, H., 1983, "Coherent Excitations in Biological Systems," Ed. by H. Fröhlich and F. Kremer, Springer-Verlag, pp. 1-6.

Furedi, A.A., and Valentine, R.C., 1962, Biochem. Biophys. Acta., 56:33.
Hartwell, L.H., Culotti, J., Pringle, J.R., and Reid, B.J., 1974, Science, 183:46.
Hess, B., and Chance, B., 1978, Theo. Chem., 4:159.
Hogetsu, T., and Shibaoka, 1978, Planta, 140:7.
Hogetsu, T., and Takeuchi, Y., 1982, Planta, 154:426.
Holzapfel, C., Vienken, J., and Zimmermann, U., 1982, J. Membrane Biol., 67:13.
Kaiser, F., 1983, International Conf. on "Nonuniform Electrodynamics in Biological Systems,: Loma Linda, CA, Edited by W.R. Adey, June 1983.
Kaiser, F., "Coherent Excitations in Biological Systems," Ed. by H. Frohlich and F. Kremer, Springer-Verlag, 1983, pp. 128-33.
Kaler, K., and Pohl, H.A., J., 1980, J. Biol. Physics, 8:18.
McCann, F.V., Cole, J.J. Greyre, P.M., and Russell, J.A.G., 1983, Science, 219:991.
Mischel, M., and Lamprecht, I., 1980, Z. Naturforsch. 35c:1111.
Mischel, M., Voss, A., and Pohl, H., 1982, J. Biol. Physics, 10:223.
Mischel, M., and Lamprécht, I., 1983, "Rotation of cells in non-uniform rotating alternating fields," J. Biol. Physics, 11: (in press).
Mischel, M., and Pohl, H., 1983, "Cellular spin resonance in rotating electric fields," J. Biol. Physics, 11: (in press).
Noyes, R.M., and Field, R.J., 1974, Ann. Rev. Phys. Chem., 25:95.
Noyes, R.M., 1977, Acc. Chem. Res., 10:214, and 273.
Okada, Y., Tsuchiya, W., and Inouye, A., 1979, J. Membrane Biol., 47:357.
Pohl, H., 1951, J. Appl. Phys., 22:869.
Pohl., H., and Crane, J.S., 1971, Biophys. J., 11:711.
Pohl, H., and Wyhof, 1972, J. Non-Cryst. Solids, 11:137.
Pohl, H.A., 1974, J. Biol. Phys., 2:113.
Pohl, H.A., and Pollak, M., 1977, J. Chem. Phys., 66:4031.
Pohl, H., 1978, "DIELECTROPHORESIS, The Behavior of Matter in Non-uniform Electric Fields," Cambridge University Press.
Pohl, H., 1980a, "Micro-dielectrophoresis of Dividing Cells," in "Bioelectrochemistry," ed. by H. Keyzer and F. Gutmann, Plenum Press, pp. 273-95.
Pohl, H., 1980b, Int. J. Quantum Chem., 7:411.
Pohl, H.A., 1980c, J. Biol. Phys., 7:1.
Pohl, H.A., 1981, Collective Phenom. 3:221.
Pohl, H., 1981a, J. Theor. Biol., 93:207.
Pohl, H., 1981b, J. Bioenerg. Biomembranes, 13, 149.
Pohl, H., Braden, T., Robinson, S., Piclardi, J., Pohl, D.G., 1981, J. Biol. Phys., 9:133.
Pohl, H., 1982, Int. J. Quantum Chem., 9:399.
Pohl, H., and Braden, T., 1982, J. Biol. Physics, 10:17.
Pohl, H., 1983, "Cellular spinning in pulsed rotating electric fields," J. Biol. Physics, 11 (in press).
Rapp, P.E., 1979, J. Exp. Biol., 81:281.

Rivera, H., 1983, private communication.
Rivera, H., Biscar, J.P., and Pohl, H.A., 1983, private communication.
Rowlands, S., Sewchand, L.S., Lovlin, R.E., Beck, J.S., and Enns, E.G., 1981, Phys. Lett., 82A:436.
Rowlands, S., Sewchand, L.S., Enns, E.G., 1982, Can. J. Physiol. Pharm., 60:52.
Rowlands, S., 1982, J. Biol. Phys., 10:199.
Rowlands, S., Eisenberg, C.P., and Sewchand, L.S., 1983, J. Biol. Phys., 11:1.
Roy, S.C., Braden, T., and Pohl, H., 1981, Phys. Lett. 83A:142.
Sargent, M.G., 1973, J. Bacteriol., 116:736.
Schmidt, S., and Ortoleva, P., 1979, J. Chem. Phys., 71:1010.
Smith, Cyril W., 1983, private communication.
Teixera-Pinta, A.A., Nejelski, L.L., Cutler, J.L., and Heller, J.H., 1960, Exp. Cell Res., 20:548.
Treherne, J.E., Foster, W.A., and Schofield, P.K., 1979, "Cellular oscillators," J. Exp. Biol., 81 (review volume).
Webb, S.J., 1980, Phys. Rep., 60:201.
Westby, M., and Kirschbaum, F., 1982, J. Compar. Physiol., 145:399.
Yano, J., Okada, Y., Tsuchiya, W., Kinoshita, M., Tominaga, T., and Nishimura, M., 1981, Acta Obst. Gyn. Japan, 33:137.
Zeuthen, E., 1958, Adv. Biol. Med. Physics, 6:37.
Zimmerman, U., Vienken, J., and Pilwat, G., 1981, Z. Naturforsch., 36c:173.

FREQUENCY - DEPENDENT BIOLOGICAL EFFECTS OF LOW INTENSITY MICROWAVES

W. Grundler

Gesellschaft für Strahlen- und Umweltforschung mbH
D-8042 Neuherberg, F.R. Germany

1. INTRODUCTION

"In the past three years, the study of the biological effects of electromagnetic fields has taken a new turn. At the beginning of the 1980s, researchers still argued about whether all effects of fields on organisms were caused simply by the heating induced by intense radiation or if some effects also occured at field strengths too low to produce even localized heating. Although laboratory results had by that time convinced some researchers that athermal effects - those that could not be explained by heating - did in fact occur, a vocal group of scientists still doubted that these laboratory results were valid and repeatable.

Now, a growing mass of evidence has virtually ended that debate. Few now question that some such weak-field effects exist" (cited from IEEE spectrum, May 1984).
The situation illustrated here about studies on bioeffect due to electromagnetic fields in general, holds especially for the mm-microwave area. The "new turn" is related to the increasing number of scientists accepting the existence of nonthermal mm-wave biological effects strongly depending on frequency. The frequency dependence of an effect, especially if showing a resonant response, is a fundamental proof of the existence of such a mechanism not to be explained within the realm of a simple thermal response.

Yet the study of the interaction of microwave fields and biological systems remains intensely controversial, with many questions not yet answered fully. For example:

- Under which conditions and at which frequencies and field strengths do interactions with organisms occur?
- What are the mechanisms by which weak fields can influence organisms?
- Which biological effects are just harmless changes, which are benefical, and which are harmful?

These questions can serve, in my opinion, as future guidelines for the scientific work in this field and they should be considered more extensively.

Today, we are standing at a starting point in attempting to give the first answers concerning for example the range of active frequencies. Questions remain in defining the important physical and biological conditions necessary for a conformation of already measured effects. This difficulty may be correlated with experimentally uncontrolled parameters, which are on the other hand important for the reaction.
This problem seems to have existed since the first spectral study of a nonthermal biological microwave effect, to my knowledge, was performed by Webb and Booth /1/ on the growth rate of E. coli. Further evidence of resonant action of millimeter waves has been established by a repetition of the above mentioned experiment /2/ and in a number of other biological systems /3-7/. Out of them, a paper should be mentioned, which reports experiments independently performed with great effort in the USSR to determine low level microwave biological effects in the frequency range from 39 - 60 GHz. Results have been reported on a dozen different experiments at different locations /3/.
Other studies report nonthermal effects on further biological systems, investigated at certain fixed millimeter wave frequencies /8-11/. On the other hand some attempts have failed to demonstrate effects /12-16/. Two fascinating aspects are common to some of the experimental results: (i) The frequency dependence is very strong and the resonance width ranges from 1 GHz to 10 MHz. (ii) The dependence on microwave intensity exhibits a threshold region below which no biological effects occur and above which they occur independently of intensity.

Otherwise the majority of research performed in the past has concerned the effects of radiation fields of intensities which were high enough to produce heating in biological specimens. These mechanisms are well understood and documented /17-19/. Effects reported in the papers /1-11/ are observed in most of the cases at intensities low enough that any mechanism based on microwave-induced heating can be excluded. A

non-thermal influence of a field, can occur at very high field strength. In order to orient a polar molecule in solution, the field must overcome the thermal motion, that means the product of dipole moment and field-strength must be greater than kT. For a dipole moment of 100 Debye, a field-strength greater than 10^5 V/cm is required.
Radiation producing such a field-strength has to deliver intensities of millions of W/cm^2, whereas the experiments were made with intensities of the order of 10 mW/cm^2.
To understand the above mentioned effects and the mechanisms involved, new concepts are necessary.

Theoretical considerations by Fröhlich have suggested that living objects might contain systems exhibiting large-amplitude oscillations at high frequencies /20-22/, a conjecture with quite far-reaching consequences for our understanding of biological processes. Excitation of such systems is thought to ordinarily occur through metabolic pumping. Radiative processes on the other hand would open such systems to spectroscopic investigations. A trigger action of resonantly applied external radiation has been discussed in a detailed nonlinear oscillator model ("limit cycle") by Kaiser /23/.

In addition to these theories able to explain such a narrow resonance like response to external radiation, a new interesting hypothesis has been developed by Keilmann /24/. It takes into account the interaction of the magnetic field of the radiation with still unknown molecules able to exist in triplet states which have different biochemical reaction rates.

In order to decide which direction further scientific work should go, a close cooperation between experiment and theory is necessary. Then, the mechanisms by which weak fields are influencing biological systems must be consistent with the biophysical laws, but due to evolution they can be very unexpected ones. The knowledge of these basic reactions on a molecular level is finally the precondition for the answer to our question, whether such nonthermal effects are beneficial, harmless or harmful.

The purpose of this contribution is

(i) to demonstrate the nonthermal resonant influence of microwaves on the growth of aqueous yeast cultures, by the use of different antenna shapes,

(ii) to show the influence of this radiation on the kinetics of single yeast cells observed microscopically and to proof the reproducibility of the results previously reported and

(iii) to report similar effects on mammalian cells for testing the transferibility of resonant frequencies to other cell systems, possibly indicating whether the mechnisms is on a molecular level common to both systems.

2. FREQUENCY-DEPENDENT YEAST GROWTH EFFECTS OF LOW INTENSITY MICROWAVES

Yeast growth under the influence of different frequencies out of the U-Band (around 42 GHz) and the E-Band (around 84 GHz) with low intensities is monitored by means of two different methods. The first one uses the proportionality between the extinction and cell density measured by a photometer, the second one enables us to define the kinetic of single cells observed microscopically. The cells are locally fixed and it is possible to construct a pedigree of every observed cell, which allows the definition of the different cell cycle times as the observation parameters.

2.1 Photometric Studies of yeast cultures: Resonant growth rate response caused by 42 GHz radiation

Entering this field some years ago, our first experiments /6/ were suggested by effects of microwaves near 42 GHz on the growth of yeast cells (*Rhodotorula rubra*) reported in /3/. Within the last two years, we have performed about 200 experiments and about 60 controls for reproduction purposes. The experimental procedures used are descripted in detail in /25-27/ and they should be briefly mentioned in this report. More emphasis will be taken on reporting results and on statistical analysis.

EXPERIMENT

We used a diploid, homozygous wild type strain of *Saccharomyces cerevisiae* (type 211). The cells were grown on agar plates for three days at 30 °C. These plates were stored at 4 °C. Cells for liquid suspension were taken from these plates after 10 to 16 days. With starting concentrations of about 3.10^5 cells/cm^3, these cultures were held in small glass cuvettes equipped both with mechanical stirrers and with submersible teflon antennas used for coupling in the microwaves.

The simultaneous growth of two yeast cultures was measured in two double-beam spectrometers, set at 550 nm /fig. 1/.

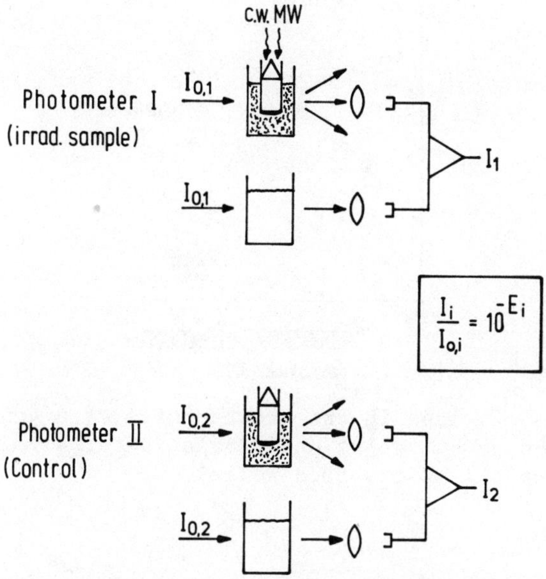

Fig. 1 Schematic drawing of the two double-beam spectrometers.

Light scattering is used to measure the growing cell concentration by recording the extinction of a transmitted light beam. In both double-beam photometers two glass cuvettes of 1 cm^2 internal cross section were inserted, one filled with medium (2.5 ml) and the other one with the cell suspension. Extinction due to the presence of yeast cells was then defined by the ratio of the transmitted and the incident light intensity.

The apparent optical signals E were amplified in logarithmic amplifiers and continuously recorded as in E vs. time t. After several hours of lag-phase the yeast cultures attain the exponential growth phase which is characterized by a linear trace of which the slope is read out to give the growth rate $\mu = t^{-1}$ ln E.

The growth rate of the unirradiated control is represented by the slope of the upper trace of fig. 2, the growth rate of the irradiated sample is read out of the lower trace after the time of switching on the microwave power.

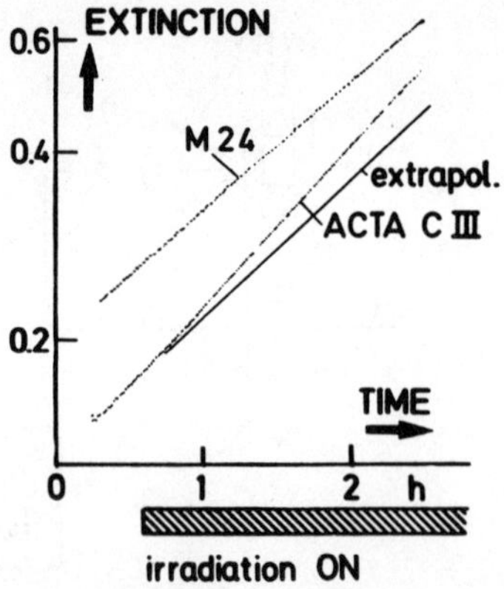

Fig. 2 Change of slope in the growth curve of a yeast culture induced by switching on irradiation at 41752 MHz (fork antenna, 21 mW absorbed power).

Two different submerged antennas were used to apply the radiation to the yeast cells through the suspension. One was a fork-shaped teflon body designed to give a large radiating surface of 8 cm^2 /fig. 3a/. This structure sustains a complex microwave field distribution. Special in situ measurements, with 1 mm spatial resolution, allow the conclusion, that for an applied power of 40 mW the maximum possible temperature difference between any two points in the stirred suspension is 0.02 °C /25/.

The second antenna was designed to have a very simple tube form: the circular waveguide was plugged with a teflon stopper with slightly curved surface to avoid trapping air bubbles. The microwave fields are now rather homogeneous but the radiating surface is smaller by nearly a factor of 10, giving the stirred cells a reduced chance of passing through the irradiation region, so that smaller effects are expected.

The very strong frequency dependence of the growth rate in our former experiments showed that a high frequency stability was needed about ± 2 MHz. The microwave sytem works in the following way:

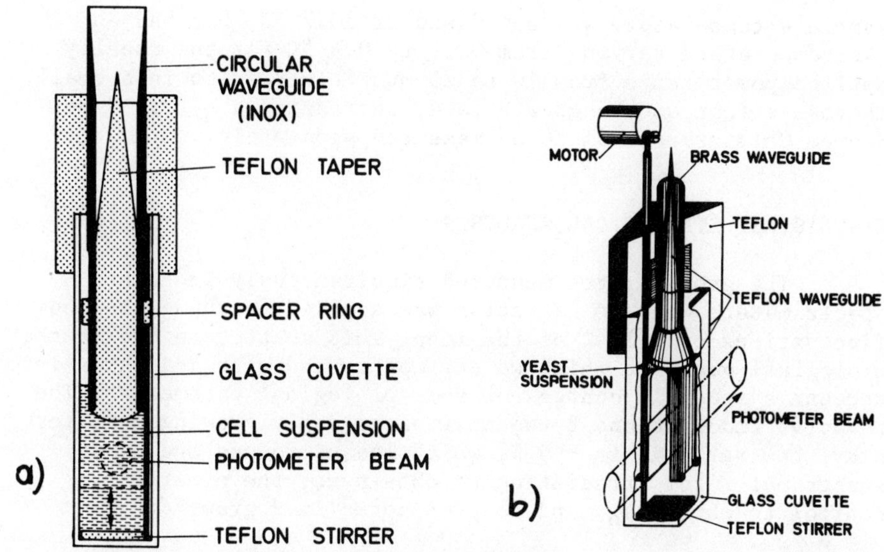

Fig. 3 Drawing of a 1 cm pathlength photometer cuvette equipped
(a) with mechanical stirrer and "tube" shaped teflon antenna.
(b) with mechanical stirrer and "fork" shaped teflon antenna.

A crystal controlled frequency standard holds for a very small systematic error in frequency (< 1 kHz) stabilizing a local oscillator, the third harmonic of which is mixed with the backward wave tube frequency (BWO 60, Siemens). A phase locked stabilisation by a Micro Now stabilizer results in a source wave stabilized at a constant frequency with an uncertainty of $\pm$ 1 MHz on an absolute scale /27/.

Care was given to temperature stabilization and power measurements. The cuvette housings were thermostatted with flowing water at 30.7 $\pm$ 0.1 °C. The microwave system included a 1-m-long waveguide which coupled source to sample via an isolator, two directional couplers to monitor forward and backward running power, and finally an impendance transformer adjusted to minimize reflections. Continuous monitoring of the power incident on the sample was supplemented by a continous calorimetric measurement of the microwave power absorbed by the sample, as already used by us /6/ whereby the temperature of the sample suspension is increased by 0.016 °C per milliwatt absorbed power. This proved that (90 $\pm$ 10)% of the incident power was absorbed in the sample. The price to pay to achieve this rather direct power measurement is that the irradiated sample's temperature was not fixed at 30.7 °C, but had an overtemperature varying from 0.16 to 0.4 °C for the usually applied power range from 10 to 25 mW. This results in a small thermal effect on the growth rate, increasing μ by 2.7 % per degree Celsius near 31 °C as measured separately.

RESULTS AND STATISTICAL ANALYSIS

The growth rates measured simultaneously in both spectrometers when no radiation was applied to the cells show fluctuations of $\pm$ 10 % of the mean. This scatter is due to the biological variance which we are confronted with and takes into account systematic changes of some biological parameters. The relative fluctuations between simultaneously growing cultures stay, however, within $\pm$ 4 %, which led us to use one spectrometer for irradiation to obtain μ_x, the other for control to obtain μ_o, and to plot normalized growth rates $\mu = \mu_x/\mu_o$.

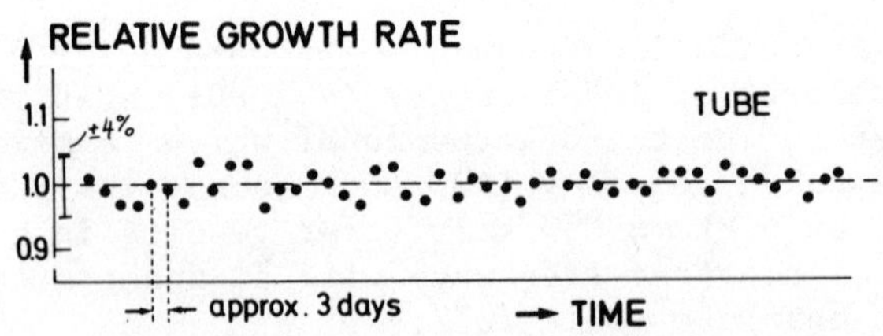

Fig. 4 Normalized growth rates μ with no irradiation applied, obtained in a long series, using tube antenna.

In fig. 4 the normalized growth rates of 48 controls are shown, with no radiation applied using tube antenna. It should be pointed out, that this scatter of $\pm$ 4 % indicates the full range within which the control data are found. Whereas the standard deviation (not used in the following data representations) of these control data is equal to $\pm$ 2 %.

Each growth experiment performed with radiation is represented in fig. 5 by a single point and an error bar of $\pm$ 4 %. On the frequency scale, an error bar of $\pm$ 1 MHz is assigned because the irradiation frequency shows residual modulation up to 0.5 MHz.

The results (Fig. 5) show that microwave effects of about 10 % are obtained, both in the directions of enhanced and of reduced growth. The effects obtained with the tube antenna do not seem to be smaller. Hence it appears that a reduced microwave intensity available quite far from the radiating surface suffices to cause the effect. At a first glance, both spectra show a great resemblance and some maxima seem to appear at same frequencies. To test this important conclusion, two types of statistical calculations were applied to our experimental data.

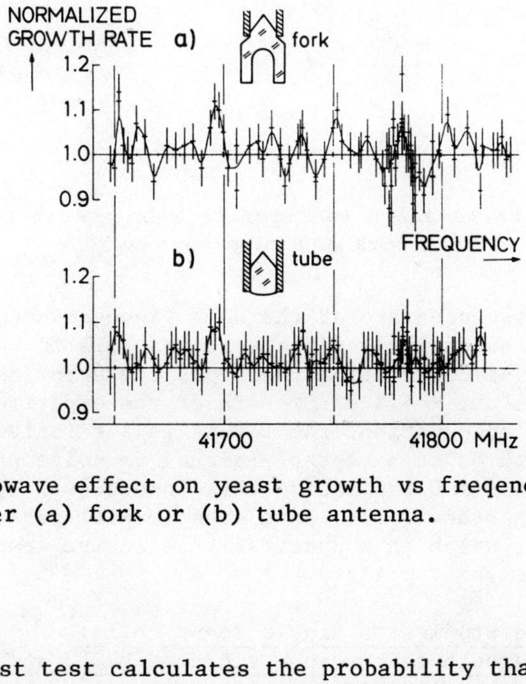

Fig. 5 Microwave effect on yeast growth vs freqency with use of either (a) fork or (b) tube antenna.

The first test calculates the probability that either spectrum in fig. 5a or fig. 5b could have been obtained as a mere chance sequence of a truly frequency-independent microwave effect. The probabilities for this hypothesis are very small and lay at $p = 2x10^{-4}$ and 10^{-7} for the fork and tube results, respectively /27/. This allows the conclusion that the existence of a frequency-dependent microwave effect is ascertained for each of the experimental spectra (fig. 5a and 5b) to a significance level of higher than 99.9 %.

The second statistical test is aimed to compare both experimental series in fig. 5a and 5b and to proof the question,

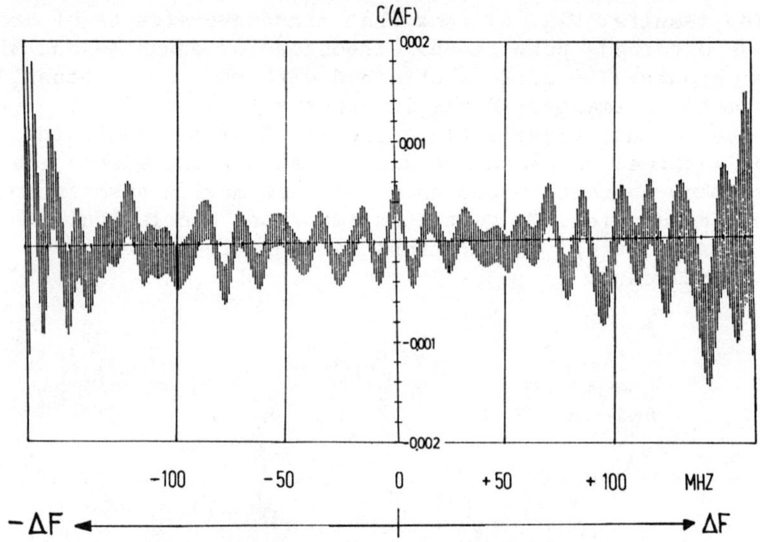

Fig. 6 Cross correlation function between growth rate spectra obtained with fork and tube antenna.

whether the resonances are at the same frequencies or not. A cross-correlation analysis clearly proves this to be true. Fig. 6 shows the cross correlation function. The error bars are calculated from the $\pm$ 4 % error bars of the experimental data by standard error propagation. The result is a relative maximum of correlation with quite acceptable signal to noise ratio at frequency shift Δ f=0 indicating that indeed both spectra overlap or with other words, both experimental series probe a common feature, which is a spectrally selective response inherent in the yeast cells.

2.2. Microscope Studies of Single Yeast cells: Resonant growth rate response caused by 42 and 84 GHz radiation

The frequency dependent growth effects photometrically measured are mean values obtained from the reaction of a population of nearly 10^6 cells.
No information can be derived for the behaviour of individual cells which are the reactive units. For example:

- A division stop of some cells would mean a slight slowing down only of the growth rate of the whole population.
- Stirring causes the cell to move through regions of different

intensities. No prediction can be made of the really applied power to an individual cell nor of the irradiation time during which a cell is affected.

These and other reasons lead to the decision of starting experiments with monolayer cells under the microscope. This method provides locally fixed cells. The intensity by which they are influenced is equal for all the observed cells. Further more, the generation times of every single cell can be defined individually for the F_1 - and F_2 generations.

EXPERIMENT:

The experimental set-up is schematically shown in fig. 7.

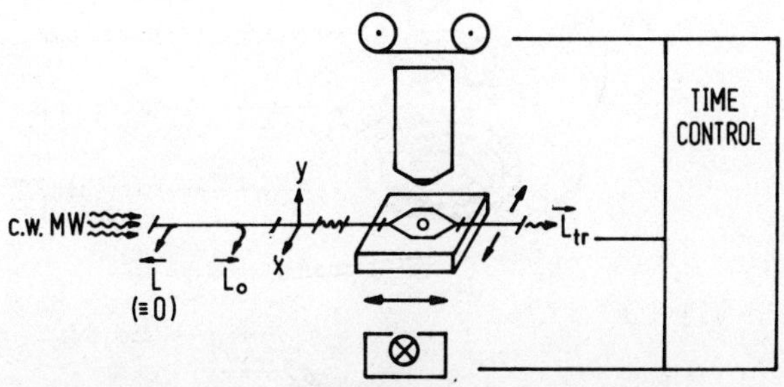

Fig. 7 Schematic drawing of the experimental set-up for observation of single cell kinetic.

The irradiation chamber, coupled by a flexible waveguide with a standardized mm-wave set-up as described in /27/, is mounted on the stage of a scanning microscope. Between the fixed illumination- and registration systems the moveable stage enables the observation of many cells (50) by taking pictures within an area of 0.2 mm^2 out of an agar area of 78 mm^2. At different time intervalls (e.g. 5 min), a set of 6 pictures is made over an observation period of about 4 cell cycle times, i.e. about 5 hours. The incoming radiation is guided through the chamber by means of a tapered sapphire plate, a dielectric substance filling the chamber.

The frequencies used are out of two regions and are lying between 83770 MHz and 83850 MHz (E-Band) and between 41772 MHz and 41800 MHz (U-Band). The scatter in frequency was $\pm$ 1 MHz. In some cases, when this stabilisation technique was not available, it rises up to ± 5 MHz on a absolute scale.

The applied power was measured just before entering the irradiation chamber and was between 1 and 3 mW for the E-Band region and between 13 and 23 mW for the U-Band, whereas the transmitted power was not exceeding 1 mW. The reflected power was set to zero by means of a transformer shown near the stage. The temperature was regulated within a scatter of ± 0.1 °C round 31 °C using a laminar air stream incubator and a thermostatted water bath connected by copper tubes with the microscope chamber.

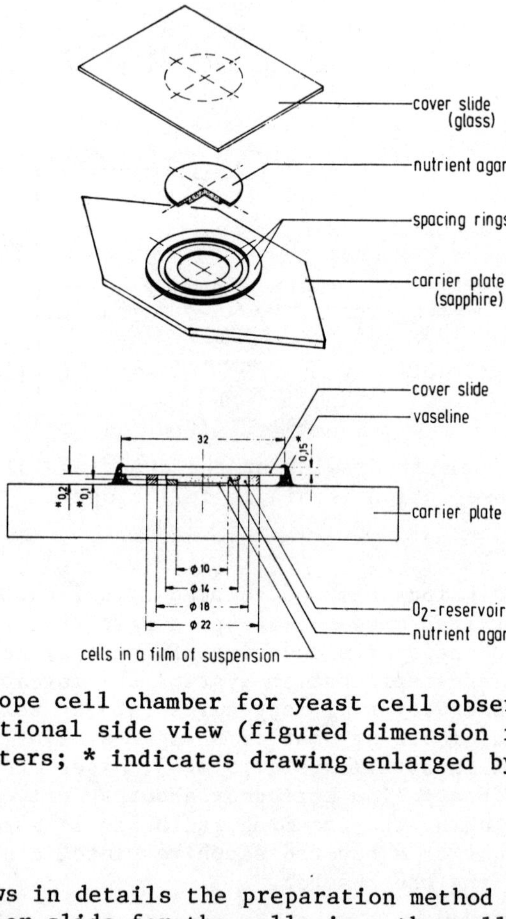

Fig. 8 Microscope cell chamber for yeast cell observation (top) and sectional side view (figured dimension is in millimeters; * indicates drawing enlarged by a factor of 10).

Fig. 8 shows in details the preparation method for the cells. The carrier slide for the cells is a thermally well conducting sapphire plate. Two plastic rings of different thickness are fixed concentrically on the sapphire plate. The inner ring is partially filled with 10 µl cell suspension (10^5

to 10^7 cells) and afterwards a agar layer and a slide are carefully placed on the outer ring. Finally the cover slide is fixed elastically to the carrier slide by means of vaseline. The cells are lying in a plane directly on the sapphire plate and are sorrounded by a thin fluid suspension layer.

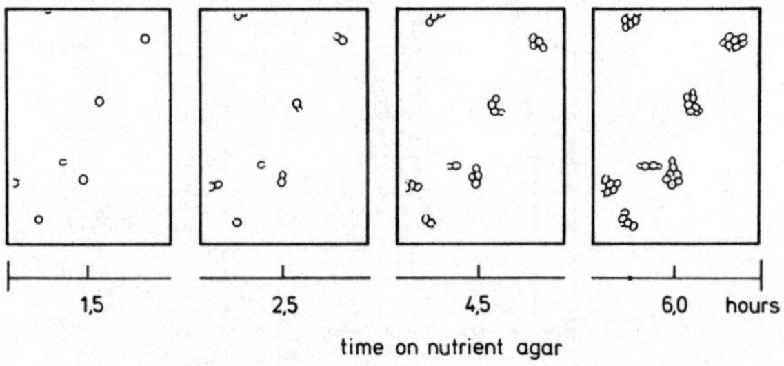

Fig. 9 Yeast growth pattern (magnification x 200) taken from a series of photographs.

The cell pattern obtained with a magnification of 200 can be seen on fig. 9. These pictures are choosen out of a long series of observations lasting up to 6 h. The cells are fixed locally within a range of $\pm$ 2 μm and there is no rotation of the whole colonies.
The cells are represented as black and whit pictures with high contrast, suitable for digitalisation and for reading by a TV camera for further treatment by pattern recognizing programmes. In this way we can compute all the important information of the cell kinetics /28/.

Let us have a look now on the biological parameters used for detecting possible mm-wave effects. Yeast is known to bud nearly at the boundary between G_1- and S-phase (fig. 10a). We can observe cell cycle times, which are defined as the time interval needed for a cell to start the S-phase, running through the G_2-, M- and G_1-phase and arrive once more in the S-phase. These cell cycle times allow the construction of a pedigree for every observed cell (Fig. 10b) and can be subdivided in two subgroups: one group (characterized by "2") describes the first division ($F_1 \rightarrow F_2$), the other one (marked by "3") the second division ($F_2 \rightarrow F_3$), when a minicolony of 8 cells is built.

All the measured time differences registrated in this way on the cells of the observed subpopulation can be pooled as shown in fig. 10c, and the mean cell cycle time and the standard deviation can be determined.

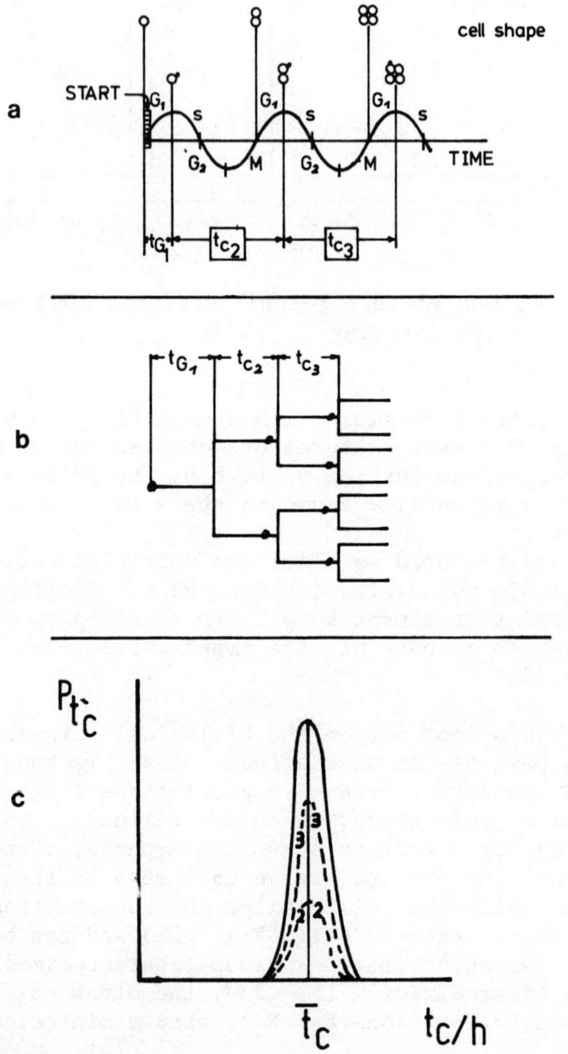

Fig. 10 Schematic diagram of the microscopically defined kinetic parameters of single yeast cells:
(a) cell shape correlated to the cell cycle phases
(b) pedigree constructed for every observed single cell
(c) cell cycle time distribution of the time intervalls defined for two generations

RESULTS AND DISCUSSION

Within the first frequency region (83770-83850 MHz), we have performed 12 control and 27 irradiation experiments, 15 of which were performed half a year later and served as tests for effect reproduction.

Fig. 11 shows the cell cycle time distribution of unirradiated control cells. The mean values of these distributions t_c are listed in fig. 11. Normalized growth rates were defined as the ratio of the mean of all controls and the mean value of every single distribution. The scatter of these relative growth rates is well within $\pm$ 8 % and give a impression of the reproducibility of the method itself. The influence of different frequencies on the cell cycle time distributions is demonstrated in fig. 12.

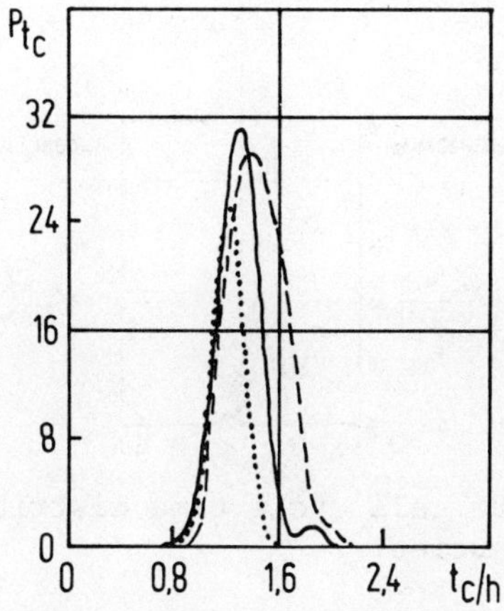

CONTROLS:

	$\bar{t}_c/h$	μ/μ_0
······	1,22	1,03
——	1,32	,95
---	1,36	,92

Fig. 11 Cell cycle time distribution of single yeast cells without irradiation (the scatter of the relative mean values is maximal $\pm$ 8 %)

The six graphs shown belong to different frequencies arranged in an increasing order. Especially, the shaded area characterizing the deviation of the mainpeaks should be looked at. One recognizes: Irradiation with 83800 MHz seems not to induce any change at all compared to the controls. The same is due to cells irradiated with 83842 MHz. One can state: Irradiation at certain frequencies does not influence the cell cycle times for the first and second division.

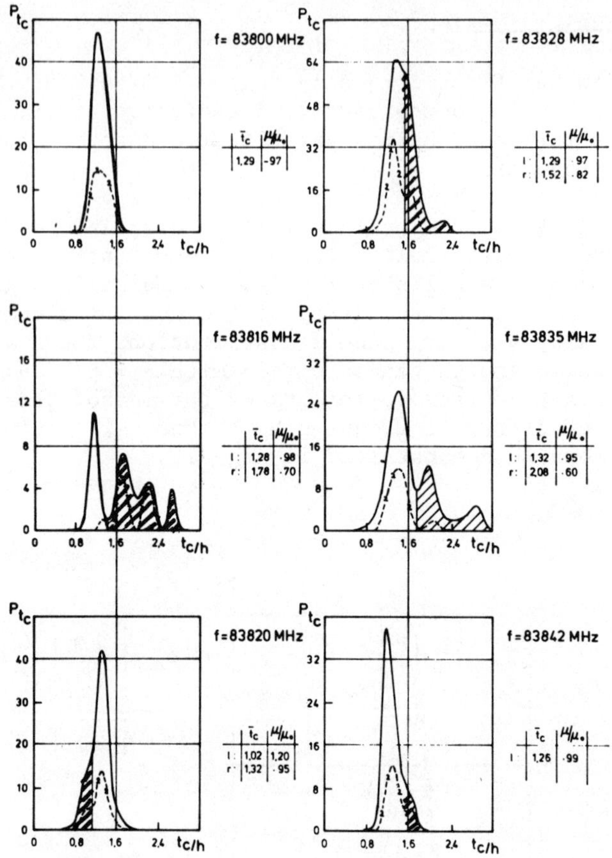

Fig. 12 Dependence of the cell cycle time distribution on different frequencies.

At other frequencies (83816 and 83835 MHz) we can in general observe a multipeaked distribution, consisting of two parts:
One part of the t_c-distribution is not altered at all, falling exactly on the control position, the other part is prolongated. The astonishing fact is that although all the cells have been treated with the same intensity (note they are all located within an area of .2 mm^2), one part of them reacts like a control, the other one has prolongated cell cycle times.

For demonstrating quantitatively a frequency dependence, the mean values of the main and the neighboured peaks have been defined (without to consider the other peaks of the shaded area). These results defining mean cell cycle times $\bar{t}_c$ and relative growth rates μ/μ_o of the two peaks are listed at the right of every graph. The main peak values are well within those of the controls.

The sidepeak values are shown in fig. 13, where the normalized growth rate vs frequency are drawn. (The dark squares correspond to the six distributions demonstrated by fig. 12).
Here, additional values are listed (open squares) and the results of the reproduction experiments (crosses) are shown.

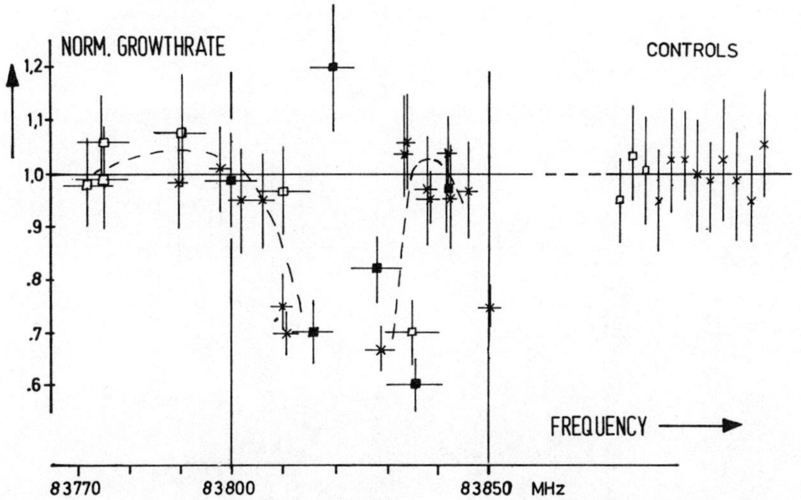

Fig. 13 Dependence of the normalized growth rates, defined by the use of cell cycle time distribution of single yeast cells, on frequency.

It should be mentioned that all runs corresponding to the region ± 10 % round unity, behave like controls without showing a multipeaked distribution, and that all the other points show, without exception, s similar abnormal shape. The overall dependence on frequency points once more to a resonance like behaviour, with a line width of around 10 MHz.

Additionally, we show the results of our first attempt of reproducing photometrically measured mm-wave effects on yeast cells in the U-Band by using this different technique.
We have analysed the growth behaviour of 6 control - and 23 irradiated cell populations microscopically. The frequency region selected ranges from 41772 MHz to 41800 MHz and covers a resonance region of the photometrically measured reaction spectrum shown in fig. 5.

The cell cycle time distributions were qualitatively identical to those of the former experiment with E-Band frequencies.

For demonstrating quantitatively a frequency dependence, the normalized mean of the cell cycle time distribution is defined by the ratio of the mean of every single distribution and of the mean control value. The inverse of this quantity is the normalized growth rate. These values are shown vs frequency in fig. 14.

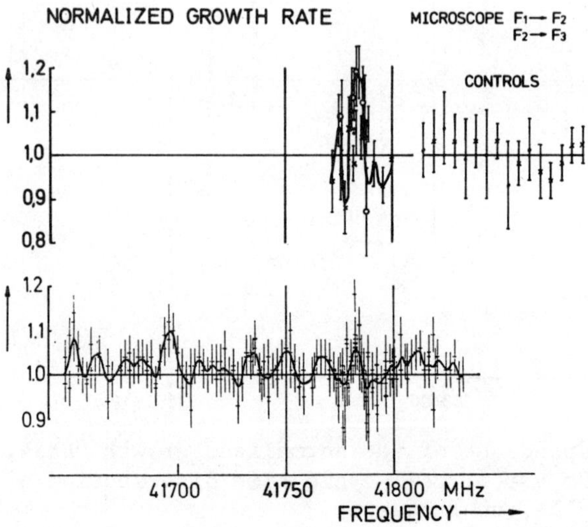

Fig. 14 Dependence of normalized growth rates on frequency. Comparison of a photometrically observed resonance region with the growth rates of microscopically observed single cells.

One recognizes clearly an irradiation effect well out of the scatter of the controls. The overall behaviour shows a great resemblance with the corresponding frequency range of the photometer experiment (lower part of figure 14). Especially, the effect maximum at 41784 MHz measured microscopically is well within the range of 41782 ± 3 MHz, where an enhanced growth was registrated photometrically. Also, the same resonance with can be defined round 8 MHz. So in general, the resonance effect on yeast growth is manifested once more, although, in our opinion, it is not clear up to now what is finally the real fine structure of such a resonance.

3. FREQUENCY-DEPENDENT MICROWAVE EFFECTS ON MAMMALIAN CELLS IN VIVO AND IN VITRO

The purpose of this chapter is to demonstrate that in higher organized mammalian cells (in comparison to yeast), similar effects can be produced and that frequencies of the mm-wave region may act in this systems on cell growth with a similar response.

Some years ago, resonance effects of mm-waves were reported /3/ on bone morrow cells of mice. The investigations deal with damage to these cells arising from exposure to X-rays and the protecting effect of microwave irradiation. When prior irradiation with microwaves has taken place, at certain frequencies and intensities the damage is then vastly reduced. These experiments have been reproduced recently /29/ and the new results are shown in fig. 15. The reduction of X-ray damaged bone marrow cells from about one half to about 80 % survial by prior microwave irradiation of certain selected frequencies is demonstrated. No effect exists, however, at certain other frequencies.

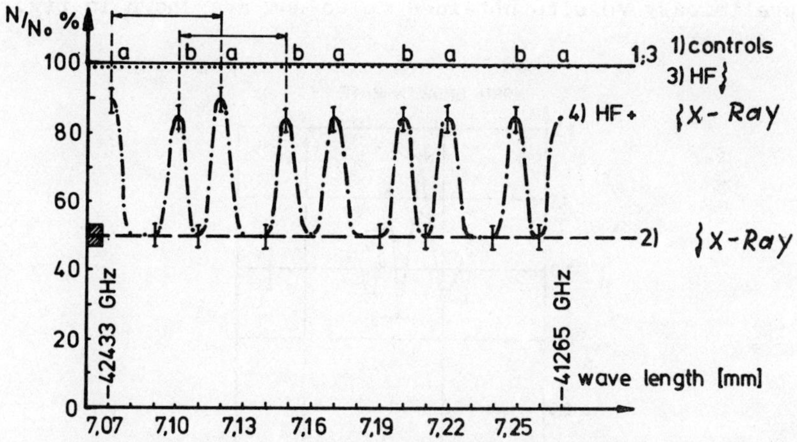

Fig. 15 Dependence of the relative number of bone marrow cells N/No on the wave length of the radiation. (N: number of surviving cells, No: number of cells without radiation; 1: control; 2: X-ray irradiated; 3: millimeter wave irradiated; 4: millimeter wave and X-ray irradiated) (Sevastyanova /23/).

Because of an absorption of about 70 % of the MW-radiation by the skin /3/, only a small fraction of the 10 mW/cm^2 used will penetrate for direct action on a deeper region. In this in vivo system, sharp resonances also occur of a width of about 80 MHz at half maximum. It seems also, that every two of the peaks are coupled where a difference in frequency of 160 MHz can be observed. The repetition periode of these double peaks itself is about 400 MHz. These results are surely not understood and point to interesting physical reactions possibly involved.

A further study on frequency dependent resonance effects of mammalian cells has been started in our laboratory. We are analysing the growth behaviour of Ehrlich Aszites Tumor cells (EAT-cells) in vitro by counting the cell density of a exponentially growing culture at different times by means of a cell counter (Symex, Colora) working similar to a coulter counter.
The cells were allowed to grow up for nearly one generation time (10 h) under controlled conditions in a chamber equipped with a control- and irradiation position. The petri dishes used have a thin gaspermeable membran at the bottom through which the radiation can be applied. By defining the cell concentration before and after every run generation times and relative growth rates for controls and irradiated cells can be calculated.
The preliminary results obtained up to now are shown in fig. 16.

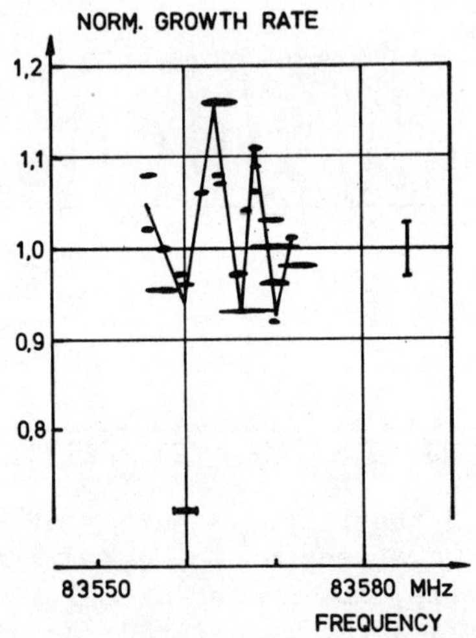

Fig. 16 Dependence of the normalized growth rates of Ehrlich Aszites Tumor cells on different frequencies.

In this mammalian cell system a change of growth in dependence on frequency occurs. At some frequencies an enhancement, at other ones a reduction is be observed. Furthermore, a resonance width of about 8 MHz can be read out well in agreement with the width observed with yeast cells. A further interesting fact is that the maxiumum of enhanced growth is caused by frequencies around 83568 MHz, which is exactly twice the former microscopically defined one of 41784 MHz, at which yeast cells are maximally forced to growth. One can state: The first harmonic of this frequency is also active in influencing cell growth, but within a different biological system.

4. CONCLUSION

We have confirmed the existence of nonthermal resonant effects of 42 GHz and 84 GHz irradiation on the growth rate of yeast cells. The resonances have narrow line widths of 8 MHz FWHM. On cells in suspension, the same resonances can be found when different antenna geometries are used. Furthermore, in the 42 GHz region the effects have been reproduced on single cells observed microscopically.
Other cell systems (mammalian cells) show also a resonant behaviour in response to microwave irradiation; here protection effects against X-rays were analysed or cell growth was monitored.

ACKNOWLEDGEMENTS

This study was supported by the Deutsche Forschungsgemeinschaft. The technical assistance of I. Stark, I. Zimmermann, O. Lock and W. Stark is gratefully acknowledged. The author is greatly indepted to Dr. M. Nüsse for support in performing the EATC experiments.

REFERENCES

/1/ S.J. Webb and D.D. Dodds, Nature 218, 374 (1968); S.J. Webb and A.D. Booth, Nature 222, 1199 (1969).
/2/ A.J. Berteaud, M. Dardalhon, N. Rebeyrotte and D. Averbeck, C.R. Acad. Sc. Paris 281, D843 (1975).
/3/ N.D. Devyatkov, L.A. Sevastyanova, R.L. Vilenskaya, A.Z. Smolyanskaya, V.F. Kondrateva, E.N. Chistyakova, I.F. Shmakova, N.B. Ivanova, A.A. Treskunov, S.E. Manoilov, N.P. Zalyubovskaya, R.I. Kiselev, V.I. Gaiduk, Yu.I. Khurgin and V.A. Kudryashova, Sov. Phys.-Usp. 16, 568 (1974).
/4/ S.J. Webb, Phys. Lett. 73A, 145 (1979).
/5/ L.A. Sevastyanova, Vestn. Akad. Med. NAUK SSSR 2, pp. 65-88 (1979), (in Russian).
/6/ W. Grundler and F. Keilmann, Z. Naturforsch. 33c, 15 (1978); W. Grundler, F. Keilmann and H. Fröhlich, Phys. Lett. 62A, 463 (1977).

/7/ B.K. Chang, A.T. Huang, W.T. Joines, Inhibition of DNA Synthesis and Enhancement of the uptake and Action of Methotrexate by Low-Power-Density Microwave Radiation in L 1210 Leukemia Cells. Cancer Res., 40, 1002-1005 (1980).
/8/ L.A. Sevastyanova, S.I. Potapov, V.G. Adamenko and R.I. Vilenskaya, Biological Sciences, No. 6, 1969 (in Russian); S.I. Potapov, L.A. Sevastyanova and R.L. Vilenskaya, Biological Sciences, No. 3, 1974 (in Russian).
/9/ M. Dardalhon and D. Averbeck, Proc. IXth Int. Congr. Soc. Francaise de Radioprotection, Effets Biologiques des Rayonnements Non Ionisants, pp. 279-299, 1978.
/10/ M. Dardalhon, A.J. Berteaud and D. Averbeck, Radioprotection 14, 145 (1979).
/11/ F. Kremer, C. Koschnitzke, L. Santo, P. Quick and A. Poglitsch, in: Coherent Excitations in Biological Systems Ed. by H. Fröhlich and F. Kremer, Springer, Heidelberg, pp. 13-24, (1983)
/12/ C.F. Blackmann, S.G. Benane, C.M. Weil and J.S. Ali, Am. New York Acad. Sci. 247, pp. 352-366 (1975).
/13/ P. Tuengler, F. Keilmann and L. Genzel, Z. Naturforsch. 34c, 60 (1979).
/14/ D.L. Jaggard and J.L. Lords, Proc. IEEE 68, 114 (1980).
/15/ L.M. Partlow, L.G. Bush, L.J. Stensaas, D.W. Hill, A. Riazi, O.P. Gandhi, P.L. Inversen and M.J. Hagman, Bioelectromagnetics 2, pp. 123-159 (1981).
/16/ G.Larson and L. Karlander, Report TRITA-TET-8201, Royal Institute of Technology, S-10044 Stockholm (1982).
/17/ H.P. Schwan, Interaction of Microwave and Radiofrequency Radiation with Biological Systems, IEEE Trans. Microwave Theory Tech. MTT-19: 146-152 (1971).
/18/ A.W. Guy, Analyses of Electromagnetic Fields induced in Biological Tissues by Thermographic Studies on Equivalent Phantom Models. IEEE Trans. Microwave Theory Tech. MTT-19: 205-214 (1971).
/19/ S.M. Michaelson, Human Exposure to Nonionizing Radiant Energy-Potential Hazards and Safety Standards, Proceedings of the IEEE, Vol. 60, No. 4, April (1972).
/20/ H. Fröhlich, Int. Quant. Chem. 2, 641 (1968).
/21/ H. Fröhlich, Nature 228, 1093 (1970)
/22/ H. Fröhlich in Advances in Electronics and Electron Physics 53 (eds. L. Marton and C. Marton), pp. 85-152, Academic Press, (1980).

/23/ F. Kaiser in: Biological Effects and Dosimetry of Nonionizing Radiation (eds. S. Michaelson, M. Grandolfo and A. Rindi) pp. 251-282, Plenum Press, New York, (1982). cf. also contribution in this volume, pp. 149-154.

/24/ F. Keilmann, Magnetic-Transition Mechanism of Resonant Microwave Bioeffects. Abstr. of Symp. on the Interaction of ELM-Fields with Biological Systems. Florence, Italy, August 27-30, (1984).
F. Keilmann, Triplet-Selective Chemistry: A Possible Cause of Biological Microwave Sensitivity, submitted to IEEE Trans., M.T.T., (1984).

/25/ F. Keilmann in: Biological Effects and Dosimetry of Nonionizing Radiation (eds. S. Michaelson, M. Grandolfo and A. Rindi), pp. 283-298, Plenum Press, New York, (1982).

/26/ W. Grundler in: Biological Effects and Dosimetry of Nonionizing Radiation (eds. S. Michaelson, M. Grandolfo and A. Rindi), pp. 299-318, Plenum Press, New York, (1982).

/27/ W. Grundler, F. Keilmann, V. Putterlik, L. Santo, D. Strube and I. Zimmermann, in: Coherent Excitations in Biological Systems, Ed. by H. Fröhlich and F. Kremer, Springer, Heidelberg, pp. 21-37 (1983).

/28/ W. Grundler, W. Abmayr, Differential Inactivation Analysis of Diploid Yeast Exposed to Radiation of Various LET, Rad. Research 94, pp. 464-479, (1983).

/29/ Nonthermal Effects of Millimeter Wave Irradiation, (ed. N.D. Devyatkov), Academ. Sci. USSR, Inst. Radiotech. Electrotech. Moscow, (1981) (in Russian).

FLUORESCENCE, ABSORPTION, PHASE-CONTRAST, HOLOGRAPHIC AND ACOUSTICAL CYTOMETRIES OF LIVING CELLS

F. Beltrame, B. Bianco, G. Castellaro, and A. Diaspro

Dipartimento di Ingegneria Biofisica ed Elettronica
dell'Università di Genova
Via all'Opera Pia 11 a, 16145 Genova - Italy

INTRODUCTION

There is in the scientific community a wide interest in the interactions between electromagnetic fields and living matter, in particular with cells. Certainly, this paper is not the right place to justify the motivations of such interest, or to present the various aspects of such interactions. On the contrary, we wish to clarify, in this brief introduction, the reasons why a work, essentially devoted to cell imaging and in particular to cytometry, is related to the general scopes of this meeting.

Broadly speaking, interactions between electromagnetic fields and cells are "weak" ones, in both senses that, if we consider the scale of observation of e. g. optical microscopy, they do not give rise to abrupt changes of physical structure inside the cell, in general, or that, when one deals with a large cellular population, their effects may even be strong on single cells, but the number of cells interested to the effects is a small fraction of the total population. As an example, electromagnetic fields of relatively low frequency and intensity can induce[1,2] cell reactivation; from an imaging point of view this phenomenon can be observed by monitoring conformational changes in chromatin, which in turns are put into evidence by changes in the optical density or in the fluorescent activity induced by suitable staining molecules. Nevertheless, these changes are not so strong as to be evident if one makes average observations on large populations, whereas analysis on single cells of a population clearly stresses the phenomena. An example of very strong interactions is cell fusion[3] induced by electrical means: in this case, dielectrophoretic motion (i. e. the motion of electrically neutral cells in a nonuniform electric field, caused by interaction between the dipole momentum created in the cell by electrical polarization, and the field itself) may put some cells of a population in membrane contact, and a sudden electric pulse causes fusion; again, however, this is a probabilistic event which happens in a fraction of the population: it can be observed only by monitoring single items.

As a consequence of this "weak" character of such interactions, their effects can be properly estimated through methods of analysis which do not treat large cell populations as a whole, since the resulting measurements of average quantities may mask the possible existence of particular, hence more interesting phenomena. This has as a further consequence that ordinary methods of biochemical analysis do not exhibit, in general, sufficient resolution in such investigations. The most suitable methods of study are those based on the observation of single cells. Modern acquisition and processing techniques can extract a very large amount of information from one cell; moreover, the present day availability of both computerized acquisition systems at relatively low cost and of mathematical methods for the treatment of large masses of data, actually put the life science researcher in the almost ideal situation, where he is able to get measurements on each cell belonging to a more or less large population. In this direction the most advanced systems are flow systems, where information is extracted from each cell from a sample flowing in a capillary tube so small in diameter to permit the passage of one cell at a time. This permits measurements on large numbers of cells in short time: this advantage in speed is paid by the fact that the information extracted from each cell is rather of global kind. Static systems are devoted to the analysis of single cells or of a small number of cells. They allow the information extraction point by point from a cell, to the resolution limit of the magnifying imaging device (microscope). In turns, the analysis of an image is relatively time expensive, so that it permits the observation of not too large populations. In this paper, we examine the major features of cell analysis based on optical or acoustical methods. We shall present a short outline of methods of analysis, each of which stresses the global value or the punctual distribution of some cell physical property. Absorption cytometry measures the absorption of the cell at the wavelength used for illumination. Fluoroscopic methods detect the activity of fluorescent dye molecules, i. e. their amount and/or distribution in the specimen. Phase-contrast imaging gives some information on the refractive properties of cells; finally, acoustical microscopy yields some map of the elastic properties of the examined cells. Moreover, a brief discussion on holographic methods is added, in view of their intrinsic "three dimensional" character. It must be stressed that most methods considered here are non destructive and, in some cases, non invasive: in such cases, it is possible to attempt to perform measurements on living cells, for example in a microchamber.

GENERAL CONSIDERATIONS ON OPTICAL METHODS

Static cell imaging by optical means (i. e. in the visible light range) is based on the magnification of the observed object through systems of lenses. The object is placed approximately in the front focal plan of an optical system, the objective, which forms a magnified real image. For visual inspection, this real image is observed through an eyepiece, or ocular, whose front focal plane coincides with the plane of that image. As a result, a final virtual image at infinity is formed. The conventional magnification G of the microscope is defined as the ratio between:

- the angle under which the observer's eye, placed in a position optically equivalent to the ocular back focus, sees the image at infinity and

- the angle under which the object is seen by the eye at a distance of 250 mm (that of distinct vision for the normal or "emmetropic" eye).

It is found that G is approximately given by:

$$G = \frac{250 \cdot L}{F_1 \cdot F_2} \tag{1}$$

where F_1, F_2 are the focal lengths of the objective and ocular respectively, and L is the distance between objective and ocular. In eq. (1), all lengths are to be expressed in millimeters. Small adjustments of L can yield a resulting image at a finite distance, instead than at infinity.

In electronic imaging systems the image formed by the objective is treated in a different manner. Some lenses and/or prisms provide to create an image focused directly on the target of a TV camera. The "electronic" image so formed, is scanned line by line and eventually displayed on a TV monitor for visual observation. For subsequent processing, the image, more exactly the set of voltage values corresponding to the target elements, is converted into digital form and stored in a computer bulk memory, via a fast buffer memory. The magnification of TV-based static systems must by no mean be confused with the previously defined G. It is typically measured in pixel per micrometer, i. e. by the number of pixels on the TV screen corresponding to a segment of the object one micrometer in length, and in a direction perpendicular to the optical axis of the system. A typical range is 3 to 30 pixels per micrometer, depending on the focal length of the objective, for a 512x512 pixel TV format. Static systems are useful for analysis of single cells or of small group of cells, more exactly of one TV frame at a time with proper magnification. The amount of information they can yield is very large. However, it must not be overestimated, on the basis of the following considerations. First of all, suppose, for example, to work in yellow light, i. e. $\lambda = 0.6$ micrometers. The resolution in a high quality microscope is some 0.3 micrometers, indeed it is given approximately by

$$\frac{\lambda}{\sin\alpha_1 + \sin\alpha_2} \qquad (2)$$

where α_1 and α_2 are respectively the angular aperture of the microscope (or, which is the same in most cases, of the objective) and of the condenser which provides illumination. For large apertures, eq. (2) yields value of about $\lambda/2$. Now, with the above said system magnification, 0.3 micrometers correspond to 1 to 9 pixels. It is clear that, at the largest magnifications, the fine distribution of pixels does not correspond to actual details in the object, since the separation between adjacent pixels, when reported on the object, is far below the microscope resolution. We have put into evidence this fact only as a warning: if one remembers it, he is guaranteed from meaningless measurements. A more serious source of troubles is the poor response of TV system at high spatial frequencies, i. e. to the fine details of the image focused on their targets. To put ourself in a neat situation, let us suppose that the object is a grating composed of pairs of "white" and "black" lines (of monotonic decreasing width), focused with negligible error onto the target of a TV tube. The image acquired by the system, and possibly presented on a TV monitor, unfortunately is far from the original item. If one plots the contrast of the image as acquired by the system (more exactly, the[4] modulation transfer function (MTF)) as a function of the grating spatial frequency, the resultant curve falls quickly at low values even for moderate spatial frequencies. As an example, for our ACTA 500 imaging system, the contrast is reduced to the 70% of its maximum value for a spatial frequency of 0.04 line pairs per pixel. Higher spatial frequencies are even more smoothed in contrast. As a consequence, care must be taken in identifying the imaging system output with the true image intensity, i. e. the intensity at the target input, irrespective of the linearity of the target itself. This kind of error could be somehow corrected by computing the optical transfer function of the system (OTF), or equivalently, the point spread function[4] (PSF). Direct measurements are not possible, while one can calculate either the OTF or the PSF from measurements on particular standard gratings. However, this point is rather delicate. In general, in the sequel of this work, we shall not consider the system error, dealing with imaging in ideal situation.

Flow systems are different. In the case of laser systems considered here, the cell population contained in an appropriate culture medium is forced to pass through a capillary tube, one cell at a time. Each cell, in a given position, is illuminated by a thin laser beam, which causes scattering of light in a fan of directions. Appropriate detectors (through the use of properly matched optical filters) measure the light intensity at fixed angles (typically in the forward and 90 degrees directions). In the case of fluorescent staining, besides scattered light, these light intensities are representative of green, red or blue emitted fluorescence from each single event (accordingly to the physico-chemical process of the dye uptake of the biological sample under investigation). So, simultaneously, multiparametric acquisition is feasible for each item: such parameters, through the use of a synchronizing electronic pulse, are collected into a file in a computer, and in short time distributions of them are displayed on a TV screen, giving rough, global, but, nevertheless, important insights on the composition of the cell population under analysis. Furthermore, in most sophisticated systems (cell sorters), different compartments of a cell population can be identified through the selection of fluorescence and scattering windows and physically sorted in different bottles for further analysis.

ABSORPTION AND FLUORESCENCE ANALYSIS

Image analysis, in absorption and in fluorescence, has been known from several years as a powerful tool to study many different problems in the field of cell biology, as for example cell cycle, virus transformation or automatic recognition and classification of leucocytes.

In absorption and fluorescence the cells are fixed and stained and presented to the microscope either on a slide or on a microchamber. Of course, fixation and staining are carried out in a quantitative and specific way, either for absorption and for fluorescence. In fact, only in such case, the dye uptake from the specimen, which is the measured physical observable, can be, in principle, related to the amount and/or conformation of a specific macromolecule, as for example DNA or RNA, inside each single cell or each single nuclei picked up by the TV sensor.

Really this is the first remarkable advantage of image analysis technique (both in absorption and fluorescence) to be pointed out with respect to other more classical biochemical and biophysical methods: it is possible, with it, to obtain a number (e. g. area, optical density, etc.) representative of a certain physical process for each single event in the field of observation, thus avoiding all the problems arising from bulk population measurements. In the attempt to quantitatively explicitate certain biological phoenomena, as cell reactivation or cell proliferation, image analysis has been often used intended as the only practical tool to look and objectively define morphological and densitometric changes of the biostructures involved in such processes (i. e. DNA synthesis, chromatin spatial rearrangement inside the nuclei) at the resolution level of light microscopy. Here are two others important features of absorption and fluorescence measurements by image analysis technique. For example, it is possible to measure in absorption, through the geometric and densitometric analysis of nuclear images obtained by differential staining with the Feulgen method, the amount and conformation of chromatin-DNA "in situ" , thus avoiding all the artifacts due to the isolation process of such macromolecule. Furthermore, the measurement is performed at a magnification level corresponding to an organization level of chromatin-DNA which plays a critical key-role in determining the cellular function. In absorption, classical studies have been carried out leading, as one of the major result, to the objective identification of cell cycle, using measured parameters as Integrated Optical Density (IOD), Area, or their "clever" combinations, such as Average Optical Density (AOD). In such example, IOD, measured in Feulgen stained nuclei, is proportional to the amount of DNA of each single nuclei, while AOD gives, as a texture parameter, a mean to distinguish two items (cells) with the same amount of material (DNA), but distributed over different surfaces. Fundamentally, measurements in absorption have been extensively carried out with the objective to reach an analytical determination of the structure-function relationship, in particular of chromatin, mainly because of the availability of a well known staining technique such as the Feulgen method, of TV cameras with good frequency response in that absorption region (i. e. around 540 nm) and good linearity in the relative optical density range (i. e. 0 to 2 OD) of interest, as the Plumbicon target. To be noticed, just in passing, that of course not only global descriptors as IOD, Area, etc. are computed with the most advanced acquisition systems, but the optical density or intensity value for each pixel of the image under investigation is measured, with obvious consequences for all the related image processing techniques available (i. e. filtering, transformations, pseudocolor etc.).

The same considerations reported for absorption could be repeated for fluorescence image analysis methods, with the following comments. As far as the staining process, fluorescence is much less a destructive technique for the cell that absorption staining methods: infact, even fixation can be sometimes avoided, for example using Acridine Orange stain, and the chemical reaction (of equilibrium type) is reversible: in particular, if the concentration of the dye is low enough, the dye itself can be washed out from the cell solution after the measurement and the cells can still survive. In a certain sense, it can be considered as a "vital" staining. Therefore, presumably, the biostructure (e. g. chromatin-DNA) under investigation is much more well preserved. Nevertheless, it is more difficult to relate the measured fluorescence (for each single cell or cell nuclei) to the amount of a specific macromolecule: it usually depends (thinking for example to an intercalation process of Acridine Orange molecules between DNA base pairs) on both the amount of the target molecule (e. g. DNA) and its organization (e. g. DNA coiling status). By fluorescence image analysis methods it is possible to look and study biochemical and biophysical processes which express themselves, simoultaneously, in different part of the cell: at membrane, cytoplasm and nuclear level, for example by using a combination of fluorescent dyes with different wavelengths of emission and, of course, proper matched optical filters. It is also worth to just remind the possibility to combine static fluorescent techniques with flow systems, as a refined or verification at an higher resolution level post-processing step.

Finally it must also be mentioned that TV cameras for fluorescent measurements, i. e. low level of illumination sensor, need usually an electronic image intensifier. This is a high cost equipment, and also the signal to noise ratio is very low if compared with a Plumbicon TV scanner, mainly because of the more complicated noise statistic involved. Nevertheless, even this problem can be partially overcome, by proper identification of the system used and with consequent restoration software design and application: this latter will only slow down the processing of each image.

PHASE-CONTRAST IMAGING

This technique is based on the concept of transforming the uneveness of refraction index into differences of image intensity from one point to another. This makes it possible to observe with good contrast even transparent objects, such living, unstained cells are in many cases. A strongly simplified model of phase-contrast microscope is shown in Fig. 1:

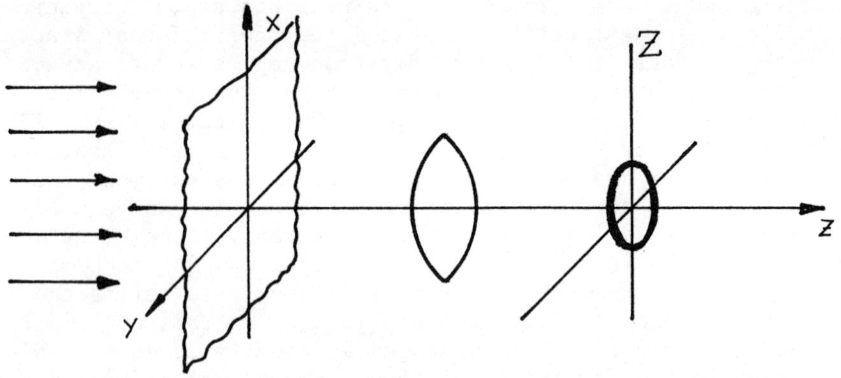

Fig. 1 - Model of phase-contrast microscope. The xy and $\mathcal{Z}$ planes are respectively the front and back focal planes of the objective.

We consider monochromatic illumination, parallel to the optical axis z. Let I_o be the amplitude of the source light; and let t(x,y) be the transmission coefficient and $\varphi(x,y)$ the phase shift of the object, considered here as a thin sheet, which actually is realized by a cell on slide. Ideally, the object can be considered as a semitransparent plane xy, having a complex transmission coefficient given by

$$T(x,y) = t(x,y)\,e^{j\varphi(x,y)} \tag{3}$$

Disregarding some fixed phase factor, the light U(x,y) in $z=0^+$ is then

$$U(x,y) = I_o\,T(x,y) \tag{4}$$

It is known [4] that a lens performs a Fourier transformation on the light between its focal planes. Hence, in the $\mathcal{Z}$-plane, the light is essentially equal to the Fourier transform $A(f_x, f_y)$ of T(x,y). In the same plane a circular disk, R in radius, η in transmittance, $\lambda/4$ in length, is centered on the optical axis.

As a consequence, the portion of the light A which falls on this disk (Zernike filter) is shifted 90 degrees in phase. The $\mathcal{Z}$-plane has a transmission coefficient given by

$$\mathcal{Z} = \begin{cases} 1 , & x^2 + y^2 > R^2 \\ j\eta , & x^2 + y^2 < R^2 \end{cases} \tag{5}$$

The light immediately on the right of the filter plane is then A$\mathcal{Z}$. It is clear that the filter effect can be actually reported at the input, by considering as imaging device a standard microscope, i. e. without filter, but where the object transmittance T(x,y) is substituted by

$$T(x,y) * \mathcal{F}^{-1}\mathcal{Z} \tag{6}$$

where $\mathcal{F}^{-1}$ is the symbol of inverse transform. With this background, the final image intensity results to be

$$I_{PC} = |U'|^2 \tag{7}$$

where

$$U' = t(x,y)\, e^{j\varphi(x,y)} * \mathcal{F}^{-1}\mathcal{Z} \tag{8}$$

Eq. (7) shows that the image intensity depends on the phase characteristics of the object. If $R \to 0$, $\eta = 1$, $t = 1$, eq. (7) reduces to the well known approximation

$$I_{PC} \cong 1 + 2\varphi(x,y) \tag{9}$$

Now, in the case of cells, the phase shift in general is not particularly small. Even if the refraction index of the cell is close to that of the surrounding medium, the thickness of the cell is usually tens of wavelength, and this fact may yield values of φ of several radians. Eq. (7) is more realistic. An analysis in simple cases shows that this simple theory of image formation may explain the occurence of halos or fringes around the interfaces between media with different refraction index such as cytoplasm and nucleus, or the whole cell and a culture liquid.

It is important to notice that I is not a function of φ, since the dependence of I from φ is established by a complicated, non linear integral operator. However, in principle, $\varphi(x,y)$ can be recovered: this is referred to as "phase reconstruction". Indeed, t(x,y) can be measured by taking an image in transmission. A phase-contrast image yields the function $I_{PC}(x,y)$: $\varphi(x,y)$ (which is contained in T(x,y)) can be recovered by solving numerically [5] the integral equation (7).

An example of application of this method is shown in Fig. 2:

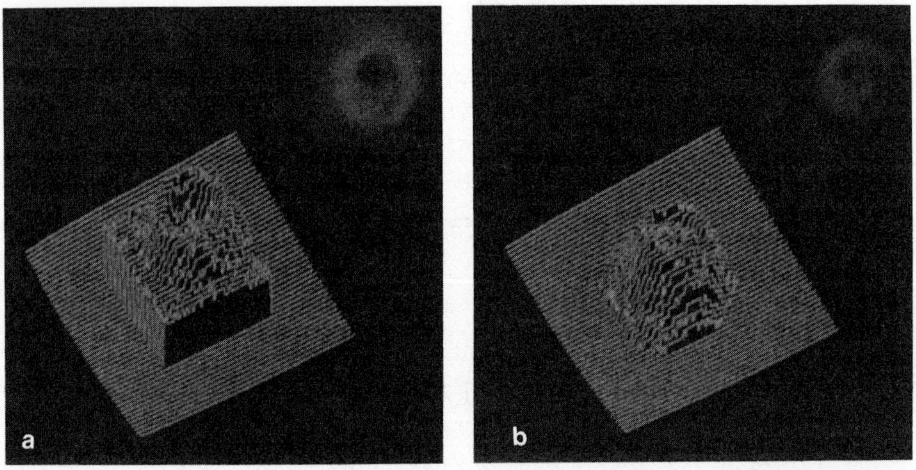

Fig. 2 - a) In the top-right corner the image of a living CHO-K1 cell, as it appears to the human eye through standard phase contrast microscopy. In the center, an axonometric view of the same image (intensity distribution) obtained by computer. b) Same cell as in a) with the quantitative information extracted after processing with a proper algorithm modelling the real physical situation of the imaging device (Zernike phase-contrast microscope). It is possible to have a number for each pixel, which is, with good approximation, proportional to the number of molecules inside the cell.

The result shown in Fig. 2 is merely a temptative one, since the microscope used was far from the model of Fig. 1: however, it seems encouraging in view of further development of this reconstruction technique, mainly based on the fact that phase information on cell is, so to say, complementary to that in absorption. Furthermore phase-contrast techniques represent almost the unique method to observe the effect of e. m. interactions on living cells in real time: it is evident the interest in transforming the simple qualitative observation (as up to now phase contrast has been used) in a quantitative measurement.

HOLOGRAPHIC MICROSCOPY

Application of holographic techniques to microscopy is not a new subject: an excellent discussion can be found in a quoted[6] paper. The obvious interest in holographic methods bears on their intrinsic three dimensional character. At the macroscopic scale, holography typically works as follows. The object is illuminated by a laser source. The scattered light falls on a photographic plate, illuminated also by a separate beam of plane waves ("reference beam"). So the plate records interference fringes; once fixed, it becomes the hologram, and the described process is the "hologram formation". It can be shown that the hologram contains, as a consequence of interference, information on both amplitude and phase, coded in some form, of the light scattered by the object. If now the latter is removed, and the hologram is illuminated by either the source or the reference beam, it is found that the scattered wavefront (in an ideal situation, i. e. disregarding the film nonlinearities and so on) is identical to the original wavefront scattered by the object, in amplitude and phase, apart some spurious light not considered here for the sake of brevity. This step is the "hologram reconstruction". As a consequence, an observer which looks at the hologram, "sees" a view of the object in its three dimensional totality, that is, if the observer moves from a standpoint to another, the object views change according to the movement.

As regards cell imaging, the matter is not so simple because of the need for magnification. A technique which could be used is the following. The slide supporting the cell to be examined is placed in front of a microscope objective, and is illuminated by a laser source aligned with the optical axis. Also a reference beam enters the objective: there is interference between this beam and the light scattered by the object (cell). The objective forms a real image of the focused plane directly on the target of a TV tube. The TV frame is converted into digital form and stored in a computer memory. In such a way, one records a magnified hologram. This process of formation is very similar to the traditional one. However, the hologram so obtained cannot obviously be reconstructed in the usual manner. It is suitable to this purpose a simulation procedure. Indeed, the wavefront reconstruction is essentially a diffraction phenomenon, and the observation of a reconstructed hologram can be described as follows. The hologram itself is a semitransparent plane xy with transmission coefficient t(x,y). Once the illuminating light is known, the diffracted light can be computed. Let $\xi\eta$ be the plane where the observer focuses his vision. If we know the light $U(\xi,\eta)$ in that plane, $|U(\xi,\eta)|^2$ yields the intensity distribution of the image seen by the observer. It can be shown that, if $A(f_x,f_y) = \mathcal{F}\, U(x,y)$, $A^*(f_\xi,f_\eta) = \mathcal{F}\, U(\xi,\eta)$, then

$$A^*(f_\xi,f_\eta)=\begin{cases} A(f_x,f_y)\dfrac{\sqrt{K^2-(a_{11}f_\xi+a_{21}f_\eta+a_{31}f_\zeta)^2+(a_{12}f_\xi+a_{22}f_\eta+a_{32}f_\zeta)^2}}{\sqrt{K^2-f_\xi^2-f_\eta^2}} \\ 0 \quad \text{if } f_\zeta \hat{=} \sqrt{K^2-f_\xi^2-f_\eta^2} \text{ is not real} \end{cases} \tag{10}$$

where $k = 1/\lambda$ and the a_{ij}'s are coefficients identifying the relative position of the xy and $\xi\eta$ planes. Simulation works as follows: apart from the usual scale factors, the hologram recorded in the computer memory yields essentially $t(x,y) = U(x,y)$. Then A is computed. One specifies the desired plane of observation $\xi\eta$, which allows to compute the a_{ij}'s coefficients; by eq. (10), A^* is found, and an inverse transformation yields $U(\xi,\eta)$, from which the intensity of the observed image is computed. This method is very effective from a computational standpoint, since is based on a direct and an inverse Fourier transformation that can be performed by[7] Fast Fourier Transform algorithms.

The technique described is a temptative one: others are possible. Generally speaking, the future of holographic microscopy is not well defined, mainly because experimental problems are heavy. Nevertheless, in comparison with other 3D imaging methods[8,9] (for example with the ones based on the acquisition of images in transmission at several different focal planes), it looks in principle so simple and so scarcely invasive as to render it attractive and worth of further research.

ACOUSTICAL MICROSCOPY

Broadly speaking, absorption and phase-contrast microscopes are sensitive to two electromagnetic properties of the cell: light attenuation and index of refraction respectively, which in turn depend on (i) the dielectric constant and (ii) the electric losses of the cell. Acoustic microscopes are sensitive to completely different properties: mainly to the elastic ones. The idea of acoustic microscopy is not recent, but instrumentation of good quality has been developed only in these years, as a consequence of the availability of appropriate ultrasonic generators. These must operate in the gigahertz range. Indeed, in order to get the acoustical image of a cell with a resolution comparable with that of optical microscopes, say 0.3 micrometers, the wavelength λ of the sound must be of the same order. Now, the velocity of sound in a cell is 1.5 - 2 x 10^3 m/sec, hence the frequency must be in a range around 1.5 x 10^3 /0.3 x 10^{-6} = 2 GHz.

The velocity v of sound waves in gaseous or liquid medium is given by:

$$v = \frac{1}{\sqrt{\varkappa \delta}} \tag{11}$$

where δ and $\varkappa$ are respectively the specific mass and the coefficient of uniform compression of the medium. For time-harmonic oscillations both the displacement of medium particles around their equilibrium positions and the pressure variations are also sinusoidal functions of time. The ratio between their amplitudes, in case of plane waves,is referred to as the (acoustical) characteristic impedance of the given material. It is found that $Z_0 = \delta v$. At the interface between two such materials, waves in general are reflected and refracted. Again in the case of plane waves, traveling from a medium with impedance Z_{01}, sound velocity v_1, toward another medium with parameters Z_{02}, v_2, the reflection coefficient (as regards the pressure amplitudes, and for normal incidence) is

$$\frac{Z_{02} - Z_{01}}{Z_{02} + Z_{01}} \tag{12}$$

If a wave is attenuated because of the medium viscous losses, the characteristic impedance is a complex quantity. Similarly, the reflection coefficient at an interface between lossy media is a complex number. Moreover, in case of oblique incidence, eq. (12) is substituted by one containing the incidence and refraction angles. The usual Snell's law holds true:

$$\frac{\sin\vartheta_1}{\sin\vartheta_2} = \frac{v_1}{v_2} \tag{13}$$

In solid materials, the situation is more involved, since they can support either longitudinal and transverse waves, having different velocities and impedances. (Fluid media do not carry transverse waves.)
A possible arrangement for coupling between sound waves and a sample in a reflection acoustical microscope is shown[10] in Fig.3.

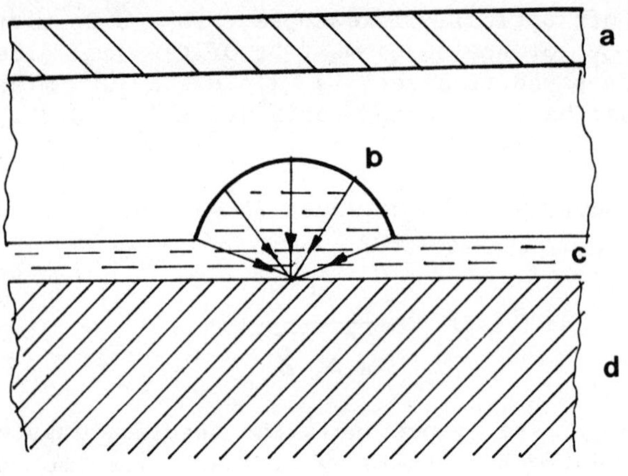

Fig. 3 - a) Piezoelectric film. b) Acoustical lens. c) Coupling medium. d) Sample.

A piezoelectric film excited by an RF source generates ultrasonic waves which are focused on some point of the sample through a spherical lens. As a general fact, when dealing with sound waves it is easy to obtain very high refraction indexes. This makes focusing of sound waves far more simple than in the optical case. (As an example, if the lens is made by a sapphire-water interface, aberrations in focusing are negligible for a portion of the spherical surface 55-60 degrees in aperture.) The reflected waves are collected by the same lens and transformed in an RF signal by the reversible piezoelectric film. Finally, the intensity of this reflected signal is recorded. The examination of the whole sample is made by a mechanical-scanning technique.

By this method, one has essentially a map of the acoustical reflection coefficient of the sample for converging spherical waves: this parameter is related, though not simply, to the elasticity and viscosity of the sample.

Some other imaging techniques have been proposed, either in reflection or in transmission. In any case, all methods yield some representation of the sample elasticity and viscosity.

Acoustical imaging can give very interesting information on the cell structure and properties, and generally speaking it is a technique complementary of optical ones. In particular it allows to detect cell structures which are unobservable, without staining, even by phase contrast techniques. It is also known that in some cases the acoustic response is different in normal and pathological conditions. The future of acoustical microscopy is oriented in the direction of imaging of living cells. For example, it will be possible to display, by coupling acoustical and interferential phase-contrast microscopy, both an image representative of the elastic properties and an image representative of optical properties of a single living cell, and to monitor their changes versus time.

CONCLUSION

In this paper we have presented a short outline of imaging techniques either of current use or still in development, which, in their totality, can give a considerable insight in the analysis of cells; in particular in non invasive analysis of unstained and even living cells. They are characterized by the fact that single cells are analyzed, in such a way as to get information able to stress the existence of weak interactions.

REFERENCES

1. A. Chiabrera, M. Hinsenkamp, A.A Pilla, J. Ryaby, D. Ponta, A. Belmont, F. Beltrame, M. Grattarola, and C. Nicolini, "Cytofluorometry of electromagnetically controlled cell dedifferentiation", Journal of Histochemistry and Cytochemistry Vol. 27(1) pp. 375-381 (Jan. 1979).

2. A. Chiabrera, R. Viviani, G. Parodi, G. Vernazza, M. Hinsenkamp, A.A Pilla, J. Ryaby, F. Beltrame, M. Grattarola, and C. Nicolini, "Automated absorption image cytometry of electromagnetically exposed frog erythrocytes", Cytometry Vol. 1(1) pp. 42-48 (1980).

3. U. Zimmermann and J. Vienken, "Electric field-induced cell-to-cell fusion", The Journal of Membrane Biology Vol. 67 pp. 165-182 (1982).

4. J. W. Goodman, Introduction to Fourier Optics, Mc Graw-Hill, New York, N. Y. (1968).

5. F. Beltrame, B. Bianco, and A. Chiabrera, "A system theory approach for phase reconstruction by optical phase contrast microscopy", Proceedings of the IEEE Vol. 71(2) pp. 270-271 (February 1983).

6. E. Feleppa, "Holography Applied to Biomedical Re", in Advances in Biomedical Engineering and Medical Physics, ed.

S. N. Levine, Interscience Publishers, New York, N. Y. (1971).

7. B. Bianco, F. Beltrame, and A. Chiabrera, "Simulazione a calcolatore della ricostruzione d'onda in olografia", Atti della 4ta Riunione Naz. di Elettromagnetismo Applicato, pp. 409-411 (). Oct. 1982 Firenze

8. F. Beltrame, G. Bonanno, D. Caviglia, A. De Gloria, and M. Grattarola, "Fluorescence image analysis techniques for the study of membrane receptors", Proc. combined Analytical Cytology and Cytometry IX and the VIth Intern. Symp. on Flow Cytometry, (Oct. 1982). Schloss Elmau W. Germany *Abstract*

9. D. A. Agard and J. W. Sedat, "Three-dimensional architecture of a polytene nucleus", Nature Vol. 302 pp. 676-681 (21 April 1983).

10. C. F. Quate, A. Atalar, and H. K. Wicksramasinghe, "Acoustical microscopy with mechanical scanning: a review", Proc. IEEE Vol. 67(8)(Aug 1979).

PERSPECTIVE ON WINDOWS AND CALCIUM EFFLUX STUDIES

Sol M. Michaelson

Department of Radiation Biology and Biophysics
School of Medicine and Dentistry, University of Rochester
Rochester, NY 14642

INTRODUCTION

When the mammalian central nervous system is exposed to low frequency (6-25 Hz) oscillating electrical and electromagnetic fields, various behavioral and electrophysiological effects can be observed (Adey, 1975, 1979; 1981; Bawin and Adey, 1976, 1977, 1978; Bawin et al., 1975, 1977; Blackman et al., 1979, 1980a and 1980b; Joines and Blackman, 1980; Kaczmarek, 1977; Lawrence et al., 1980). Fields at 5-15 Hz with a peak amplitude of 2-10 V/m in air produce changes in human circadian rhythms and shorten reaction times in both the human and monkey (Adey, 1975). Furthermore, characteristic changes occur in the intensity of certain frequency components of the EEG. Similar effects are observed when the brain is exposed to weak, very high frequency fields (50-450 MHz) amplitude-modulated at frequencies below about 30 Hz. Such VHF fields, amplitude-modulated at brain wave frequencies, strongly influence both spontaneous and conditioned EEG patterns in the cat.

Adey (1975) hypothesizes that local variations in the electrical field surrounding the external neuronal membrane produce slight modifications of negative binding sites in the membrane as a result of conformational changes of the surface macromolecules or of small displacements of the surface-bound cations. These alterations, in turn, would influence the activity of adjacent ions, which would further disturb the molecular arrangement of the surface macromolecules, thus propagating and multiplying the initial electrical disturbance. A key role is postulated for the calcium ion.

PHYSIOLOGICAL AND BIOCHEMICAL ASPECTS OF Ca^{++} ION FLUX

Physiological Aspects of Calcium Function in the Brain

Calcium plays a role in the physiology of the nervous system in relation to:

(a) electrophysiological or biophysical properties of the neuronal membrane,
(b) mediating and modulating the release of neurotransmitter from presynaptic nerve terminals,
(c) functioning of the neuronal cytoskeleton as it relates to axoplasmic transport, cell shape, and mobility,
(d) brain energy metabolism,
(e) thermoregulation,
(f) the blood brain barrier.

Molecular and Chemical Aspects of Calcium Metabolism

Many hormones, neurotransmitters and peptide ligands can produce a specific effector response in a biological system without penetrating the cell membrane. These effector molecules produce their response by activation of externally faced receptor sites. These sites are in principle coupled to effector or amplification mechanisms which then relay the signals intracellularly to complete the pharmacological or metabolic response (Myers and Ross, 1981).

By virtue of calcium's unique role in binding to the cell membrane, a rapid conformational change in membrane proteins, in the form of cation channels and enzyme-receptor reactions, may be influenced directly by the regulation of the amount of calcium bound to and transported by the cell membrane. This rapid control of cations and other co-factors over membrane permeability is very pertinent to the present review. The reason is that any change in electronegativity or net negative surface charge on the surface of the membrane will influence Ca^{++} ion-membrane interactions which in turn may alter the functional activity of the cell (Myers and Ross, 1981).

The involvement of calcium with many of the processes involved in neurotransmission including binding, uptake and release of neurotransmitters, activation of enzymes and opening and closing of ion channels, suggests that much of the calcium implicated in nerve activity is probably in the bound form. The question still to be answered is what is the functional relevance of these bound Ca^{++} ions? Conceivably, any change in the properties of the membrane surface which would reduce the net negative surface charge would contribute to a decrease in the number of binding sites. Similar changes may occur as a function of the dielectric constant of the solution used for the suspension of the membrane (Myers and Ross, 1981).

Electrophysiological Functions of Calcium

Calcium plays several essential roles in the electrophysiological processes of nerves: (1) external calcium stabilizes excitable membranes; (2) external calcium enters excitable cells in a voltage-dependent manner, thereby providing some of the current for the action potential; and (3) internal calcium regulates the permeability of the neuronal membrane to other cations.

Calcium and Neurotransmitter Release

Calcium ions (Ca^{++}) regulate the release from neurons of chemical transmitters, neurohormones, and other substances of ill-defined modulator, tropic, or other function. In their dependence on calcium ions to regulate secretion, neurons can be recognized as behaving much like many other secretory cells where calcium ions participate critically in transducing stimuli into secretory responses (Greengard et al., 1982). Calcium is not only the central element in the physiological regulation of the secretion of nerve cells, but it is the link that is affected, directly or indirectly, by a host of physiological factors, drugs, and toxins.

The excitation of the peripheral nerve is dependent on Ca^{++} ion flux. Ca^{++} ions play an important role in the release of neurotransmitters and other humoral substances within the CNS. The extracellular level of free Ca^{++} ions declines at the same time that paroxysmal discharges are recorded from the cerebral cortex during either a seizure episode produced by metrazol or during the cyclical spiking of a penicillin focus. Ca^{++} ions induce the phosphorylation of protein in synaptosomes of the brain, facilitate the incorporation of amino acids into protein and mediate the secretory events involving dopamine and other substrates from the cerebral cortex and corpus striatum. Calcium is one of the crucial cations involved in the control of the vasomotor tonus of the vessels in the cerebral meninges. During conditions of Ca^{++} ion deficiency, a static hypertension can ensue which could be related to the neccessity of a functional pool of Ca^{++} ions within the brain stem.

In relation to the functional importance of a physiological response to Ca^{++} ions particularly, it is noteworthy that a mMolar concentration of Ca^{++} ions from 2 to 40 times higher than that in the CSF or extracellular fluid is required to produce any observable change in the hypothalamic mechanism underlying motivational metabolic processes. Thus, a rather substantial shift in Ca^{++} ion kinetics is required in order for any external stimulus to change nerve cell function and thereby produce a consequent physiological and behavioral reaction (Myers and Ross, 1981).

Calcium Homeostasis

The involvement of calcium in the secretion of neurotransmitters from nerves is but one example of the more general mediator function of calcium ions in stimulus-secretion coupling in many cells (Douglas, 1968; Rubin, 1974). Moreover, calcium ions are used in corresponding manner in a host of nonsecretory cells of widely different functions as a signal to initiate or regulate diverse cellular responses, of which the most familiar is muscular contraction. The ability of calcium to function in such mediator roles hinges on the fact that cells generally maintain the intracellular concentration of free calcium ions in their cytoplasm at extremely low levels (about 10^{-7} M or less) that are about four orders of magnitude lower than found extracellularly (about 10^{-3} M). This enormous concentration gradient directed inward across the plasma membrane, together with the similarly directed high electromotive force across the membrane (positive outside), permits a prompt influx of calcium ions in response to various membrane perturbations by chemicals, hormones, and mechanical or immunological means that tend to increase the permeability of the membrane to calcium ions. Moreover, since cells commonly have large stores or reservoirs of sequestered calcium (in mitochondria and sarcoplasmic reticulum, for example), various stimuli can mobilize calcium from such stores and thus raise the intracellular concentration of free calcium ions by this alternative mechanism (Greengard et al., 1982).

Calcium and Brain Energy Metabolism

Calcium plays an important role in brain energy metabolism. Respiration in brain slices is stimulated under calcium-free conditions (McIlwain, 1952), whereas incorporation of amino acids into protein is decreased under these conditions (Dickens and Greville, 1935). One possible explanation for this observation relates to the membrane stabilizing action of extracellular calcium. Briefly, as extracellular calcium is reduced, the permeability of the neuronal membrane to cations increases, thus leading to increased neuronal excitability and membrane depolarization. Spontaneous shifts in the state of arousal, with associated changes in metabolic CO_2 production, produce corresponding changes in the ^{45}Ca efflux.

The efflux of ^{45}Ca from preincubated slices of cerebral tissue is characterized first by an early phase (within 5 to 10 minutes of washing in isotope-free solution) presumably representing washout of rapidly exchanging calcium from the extracellular space (Bawin and Adey, 1976, 1977). Thereafter, there is a slower rate of radiocalcium efflux, which could result from exchange with more sheltered calcium pools.

Since the classical experiments of Ringer (1883), calcium has been shown to play a considerable role in a variety of neuronal processes such as cellular respiration, excitation-secretion coupling in nerve endings, coupling and uncoupling of intercallular connections, functioning of the cytoskeleton, and general maintenance of membrane stability (Blaustein, 1974). The diverse spectrum of calcium action suggests that hemeostatic mechanisms must exist to maintain relatively constant levels of calcium in the extracellular fluid and that, in addition, the level of calcium within regions of neurons must be under fine local control (Greengard et al., 1982).

The homeostatic mechanisms of the body achieve remarkable control of the plasma concentrations of most substances, including free calcium. The degree of control achieved is sufficient for most organs, but not for the brain. The blood-brain barrier, then, represents an additional homeostatic mechanism for strictly maintaining the composition of brain extracellular fluid (Greengard et al., 1982).

About half the extracellular calcium in a mammal is bound to plasma proteins so that the free-calcium concentration extracellularly is about 1 mM. Intracellularly the concentration is several orders of magnitude lower and is of the order of 0.1 μM. In neurons, with a resting membrane potential of the order of -60 mV (the plasma being negative), one would expect the intracellular activity of calcium to be about two orders of magnitude greater than the extracellular activity rather than being several orders below it. This means that the steady-state distribution of calcium is far from being in thermodynamic equilibrium, and that the influx and passive accumulation of calcium by cells is strongly favored. This thermodynamically unfavorable concentration gradient is maintained by (a) a neuronal membrane which at rest is relatively impermeable to calcium; (b) a transport process(es) contained in the neuronal membrane which actively pumps calcium out of the cell; and (c) intracellular buffers by which excess internal free calcium is sequestered (Henkart, 1980a,b).

That calcium is extruded from nerve cells has been demonstrated by observing the efflux of ^{45}Ca from nervous tissue preloaded with radiolabeled Ca^{++}. The source of the efflux of radiolabeled calcium from excitable tissue and the mechanism of its extrusion are not entirely clear yet. One major component appears to involve a Ca-Ca exchange. Probably of equal importance is a Na-Ca exchange, which derives its energy from the sodium concentration gradient actively maintained by the Na-K ATP-dependent pump. Such a role for sodium is indicated by observations that replacement of external sodium by lanthanum produces as great a fall in ^{45}Ca efflux as does removal of external Ca^{++} (Chen, 1974, Blaustein, 1974). Removal of external calcium and sodium has resulted in an 85 percent reduction in the efflux of calcium. The remaining efflux could be attributed to exchange with external magnesium, since in the absence of external calcium, the addition of magnesium to the external solution causes an increase in calcium efflux (Chen, 1974; Blaustein and Hodgkin, 1969).

THERMAL CONSIDERATIONS

Calcium and Thermoregulation

A role for Ca^{++} regulation of several autonomic nervous system functions has been postulated, e.g., feeding and drinking (Myers et al., 1972). The large amount of evidence accumulated concerns thermoregulation.

Three major reasons for a detailed discussion of the role of Ca^{++} in thermoregulation are:

(a) A large amount of supportive evidence.
(b) The release of Ca^{++} found in the brain after radiofrequency (rf) irradiation may have a direct effect on the thermoregulation mechanisms.
(c) The fluxes of Ca^{++} in brain after rf irradiation may be the result of heating.

The Role of Ca^{++} in Control of Thermoregulation

Freund (1911) and Schultz (1916) reported that changes in the systemic concentration of calcium ions caused a change in body temperature. One of the very first observations of the powerful effect of calcium on the brain stem was the discovery of a precipitous fall in body temperature following an infusion of Ca^{++} ions directly into the basal forebrain (Hasama, 1930). Myers and Veale (1970) have proposed that the ratio of sodium to calcium ions within the posterior hypothalamus constitutes the mechanism whereby the set point for body temperature is established and maintained. They suggested that "a temperature of 37°C is established and sustained by an inborn constancy in the ratio of calcium to sodium ions which leads to a stabilization of the local firing rate of neurons in the caudal hypothalamus that are part of the efferent temperature pathways" (Myers et al., 1976a,b). The evidence for this came from the observation that perfusion of calcium, in excess of its normal concentration in interstitial fluid, into the posterior hypothalamus of the unrestrained cat induced a fall in the core temperature of the animal. Perfusion of excess Na^+ gave the opposite results. Alteration of Na^+ and Ca^{++} concentrations in constant ratio had no effect, thus indicating that the relative ratio, not the absolute levels, of the two cations was important. The temperature shift could be changed upwards or downwards until the point of death, or a new set point could be established around which the animal would thermoregulate (Myers and Yaksh, 1971). The effects of sodium and calcium were specific since no similar effects of potassium or magnesium were recorded. Further, the effects were anatomically specific in all animals tested; e.g., monkey, rat, and rabbit (Myers and Yaksh, 1971; Myers et al., 1976a; Veale et al., 1977).

Further experiments (Myers et al., 1976b) measuring $^{45}Ca^{++}$ release from the posterior hypothalamus showed that an increase in ambient temperature caused a relative retention of Ca^{++} and that a decrease in temperature caused a release of Ca^{++}. Similarly, raising the temperature of the anterior hypothalamus, the probable site of temperature-sensitive neurons, caused a relative decrease in Ca^{++} release, whereas a decrease in temperature caused an increased release of Ca^{++}. In a more complex set of experiments, Myers et al. (1976b) examined the effect of ambient temperature changes during perfusion of Ca^{++}. If the animal was warmed prior to and during infusion of Ca^{++}, a Ca^{++}-induced fall in temperature was prevented. The Ca^{++}-induced fall did occur, however, if the animal was cooled before infusion of Ca^{++}, suggesting that the response to an increased ambient temperature was mimicked by the infusion of Ca^{++}.

The results of Myers and other workers (Myers et al., 1977; Gisolfi et al., 1977; Wilson et al., 1978) show changes in Ca^{++} movement during exercise, which is opposite to the results when they only increased ambient temperature. Their results also show that Ca^{++} is released during exercise and that this release is unaffected by an increase, but stimulated by a decrease, in ambient temperature. The release of Ca^{++} from the posterior hypothalamus during hyperthermia brought about by exercise or fever is the opposite of that brought about by an increase in the ambient temperature, when a relative retention of Ca^{++} is measured.

In the case of a temperature perturbation brought about by external heating of the body or by a bacterial challenge, the actual measured deflection in body temperature does not occur until well after calcium efflux ensues. Unbinding or release of Ca^{++} may signal either the onset of a temperature change or the occurrence of a thermal challenge in response to which compensatory physiological reaction keeps the animal's temperature stable. Therefore, a thermal effect may indeed cause a change in calcium kinetics without eliciting measurable temperature increase (Greengard et al., 1982). It is particularly significant that the efflux of Ca^{++} ions from circumscribed tissue in the brain typically forewarns or "signals" the onset of the thermogenic stimulus. This "signal" occurs prior to any recordable change in the tissue temperature of the animal (Myers and Ross, 1981).

RADIOFREQUENCY IRRADIATION AND CALCIUM EFFLUX

Over the last decade, Adey and Bawin and their colleagues have reported extensively on their studies of changes in radioactive-calcium ion (^{45}Ca) efflux from neonate chick brain preparations and isolated samples of cat cortex under very specific frequency and power density regimes of unmodulated ELF fields and of amplitude-modulated VHF and UHF fields. Reviews of this work have been presented by Adey (1979, 1980, 1981), and some details of the experimental protocol are given by Bawin and Adey (1976, 1977).

Blackman et al. (1979) conducted experiments that extended Bawin and Adey's findings for chick brain at 147 MHz and later at other modulations and frequencies. Treated tissues were exposed in a Crawford cell to power densities between 0.5 and 2.0 mW/cm^2 and amplitude modulation of the carrier at selected frequencies between 3 and 30 Hz. They found a statistically significant increase in calcium efflux when the frequency modulation was 16 Hz and the power density was 0.750 mW/cm^2. The apparent existence of a power window, in addition to the frequency window suggested by Adey and Bawin, was reported by Blackman et al., 1980a).

In studies at 9 Hz or 16 Hz, a potential artifact apparently arose from differences in the number of samples simultaneously exposed. Statistically significant differences in normalized calcium efflux were found between exposed and sham-exposed tissues for eight of the combinations of power density, modulation frequency, and number of tubes (brains) per dummy load.

To determine whether changes in carrier frequency altered the range of power densities effective in producing statistically significant alterations in calcium efflux, Blackman et al. (1980b) conducted experiments at 50 MHz, amplitude-modulated at 16-Hz. Exposure conditions and protocol were similar to those previously used by these investigators. The results of a power density series demonstrated three effective power densities: 1.44, 1.67, and 3.64 mW/cm^2. No statistically significant effects were found at 0.72, 2.17, and 4.32 mW/cm^2. Of interest was the calculation that peaks of positive findings were associated with nearly identical rates of energy absorption, 1.4 W/kg at 147 MHz and 1.3 W/kg at 50 MHz.

Direct Electrical Stimulation

Very weak electrical stimulation of the cat cerebral cortex has been reported to elicit significant increases in the efflux of ^{45}Ca and GABA (Kaczmarek, 1977). Stimulation at 200 pulses per second (1 msec duration) increased ^{45}Ca release by 130 percent and ^{3}H-GABA release by 12 percent. These results suggest that calcium binding in the neurons of the cerebral cortex is highly sensitive to relatively small perturbations in the local electric field. The calculated electric gradient induced by the stimulation would be, at most, 2.5 μV across a typical synaptic terminal.

Using the cat with electrodes implanted in the cerebral cortex, amygdala or other subcortical structure, Bawin et al. (1973) reported that low level VHF fields, which are amplitude-modulated at specific frequencies, altered the rhythms from the brain. Since the rate of extinction was markedly enhanced when the animal was exposed to 147 MHz electrical field, Bawin et al. (1973) presented the hypothesis "...that these fields were acting as effective contingent reinforcers." In this paper, the possibility was rejected of any possible artifact influencing their results. These include the "antenna effect" occasioned by electrodes positioned in the brain; thermal effect because of the low field intensity of 1.0 mW/cm^2; and the improbability that the observed effect could be due to peripheral receptors.

Effect of Low Frequency (LF) Fields on Calcium Exchanges

The responses to direct stimulation prompted an examination of the effects of low frequency fields on the calcium exchange in isolated cerebral hemispheres of cats and neonatal chicks (Bawin and Adey, 1976, 1977; Kaczmarek, 1977). The brains were bathed in a small volume of physiological solution and loaded with labeled calcium, then transferred to a nonradioactive solution; the efflux of calcium subsequently followed. The chamber was placed between two parallel metal plates 1 m^2 in area, 50 cm apart, and was exposed to electrical fields applied at levels of 5-100 V/m and at frequencies of 1-75 Hz. Exposure to such a field generally reduced the release of the preincubated labeled calcium. This release was sensitive to both the frequency and amplitude of the applied field. Thus, maximum decreases (Bawin and Adey, 1976, 1977) occurred at 6 and 16 Hz (12-15 percent), and little or no effect was observed below 3 or above 25 Hz (Bawin and Adey, 1977). Furthermore, no significant effects were observed when the field strength was lower than 10 V/m or higher than 100 V/m. Although the actual field gradients in the tissue could not be measured directly, they were estimated to be of the order to 0.1 μV/cm (Bawin and Adey, 1976).

The Effect of Modulated Very High Frequency (VHF) Fields on Calcium Efflux

Experiments similar to those just described have been performed on the cerebral hemispheres of neonatal chicks irradiated with low frequency, sine-wave-modulated VHF fields (147 MHZ 1-2 mW/cm^2). Exposure of the cerebral hemispheres to unmodulated VHF fields had no significant effect on the calcium efflux. However, when the modulation frequency was between 6 and 32 Hz, calcium efflux increased, in contrast to the decrease in calcium efflux produced by the low frequency fields. The maximum effect was produced at 16 Hz, which led to a 20 percent increase in calcium efflux (Bawin et al., 1975).

The effects obtained with a 147-MHZ carrier frequency have been repeated using both lower frequency VHF radiation (50 MHZ) (Blackman et al., 1980b) and higher radiofrequencies (450 MHZ) (Bawin et al., 1977).

In Vivo Experiments

The sensitivity of calcium efflux to amplitude-modulated high frequency fields has been primarily revealed by experiments on freshly isolated chicken or cat cerebral tissue. A similar effect of high frequency electromagnetic radiation, however, has been demonstrated in the intact awake cerebral cortex of the cat (Bawin et al., 1977). In these experiments the cerebral cortex of a cat was exposed through a trephine hole, loaded with radioactive calcium, whose washout into isotope-free solution was then studied. Field stimulation with a high frequency electromagnetic field (450 MHz) amplitude-modulated at a low frequency (16 Hz) for 20 minutes lead to a marked (> 30 percent) increase in calcium efflux, which slowly returned to baseline over 20-40 minutes. These changes are similar to those found for both calcium efflux and GABA efflux with low-level electrical stimulation of the cat cerebral cortex with gradients of 20-50 mV/cm. They are also similar to changes associated with the electroencephalogram (Kaczmarek and Adey, 1973). Neither the resting nor field-stimulated increases in efflux are affected by altering the extracellular calcium concentration (by removing calcium or by adding an extra 2.0 mM Ca^{++}). Omission of bicarbonate from the bathing medium results in a consistent decrease in the resting efflux (by 8.5-11 percent). The response to field stimulation in this zero bicarbonate solution is abolished.

When the pH of the bathing medium is reduced from 7.4 to 6.8 (by adding hydrochloric acid), the resting calcium efflux changes little. However, the field-stimulated efflux under these conditions is almost twice that normally obtained.

Summary of Experimental Findings

The major findings in this work are that exposure of mammalian and avian central nervous tissue to weak electromagnetic fields causes a decrease in calcium efflux from the tissues when the frequency is low (6-32 Hz), but an increase when these same low frequencies are used to modulate a radiofrequency carrier (50-450 MHz). This has suggested both a frequency window (the effect not being observed at frequencies lower than 3 Hz or greater than 25 Hz) and a power window which is relatively narrow, covering about an order of magnitude (the effect being obtained only between 0.1 and 1.0 mW/cm^2). For the cat cortex, tissue gradients are of the order of 100 mV/cm and are thus comparable with the gradients of 20-50 mV/cm associated with the EEG. Weak direct electrical stimulation of the cortex leads to an increase in labeled calcium efflux.

Shelton and Merritt (1981) examined whether pulsed rf (instead of sinusoidally amplitude-modulated rf) might also elicit alterations in calcium efflux from cerebral tissue. Rat cerebral tissue preloaded with $^{45}Ca^{++}$ was exposed to pulsed rf (1-GHz carrier, 0.32 duty factor) at one of several pulse-width, power-density exposure schemes: 16-Hz PRF at 0.5, 1.0, 2.0, or 15 mW/cm^2 average power density; or 32 Hz PRF at 1.0 or 2.0 mW/cm^2 average density. Measurements of radioactivity (CPM) in the efflux medium and in the tissue sample were used to calculate an efflux value for each sample. (This measure of efflux differs from that used by other investigators.) The results of exposing or sham-exposing cerebral tissue from 167 rats showed no statistically significant differences between any of the $^{45}Ca^{++}$ efflux values when irradiated samples were compared with the accompanying sham-irradiated control samples. Shelton and Merritt pointed out that no direct comparisons can be made between their results and those of the other investigators because of the differences in radiation exposure parameters, carrier frequency and modulations.

THE CONCEPT OF "WINDOWS"

Does a frequency or power "window" exist? The dictionary definition by Webster of a "window" is an "opening in a structure or wall." Some researchers use the term to describe a specific EMR frequency or amplitude which has the potentially unique characteristic of producing a peak rise or decline in Ca^{++} ion activity in brain tissue. Actually, the highest point or value of efflux of Ca^{++} ions, along a continuum of radiation frequency or power parameters, would constitute a peak or spike in the cation's activity rather than a "window" (Myers and Ross, 1981).

Bawin and colleagues have shown a Gaussian type of curve in $^{45}Ca^{++}$ activity as related to frequency of amplitude modulation of a 145 MHz field. The higher values along this curve were 14, 15 and 18 percent elevation above control Ca^{++} efflux at 9 Hz, 11 Hz and 16 Hz frequencies, respectively. According to Myers and Ross (1981), for one value to be classified as a unitary, functionally significant "peak" a statistically significant difference between the 16 Hz and 11 Hz values (i.e., 18 and 15 percent) would have to be demonstrated. An enhanced efflux of Ca^{++} ions at one frequency parameter, but a supressed efflux at a specific ELF field intensity appears to be physiologically inconsistent, even if the phenomenon were incontrovertibly documented, in terms of the neurochemical mechanisms of calcium's metabolism, binding and transport.

Although the data collected by Blackman and colleagues are thought to replicate the "window" phenomenon, close scrutiny of the published values of $^{45}Ca^{++}$ efflux do not provide incontrovertible evidence of its existence. In reality, the "window" exists only insofar as the differences seen in the published Tables and Figures, are due to fluctuations in the control "sham-exposed" samples of the chick's brain tissue (Myers and Ross, 1981).

CRITICAL ANALYSIS OF Ca^{++} FLUX STUDIES

Source of Increased Calcium Efflux

The increased calcium efflux produced by low frequency amplitude-modulated VHF electromagnetic fields could either be on the external neuronal membrane or from some intracellular site. The trivalent lanthanum ion, which is presumed not to cross cell membranes, has been found in many studies to be extremely effective in blocking calcium fluxes across excitable membranes (Adams et al., 1980). If the source of the increased calcium efflux were intracellular, one would expect it to be blocked by lanthanum. Although lanthanum does decrease the resting efflux, it seems that modulated field stimulation still has an effect (Bawin and Adey, 1976, 1977, 1978). However, this effect is in the opposite direction of that usually seen with stimulation with a modulated 450-MHz field--in fact, a 7 percent decrease in calcium efflux.

Few studies on calcium efflux have sought to determine the source of the calcium involved in the observed efflux, the mechanism by which energy deposition alters calcium efflux, and the functional implications of the induced changes in calcium efflux. While these studies on calcium efflux are interesting and potentially important, they fail to demonstrate physiologically significant effects of electric, magnetic or electromagnetic fields on calcium function in the brain. Adey suggests that electromagnetic fields may induce conformational changes of the neuronal membrane resulting in displacement of the surface bound cations.

Regarding the change in resting calcium flux produced by "weak" electromagnetic fields, Blackman et al. (1980a) state that "both the mechanism of interaction response for this change and the consequences of the biochemical change for the intact animal have yet to be determined."

Influence of Temperature

Blackman et al. (1980b) reported that exposure of the chick's brain kept in a water bath at a low ambient temperature of 32°C depressed the efflux of $^{45}Ca^{++}$ by 9 percent. On the other hand, exposure of the tissue to an ambient temperature of 41°C augmented the $^{45}Ca^{++}$ efflux by 15 percent. These results coincide generally with the increased efflux of Ca^{++} from the hypothalamus of the live, unanesthetized mammal undergoing thermogenesis as a result of an injection of a bacterial endotoxin (Myers, 1980) or other hyperthermia-inducing substance (Myers et al., 1976b).

A more thorough examination of this thermal phenomenon in the chick's brain at much smaller increments in deviations of temperature around 37°C would also be desirable. The reason for this is that the threshold for Ca^{++} ion efflux from brain tissue following a thermal challenge exhibits the characteristics of an "all-or-none" phenomenon. Actually, the kinetics of Ca^{++} efflux change very abruptly once a given ambient temperature, tissue temperature, or chemical stimulus is imposed (Myers et al., 1976b).

Greengard et al. (1982) have also stated "cooperative phenomena, long-range interactions, nonlinear dynamic phenomena in biological membranes, long-range coherence and energy storage in biological membranes, functional analysis of antiferromagnetic crystals in external magnetic fields, macroscopic quantal amplification, and flop configuration are terms frequently used to account for the observed effects. However, until the use of these concepts can be shown to lead to further predictions that can be tested experimentally discussing them in any detail here seems unprofitable. Facts and theory should be natural enemies, each pitting their strength against the other. In this particular case, an extremely interesting set of facts and a highly speculative set of theories do not seem to impinge on each other to any reasonable extent. Perhaps experiments with drugs such as local anesthetics, or calcium channel blockers of a diverse group (since they are often nonspecific in their action) that also affect calcium turnover in various compartments of nerve, would yield interesting results. Why not look at the effects of temperature on the calcium efflux? In addition, the effects of radiation on well-described physiological or biochemical processes in which a role for calcium is well known could be studied. Rather than repeatedly studying calcium efflux under essentially the same conditions and postulating an array of untested theories, why not study the effects of radiation on the electrical stability of neuronal membranes or on the activity of calmodulin-dependent processes?"

"In general, therefore, the effect of electromagnetic radiation on calcium efflux is potentially interesting, but further progress should be along more imaginative lines where the functional significance, if any, of this observation can be studied" (Greengard et al., 1982).

CONCLUSION AND PERSPECTIVE

Over the last decade, Adey and Bawin and their colleagues have reported extensively on their studies of changes in radioactive-calcium ion ($^{45}Ca^{++}$) efflux from neonate chick brain preparations and isolated samples of cat cortex under very specific frequency and power density regimes of unmodulated sub-ELF fields and of amplitude-modulated VHF and UHF fields. Blackman conducted experiments that extended Adey's findings for chick brain at 147 MHz and later at other modulations and frequencies. Few studies on calcium efflux have sought to determine the source of the calcium involved in the observed efflux, the mechanism by which energy deposition alters calcium efflux, and the functional implications of the induced changes in calcium efflux. It should be recognized that unbinding or release of Ca^{++} may signal either the onset of a temperature change or the occurrence of a thermal challenge in response to which compensatory physiological reaction keeps the animal's temperature stable. Therefore, a thermal effect may indeed cause a change in calcium kinetics without eliciting measurable temperature increase. The concept of "windows" which has evolved from such studies may be reflective of compensatory physiological responses or rate changes in biochemical reactions. Adey suggests that electromagnetic fields may induce conformational changes of the neuronal membrane resulting in displacement of the surface bound cations. While these studies on calcium efflux are interesting, and potentially important they fail to demonstrate physiologically significant effects of electric, magnetic or electromagnetic fields on calcium function in the brain.

Myers and Ross (1981) noted three distinct classes of data have arisen in experiments on the effect of an EM radiation source on the CNS or on an organ system which influences the brain and spinal cord. In the first class, there is a defined, quantitative and reproducible effect of an EM signal contingent on its arrival into the tissue. The biological change in the tissue's characteristic(s) is due to the thermal effect produced by it rather than to the radiation source per se. Thus, in virtually every speciality reviewed in this subfield of EM-nervous system interaction, from cochlear microphonics to neuroendocrine phenomena, a focal temperature alteration is implicated in the induction of the effect. Secondly, no consistently reproducible nor significant effect is produced by the EM source on an animal's behavior or on an organ system or fragment of tissue even though the physical parameter of the source is varied widely.

The third class of data falls into the category of variability designated by Graves et al. (1979) as the "Cheshire Cat Phenomenon." Here we find that a transitory effect may be seen once during a block of repeated attempts, but the observation is not necessarily replicable even by the same laboratory group. Alternatively, the phenomenon is witnessed inconsistently from laboratory to laboratory. In this class of finding fits also the result which is due apparently to an uncontrolled environmental factor which is unwittingly or unintentionally introduced into an experimental situation and for which no plausible explanation exists.

Falling outside of these three classes are still other data for which a conclusive evaluation cannot be made. There are several reasons for this which include: (1) insufficient information is provided by the experimenter about techniques and procedures; (2) the statistical or graphical treatment of the data is incomplete, invalid or otherwise open to question; or (3) uncertainty surrounds one or more of the unknown conditions which influence a published result, thereby precluding a decisive conclusion one way or another (Michaelson, 1980).

In the case of the newly-hatched chick, not only is the blood brain barrier absent (Dewhurst and Marley, 1965) but because the CNS of the bird develops so rapidly, the brain is unstable morphologically and neurochemically.

Cerebral lateralization in the brain (Pediconi and Rodriguez-deTurco, 1984) may also be implicated. In the adult brain, Ca^{++} ions bind to anionic sites on the membrane, comprised of acidic lipids and sialic acid residues on glycoprotein and glycolipids. The membrane constituents which are functionally well documented in adult tissue are in a state of constant transformation in the newborn CNS. Thus, any change in Ca^{++} ion efflux which might occur in the immature chick's brain may be unique to the time period of the bird's development. Further, the factor of age may contribute partially to the unusual amount of reported variance in experiments on calcium (Bawin and Adey, 1976), particularly as noted in "sham-exposed" control tissue (Blackman et al., 1980a).

If a given EM source does exert an effect on the activity of a cation or other factors in brain tissue, the phenomenon should be readily demonstrable in the adult brain. Further, the effect should be evident in a higher form of mammal since phylogenetic proximity of the test animal to the human yields a more germane observation in reference to an applied question (Myers and Ross, 1981).

Although many neurophysiological events can be studied in an animal under anesthesia, one nagging question always persists: What is the effect of the anesthetic agent on the intact nervous system? All surgical anesthetics and other drugs such as alcohol cause poikilothermia in an animal which maintains its set point temperature at approximately 37°C (Myers, 1981a). Thus, at a typical laboratory temperature of 22°C, external heating must be applied to the animal to prevent heat dissipation and a consequent and potentially lethal decline in core temperature (Myers, 1981b).

Whenever excised cerebral tissue is used, the in vitro information obtained may or may not be readily applicable to the in vivo events that occur in the brain of the conscious animal. In other words, the further one's observation is removed from the intact functioning brain, the greater the degree of caution is necessary in an interpretation. In an anoxic brain lacking a blood supply and kept in a medium which itself is not oxygenated, the significance of a finding must be viewed in the context of the state of the tissue (Myers and Ross, 1981).

The activity of a neurotransmitter and probably other neurohumoral substances reaches its peak in the brain within 10 minutes or less, post-mortem, as measured by intracellular extrusion of substrates into extracellular fluid (Feldberg and Myers, 1966). Presumably this occurs because the dead tissue is devoid of the vital nutrients that are carried by the circulatory system to the constituent cells. Because it is disconnected from the rest of the animal's nervous system, the cellular material of an isolated brain is not physiologically viable (Myers and Ross, 1981).

Depending upon the inhibitory or excitatory characteristics of an individual region or population of cells, a class of ions may be retained or alternatively unbound by the membranes. Conceivably, the net effect of a specific EMR field could be to mutually self-cancel a simultaneous increase and decrease in cation flux from excitatory and inhibitory systems. Hence, no effect would be seen in that particular sample of cerebral tissue. Should a slight difference occur in the spatial orientation of the brain relative to the EMR field, a thermal effect resulting in excitation could unbind the ions from tissue: conversely, an effect on an inhibitory system could prevent the uncoupling. In either case, variation in the experimental data from sample to sample would simply be enhanced (Myers and Ross, 1981).

Within the living brain, large differences in temperature gradients exist, (Hayward and Baker, 1968), in the coronal aspect of cerebral tissue. Temperature differentials vary from the superficial cortex to the depth of the diencephalon, dorsal to the circle of Willis. Because of locally unequal blood flow, distribution and the variations in distance to the cranial surface, a gradient of as much as 0.5°C exists between the regions encompassing the corpus striatum and the anterior hypothalamus. If thermodetector neurons in the brain are warmed, the animal will respond as if it were thermally stressed even though the temperature recorded from the colon or other structure in the periphery exhibits no change whatsoever. In this connection, it should be reiterated that a major physiological effect of a thermal challenge on an animal is on the kinetics of Ca^{++} ions within its brain-stem. The onset of of a febrile response, engendered by a bacterial pyrogen, is signalled by a sudden but transient efflux of Ca^{++} ions within circumscribed tissue of the diencephalon (Myers et al., 1976a).

The absence of any electroencephalographic activity as recorded from a dead brain reflects the nonfunctional state of the tissue. Thus, it is difficult to envisage a Ca^{++} ion response to a physical stimulus of excitable cells which are biologically inactive for a period of time. Electric current or thermal energy could possibly elicit a physical change in the properties of the dead tissue just as it may do in any viscous, nonliving, protein-like material.

Myers and Ross (1981) note in the discipline of neuropharmacology, the effect of a single substance in the brain on an isolated physiological event is often examined; however, a second compound is tested to control for specificity of the first (Myers, 1974). In an experiment dealing with the flux of a given cation a control for another cation is similarly essential. Frequently, the activity of two or more cations changes in harmony with one another (Nicholson et al., 1978). For example, current passed across the surface of the cortex depletes or augments the level of extracellular K^{+} ions while not affecting the level of Ca^{++} ions (Gardner-Medwin and Nicholson, 1977). If an effect on one ion would be produced by an EMR field, the obvious question is whether the effect is limited to that ion, or instead is a nonspecific action of EMR on all monovalent, divalent and other ion groups.

The functional significance of a 5 percent to 18 percent fall or rise in the extracellular concentration of 1.2 mM calcium, would constitute an excess or depletion of 0.12 mM Ca^{++} or less to which the cells would be exposed. Because of the remarkable buffering capacity of the extracellular fluid of the brain's parenchyma, such an excess in Ca^{++} ions, at least from pharmacological evidence, is inconsequential.

Is the efflux of Ca^{++} ions from the brain biologically significant? Does a radiation field induce a change in the level of extracellular Ca^{++} ions? If so, what is the functional significance of this change? Is the effect transient and is it reproducible across different species? The answers to these questions are presently unknown. However, in view of the controversy associated with the published accounts of Ca^{++} efflux in response to EMR exposure, these points should be emphasized (Myers and Ross, 1981).

ACKNOWLEDGEMENT

This paper is based on work performed under contract No. DE-AC02-76EV03490 with the U.S. Department of Energy and PHS Grant No. ES 03239 at the University of Rochester Department of Radiation Biology and Biophysics and has been assigned report No. DOE/EV/03490-2440.

REFERENCES

Adams, D.J., Smith, S.J., and Thompson, S.H., 1980, Ionic currents in molluscan soma, Annu. Rev. Neurosci., 3:141.

Adey, W.R., 1975, Introduction: Effects of electromagnetic radiation on the nervous system, Ann. N.Y. Acad. Sci., 247:15.

Adey, W.R., 1979, Neurophysiologic effects of radiofrequency and microwave radiation, Bull. N.Y. Acad. Med., 55:1079.

Adey, W.R., 1981, Frequency and power windowing in tissue interactions with nonionizing electromagnetic fields, Proc. IEEE, 58:119.

Adey, W.R., 1981, Tissue interactions with nonionizing electromagnetic fields, Physiol. Rev. 61:435.

Bawin, S.M., and Adey, W.R., 1976, Sensitivity of calcium binding in cerebral tissue to weak environmental electric fields oscillating at low frequency, Proc. Natl. Acad. Sci. U.S.A., 73:1999.

Bawin, S.M., and Adey, W.R., 1977, Calcium binding on cerebral tissue, in: "Symposium on Biological Effects and Measurement of Radio Frequency/Microwaves," Rockville, MD.

Bawin, S.M., Adey, W.R., and Sabbot, I.M., 1978, Ionic factors in release of $^{45}Ca^{2+}$ from chicken cerebral tissue by electromagnetic fields, Proc. Natl. Acad. Sci. USA, 75:6314.

Bawin, S.M., Gavalas-Medici, R.J., and Adey, W.R., 1973, Effects of modulated very high frequency fields on specific brain rhythms in cats, Brain Res., 58:365.

Bawin, S.M., Kaczmarek, L.K., and Adey, W.R., 1975, Effects of modulated VHF fields on the central nervous system, Ann. N.Y. Acad. Sci., 247:74.

Bawin, S.M., Sheppard, A., and Adey, W.R., 1977, Possible mechanisms of weak electromagnetic field coupling in brain tissue, in: "The Physical Basis of Electromagnetic Interactions with Biological Systems," L.S. Taylor, and A.Y. Cheung, eds., Proc. Workshop held at Univ. of Maryland, College Park, MD.
Blackman, C.F., Elder, J.A., Weil, C.M., Benane, S.G., Eichinger, D.C., and House, D.E., 1979, Induction of calcium-ion efflux from brain tissue by radio-frequency radiation: Effects of modulation frequency and field strength, Radio Sci., 14:93.
Blackman, C.F., Benane, S.G., Elder, J.A., House, D.E., Lampe, J.A., and Faulk, J.M., 1980a, Induction of calcium-ion efflux from brain tissue by radio-frequency radiation: Effect of sample number and modulation frequency on the power density window, Bioelectromagnetics, 1:35.
Blackman, C.F., Benane, S.G., Joines, W.T., Hollis, M.A., and House, D.E., 1980b, Calcium-ion efflux from brain tissue: Power density versus internal field-intensity dependencies at 50-MHz RF radiation, Bioelectromagnetics, 1:277.
Blaustein, M.P., 1974, The interrelationship between sodium and calcium fluxes across cell membranes, Rev. Physiol. Biochem. Pharmacol., 70:33.
Blaustein, M.P., and Hodgkin, A.L., 1969, The effect of cyanide on the efflux of calcium from squid axon, J. Physiol. (Lond.). 200:497.
Chen, S.S., 1974, Studies of radiocalcium efflux in single barnacle muscle fibers: Effects of procaine and external divalent cations, J. Physiol., 238:313.
Dewhurst, W.G, and Marley, E., 1965, The effects of #-methyl derivative of noradrenaline, phenylethylamine and tryptamine on the central nervous system of the chicken, Br. J. Pharmac., 25:682.
Dickens, F., and Greville, G.D., 1935, The metabolism of normal and tumor tissue, Biochem. J., 29:1468.
Douglas, W.W., 1968, Stimulus-secretion coupling: the concept and clues from chromaffin and other cells. The First Gaddum Memorial Lecture, Br. J. Pharmacol., 34:451.
Feldberg W., and Myers, R.D., 1966, Appearance of 5-hydroxy-tryptamine and an unidentified pharmacologically active lipid acid in effluent from perfused cerebral ventricles, J. Physiol., 184:837.
Freund, H., 1911, Über das Kochsalzfieber, Arch. Exp. Path. Pharmak., 65:225.
Gardner-Medwin, A.R., and Nicholson, C., 1977, Measurements of extracellular potassium and calcium concentration during passage of current across the surface of the brain, Proc. Physiol. Soc., Nov:66P.
Gisolfi, C.V., Mora, F., and Myers, R.D., 1977, Diencephalic efflux of calcium ions in the monkey during exercise, thermal stress and feeding, J. Physiol. (Lond.), 273:617.

Graves, H.B., Long, P.D., and Poznaniak, D., 1979, Biological effects of 60-Hz alternating-current fields: A Cheshire cat phenomenon , in: "Biological Effects of Extremely Low Frequency Electromagnetic Fields," R.D. Phillips, M.F. Gillis, W.T. Kaune, and D.D. Mahlum, eds., U.S. Dept. of Energy, Washington.

Greengard, P., Douglas, W.W., Nairn, A.C., Nestler, E.J., and Ritchie, J.M., 1982, Effects of Electromagnetic Radiation on Calcium in the Brain, Aeromedical Rev., 2-82, USAF/SAM, TX.

Hasama, B., 1930, Pharmakologische und physiologische Studien über die Schweisszentren, Arch. exp. Path. Pharmak., 153:291.

Hayward, J.N., and Baker, M.A., 1968, Role of cerebral arterial blood in the regulation of brain temperature in the monkey, Am. J. Physiol., 215:389.

Henkart, M., 1980a, Identification and function of intracellular calcium stores in neurons, Fed. Proc., 39:2776.

Henkart, M., 1980b, Identification and function of intracellular calcium stores in axons and cell bodies of neurons, Fed. Proc., 39:2783.

Joines, W.T., and Blackman, C.F., 1980, Power density field intensity, and carrier frequency determinants of RF-energy-induced calcium-ion efflux from brain tissue, Bioelectromagnetics 1:271.

Kaczmarek, L.K., 1977, Cation binding models for the interaction of membranes with EM fields, Neurosci. Res. Program Bull., 15:54.

Kaczmarek, L.K., and Adey, W.R., 1973, The efflux of $^{45}Ca^{2+}$ and (^{3}H)γ-ammionbutyne acid from cat cerebral cortex, Brain Res., 63:331.

Lawrence, A.F., Bawin, S.M., and Adey, W.R., 1980, Reaction-diffusion models for calcium fluxes in nervous tissue, Soc. Neurosci. Abstr., 137A.8.

McIlwain, H., 1952, Phosphates of brain during in vitro metabolism, J. Biochem., 52:289.

Michaelson, S.M., 1980, Microwave biological effects: An overview, Proc. IEEE, 68:40.

Myers, R.D., 1974, "Handbook of Drugs and Chemical Stimulation of the Brain," Van Nostrand Reinhold, New York.

Myers, R.D., 1980, Hypothalamic control of thermoregulation: Neurochemical mechanisms, in: "Handbook of the Hypothalamus," P.J. Moragane, and J. Pankseep, eds., Marcel Dekker, New York.

Myers, R.D., 1981a, The role of ions in thermoregulation and fever, in: "Handbook of Experimental Pharmacology," A.S. Milton, ed., Springer Verlag, Berlin.

Myers, R.D., 1981b, Alcohol's effect on body temperature: Hypothermia, hyperthermia or poikilothermia?, Brain Res. Bull., 7:209.

Myers, R.D, Bender, S.A., Kristic, M.K., and Brophy, P.D., 1972, Feeding produced in the satiated rat by elevating the concentration of calcium in the brain, Science, 176:1124.

Myers, R.D., Gisolfi, C.V., and Mora, F., 1977, Role of brain Ca^{2+} in central control of body temperature during exercise in the monkey, J. Appl. Physiol., 43:689.
Myers, R.D., Melchior, C.L., and Gisolfi, C.V., 1976a, Feeding and body temperature in the rat: Diencephalic localization of changes produced by excess calcium ions, Brain Res. Bull., 1:33.
Myers, R.D. and Ross, D.H., 1981, Radiation and brain calcium: A review and critique. Neurosci. Biobehav. Rev. 5:503.
Myers, R.D., Simpson, C.W., Higgins, D., Nattermann, R.A., Rice, J.C., Redgrave, P., and Metcalf, G., 1976b, Hypothalamic Na^+ and Ca^{++} ions and temperature set-point: New mechanisms of action of a central or peripheral thermal challenge and intrahypothalamic 5-HT, NE, PGE_1, and pyrogen, Brain Res. Bull., 1:301.
Myers, R.D., and Veale, W.L., 1970, Body temperature: Possible ionic mechanism in the hypothalamus controlling the set point, Science, 170:95.
Myers, R.D., and Yaksh, T.L., 1971, Thermoregulation around a new 'set point' established in the monkey by altering the ratio of sodium to calcium ions within the hypothalamus, J. Physiol. (Lond.), 218:609.
Nicholson, C., Ten Bruggencate, G., Steinberg, R., and Stockle, H., 1978, Calcium and potassium changes in exracellular microenvironment of cat cerebellar cortex, J. Neurophysiol., 41:1026.
Pediconi, M.F., and Rodriguez, de Turco, E.B., 1984, Free fatty acid content and release kinetics as manifestations of cerebral lateralization in mouse liver, J. Neurochem., 43:1.
Ringer, S., 1883, A further contribution regarding the influence of the different constituents of the blood on the contraction of the heart, J. Physiol. (Lond.), 4:29.
Rubin, R.P., 1974, "Calcium and the Secretory Process," Plenum Press, New York.
Schultz, J., 1916, Zur Kenntnis der Wirkung des Magnesiums auf die Korportmeparatur, Arch. Exp. Path. Pharmak., 79:285.
Shelton, W.W., and Merrit, J.H., 1980, Efflux of $^{45}Ca^{++}$ from rat cortex tissue under microwave radiation, Bioelectromagnetics, 1:250.
Veale, W.L., Benson, M.J., and Malkinson, T., 1977, Brain calcium in the rabbit: Site of action for the alteration of body temperature, Brain Res. Bull., 2:67.
Wilson, H.C., Gisolfi, C.V., and Phillips, M.I., 1978, Influence of EGTA on an exercise-induced elevation in the colonic temperature of the rat, Brain Res. Bull., 3:97.

THE BIOLOGICAL INFLUENCES OF LOW-FREQUENCY SINUSOIDAL ELECTROMAGNETIC SIGNALS ALONE AND SUPERIMPOSED ON RF CARRIER WAVES

Carl F. Blackman

Health Effects Research Laboratory
U.S. Environmental Protection Agency
Research Triangle Park, NC 27711

Interest in the biological effects of low frequency modulation of RF radiation stems from reports of changes caused by exposure to electric and magnetic fields in the sub-ELF range (0-30 Hz). It has been reported that exposure to low-frequency electric fields changes the reaction time in humans (Konig and Ankermuller 1960, Hamer 1968, Konig 1971) and in monkeys (Gavalas et al. 1970; Gavalas-Medici and Day-Magdaleno 1976), and alters circadian activity in human beings (Wever 1973). Friedman et al. (1967) observed that magnetic fields modulated at low frequencies also change reaction time in human beings.

Two other studies that provide important background information are Kaczmarek and Adey (1973, 1974). In the first report, they described release of calcium ions and gamma-aminobutyric acid (GABA) from the cerebral cortex of cats in response to small changes in the extracellular concentration of calcium. In 1974, they demonstrated release of calcium ions and GABA from the cat cortex in response to low intensity electric currents, pulsed at 200 Hz, applied directly to the cerebral cortex. Thus, extracellular calcium and electric current have similar effects on the release of GABA and calcium ions from brain tissue.

The studies of (1) behavioral changes in animals and human beings induced by low frequency signals and (2) biochemical changes in the cat brain caused by electric currents led to a study of the influence of electric fields on EEG patterns associated with a conditioned behavioral response in cats (Bawin et al. 1973). To increase the penetration of the signals into the tissue, they chose an RF carrier wave of 147 MHz, which was amplitude modulated in the sub-ELF range (e.g., 3-14 Hz). Alterations were observed in the rate of performance, accuracy of reinforced patterns, and resistance to extinction in learned behavior of the exposed animals compared to controls, indicating that the fields were acting as reinforcers. In order to determine whether these effects were mediated via peripheral receptors or occurred as a result of changes induced directly on the CNS, experiments were designed to examine the effects of modulated RF carrier waves on brain tissue *in vitro*.

CALCIUM ION EFFLUX *IN VITRO*: A FUNDAMENTAL FINDING

The association of calcium ions with brain tissue was selected as the biochemical marker to examine the influence of modulated RF fields because calcium ion efflux had been shown to be sensitive to electric currents applied directly to brain tissue *in vitro*, and because calcium ions have a prominent role in many biochemical and biophysical processes (e.g., cellular membrane integrity and function, enzyme cofactor, putative second messenger for the conduction of extracellular signals to the nucleus of the cell, neural tissue excitation and secretion of transmitter substances at synapses). The first report of the influence of modulated RF fields on excised brain tissue was Bawin et al. (1975), who showed that a 20-min exposure of chick brain tissue *in vitro* to a 147-MHz field at 1 to 2 mW/cm^2 (SAR estimated at 0.002 W/kg) caused enhanced efflux of calcium ions, but only if the field was sinusoidally amplitude modulated at frequencies of 6, 9, 11, 16, or 20 Hz. Maximal efflux was measured at 16 Hz. Modulation frequencies of 0, 0.5, 3, 25, and 35 Hz were ineffective. This frequency-specific response, which occurred while the 147-MHz carrier field was maintained at the same power density, indicates that the field-induced efflux of calcium ions was not due to heating of the samples.

In another report, Bawin et al. (1978a) exposed chick brain tissue for 20 min to 450-MHz fields, amplitude modulated at 16 Hz, at 0.75 mW/cm^2 (SAR estimated at 0.0035 W/kg) under a variety of chemical conditions. The results demonstrated that a) the enhanced efflux of calcium ions is not highly sensitive to the external calcium concentration, b) bicarbonate appears to be important for enhanced efflux, c) lowering the pH from 7.6 to 6.8 in the presence of bicarbonate may enhance the magnitude of efflux, and d) lanthanum causes a reversal to field-induced retardation of calcium-ion efflux.

Corroboration of the frequency-specific response described by Bawin and co-workers was provided by Blackman et al. (1979) who showed that 16-Hz amplitude modulation of 147-MHz carrier waves caused enhanced efflux in chick brain tissue *in vitro*, while modulation frequencies of 3, 9, and 30 Hz did not. Although the data had large variances, an unusual intensity response was described, i.e., only 0.83 mW/cm^2 (SAR estimated at 0.0014 W/kg) produced a statistically significant efflux enhancement (intensity values are corrected based on discussion in Blackman et al. 1980a) while power densities (0.11, 0.55, 1.11 and 1.38 mW/cm^2) below and above the effective value did not cause efflux. In a later report, Blackman et al. (1980a) used a revised statistical model and experimental procedure to reduce the influence of the large sample variance. An intensity response identical to their earlier result was found. However, when the distance between samples was halved, the range of intensities that produced enhanced efflux increased to include 0.55, 0.83, 1.11 and 1.38 mW/cm^2, whereas lower and higher values of 0.11 and 1.66 mW/cm^2 were ineffective. In addition, an intensity region from 0.55 to 1.11 mW/cm^2 caused enhanced efflux when 9 Hz was used as the modulation frequency. These data, obtained with a more rigorous experimental protocol, provided additional support for the results of Bawin et al. (1975) and Blackman et al. (1979); however, the explanation for the dependence on sample spacing awaited further developments.

Joines et al. (1981) examined the dependence on sample spacing by calculation of the electrical coupling between the samples; for simplicity the samples were modeled as spheres. They found that increased electrical interaction between the more closely-packed spheres produced a broader range of internal field strengths within each sphere. Thus, if a given internal field strength were necessary to cause enhanced efflux, there is a greater chance for that internal field strength to occur in closely-coupled samples exposed to a specific range of incident intensities. Joines et al. (1981) found this result to be consistent with the experimental findings in Blackman et al. (1980a). Thus a potential artifact was shown to be a logical result of the experimental procedures.

The intensity response observed by Blackman et al. (1979) with modulated 147-MHz carrier waves was confirmed by Sheppard et al. (1979) with 450-MHz carrier waves, modulated at 16 Hz; calcium-ion efflux was enhanced at 0.1 and 1.0 mW/cm^2 but not at 0.05, 2.0 or 5.0 mW/cm^2 (the estimated SAR at 1.0 mW/cm^2 is 0.0047 W/kg). The results of these two reports show that the intensities producing calcium-ion efflux from chick brain tissue _in vitro_ are within the range of 0.1 to 1.38 mW/cm^2 for modulated 147-MHz and 450-MHz carrier waves.

The apparent carrier-frequency independence of effective intensities was tested with a 50-MHz carrier wave, amplitude modulated at 16 Hz. Enhanced efflux of calcium ions from brain tissue occurred within two intensity regions (between 1.44 and 1.67, and at 3.64 mW/cm^2; SAR's $<$0.0035 W/kg) separated by intensities of no effect including 0.72 mW/cm^2 (Blackman et al. 1980b). These effective intensity values were different from the corresponding values of 147-MHz radiation; thereby indicating a dependence on carrier frequency. In addition this result revealed the existence of more than one range of effective intensities.

The apparent discrepancy in effective power densities at the three different carrier frequencies (50, 147 and 450 MHz) has been resolved by the finding that efflux is dependent on the electric field strength within the tissue and not on incident intensity (Joines and Blackman 1980). The calculation to transform the incident intensity to internal field strength was based on an empirical model described by Joines et al. (1981). With the data available at 50 and 147 MHz, the model was used to predict intensities that would produce both alterations and no alterations in calcium-ion efflux; some predictions were tested and found to be valid (Blackman et al. 1981). These reports described two intensity ranges that appear effective for enhanced efflux at both 50 and 147 MHz, identified the internal electric field strength rather than incident intensity as the important exposure parameter, and showed the importance of frequency-dependent complex permittivity values of brain tissue in the conversion of incident intensity to internal field strength. The exposures at 50 and 147 MHz caused no generalized heating of the sample. The maximum temperature rise was calculated to be <0.0004 °C, and calculated SAR's at each carrier frequency were ≤ 0.0014 W/kg (Blackman et al. 1980b).

Subsequent to the critique by Athey (1981) that the simple spherical model used by Joines and Blackman (1980) was too idealized, these authors showed that a layered sphere model produced relationships between incident intensities at 50, 147 and 450 MHz and internal field strengths that were also consistent with the experimental results (Joines and Blackman 1981). The success of the initial, simple models to predict intensity regions of both field-induced efflux enhancement and no enhancement demonstrates the utility of the approach. More refinements in the models are necessary before the experimental situation is realistically described. For example, Blackman and Wilson (1983) using autoradiographic techniques established the distribution and penetration of radioactive calcium ions in chick brain tissue immediately prior to exposure to electromagnetic fields. They concluded that the radioactive label penetrated by diffusion and that there was no label greater than 1 mm from the surface. This result agrees with previous work indicating that the radioactive label only penetrated to the outer 1 mm of the tissue (Adey et al. 1982). Thus, refinements to the mathematical models should focus on conditions that occur at or near the tissue surface.

Two ancillary experiments were performed by Blackman et al. (1981). In one, they reported that the presence of the bicarbonate buffer component was essential in potentiating enhanced calcium-ion efflux due to electromagnetic radiation. When Tris or HEPES buffers were substituted for bicarbonate, they failed to observe changes in radiation-induced efflux. They also reported that Bawin, in a personal communication, had failed to detect radiation-induced changes when bicarbonate was replaced with imidazole or MOPS buffers. In the second experiment, they examined the metabolic state of the brain tissue and found that substantial respiratory activity lasted through the 55- to 60-minute experimental period, although freshly excised tissue exhibited slightly higher respiratory rates.

Shelton and Merritt (1981), who used different procedures from those described by Bawin et al. (1975), Blackman et al. (1979, 1980a,b), and Sheppard et al. (1979) reported no change in calcium ion efflux from rat brain. Brain tissue slices, labeled *in vitro* with radioactive calcium, was irradiated at 1-GHz, pulse-modulated with square waves at 16 or 32 Hz (0.5, 1.0, 2.0 and 15 mW/cm^2). In a second report, Merritt et al. (1982) exposed rat brain tissue labeled *in vivo* with radioactive calcium to microwave radiation, pulse modulated at 16 Hz (20-ms pulse width). The intensities for the 1-GHz carrier frequency were 1 mW/cm^2 (SAR = 0.29 W/kg) and 10 mW/cm^2 (SAR = 2.9 W/kg); and for the 2.45-GHz carrier frequency, 1 mW/cm^2 (SAR = 0.3 W/kg). In addition, animals labeled with radioactive calcium were exposed for 20 min to 2.06-GHz radiation at one of 17 different combinations of intensity and pulse repetition rate: 0, 0.5, 1.0, 5.0, 10.0 mW/cm^2 (SAR was 0.24 W/kg per mW/cm^2), and 0, 8, 16, 32 Hz (pulse width was 10 ms). Following exposure, brain tissue was analyzed for radioactivity. No statistically significant field-induced enhancement of calcium-ion efflux or change of calcium content in the brain tissue was found. The reason for these negative findings is not known; however, the use of square wave rather than sine wave modulation, the different biological preparation, and different medium composition are factors that may have influenced the outcome.

ADDITIONAL CNS STUDIES

The reports of field-induced calcium-ion efflux from chick brain tissue *in vitro* have led to other CNS studies. Synaptosomes, prepared from rat cerebri and labelled with radioactive calcium, were exposed for 10 min at 0.5 mW/cm^2 to 450-MHz fields, amplitude modulated at 0, 16, or 60 Hz (Lin-Liu and Adey 1982). Only 16 Hz affected the efflux kinetics of calcium ions. The exact SAR while low, can not be unequivocally established because the exposure chamber may have been operated in a multi-modal condition (see Weil et al. 1981). Nevertheless, this result is modulation dependent and it is unlikely that heating is involved as a causative agent.

Similar field-induced efflux enhancement has been reported in a live animal. Adey et al. (1982) exposed awake, immobilized cats to 450-MHz fields, amplitude modulated at 16 Hz, at 3.0 mW/cm^2 (SAR = 0.29 W/kg). The release of calcium ions from the cortex was observed as a function of time. Irradiation for 60 min caused episodes of enhanced efflux lasting 20 to 30 min and extending into the post-exposure period. Although focusing on a different component of the efflux kinetics than studied by Lin-Liu and Adey (1982), these results demonstrate that RF fields modulated at 16 Hz can cause changes in both a subcellular membrane system and in the live mammal. Thus, the field-induced phenomenon is not restricted to an avian species nor to *in vitro* preparations.

Recently, Dutta et al. (1984) observed field-induced enhancement of calcium ions from cells of human origin. Monolayer cultures of human neuroblastoma cells were exposed for 30 min at ten SAR's from 0.01 to 5.0 W/kg to 915-MHz fields, with or without sinusoidal amplitude modulation (80%) at frequencies between 3 and 30 Hz. Significant increases in the efflux of calcium ions occurred at two SAR's (0.05 and 1.0 W/kg). The increased efflux at 0.05 W/kg was dependent on the presence of 16-Hz modulation but not at the higher value. Exposure at modulation frequencies between 3 and 30 Hz (SAR = 0.05 W/kg) revealed a peak in the response at 16 Hz. Although the effective SAR (0.05 W/kg) for 16 Hz modulation is over 38 times larger than the SAR's for enhanced efflux of calcium ions from chick brain tissue *in vitro*, the low frequency response pattern was similar to that reported by Bawin et al. (1975) and Blackman et al. (1979). The relation of enhanced efflux with unmodulated fields at 1.0 W/kg with the effects of modulated fields is not known at this time; however, it is not due to a temperature increase in the sample because enhancement was not found at SAR's of 2.0 and 5.0 W/kg.

The effect of modulated RF fields on the EEG was investigated by Takashima et al. (1979). Rabbits were exposed 2 h daily for six weeks to 1.2 MHz, amplitude modulated at 15 Hz, or 5 MHz amplitude modulated at 14 Hz. Following exposure, the EEG was recorded with scalp electrodes and, when compared to the pre-treatment EEG pattern, was found to be altered with enhanced low-frequency components and decreased high-frequency components. The EEG pattern returned to the pre-treatment pattern after several weeks post-exposure. Although the electric field intensity was given as 500 V/m, with an error factor as large as two, the important aspect of the results was that unmodulated fields of similar intensity had no effect on the EEG pattern. The authors have recently stated that this field-induced change in the EEG pattern proved to be elusive and further study was discontinued because it was beyond the resources of the group. Nevertheless, the absence of metallic electrodes in the animal during exposure avoids the major criticism of earlier studies that reported field-induced changes in EEG patterns (Gavalas et al. 1970, Bawin et al. 1973). Thus, this report stands as an important adjunct to the previous work.

Sagan and Medici (1979) studied the influence of 450-MHz fields, sinusoidally amplitude modulated at either 3 or 16 Hz, on locomotor activity in young chickens. The experiments were performed in a plastic, modified Skinner box with light beams to monitor activity; the complete apparatus was placed in an anechoic chamber and exposed in the far field. The authors found no statistically significant change in performance during or immediately after a 23-min exposure at 1 or 5 mW/cm^2 (SAR estimated at 0.2 and 1.0 W/kg). They concluded that the lack of a field-induced response could be due to the use of modulation frequencies not present in the chicken's EEG during performance on the particular (fixed-time schedule) task. Two alternative possibilities exist: (1) based on the multiple-intensity ranges observed for field-induced calcium-ion efflux, the two intensities used in this study may have been outside the effective ranges, and (2) the appropriate combination of frequency and steady-state magnetic field intensity may not have been used (see below for a more detailed discussion).

In summary, four groups (Adey et al., Blackman et al., Dutta et al., and Takashima et al.) have shown that RF fields, sinusoidally modulated at sub-ELF frequencies, especially 16 Hz, cause changes in different CNS preparations including the live animal. Many of these studies have been analyzed in reviews (Adey 1981, Blackman et al. 1981, Greengard et al. 1982, Myers and Ross 1981, Postow and Swicord 1985). It is generally agreed that both the mechanism of interaction and the physiological consequences of these changes are yet to be established.

NON-CNS STUDIES

The effect of exposure of pancreatic tissue and T-lymphocytes to RF fields, sinusoidally amplitude modulated at low frequencies, have been examined. An increase of calcium ion efflux from rat pancreatic tissue exposed *in vivo* at 2 mW/cm^2 for 1 to 2.5 h at 147 MHz, modulated at 16 Hz (estimated SAR <0.075 W/kg), has been reported by Albert et al. (1980). However, the efflux was not accompanied by a change in protein secretion which is normally associated with calcium mobilization in the pancreas. The authors attributed the lack of protein secretion to a limitation imposed by the exposure conditions, i.e., a relatively small volume of medium was available to the tissue for normal metabolic activity.

In another *in vitro* assay, the cytotoxic activity of mouse T-lymphocytes was suppressed by a 2-h exposure (1.5 mW/cm^2) to 450-MHz fields, modulated at frequencies between 16 and 100 Hz (Lyle et al. 1983). Peak suppression occurred at 60 Hz modulation, with smaller effects at 16, 40, 80 and 100 Hz. The exposed cells recovered full cytotoxic activity 12.5 h after the termination of exposure. This result demonstrated an inhibitory but reversible effect on a cell-mediated immune response by modulation frequencies.

SINUSOIDAL ELF AND SUB-ELF SIGNALS

Most of the studies reviewed above demonstrate an absolute requirement for low frequency sinusoidal modulation of the RF carrier wave in order for the signal to be effective biologically. There are also reports that describe biological effects of exposure to low frequency electromagnetic signals in the absence of an RF carrier wave. Bawin and Adey (1976, 1977) exposed chick and cat cerebral tissue for 20 min to 1, 6, 16, 32 or 75 Hz at electric field gradients of 5, 10, 56 and 100 Vp-p/m in air. Only two frequencies, 6 and 16 Hz, caused a reduction in calcium ion efflux at 10 and 56 V/m for the chick tissue, and 56 V/m for the cat tissue. Because all other combinations produced no field-induced responses, the authors described "amplitude and frequency windows" for calcium-ion efflux. Electric field gradients within the tissue were estimated to be 10^{-5} V/m. The field-induced reduction in efflux is in contrast to the enhancement caused by modulated RF carrier waves. Nevertheless, the frequency dependence observed in the two studies was similar which suggests an interaction with a common substrate as the site of interaction.

Blackman et al. (1982) used chick brain to study the influence of 16-Hz signals at 15 intensities between 1 and 70 Vp-p/m on the efflux of calcium ions. Two intensity regions that included 5, 6, and 7.5 V/m and 35, 40, 45, and 50 V/m caused enhanced efflux. No field-induced effects were seen below (1, 2, and 3.5 V/m), between (10, 20 and 30 V/m), or above (60 and 70 V/m), the two effective intensity regions. Moreover, 1 and 30-Hz signals at 40 V/m caused no change in efflux. This finding is consistent with the reports of multiple-intensity regions of enhanced efflux caused by modulated RF radiation (Blackman et al. 1980b, 1981). In addition to the intensity response, the frequency dependence corroborated reports by Bawin and Adey (1976) for low frequency signals, and by Bawin et al. (1975) and Blackman et al. (1979) for modulated RF fields.

In these two low-frequency studies, the cause of the slight difference in effective intensities is unknown. The major disagreement in the results of Bawin and Adey (1976) and Blackman et al. (1982) is the direction of the change in efflux; the latter authors state that the "cause may be found in the slightly different preparations and procedures used in the two laboratories."

Blackman et al. (1983, 1985) have extended their low frequency studies to cover the range from 1 to 120 Hz at an intensity of 42.5 Vp-p/m, and over a series of intensities at single frequencies of 42, 45, 50, and 60 Hz. They found intensity ranges at 45, 50 and 60 Hz that caused enhanced calcium-ion efflux from chick brain tissues, while the four intensities tested at 42 Hz were ineffective. In the range from 1 to 120 Hz, two frequency ranges elicited enhanced efflux--one centered on 15 Hz and the other extending from 45 to 105

Hz. The results in this report emphasize the importance of the interrelation between frequency and intensity in establishing exposure conditions that can elicit the biological response. It should serve as a guide for designing future experiments to investigate the biological influence of low frequency electromagnetic signals.

Recently Blackman and co-workers have shown that the intensity of the steady-state (DC) magnetic field of the earth is directly related to the frequencies of the oscillating (AC) EM signal that cause enhanced efflux of calcium ions from the chick brain tissue. The effective frequency was directly proportional to the intensity of the DC magnetic field times an index, 2n + 1, where n = 0,1. A 40 V/m EM signal at 15 Hz, was effective in causing enhanced efflux of calcium ions when the intensity of the DC magnetic field was 38 μT, but not when it was reduced to 19 μT; similarly a 30 Hz signal that was ineffective at 38 μT was effective at ±25.3 μT and at ±76 μT. In all the previous work published by Blackman and co-workers the intensity of the DC magnetic field was 38 μT. Blackman et al. (1984) have shown that the AC magnetic component must be present for a 16-Hz, 40-V/m EM signal to cause enhanced efflux. Because the orientation of the DC magnetic field was always perpendicular to the plane formed by the electric and magnetic AC vectors, there is insufficient information to distinguish between a cyclotron resonance-like mechanism and a magnetic resonance-like mechanism. In the former case, the electric component of the AC signal must be perpendicular to the DC field for maximum coupling; in the latter case, it is the magnetic component that must be perpendicular. The experimental situation at the transduction sites may be too complicated for simple manipulations of the field orientations to resolve this ambiguity. Nevertheless, while these results complicate any theoretical analysis of the calcium-ion efflux phenomenon, they may provide an essential clue not only to the underlying mechanism of action but also to explain why otherwise identical experiments with low frequency electromagnetic signals can produce different results.

Several research groups have reported biological changes induced by low frequency, sinusoidally oscillating magnetic fields. The myxomycete Physarum polycephalum has a longer mitotic cycle and reduced respiration rate after chronic exposure to 2.0 gauss magnetic fields at 75 Hz (Goodman et al. 1979). Human fibroblasts in culture exposed to sinusoidally varying magnetic fields for a wide range of frequencies (15 Hz to 4 kHz) and amplitudes (0.25 to 5.6 gauss) exhibit enhanced DNA synthesis (Liboff et al. 1984). Fruit flies (Drosophila melanogaster) preferred not to deposit eggs in a 10-gauss, sinusoidally varying 50-Hz magnetic field; similar exposure during development of the egg produced less viable eggs and pupae in the exposed samples than in controls (Ramirez et al. 1983). These results suggest that low frequency, sinusoidally varying fields may alter fundamental biological processes.

Low frequency, pulsed magnetic fields have also been reported to produce alterations in diverse biological systems. These systems include the developing chick embryo (Delgado et al. 1982, Ubeda et al. 1983), Drosophila egg laying and mortality (Ramirez et al. 1983), the de-differentiating amphibian red blood cell (Chiabrera et al. 1979), transcription in the Dipteran chromosome (Goodman et al. 1983), nerve cells in culture (Dixey and Rein 1982), and mouse bone cells in culture (Luben et al. 1982). Many of these studies used an intricate pulsed waveform, which has been used in therapeutic devices for bone nonunions. All the studies used pulse repetition rates below 500 Hz, with most below 100 Hz. It is not known whether these pulsed fields cause alterations via the same mechanism(s) as sinusoidal fields. Recently, Liboff et al. (1984) questioned the need for the particular wave shapes; it appears that the essential element is the low frequency field.

ESSENTIAL STEPS IN A MECHANISM OF ACTION

The field-induced calcium-ion efflux phenomenon can be conceptualized as occurring in a multistep process. The first step is the initial transduction of electromagnetic energy into a chemical change at the primary reaction site. This chemical change would be expected to be small because the electromagnetic signal contains very little energy when compared to thermal energy at 37 °C. Identification of this primary transduction site and the nature of the chemical change are presently unanswered issues. The second step is the amplification of the small chemical change by utilizing energy already stored in the biological system. This amplification process could occur through a cooperative transition, such as a phase change, in which the biological system is energetically poised and needs only a minute change to initiate the transition. The third step is the secondary reactions that ensue as a result of the transition, that cause changes in observable properties, such as calcium-ion association with the tissue surface. These processes may occur within the same molecular species and be experimentally inseparable steps or, in the other extreme, they may occur between different molecular species that are chemically coupled and distinguishable under appropriate conditions. By viewing phenomena induced by ELF signals in this way, it may be possible to resolve apparent similarities and devise appropriate tests to determine whether those similarities represent the products of the same underlying mechanism.

SUMMARY

Many reports of effects of RF fields, amplitude modulated at very low frequencies, have not been independently corroborated. The major exception is calcium-ion efflux from chick brain tissue in vitro at intensity levels far below those that cause heating. This exception, combined with the results of studies of brain biochemistry and EEGs in animals and with synaptosomes and human neuroblastoma cells in culture, provides evidence that CNS tissue from several species, including human beings, are affected by low-intensity RF fields sinusoidally amplitude modulated at specific low frequencies or by those low frequencies directly. The responses over several distinct intensity ranges and the recently demonstrated influence of the earth's steady-state magnetic field in the calcium-ion efflux experiments further emphasize the unusual nature of these findings in chick brain tissue. The physiological significance of these field-induced effects is not established.

In addition to the CNS-related changes, amplitude-modulated RF fields have been reported to alter an immune response and a pancreatic tissue function. These reports with diverse biological systems are without apparent connection to each other except for the physical agent causing the change. The critical parameter common to all these experiments is the specificity of the low frequency electromagnetic signal. However, in most cases dose response curves and action spectra, viz. magnitude of the response as a function of frequency, have not been obtained. More detailed parametric manipulation of these factors, with appropriate allowances for the steady-state magnetic field vector, would permit comparisons and correlations between widely differing biological endpoints to assist in focusing on potential common underlying mechanism(s) of action. Unfortunately, this viewpoint is currently not appreciated by researchers who narrowly focus on the biological ramifications in their own specialized systems, in the hunt for the underlying mechanisms and physiological significance. By adopting this approach, they may be missing an opportunity to coalesce the individual efforts and to identify the critical common parameters. Perhaps the discovery of the influence of the steady-state magnetic field will provide a unifying guide for future research.

No report has yet described a mechanism of action in sufficient detail to identify the conditions necessary and sufficient to explain unequivocally calcium ion efflux in the brain or the other biological changes caused by low frequency signals or by modulated RF fields. The response to specific frequencies and intensities is unusual and at present unexplained. This response to amplitude modulated RF radiation or to sub-ELF signals alone may be a true field effect at a very low SAR and at biologically relevant frequencies, i.e., in the range of frequencies normally present in the EEG. The frequency-specific nature of the responses and the involvement of the earth's steady-state magnetic field provide evidence against heat as the underlying cause. The unusual, multiple-intensity-range response challenges standard dose-response analyses, and by its very nature, may prohibit the invocation of threshold levels.

ACKNOWLEDGMENTS

The author wishes to acknowledge and thank S. G. Benane, J. A. Elder, D. E. House, W. T. Joines, and J. R. Rabinowitz for their research collaboration in various aspects of this project; J. A. Elder for his contributions in editing this report; and T. D. Wall and N. C. Sawaya for typing.

REFERENCES

Adey, W. R., 1981, Ionic nonequilibrium phenomena in tissue interactions with electromagnetic fields, in: Biological Effects of Nonionizing Radiation, K. H. Illinger, ed., ACS Symposium Series, 157:271-297.

Adey, W. R., S. M. Bawin, and A. F. Lawrence, 1982, Effects of weak amplitude-modulated microwave fields on calcium efflux from awake cat cerebral cortex, Bioelectromagnetics, 3:295-307.

Albert, E., C. Blackman, and F. Slaby, 1980, Calcium dependent secretory protein release and calcium efflux during RF irradiation of rat pancreatic tissue slices, in: Ondes Electromagnetiques et Biologie, A. J. Berteaud and B. Servantie, eds, Paris, France, pp. 325-329.

Athey, T. W., 1981, Comparison of RF-induced calcium efflux from chick brain tissue at different frequencies: do the scaled power density windows align?, Bioelectromagnetics, 2:407-409.

Bawin, S. M., R. J. Gavalas-Medici, and W. R. Adey, 1973, Effects of modulated very high frequency fields on specific brain rhythms in cats, Brain Res., 58:365-384.

Bawin, S. M., L. K. Kaczmarek, and W. R. Adey, 1975, Effects of modulated VHF fields on the central nervous system, Ann. N. Y. Acad. Sci., 247:74-81.

Bawin, S. M., and W. R. Adey, 1976, Sensitivity of calcium binding in cerebral tissue to weak environmental electric fields oscillating at low frequency, Proc. Natl. Acad. Sci. USA, 73:1999-2003.

Bawin, S. M., and W. R. Adey, 1977, Calcium binding in cerebral tissues, in: Symposium on Biological Effects and Measurement of Radio Frequency/Microwaves, D. G. Hazzard, ed., HEW Publication (FDA) 77-8026, Rockville, Maryland, pp. 305-313.

Bawin, S. M., W. R. Adey, and I. M. Sabbot, 1978, Ionic factors in release of $^{45}Ca^{2+}$ from chicken cerebral tissue by electromagnetic fields, Proc. Natl. Acad. Sci., USA, 75:6314-6318.

Blackman, C. F., J. A. Elder, C. M. Weil, S. G. Benane, D. C. Eichinger, and D. E. House, 1979, Induction of calcium-ion efflux from brain tissue by radio-frequency radiation: Effects of modulation frequency and field strength, Radio Sci., 14(6S): 93-98.

Blackman, C. F., S. G. Benane, J. A. Elder, D. E. House, J. A. Lampe, and J. M. Faulk, 1980a, Induction of calcium-ion efflux

from brain tissue by radio-frequency radiation: Effect of sample number and modulation frequency on the power-density window, Bioelectromagnetics, 1:35-43.
Blackman, C. F., S. G. Benane, W. T. Joines, M. A. Hollis, and D. E. House, 1980b, Calcium-ion efflux from brain tissue: Power-density versus internal field-intensity dependencies at 50 MHz RF radiation, Bioelectromagnetics, 1:277-283.
Blackman, C. F., W. T. Joines, and J. A. Elder, 1981, Calcium ion efflux induction in brain tissue by radiofrequency radiation. In: Biological Effects of Nonionizing Radiation, K. H. Illinger, ed., ACS Symposium Series, 157:299-314.
Blackman, C. F., S. G. Benane, L. S. Kinney, W. T. Joines, and D. E. House, 1982, Effects of ELF fields on calcium-ion efflux from brain tissue in vitro, Radiat. Res., 92:510-520.
Blackman, C. F., B. S. Wilson, 1983, Distribution of label in studies on the effects of nonionizing radiation on the association of calcium ions with brain tissue, Abstract #GJ-24, 5th Annual Scientific Session, The Bioelectromagnetics Society, 12-17 July, Boulder, Colorado (available from The Bioelectromagnetics Society, One Bank Street, Suite 307, Gaithersburg, Maryland 20878), p. 73.
Blackman, C. F., S. G. Benane, W. T. Joines, D. E. House, 1983, Effects of ELF fields between 1 and 120 Hz on the efflux of calcium ions from brain tissue, in vitro, Abstract #E-8, 5th Annual Scientific Session, The Bioelectromagnetics Society, 12-17 July, Boulder, Colorado (available from The Bioelectromagnetics Society, One Bank Street, Suite 307, Gaithersburg, Maryland 20878), p. 28.
Blackman, C. F., S. G. Benane, D. E. House, J. R. Rabinowitz, W. T. Joines, 1984, A role for the magnetic component in the field-induced efflux of calcium ions from brain tissue, Abstract # SA-4, 6th Annual Scientific Session, The Bioelectromagnetics Society, 10-14 June, Atlanta, Georgia (available from The Bioelectromagnetics Society, One Bank Street, Suite 307, Gaithersburg, Maryland 20878), p. 2.
Blackman, C. F., S. G. Benane, D. E. House, W. T. Joines, 1985, Effects of ELF (1-120 Hz) and modulated (50 Hz) RF Fields on the efflux of calcium ions from brain tissue, in vitro, Bioelectromagnetics 6(1) in press.
Chiabrera, A., M. Hinsenkamp, A. A. Pilla, J. Ryaby, D. Ponta, A. Belmont, F. Beltrame, M. Grattarola, and C. Nicolini, 1979, Cytofluorometry of electromagnetically controlled cell dedifferentiation, J. Histochem. Cytochem., 27:375-381.
Delgado, J. M. R., J. Leal, J. L. Monteagudo, and M. Garcia-Gracia, 1982, Embryological changes induced by weak, extremely low frequency electromagnetic fields, J. Anat., 134:533-551.
Dixey, R. and G. Rein, 1982, ^{3}H-noradrenaline release potentiated in a clonal nerve cell line by low-intensity pulsed magnetic fields, Nature, 296:253-256.
Dutta, S. K., A. Subramoniam, B. Ghosh, and R. Parshad, 1984, Micro-

wave radiation-induced calcium ion efflux from human neuroblastoma cells in culture, Bioelectromagnetics, 5:71-78.

Friedman, H., R. O. Becker, and C. H. Bachman, 1967, Effect of magnetic fields on reaction time performance, Nature, 213:949-950.

Gavalas, R. J., D. O. Walter, J. Hamer, and W. R. Adey, 1970, Effect of low-level, low-frequency electric fields on EEG and behavior in Macaca Nemestrina, Brain Res., 18:491-501.

Gavalas-Medici, R., and S. R. Day-Magdaleno, 1976, Extremely low frequency weak electric fields affect schedule-controlled behavior of monkeys, Nature, 261:256-258.

Goodman, E. M., B. Greenebaum, and M. T. Marron, 1979, Bioeffects of extremely low-frequency electromagnetic fields, Radiat. Res., 78:485-501.

Goodman, R., C. A. L. Bassett, and A. S. Henderson, 1983, Pulsing electromagnetic fields induce cellular transcription. Science, 220:1283-1285.

Greengard, P., W. W. Douglas, A. C. Nairn, E. J. Nestler, and J. M. Ritchie, 1982, Effects of electromagnetic radiation on calcium in the Brain, USAF School of Aerospace Medicine, Report Number SAM-TR-82-15, 113 pp.

Hamer, J., 1968, Effects of low level, low frequency electric fields on human reaction time. Commun. Behav. Biol., 2(5) Part A:217-222.

Joines, W. T., and C. F. Blackman, 1980, Power density, field intensity, and carrier frequency determinants of RF-energy-induced calcium ion efflux from brain tissue, Bioelectromagnetics 1:271-275.

Joines, W. T. and C. F. Blackman, 1981, Equalizing the electric field intensity within chick brain immersed in buffer solution at different carrier frequencies, Bioelectromagnetics, 2:271-275.

Joines, W. T., C. F. Blackman, and M. A. Hollis, 1981, Broadening of the RF power-density window for calcium-ion efflux from brain tissue, IEEE Trans. Biomed. Eng., BME-28:568-573.

Kaczmarek, L. K., and W. R. Adey, 1973, The efflux of $^{45}Ca^{2+}$ and $[^{3}H]\gamma$-aminobutyric acid from cat cerebral cortex, Brain Res., 63:331-342.

Kaczmarek, L. K. and W. R. Adey, 1974, Weak electric gradients change ionic and transmitter fluxes in cortex, Brain Res., 66:537-540.

Konig, H. and F. Ankermuller, 1960, Uber den einfluss besonders niederfrequenter elektrischer vorgange in der atmosphare auf den menschen. Naturwissenschaften, 21:486-490.

Konig, H, 1971, Biological effects of extremely low frequency electrical phenomena in the atmosphere, J. Interdiscipl. Cycle Res. 2:317-323.

Liboff, A. R., T. Williams, Jr., D. M. Strong, and R. Wistar, Jr., 1984, Time-varying magnetic fields: effect on DNA synthesis, Science, 223:818-820.

Lin-Liu, S. and W. R. Adey, 1982, Low frequency amplitude modulated microwave fields change calcium efflux rates from synaptosomes, Bioelectromagnetics, 3:309-322.

Luben, R. A., C. D. Cain, M. C.-Y. Chen, D. M. Rosen, and W. R. Adey, 1982, Effects of electromagnetic stimuli on bone and bone cells in vitro: Inhibition of responses to parathyroid hormone by low-energy low-frequency fields, Proc. Natl. Acad. Sci. USA, 79:4180-4184.

Lyle, D. B., P. Schechter, W. R. Adey, and R. L. Lundak, 1983, Suppression of T-lymphocyte cytotoxicity following exposure to sinusoidally amplitude modulated fields, Bioelectromagnetics, 4:281-292.

Merritt, J. H., W. W. Shelton, and A. F. Chamness, 1982, Attempts to alter $^{45}Ca^{2+}$ binding to brain tissue with pulse-modulated microwave energy, Bioelectromagnetics, 3:475-478.

Myers, R. D. and D. H. Ross, 1981, Radiation and brain calcium: A review and critique, Neurosc. and Biobehav. Rev., 5:503-543.

Postow, E., and M. L. Swicord, 1985, Modulated fields and "window" effects, in: CRC Handbook of Biological Effects of Electromagnetic fields, edt. C. Polk and E. Postow, CRC Press, Boca Raton, FL.

Ramirez, E., J. L. Monteagudo, M. Garcia-Gracia, and J. M. R. Delgado, 1983, Oviposition and development of drosophila modified by magnetic fields, Bioelectromagnetics, 4:315-326.

Sagan, P. M., and R. G. Medici, 1979, Behavior of chicks exposed to low-power 450-MHz fields sinusoidally modulated at EEG frequencies, Radio Sci., 14(6S):239-245.

Shelton, W. W. and J. H. Merritt, 1981, In vitro study of microwave effects on calcium efflux in rat brain tissue, Bioelectromagnetics, 2:161-167.

Sheppard, A. R., S. M. Bawin, and W. R. Adey, 1979, Models of long-range order in cerebral macromolecules: effect of sub-ELF and of modulated VHF and UHF fields, Radio Sci., 14(6S):141-145.

Takashima, S., B. Onaral, and H. P. Schwan, 1979, Effects of modulated RF energy on the EEG of mammalian brains, Radiat. Environ. Biophys., 16:15-27.

Ubeda, A., J. Leal, M. A. Trillo, M. A. Jimenez, and J. M. R. Delgado, 1983, Pulse shape of magnetic fields influences chick embryogenesis, J. Anat., 137:513-536.

Weil, C. M., W. T. Joines, and J. B. Kinn, 1981, Frequency range of large-scale TEM mode rectangular strip lines, Microwave J., 24(11):93-100.

Wever, R., 1973, Human circadian rhythms under the influence of weak electric fields and the different aspects of these studies. Int. J. Biometeor., 17:227-232.

BIOPHYSICAL STUDIES OF PULSED MAGNETIC FIELD INTERACTION WITH BIOLOGICAL SYSTEMS: PART 1 - BIOPHYSICAL INTERACTIONS

Sylvia Fitton Jackson

Strangeways Research Laboratory
Cambridge, UK

The background to the research is based on the premise that pulsed magnetic fields (PMF) induce electromagnetic fields which may affect recalcitrant fractures of long bone, particularly "pseudarthrosis" and non-unions[1-4]. Use of such a method originated from the observation that stress generated potentials are present in bone at all times[5]. For example, an applied load of 500 grams will produce a measurable field of about 20V/m; release from the load, after say 10 seconds, will result in an opposite polarity signal of similar strength. The piezoelectrical properties of the bone may transduce such mechanical stress into an electrical signal which may then initiate a metabolic response essential to the maintenance of the bony tissue. This normal physiological flow of events will be interrupted when bone is fractured; hence non-invasive PMF may induce small fields that "replace" those missing because of the lesion, and thus act as an aid to the repair process.

In the laboratory, PMF may be regarded as a tool for studying factors involved in repair and regeneration, and the work to be described here bears special reference to connective and skeletal tissues, that are, by their very nature highly complex entities that undergo constant cycles of formation and resorption. The various processes involved in these opposing activities are numerous (Table 1) and must be maintained in physiological balance.

Whether wide-band ELF signals affect a specific process or whether their action may be more generalized is open to question, and whether such activities occur simultaneously or in sequence depending on requirements. The repair process evoked by trauma probably proceed at rates that are limited by the inherent control systems in the cell/tissue. Hence it is necessary for PMF to work within these

Table 1. Factors Concerned in Skeletal Formation.

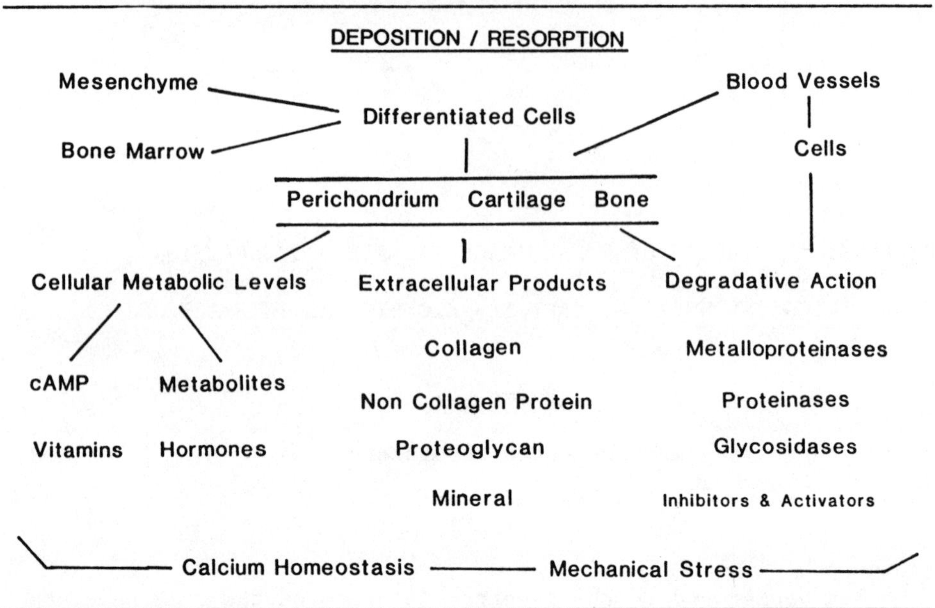

limits; a sudden rise in a specific action would lead to a "cut-off" whereas a drop below a certain level would "cut in" an appropriate response, until the normal rate was reestablished[6]. If this is so, measurable responses to PMF are likely to be small.

EXPERIMENTAL

Research of the group at Strangeways has used cell and tissue culture systems so that not only biological requirements can be precisely controlled, but a reproducible physical environment with the use of PMF can be imposed[7]. All cultures have been grown under controlled atmosphere within perspex jars held between 'Helmholtz-aiding' coils, so that the experimental field parameters do not vary by more than 4%. Culture temperature levels have been maintained either by use of a hot room or matched incubators similarly orientated with respect to the earth's magnetic field. The PMF treatment regimes usually involve 12 hr cycles of equal ON/OFF periods, care being taken to employ similarly staged cultures from one experiment to the next.

The studies to be described have, essentially, involved the use of two patient type signals+, the single pulse (SPP) and the pulse train (PTP), the former repeated at 72 Hz and the latter at 15Hz (Figure 1 and 2.) For clarity, a more detailed account of the

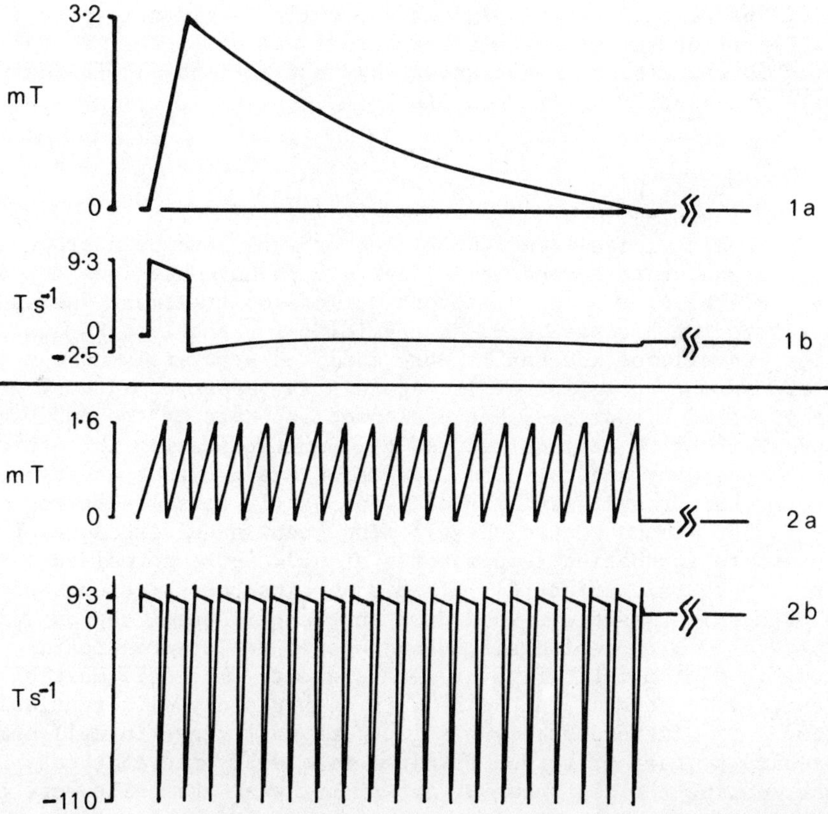

Fig. 1 and 2. Diagrams of the SPP (Figure 1) and the PTP (Figure 2) waveforms showing the initial 5ms phase of each pulse of the respective signals, down to illustrate the marked differences between the waveforms. The peak magnetic flux density B of SPP (Figure 1a) while double that of PTP (Figure 2a) falls off very slowly as compared with the rapid peak to base fluctuation (x 21) present in the PTP waveform. Hence the rate of change of flux density dB/dt of SPP (Figure 1b) contains a very slow time derivative against the rapid and much larger fluctuation (peak x 11 greater) of PTP. These differences and the consequential changes they will impose on e.g. the frequency content and the average power of the signals may profoundly affect different biological responses in different ways. See Part II of this contribution for a more detailed account of the waveforms.*

*Biosteogen system, kindly loaned by ElectroBiology INC of New Jersey, USA, and Reading, UK.

physical parameters of the signals and coils is given in part II. It is important to stress that the actual values of the fields induced in the tissue is not known, but estimations put the current density at about 0.6uA/cm^2 per T/s.

CELL MULTIPLICATION

Embrhonic tibiae from 14d chicken embryos have been grown for 8 days in organ culture, and individual groups have been subjected to SPP and PIP waveforms for different periods of treatment during the last 7 days, with exposure to ^{3}H thymidine for the last 24 hrs. Terminal cartilages and shafts were analyzed separately as cpm/ugDNA. The rate of incorporation of isotope was not affected in most groups as compared with controls, but prolonged exposure of the rudiments for cyclic 9 hr ON periods reduced cell multiplication in both cartilage and the bony shafts. The depression was greatest under the SPP waveform; this result may have been due partly to self-heating of the coils (0.75C around control level) since continuous exposure of rudiments to incubation temperatures of 0.7C above normal have been shown to have this effect[8]. In histological preparations, however, much ostegenic tissue was laid down in the periosteal region during culture, and in the epiphysis, chondrogenic cells were clearly active, in that mitotic figures were present. The cell multiplication rate is normally about 3-5% in such rapidly differentiating systems. Dr Clifford Murray has found no difference in cell numbers in primary culture of tendon obtained from embryonic chickens, neither during the log phase of growth nor when the cells were confluent; PMF treatment was with the most energetic pulse train signal (PT2) for 6-7 days[9]. Dr Richard Farndale has also failed to find any differences in primary and passaged cells derived from bone marrow of young rabbits, grown for considerable periods under the influence of PTP waveform[10].

COLLAGEN PRODUCTION

A significant dose-dependent accumulation of collagen, measured as protein-bound hydroxyproline was present in both cartilage and shafts of tibiae grown in organ culture for 8 days. Under the PTP signal, osteogenic levels rose by 45%, and those in cartilage by about 30% when the treatment involved 9 hrs ON and 3 hrs OFF cycles. Shorter repetitive doeses of 6 hrs ON/OFF failed to increase the amounts of collagen in cartilage but elevated the content of the shafts by about 25%. The SPP waveform was less efficient in raising the collagen moiety, but prolonged exposure elevated the shaft content by close to 35%, while the chondrogenic response was smaller. The fact that PMF treatment of cultures can enhance the accumulation of collagen in a rapidly differentiating system suggests that similar activity might occur in tissues undergoing repair.

Detailed studies of collagen metabolism have been made in both the tendon and bone marrow fibroblast systems. By measuring the amount of incorporation of ^3H proline and subsequent conversion of some of the radiolabel to ^3H hydroxyproline in macromolecular material the proportion of newly synthesized collagen actually released by the cells and so available for matrix formation has been evaluated over 24 hr periods after various days of exposure of the cells to PMF. In primary cultures of tendon cells the proportion of macromolecular ^3H hydroxyproline doubled from 12% to about 25%, and the distribution of label clearly showed that the net increase in collagen arose because of diminished intracellular breakdown of newly formed collagen. These results only occurred with 6-7 days of treatment when the cells were confluent and essentially non-proliferative. Use of bone marrow cells also showed that collagen production was elevated in postconfluent cultures treated with PTP, but about 65% of the ^3H hydroxyproline was in low molecular weight form.

Measurements of cAMP levels in tendon fibroblasts treated with PT2 immediately at the end of a 6 day treatment period showed that the intercellular cAMP levels were significantly reduced. These results, however, apply only to a single time point and do not necessarily indicate that cAMP is chronically reduced. A smaller level of cAMP suppression was found in bone marrow cells treated with PTP. Intercellular degradation of collagen has been identified as a cAMP-dependent pathway by which the cells modulate collagen production prior to its secretion from the cell[11]. The observations of Drs Murray and Farndale support the view that PMF may affect collagen production by modification of cAMP metabolism.

In the light of the above results, it is of interest that a preliminary study of adult horse tendon, grown on primary culture, has shown that during a 24 hr period, macromolecular collagen is significantly increased to 71% of the total present in the treated cultures compared with 59% in control. Furthermore , well over half of this fraction was localized in the cell layer when the cells were treated with PMF. The culture had been grown for 14 days and were exposed to PTP for a quarter of an hour every hour throughout the last 7 days. These results uphold the idea that breakdown of intracellular collagen has been reduced, even in cells from adult tissue. The data reported here shown that PMF may swing the balance of the cell towards formation of the major structural protein of connective and skeletal tissue. The fact that only confluent cells, which perhaps bear a closer resemblance to the fully differentiated state, responded to PMF, and that there was no effect on the types of collagen synthesized, reinforces the hypothesis that PMF modifies cellular physiology subtly, remaining within normal control limits.

MINERALIZATION

The effect of PMF on mineralization has been studied in the shafts of rudiments grown in organ culture. Two methods have been used. Firstly, protein bound ^{45}Ca has been employed to assess the initial stages including the seeding of the bone mineral in relation to newly laid down osteoid material. This has been achieved by appropriate treatment of the shafts with TCA solutions, which also insures that all the free isotope has been removed. Comparison of the differently treated groups of rudiments showed that the PTP waveform at a 6 hr dose for 7 days increased the rate of uptake by 40% or more but incorporation dropped back to the control level with prolonged exposure, suggesting that inherent control mechanism may have been brought into play. Interestingly, the SPP signal only increased uptake under the 9 hr ON dose; if the repetition rate of the signal was reduced from 72 to 15 Hz (SP15), the incorporation of isotope at both 6 hr and 9 hr ON was raised between 75% and 100% as compared with controls. It should be noted that the 15 Hz repetition rate is close to the peak found by Adey and colleagues[12] for the efflux of calcium from brain tissue when subjective to modulated microwave irradiation. An additional factor was found to influence the rate of uptake of ^{45}Ca. If the rudiments, rather than lying at random within the petri-dishes, were held throughout culture with their long axis parallel to the magnetic flux lines (coils vertical to the plane of the dish) rather than at right angles (coils either vertical or horizontal) incorporation of isotope was nearly tripled. It is perhaps appropriate to mention that the orientation of the collagen fibrilsand the long (c-axis) of the mineral crystals are also aligned in parallel array with the long axis of the rudiment, and hence with the magnetic flux lines.

The second method of measurement for mineralization has involved absolute calcium analyses. The result of estimations made at the end of culture showed that elevated values for uptake of calcium isotope were paralleled by increases of total calcium present in the shafts, thus demonstrating that additional mineral had indeed been laid down and that the isotopic incorporation was not solely a measurement of turnover during the final day of treatment with PMF. The use of the von Kossa technique to stain mineral in histological sections of the rudiments has shown that some 60% of those examined after PMF treatment contained small centers of calcification in the hypertrophic cartilage, located exactly in the region that starts to calcify in tibiae of much older embryos than those used for culture. No such deposits were found in controls. This result suggests that PMF treatment is capable of switching on a process inherent within the cell.

The body of evidence from the detailed work briefly described in this section, indicates that PMF treatment plays an active role in certain mineralization processes, and hence may indeed be capable of influencing these events during fracture repair. Furthermore, it is evident that certain waveforms are more efficient in this respect than others. In studies designed to test factors involved in the interaction of the various waveforms with mineralization processes, sinewave excitation has been used to test certain aspects. The sinewaves have been used in conjunction with culture in the same way as used with the wideband signals discussed above. The apparatus used is described briefly in Part II of this contribution. The results of some experiments showed that a carrier wave of 2.7 MHz, ON for a 65us burst at 15 Hz repetition rate, effectively raised the level of Ca uptake by 50%, while a continuous sinewave, or one with a pulse burst of 5 ms at the same frequency and amplitude did not[13]; this pulsed sinewave as compared with the SP15 mentioned previously, lowers the B field by 2 orders of magnitude, and the time average power by 3 orders. The effects produced by such low levels of energy input suggest that PMF action may involve the triggering of cascade-like mechanism.

PMF AND BREAKDOWN

The maintenance of the correct physiological balance in skeletal tissues, even when undergoing repair also involves degradative actions (Table 1). It is, therefore, of interest that studies of the effects of PMF on the release of enzymes responsible for connective tissue resorption have indicated that PMF may modify some degradation pathways. Synovial fibroblasts respond to factors produced by monocytes by releasing high levels of collagenase and prostaglandin E_2 (PGE_2), both of which are directly involved in connective tissue breakdown. Dr Murray has mimicked this pathway in vitro and examined the effects of PMF. The results showed that PMF did not activate such processes, but the treatment produced a reduction in the release of two lysosomal enzymes, -glucuronidase and N-acetyl glucosaminidase, by both synoxial fibroblasts and the monocyte/macrophage system[14]. Long term treatment with PMF reduced the levels of enzymes in the culture medium by about 50%. Metalloproteinases, measured as collagenase, remained unaffected. It is worth noting that lysosomal enzymes release is controlled in part by cyclic nucleotide levels whereas collagenase release is not. Interestingly, use of organ culture of pig articular cartilage by Dr Farndale[10] has shown that 14 day treatment with PTP, prevented the loss of glycosaminglycans from the cartilage into the medium, as occurred in control cultures. The data showed that there was no additional synthesis of the GAGS during treatment, but rather there was a redistribution of the material in favor of retention by the tissues. This finding possibly indicates that PTP had limited the self-distributive mechanism of the cells[15]. These studies viewed in conjunction with the effects of PMF on collagen production suggest that PMF may have a general effect on connective and skeletal tissue turnover, reducing degradative enzyme levels while increasing net collagen production.

CONCLUSION

The experimental results briefly referred to in this presentation indicate that PMF measurably affects a number of different cellular events, and that different responses are obtained by different signals. The mechanism involved in these reactions are not yet clear. There has been much speculation that ion fluxes/gradients may be directly involved (e.g.[16]); careful investigations under physiological conditions by Dr Farndale[10], however, have failed to demonstrate any such transport in a number of different systems used classically for such studies. But such experiments do not register very rapid changes (i.e. less than 1 minute) in, for example, internal cellular calcium levels that may be involved in the regulation of various metabolic events. Indeed the present observations, in line with those of Luben et al.[17], that cAMP metabolism is suppressed, imply that PMF may act through the opposing activation of both adenylate cylase and phosphodiesterase, as noted by Dr Jones[18]; both of these enzymes are under the control of calcium levels. The evidence presented here, that certain PMF enhance mineralization in osteogenic tissues, and may trigger calcification in cartilage, also implicates calcium. Since many different biological processes are controlled by transient changes in the level of calcium concentrations, particularly within the cell, the possible capability of PMF targeting on the calcium ion, may explain their effectiveness in the activation/regulation of certain cellular events.

REFERENCES

1. J. K. Kort, M. M. Schink, S. N. Mitchel, and C. A. L. Bassett, Clin.Orthop., 166:124-137 (1982).
2. M. L. Sutcliffe and A. A. J. Goldberg, Clin.Orthop., 166:45-57 (1982).
3. C. A. L. Bassett, S. N. Mitchel, and S. R. Gaston, JAMA, 247:623-628 (1982).
4. J. C. Mulier and F. Spass, Arch.Orthop.Ir., 97:293-297 (1980).
5. C. A. L. Bassett and R. O. Becker, Science, 137:1063-1064 (1962).
6. S. Fitton Jackson, Proc.Roy.Soc.Lond., B. 175:405-453 (1970).
7. S. Fitton Jackson and C. A. L. Bassett, "Tissue Culture in Medical Research II", R.J. Richards and K.T. Rajan, eds., Pergamon Press, Oxford and New York, p.21-28 (1980).
8. S. Fitton Jackson and T. P. Marsland, Trans.Bioelect.Repair & Growth Soc., 3:38 (1982).
9. J. C. Murray and R. W. Farndale, Biochim.Biophys.Acta., in press (1985).
10. R. W. Farndale, PhD Thesis CNAA (1984).
11. S. I. Rennard, L. E. Stier, and R. G. Crystal, J.Inv.Dermatol., 79:77-82 (1982).

12. S. M. Bawin, A. Sheppard, and W. R. Adey, Bioelectrochem. Bioeng.., 5:67-76 (1978).
13. S. Fitton Jackson and T. P. Marsland, Manuscript in preparation (1985).
14. J. C. Murray and M. Lacy, Trans Bioelect.Repair & Growth Soc., 3:7 (1983).
15. H. B. Fell and R. W. Jubb, Arth.Rheum., 20:1359-1371 (1977).
16. A. A. Pilla, "Bioelectrochemistry", H. Keyzer and F. Gutman, eds., Plenum Press, New York, p.353-392 (1980).
17. R. A. Luben, C. D. Cain, M. C.-Y. Chen., D. M. Rosen, and W. R. Adey, Proc.Natl.Acad.Sci., 79:4180-4184 (1982).
18. D. B. Jones, Trans.Bioelect.Repair & Growth Soc., 2:29 (1982).

BIOPHYSICAL STUDIES OF PULSED MAGNETIC FIELD INTERACTION WITH BIOLOGICAL SYSTEMS: PART II - PHYSICAL ASPECTS

T. P. Marsland

Strangeways Research Laboratory, and
Cambridge University Engineering Department
Cambridge, UK

INTRODUCTION

The purpose of this chapter is to give some pertinent physical data concerning the parameters of the electromagnetic fields used for pulsed magnetic field (PMF) therapy. In addition it describes the apparatus and field parameters used for experimental investigation of the effects of time varying magnetic fields on *in vitro* biological systems.

The Electro-Biology Inc. (EBI) Biosteogen system has been widely used both in the clinic[1,2] and by ourselves and others[3], on *in vitro* biological systems. Several other types of PMF apparatus have been developed but are not discussed here[4,5].

The EBI system consists of a set of treatment coils and a special purpose pulse generator to drive them generating a time-varying electromagnetic field in the region to be treated. There are two major types of signal produced by the EBI pulse generator. The first signal is a straightforward repetitive single pulse (SP waveform), and the second is a more complex repetitive burst of pulses (PT waveform).

Exposure Apparatus used for Experimental Work

The biological system under investigation is exposed to the time-varying fields produced by an SP or PT pulse generator driving current through a pair of coils, as in the clinical environment. However, we use the true Helmholtz-aiding coil configuration[6] in order to produce a uniform axial magnetic field in as large a volume as possible. Each of the two coils, which are normally connected in parallel, are 18cm in diameter and consist of 40 (SP) or 32 (PT) turns of insulated 1.3mm copper wire. The coils are attached to a plexiglas stand which determines their orientation with respect to the biological system under investigation. The measured magnetic field strength in the exposure volume ($7 \times 7 \times 9cm^3$) varies by less than 4%.

Fields in air

The Helmholtz coils generate a time-varying magnetic flux density B. The flux density on the coil axis is given by

$$B = \mu_0 \frac{NI}{a} \left(\frac{4}{5}\right)^{3/2} \tag{1}$$

where N is the number of turns on each coil, I is the current that flows through each coil and a is the coil radius and spacing. The flux density was measured directly with a Hall-effect probe; the rate of change of flux density, dB/dt, was determined from measurements of the voltage across a resistively loaded search coil. The search coil voltage is given by

$$u_{sc} = -\pi c^2 M \frac{dB}{dt} \tag{2}$$

where M is the number of turns on the search coil and c is the search coil radius. For our search coil (supplied by EBI) the calibration constant was determined both from (2) and from measurements of known fields so that

$$\frac{dB}{dt} = -0.62\, u_{sc} \tag{3}$$

These two measurements give the time domain form of the magnetic flux density and its time derivative. Now the current density induced in media which is predominantly conductive will have the same time dependence as the rate of change of flux density, dB/dt. Consequently, most workers describe their waveforms, either directly or indirectly, in terms of this time derivative. We have, therefore, determined the frequency domain amplitude spectra of dB/dt for the different signals experimentally. The search coil waveforms were sampled and processed with a microprocessor based waveform analyzer[7], and the fast Fourier transform algorithm was used to calculate the amplitude spectra of the sampled waveforms. We have not attempted any interpretation of the phase spectra.

SP and PT signal parameters

An example SP signal is shown in Figure 1(a), and its corresponding magnitude spectrum is shown in Figure 1(b). Signals of this general form have been used widely in the clinic and we

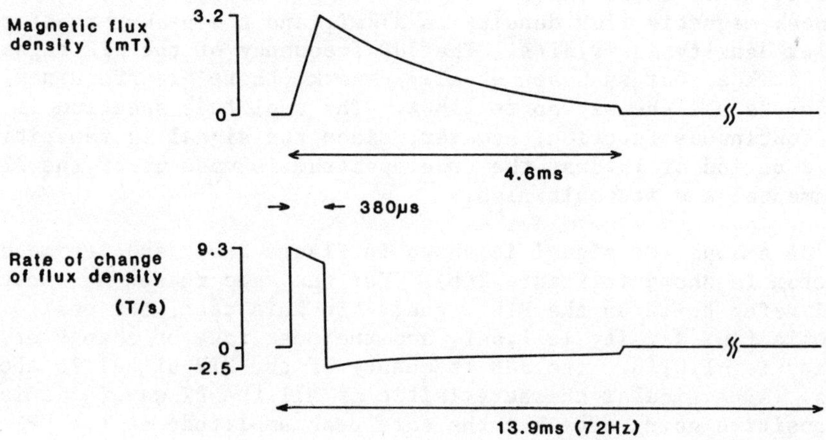

Figure 1. (a) Magnetic flux density B, and rate of change of flux density dB/dt for the SPP waveform, as measured by a Hall-effect probe and a search coil respectively.

Figure 1. (b) Magnitude spectrum of dB/dt for the SPP waveform. The spectrum extends to 11MHz, yet possesses harmonics at multiples of the 72Hz repetition rate.

shall refer to it as the SPP signal (Single Pulse, Patient type). The peak magnetic flux density is 3.2mT, and the peak rate of change of flux density is +9.3T/s. The 3dB frequency of the SPP signal is about 1.3kHz, but as Luben et al.[3] remark, there are frequency components all the way up to 11MHz. The amplitude spectrum is shown as a continuous function, however, since the signal is repetitive with a period of 13.8ms, the true spectrum is made up of the 72Hz fundamental and its multiples.

An example PT signal is shown in Figure 2(a), and its magnitude spectrum is shown in Figure 2(b). For the same reason as above, we shall refer to it as the PTP signal. In this case, the peak magnetic flux density is 1.6mT, and the peak rate of change of flux density is -110T/s. The 3dB frequency of the PTP signal is about 21kHz. A particular characterisitic of all the PT waveforms is that the positive going edge has the same peak amplitude as the SPP signal i.e. +9.3T/s, however, the signal spectrum is vastly dissimilar due to the narrow high amplitude negative going portion of the pulse (about 20μs FWHM). The harmonics in the PTP waveform are also multiples of the repetition rate (15Hz) but the spectrum is further complicated by the 'pulse-burst' nature of the waveform.

The envelope of the signal spectrum is determined by the spectrum of an individual pulse from the burst which can be seen in Figure 2(b). These pulses are repeated every 220μs, producing the dominant spikes in the spectrum at multiples of 4.4kHz. The PT waveform is synthesized by gating these pulses with a signal that has an ON period of 4.8ms, and a 67ms repetition period. This gating can be viewed as time domain multiplication of the train of pulses by a repetitive gating function, or equivalently as frequency domain convolution of the pulse spectrum with the gate function spectrum - a sinc function with harmonics spaced at 15Hz intervals. It is this frequency domain convolution that produces the fine structure in the overall spectrum, i.e. a sinc function convolved around each 4.4kHz spike.

Experimental PMF Signals

Several modifications have been made to the standard EBI signals described above, changing the amplitude or duration of various components of these waveforms. For example, the 'SP15' signal has a repetition rate of 15Hz, but it is otherwise identical to the SPP signal. This multiplies the signal energy by a factor of 15/72, and changes the spacing between the harmonics in the frequency domain from 72Hz to 15Hz. Experimental PT signals have also been used; 'PT2' is our most energetic signal - it has a peak dB/dt of -310T/s and a narrower negative going portion (7us FWHM) which leads to a broader spectrum in the frequency domain.

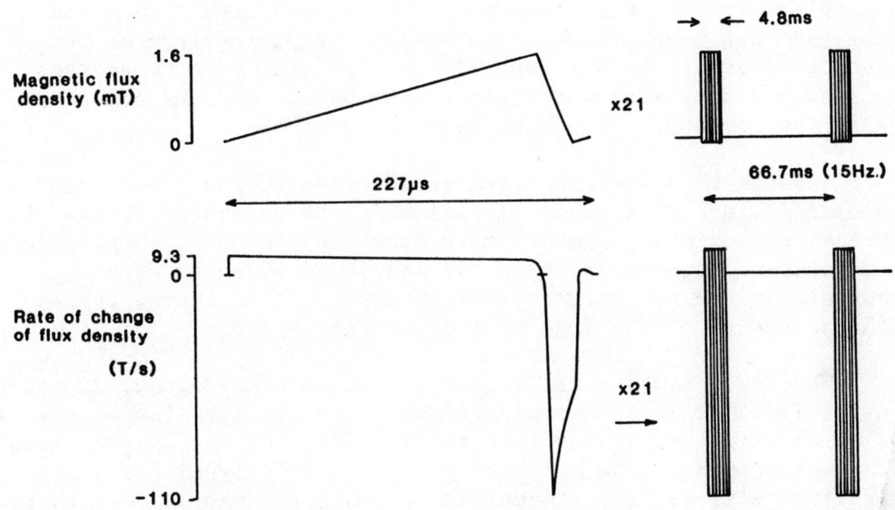

Figure 2. (a) Magnetic flux density B, and rate of change of flux density dB/dt for the PTP waveform, as measured by a Hall-effect probe and a search coil respectively.

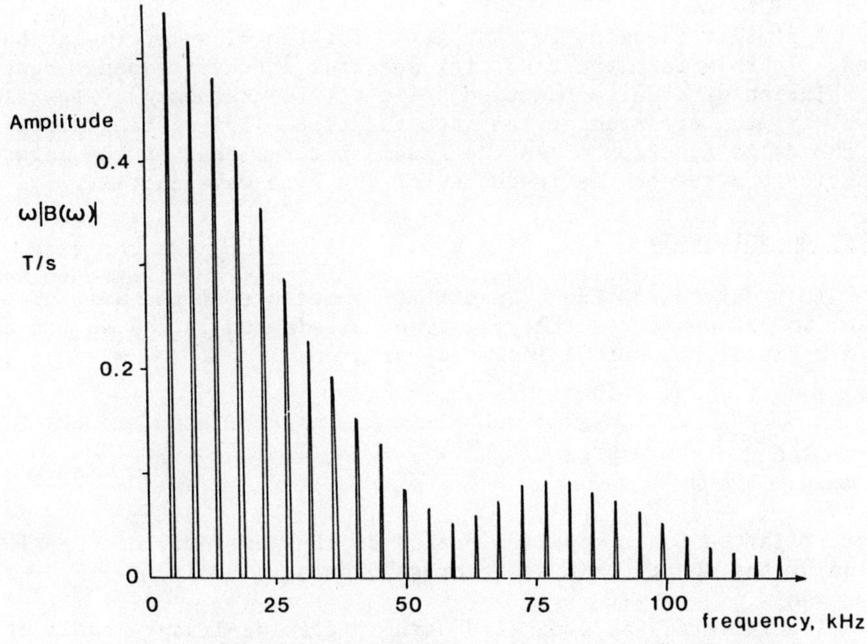

Figure 2. (b) Magnitude spectrum of dB/dt for the PTP waveform. The dominant harmonics are spaced at 4.4kHz intervals, note however the 15Hz fine structure of the spectrum (see text).

The EBI waveforms are quite complex, both in the time domain and in the frequency domain, yet it seems that their form and amplitude has been arrived upon somewhat empirically - do the waveforms need to be so complex? We have attempted to address this problem by using single frequency sine waves, both CW and gated, yet using the same (PT) Helmholtz coils.

Wavetek 182A function generators (Wavetek Inc.) were used to generate 2.7MHz CW sinusoidal fields in the Helmholtz coils. The 182A also provides a ´gated´ mode in which the output amplitude of the sinewave signals can be pulse modulated by an external generator. The generator circuitry ensures that the switching is done at the zero crossings of the waveform.

We chose, somewhat arbritrarily, to take 1mV (peak) as measured in the Search Coil as a standard amplitude for the sine waves. This value is slightly greater than the magnitude of the largest single frequency component in our most energetic EBI signal (PT2). Unfortunately, at high frequencies (>1MHz) the search coil becomes inaccurate, so the voltage applied to the Helmholtz coil is measured and scaled to the equivalent search coil voltage using a calibration factor derived at 20kHz.

The Spectra of Gated Sine Waves

A similar argument to that given for the PT waveform can be used. In this case the resulting spectrum is the frequency domain convolution of a delta function spike at the frequency of the sine wave with the spectrum of the gate function. The sifting property[8] of the delta function makes the final spectrum that of the gate function centred on the frequency of the sine wave carrier.

Power measurements

In order to provide a quantitative measure of the available power in PMF waveforms the real power dissipated in the search coil load resistor has been calculated using

$$P_{sc} = \frac{1}{TR_{sc}} \int_0^T u_{sc}^2 \, dt \qquad (4)$$

where T is the basic repetion period of the waveform, and R_{sc} is the value of the search coil resistor (470 ohms).

The integral was evaluated using piecewise-linear models of the SP and PT waveforms, and have been confirmed experimentally for the PTP signal. Approximately 260nW is dissipated in the resistor by the PTP signal, and 11nW by the SPP signal. The integral is readily evaluated for sinusoidal signals, and has the value of 1.06nW for a 1mV sine wave, independent of frequency. One obvious interpretation of this value is that it is a scaled version of the power available to produce heating by electromagnetic induction. However, the relatively poor electrical conductivity of biological media together with the low frequencies involved mean that bulk temperature rises from this mechanism are insignificant.

The average power supplied to the coils has also been determined for the PTP signal. Measurements of both the current and voltage waveforms were made using the aforementioned waveform analyser, the power was evaluated numerically as

$$P_s = \frac{1}{T}\int_0^T V_s I_s \, dt \qquad (5)$$

where V_s is the coil voltage, and I_s is coil current.

Approximately 0.4W is delivered to the PT coils. Almost all of the power is converted to heat by current flowing in the resistive component of the coil impedance. Naturally, this leads to heating of the Helmholtz coils, and by conduction and convection, of the immediate surroundings. Indeed temperature changes of the order of 0.4°C in cultures which were exposed to PMF signals have been observed. Clearly it is important to establish the temperature sensitivity of any biological system when using such signals.

Fields in Biological Media

There have been many attempts to calculate the electric and magnetic fields induced in biological media by pulsed magnetic fields. Lunt[9], McLeod and Parker,[10] and Freeston[11] have calculated the macroscopic fields generated by PMFs around an electrical model of a clinical non-union. McLeod et al.[12] have determined an analytical solution for the induced current density in culture medium in a culture dish between two Helmholtz aiding coils.

Most authors conclude that macroscopically at least, the magnetic field in air is the same as that in tissue, and that the induced currents are so small that they do not contribute significantly to the magnetic field. This indeed simplifies the electromagnetic problem somewhat, but major difficulties remain concerning the spatial inhomogeneity of biological tissue, together with the vast range of scale lengths necessary to describe structures ranging from single cells to bones. Even if the electromagnetic problem were solved, field computation would still be difficult because of the lack of relevant dielectric property data over the wide frequency range encompassed by the signals.

We have also investigated the induced current problem, with particular reference to the culture systems used by my colleagues in this Laboratory. Anticipating spatial inhomogeneity from the start, a numerical approach was adopted which allows the specification of arbitrary conductivity and permittivity over the region to be modelled. The technique used was suggested to us by Dr. R. Pethig, and a concise description is given by Armitage et al.[13] The technique sets up a network analogue to the field problem which is then solved by a relaxation method. Figure 3(a) shows a cross-sectional view of a 35mm diameter culture dish filled with culture medium whose conductivity is 0.67S/m. We assume that $\sigma >> w\varepsilon$, strictly inadequate for biological media, in order to get some feel for the problem without attempting a more complex transient analysis of the network analogue. The magnetic field is into the paper, and the arrows represent the current density vectors. The largest current density is experienced at the top and bottom of the culture medium, which is $0.65\mu A/cm^2$ for for dB/dt=1T/s. Figure 3(b) shows a similar view of another culture dish this time containing a small region of much lower conductivity.

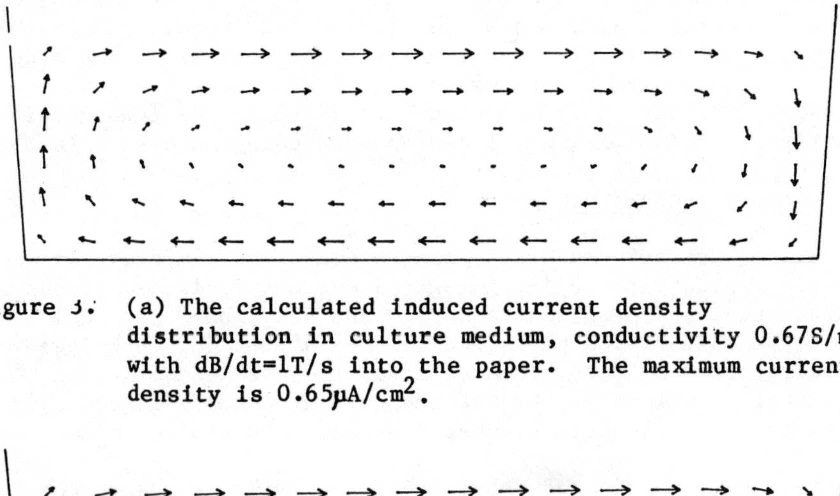

Figure 3. (a) The calculated induced current density distribution in culture medium, conductivity 0.67S/m, with dB/dt=1T/s into the paper. The maximum current density is $0.65\mu A/cm^2$.

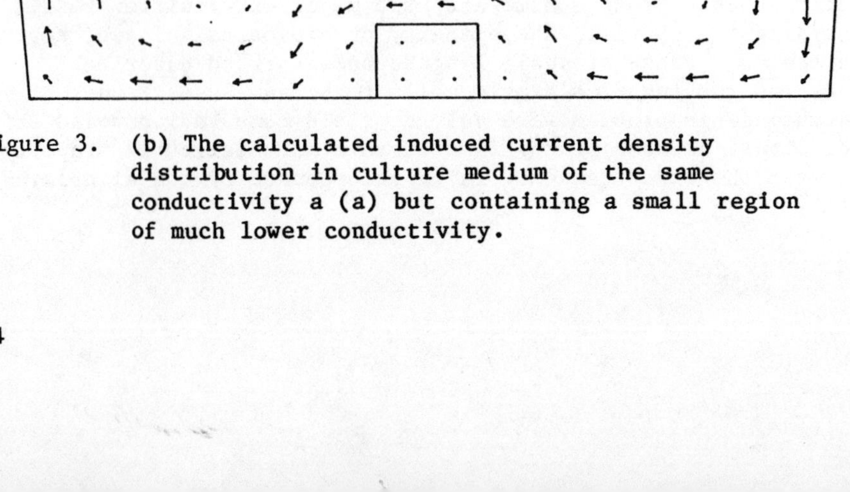

Figure 3. (b) The calculated induced current density distribution in culture medium of the same conductivity a (a) but containing a small region of much lower conductivity.

ACKNOWLEDGEMENTS

Thanks are due to Mr. R. D. Jackson for the use of his waveform analyzer system and for many useful discussions during the course of this work. I should also like to thank my colleagues in the Biophysics group - Dr. S. Fitton Jackson, Dr. R. W. Farndale, and Dr. J. C. Murray for numerous discussions and ideas.

REFERENCES

1. Bassett C. A. L., Mitchell S. N., Gaston S. R., 1981, J. Bone Jt. Surgery, 63A:511-523
2. Hinsenkamp M. G., 1982, Acta Orthop. Scand., 53, suppl. 196, 63-79
3. Luben R. A., Cain C. D., Chi-Yun Chen M., Rosen D., Adey W. R., 1982, Proc. Natl. Acad. Sci., 79:4180-4184
4. Watson J., and Downes M., 1979, Med. & Biol. Eng. & Comput., 17:161-169
5. Barker A. T., 1981, Clin. Phys. Physiol. Meas., 2:9-16
6. Reitz J. R., Milford F. J., Christy R. W., 1979, "Foundations of Electromagnetic Theory", 3rd Ed., Addison-Wesley
7. Jackson R. D. and Prendergast E. J., 1983, Chapter 5 in "Microprocessors in Signal Processing, Measurement and Control'", S G Tzafestas, ed., D Reidel.
8. Bracewell R. N., 1978, "The Fourier Transform and its Applications", 2nd Ed., McGraw-Hill
9. Lunt M. J., 1982, Med. & Biol. Eng. & Comput., 20:501-511
10. McLeod B. R. and Parker R. A., 1984, J. Bioelec., 3(1&2):41-56
11. Freeston I. L., 1982, IEE Colloquium, Digest No. 1982/61
12. McLeod B. R., Pilla A. A., Sampsel M. W., 1983, Bioelectromagnetics, 4:357-370
13. Armitage D. W., LeVeen H. H., Pethig R., 1983, Phys. Med. Biol., 28:31-42

TOPOGRAPHY OF EM EXPOSURE AND ITS RELATIONSHIP TO BIOLOGICAL EFFECTS ON TISSUES

M. Hinsenkamp[1], M. Rooze[2], M. Noorbergen[1], B. Tuerlinckx[3], and E. Coussaert[4]

Brussels University

(1) Service d'Orthopédie-Traumatologie, Hôpital Erasme (Prof. F. Burny)
(2) Service d'Anatomie et d'Embryologie Humaines (Prof. J. Mulnard)
(3) Service d'Electricité Générale (Prof. L. Rybowski)
(4) Département de Calcul Scientifique (Prof. F. Cantraine)

INTRODUCTION

The biological effects of time-varying electromagnetic fields are the less known because of the multiplicity of the electrical parameters.

In order to precise our knowledge in this field of research, we have to define at least three groups of electrical variables: the induction waveform, the topography of the electromagnetic fields and the dielectrical characteristics of the living tissues.

Previous papers analysed the effects of different induction waveforms on the frequency range of magnetic and electric fields (Hinsenkamp, 1984) and the interaction of frequency and dielectric anisotropy of biological tissue on the electric field in a three dimensional model of a limb (Tuerlinckx, 1984).

The significance of this interaction is discussed further in case of fresh fractures and non-unions (Hinsenkamp, 1984).

The present paper investigates the variation of the magnetic and electric fields between Helmholtz coils and its effect on the biological response.

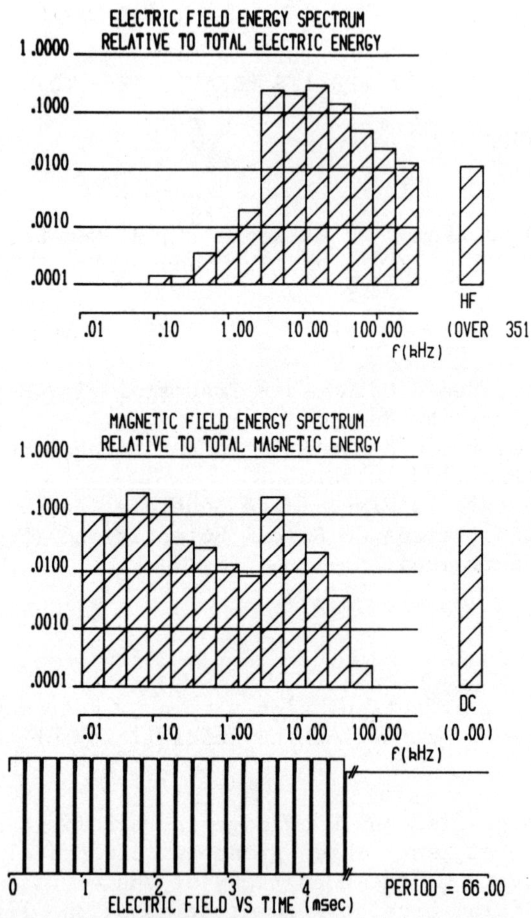

Fig. 1. Schematic representation of the electric signal used (below) and the dimensionless spectral analysis of the magnetic (center) and electric field (Top).

The magnetic and electric fields are generated in our experiments by Helmholtz coils. Their respective values can be calculated following Maxwell's equations for a homogeneous medium like air. The magnetic field values show a relative homogeneity in the center of the coils though the electric field is zero right in the center and increases linearly nearly up to the edges of the coils (Dierickx, 1982).

If this is true for a homogenous medium, these values have to be redefined when heterogenous medium like cells or tissue cultures, animals or limbs are introduced between the coils.

For this reasons in all our previous protocols,the target areas having to be stimulated are located halfway between the center and the edges of the coils.

Previously, using chicken embryos, we compared the total length of the bone rudiments and of the primary ossification point after electromagnetic exposure (Rooze, 1982, 1984).

The results showed a significant increase of both values. In the same experiment few embryos were stimulated in the center of the coils where no or weak effects of the electromagnetic stimulation were observed.

In order to precise this effect, we design the present experiment using smaller eggs of quail embryos to obtain a better topographical definition of the fields response.

MATERIAL AND METHODS

Quail eggs are incubated from day 4 during 100 hours under time varying electromagnetic field. Simultaneously, control eggs are located in the same incubator far enough to be only exposed to very weak residual electromagnetic field.

The stimulation devices, provided by EBI[★],are composed of a generator unit delivering an induction pattern described on Fig. 1 and two square shaped parallel coils.

The eggs are set on a tray having two alternative locations: A plane, 5 cm from each coils, and B plane at 1,4 cm from one coil. On the trays, eggs may lay near the corners, along the edges or near the center (Fig. 2).

[★]Electro Biology Inc., 300 Fairfield Road, Fairfield, New Jersey.

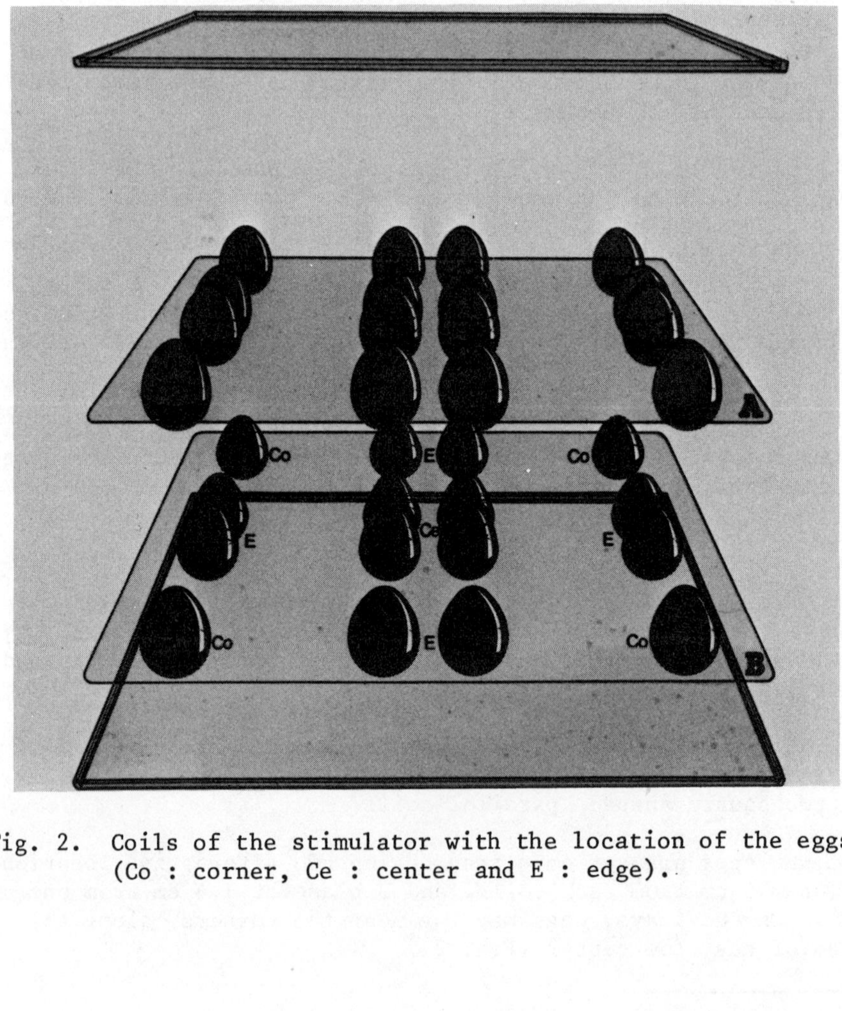

Fig. 2. Coils of the stimulator with the location of the eggs (Co : corner, Ce : center and E : edge).

For each of the six stimulated locations and for the control position, the electric field is calculated with the following approximations.

The eggs are assimilated to spheres of 2 cm in diameter and E max was calculated in a concentric sphere of 2 mm in diameter.

The basic assumption is that both electrical skin depth and wavelength are much larger than a characteristic length, say the egg radius.

Mathematically, we write :

$$\frac{1}{\sqrt{\omega \sigma \mu}} \gg a \qquad (1)$$

$$\frac{1}{\sqrt{\omega^2 \epsilon \mu}} \gg a \qquad (2)$$

with ω = pulsation
σ = conductivity
μ = permeability
ϵ = permittivity
a = characteristic length (egg diameter)

In a first step, the vector potential is computed explicitly, given the amplitude of the current in the coils and the previous estimate of the induced current density (initially set to zero).

Then the scalar potential is expressed as the solution of a Laplace equation, obtained by means of the Boundary Element Method. In a third step, knowing A (vector potential) and V (scalar potential), E (electric field) and J (current density) can be computed at any point. If the induced current density J differs significantly from the one of the previous estimation, we return to the first step.

At the end of the 100 hours exposure, after macroscopic examination, the quail embryos are weighted. Then bulk skeletal double staining following Watson's method (1977) is performed to allow skeletal observation.

After treatment, the cartilaginous matrix is stained blue and the ossified matrix appears to be stained red. The different skeletal segments are isolated by dissection. The total length of the bone rudiments and the length of the primary ossification points are measured on calibrated pictures.

The results are presented for the tibiae and expressed in millimeters. The ratio of the length of the primary ossification point on the total length (here after called : ossification ratio) is computed to minimize the interindividual variations.

The statistical analysis is realized using t test to compare the significance of the differences between the mean values of pairs of groups, variance analysis and polymomial regression to investigate the relation between the E max values and the ossification ratios.

One hundred sixty eight embryos entered in this study in 8 successive runs, four for each A and B plane. All runs include control and stimulated embryos and are separated by an interval of about one month.

RESULTS

Table 1. Mean values of the total length of the tibial rudiments, of the primary ossification points, of the ossification ratio and of the local electric field.

TIBIA		A PLANE			B PLANE		
	CONTROL N = 76	CENTER N = 13	CORNER N = 12	EDGES N= 25	CENTER N = 11	CORNER N = 10	EDGES N= 21
TOTAL LENGTH (MM.)	2.05	2.09	2.23	2.12	2.18	2.16	2.19
LENGTH OF THE OSSIFICATION POINT (MM.)	.60	.64	.71	.66	.70	.74	.86
RATIO OSSIFICATION POINT / TOTAL LENGTH	.285	.303	.315	.304	.318	.332	.360
Bmax (mT)	.3	2.1	.9	1.5	2.0	10.6	6.2
E MAX. (V/M.)	.02	.10	.06	.085	.10	.21	.67

Neither edema nor malformation is observed with significant frequency in stimulated and control groups after the 100 hours of experiments. The embryonic mortality is the same in each groups.

The mean of the total length of the tibiae and of their primary ossification points for each groups are expressed in real value on the Table 1.

On the same table are given the values of the ossification ratio and of the local electric field.

The total length of the tibiae appears to be not significantly modified after 100 hours stimulation. The length of the primary ossification points shows a significant increase mainly for the highest value of the electric field in B plane along the edges of the coils.

The ossification ratio is also significantly increase for the highest electric field values. Table 2 and 3 compares the differences between pairs of groups respectively in A and B plane.

Table 2. Statistical analysis of the differences between the ossification ratio of groups stimulated in A plane (t test)

<u>Ossification point</u> Total length	Centers	Corners	Edges $\bar{M}$ = .304 N = 25
Controls $\bar{M}$ = .285 N = 76	N.S.	N.S.	N.S.
Centers $\bar{M}$ = .303 N = 13		N.S.	N.S.
Corners $\bar{M}$ = .315 N = 12			N.S.

Table 3. Statistical analysis of the differences between the ossification ratio of groups stimulated in B plane (t test)

OSSIFICATION POINT TOTAL LENGTH	CENTERS	CORNERS	EDGES $\bar{M}$ = .360 N = 21
CONTROLS $\bar{M}$ = .285 N = 76	N.S.	$P < 0.01$	$P < 0.001$
CENTERS $\bar{M}$ = .318 N = 11		N.S.	$P < 0.02$
CORNERS $\bar{M}$ = .332 N = 10			N.S.

Table 4. Relative increase of the ossification ratio over the control value

OSSIFICATION POINT TOTAL LENGTH	A PLANE	B PLANE
CENTER	6.3 %	11.5 %
CORNER	10.5 %	16.5 %
EDGES	6.6 %	26.3 %

In A plane no significant difference is found between control and stimulated samples. It is interesting to note that the embryos located in the center on A plane are in the area of the most homogenous part of the magnetic field.

In B plane a hightly significant increase of the ossification ratio is observed for the embryos stimulated along the edges of the coils though no significant differences is found near the center.

The table 4 summarizes the relative increase of the ossification ratio of each stimulated group over the control value.

The variance analysis of the ossification ratio of all groups and their respective electric field value confirms the hightly significant difference (F = 8.024 for 5/162 degree of freedom, $p < 0.001$).

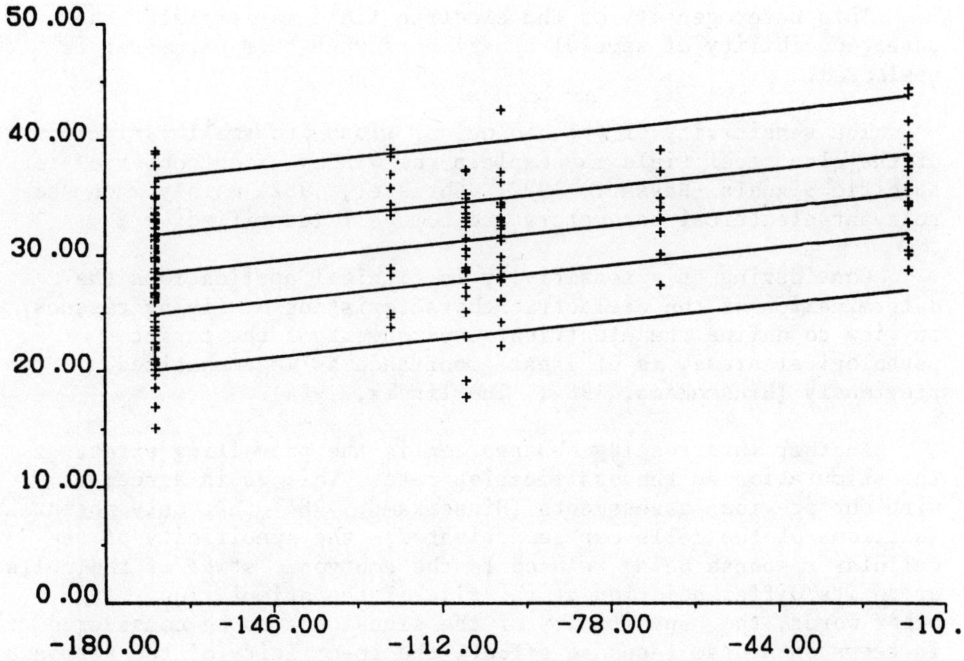

Fig. 3. Linear relation between the local electric field values after logarithmic transformation (abcissa) and the ossification ratio (ordinate).

Then the relation between the local electric field and the ossification ratio is studied. The polynomial regression reveals a linear relation between the logarithmic values of the electric field and the ratios ($Y = 36.26 + 0.04638.\ln E$, where $p < 0.01$ for the angular coefficient) (figure 3).

CONCLUSION

The effect of the heterogeneity of the electric field on biological response is confirmed by this experiment. It appears that though the magnetic field is the most homogenous in the central area between two Helmolthz coils, the biological response is nevertheless weaker.

The most significant effects occur at the periphery of this area where the local electric field reaches its hightest values.

A linear dose response relation exists between the local electric field and the biological response (in the present model : the ossification rate).

This heterogeneity of the electric field may explain the unreproducibility of several experiments when this parameter is neglected.

The sensitivity of the biological tissues to small variations of the electrical field may explain the window effect observed for specific signals (Bassett, 1982; Christel, 1982) mainly when the relevant electrical parameters are not well identified.

Considering this sensitivity in clinical applications the determination of the dielectric characteristics of living tissues in view to define the electrical environment of the target pathological areas, is of great importance as we emphasized previously (Hinsenkamp, 1984; Tuerlinckx, 1984).

Another interresting observation is the prevailing effect of the stimulation on the ossification rate. This is in agreement with our previous assessments (Hinsenkamp, 1984) that only potential functions of the cells can be activated; the specificity of the cellular response being related to the embryonic stage of the cells or to its differenciation at the time of the stimulation. In other words, the specificity of the signal has to be considered in terms of a dose response effect, the specificity of the response appears to be related to the target cell types.

REFERENCES

Bassett, C.A.L., Valdes, M.G. and Hernandez, E., 1982, Modification of fracture repair with selected pulsing electromagnetic fields, J. Bone Joint Surg., 64A : 888.

Christel, P., Cerf, G. and Pilla, A., 1982, Modulation of rat radial osteotomy repair using electromagnetic current induction, in : "Mechanisms of Growth Control", R. Becker, ed., Thomas Press, New York.

Dierickx, M., Hinsenkamp, M., Rybowski, L. and Burny, F., 1982, Electromagnetic and electric field configuration produced by two coils, Acta Orthop. Scand., Suppl. 196: 19.

Hinsenkamp, M., Burny, F., Donkerwolcke, M. and Coussaert, E., 1984, Electromagnetic stimulation of fresh fractures treated with Hoffmann external fixation, Orthopedics, 7 (3) : 411.

Hinsenkamp, M., Tuerlinckx, B. and Rooze, M., 1984, Effect of ELF fields on bone growth and fracture repair, in "Proceedings of the Course on "Biological effects and dosimetry of non ionizing radiation : Static and ELF electromagnetic fields, November 15-25, 1983 - Erice, Sicily, Plenum (in press).

Hinsenkamp, M., Rooze, M. and Tuerlinckx, B., 1984, Biophysical and Biological Intervention. Future Directions, in "Proceedings of the Symposium on "Fracture Repair : Conflicts and Consensus", May 9-11, 1984, New York, Churchill Livingstone (in press).

Rooze, M., Hinsenkamp, M. and Duchateau J., 1983, In vivo skeletal modifications of chicken embryos induced by electromagnetic fields, in "Transactions of the Second Annual Meeting of the Bioelectrical Repair and Growth Society, September 20-22, Oxford, Abstract 73.

Rooze, M. and Hinsenkamp, M., 1984, In vivo modifications induced by electromagnetic stimulation of chicken embryos, in "Reconstructive Surgery and Traumatology", S. Karger AG, Basel, vol. 19 (in press).

Tuerlinckx, B., Boido, M. and Hinsenkamp, M., 1984, Analysis of the electromagnetic fields caused by bone growth stimulators. A boundary-element simulation program, in "Reconstructive Surgery and Traumatology", S. Karger, Basel, vol. 19 (in press).

Watson, A.G., 1977, In toto alcian blue staining of the cartilaginous skeleton in mammalian embryos, Anat. Rec., 187 : 743.

A CRITIQUE OF MEDICAL USES OF WEAK PULSING ELECTROMAGNETIC FIELDS

C. Andrew L. Bassett* and Sylvia Fitton Jackson+

College of Physicians and Surgeons of Columbia University
New York 10032, USA* and Strangeways Research Laboratory
Cambridge. UK +

This presentation sets forth the current clinical applications of pulsed electromagnetic fields (PEMF), a non-invasive method proven to be essentially riskless, and discusses factors involved in their use. To date, present patient trials have demonstrated that certain pulsed waveforms+ can be used to treat congenital "pseudarthrosis" (1), ununited fractures (2), failed fusions (3), avascular necrosis (4), Legg Perthes disease (5) and chronic refractory tendinitis (6). Although each of these entities affects a skeletal tissue, their etiologies and pathological details are quite different. PEMF do not have panacea qualities; any therapeutic specificity must derive from complex interactions between the physical parameters of the waveform and the heterogeneous properties of the normal or pathological target tissue (7,8). The events that occur in any healing process are numerous, and vary in both timing and rate. The mechanisms that may be involved in PEMF action on tissues are not yet understood in detail, but, as stressed in the first presentation, present experience suggests that PEMF must work within the normal levels of cellular responses that are under their own physiological control systems.

In considering PEMF usage in the treatment of bony union when it has not yet occurred, it is important to define briefly fundamental differences between fresh and ununited fractures (7,9). Normally, fresh fracture repair in mobile long bones proceeds through a series of time dependent stages which span several weeks or months, depending upon site and species. These are listed, sequentially, in Table 1.

+Specifically those provided by Electro-Biology Inc, New Jersey, USA and Reading UK.

Table 1. Stages in Fracture Repair

	Stages	Time span	Strength
I	Anchoring	0 - 14 days	0
II	Bridging	7 - 21 "	$0 - 1^+$
III	Calcifying	10 - 30 "	1 - 2
IV	Vascularizing	10 - 40 "	$2 - 2^+$
V	Remodeling	40 - 500 "	$2 - 4^+$

The first stage involves periosteal activation leading to fibre bone formation, followed by the laying down of fibrocartilage at the fracture gap, with some early calcification; proteoglycan and collagen synthesis are essential features of these two stages of repair. Stage III involves progressive calcification, from the anchoring fronts of fibre bone across the edge of the fracture gap. Blood vessel invasion (Stage IV) leads to endochondral ossification with some bridging within the gap, while the final stage involves lamellar bone and osteone formation and full mineralization. Of the five intermediate steps in fresh fracture repair, between the original injury and final remodelling, stage III - calcification appears to be critically important to the repair process. If the fibrocartilage fails to calcify, penetration of blood vessels does not occur and the early stages of final bony union bridging by endochondral ossification is prevented. In fact, in many non-unions of long bone, steps I and II are completed, but normal progression to subsequent stages ceases.

The first PEMF waveforms were derived, originally, from pulses found to occur in physiologically stressed bone (10,11). Both the amplitude and shape characteristics of the applied fields came from this source and resulted in the development of quasi-triangular and quasi-rectangular assymetrical, wide band single repetitive pulsed signals described in earlier publications (7,12). They were shown to be effective in the treatment of experimental osteotomies (13) with a signal repetition rate set at about 72 Hz. The actual fields induced by signals would, of course be strongly influenced by the local dielectric properties of the tissues, and, due to inhomogeneities would vary from point to point. For purposes of clarity, the detailed physical assessment of the waveforms that is now available have been set forth in the previous presentation (Part II).

These 'first generation' single pulse waveforms were used, initially, in the treatment of children facing amputation of their leg because of surgically-resistant, congenital "psudarthrosis" of the tibia. In the early days the fields were inductively coupled to the tissues in the affected limb by a single external coil, and results from this treatment were published in 1974 (12) and 1977 (14). In an attempt to improve field uniformity, a quasi-Helmholtz double coil system was then adopted. In the main, the single pulse waveform has been retained, on the basis of its clinical success,

Table 2 PEMF Treated Congenital "Pseudarthrosis" of Tibia

Type	Healed	Failed
I	12	7
II	28	10
III	7	23

as the PEMF treatment of choice for the child with congenital "pseudarthrosis". The entire clinical experience between 1972 and early 1984 for cases supervised at Columbia is summarized in Table 2. Several points deserve emphasis. The average follow-up time after union has been 3 years, and the average age of the patients is presently 11.7 years old. Five of the 49 patients that initially achieved union, refractured, but healed after a further 14.4 months (average) of treatment. As pointed out by Sutcliffe and Goldberg (15) in a report of 52 cases in Europe an overall healing rate of 55% with PEMF treatment represents a significant advance over normal conservative treatment and routine forms of bone grafting. The success rate of 70% for types I and II lesions stands in sharp contrast to the 25% for type III cases. These latter patients tend to have their initial fractures at a very early age (i.e. av. 1.3yrs); they, now, are reclaimed at a far higher level of success by a combination of PEMF and Boyd dual onlay cortical grafts or by fibular grafting with microvascular anastamoses (16). It is clear that PEMF, alone, cannot restore length, correct deformity, bridge very wide gaps or heal true synovial pseudarthrosis. On the other hand, for the child with a congenital lesion, PEMF should be used as the primary form of treatment unless the limb will be non-functional or too short after union, or if union has not resulted after 3 years of PEMF treatment. In general, the use of PEMF alone, or in conjunction with bone grafts, has virtually reduced the need to amputate by 90%.

Since difficult "pseudarthrosis" cases in children were yielding to PEMF treatment by 1975 their application to ununited fractures of long standing nonunions in adults was initiated. Simultaneously, the double coil system (Helmholtz aiding) was adopted. The pulse-burst waveform at 15Hz was developed by grouping single pulses together in an attempt to couple to certain cellular 'relaxation times' (17, 18). Inevitably the coils have to be distorted somewhat to conform to the surface of the limb, thus leading to some spatial field inhomogeneity in the treatment site, and although different sized coils are required to treat different sized limbs, dB/dt is kept constant. Details of the pulse-burst signal parameters, frequency content and average power are given in Part II of the accompanying contribution. (see T.P. Marsland "Biophysical studies of pulsed magnetic field interaction with biological systems: Part II-physical aspects in this book).

The treatment of difficult ununited fractures is now well documented in many publications from different clinical centres (eg 19-22); these include substantial numbers of very long duration post-fracture (i.e. greater than 4 years) surgically resistant

Table 3 Treatment of Ununited Fractures with PEMF

Site	Patients	Healed	%
Tibia	177	158	89
Femur	205	150	73
Radius/Ulna	73	52	71
Humerus	68	41	61
TOTALS	521	401	77 (mean)

infected cases in which the chance for spontaneous union is remote yet in which the union rate is in the 80% range. Results of PEMF treatment taken from recently documented case histories of 521 patients are summarized in Table 3. As can be seen, the ununited tibia responds to PEMF treatment better than the femur, which in turn responds a little better than the radius and ulna, with the humerus the most difficult to mend. There appears to be a marginal correlation between the length of disability time and between bony lesion reactivity and PEMF-triggered union. The longer the time since the original fracture, the more severe the therapeutic challenge, within a 10-15% variation. Hypertrophic lesions, which contain largely fibrocartilage, are easier to treat than atrophic lesions which are largely fibrous tissue. The humerus presents the severest single therapeutic problem, probably because it is so mobile; however, when substantial numbers of long-standing atrophic, actively infected lesions respond under PEMF, it is evident that such treatment has significant clinical importance.

The clinical problems arising from congenital "pseudarthrosis" are generally far greater than for the acquired non-union in the adult for two main reasons. First, acquired lesion characteristics are more amenable, since fracture gaps in adults frequently contain a large proportion of fibrocartilage, a tissue more likely to calcify than the fibrous tissue which often is present in the congenital "pseudarthrosis". Secondly, motion can be more easily controlled in the adult than the child; immobilization of bony fragments appears to work synergistically with PEMF to produce a higher percentage of success. Attention to this biomechanical fact of life has been an important aspect of clinical experience and may derive its importance from the observation that fibrous tissue, protected from repeated tensile loading, converts to fibrocartilage (23) and hence may calcify much more readily than fibrous tissue. Furthermore PEMF appear to specifically affect the inability of non unions to calcify. Once stage III of normal fracture repair (Table I) has been triggered by PEMF and the blockage to continuing repair removed, stages IV and V proceed in normal fashion and at normal rate. This process has been activated by PEMF as many as 45 years after the initial fracture and has been documented by X-ray and by biopsy.

In a recent detailed retrospective study of the interaction between PEMF and immobilization, 50% of all ununited long bone fractures were found to unite even when motion was uncontrolled, this rate could be nearly doubled if motion were controlled. The overall success in the 70-80% range quoted for more than 30,000 cases reflects the average of these two extremes of patient management. The recent preliminary report of a double-blind pulsed magnetic field trial of 17 tibial non-unions gave indefinite results, but stressed the importance of conservative treatment in fracture repair (24). Subsequent correspondence in the Lancet (25-27) however indicated that the patients were not matched, always a problem with such small numbers, and also stressed the need for at least 200 clinical cases in such a trial so that the variations in types of fracture and other factors could be adequately covered. An ongoing double-blind study, started 3 years ago in England of delayed-unions (ie 6-9 mths post fracture) may help to throw more light on the effectiveness of PEMF treatment.

Emphasis has been placed on the role of PEMF in triggering calcification of fibrocartilage in the old fracture gap as a prelud to bony bridging. It should not be surprising, therefore, that pulse characteristics, changed to achieve greater calcification efficiency, would improve the clinical effectiveness for non-unions. Results in the laboratory have indicated that a 15Hz, 200 us version of the original 72Hz, 380 us single pulse waveform improved, considerably the calcium content of chick tibial rudiments grown in organ culture (28).The change in signal parameters increased the bandwidth, altered the harmonic spacings and reduced the average power by over two orders of magnitude. Application of this signal to 22 non-union patients who had failed to respond to the standard pulse-burst signal resulted in union of 3 out of 4 cases. This response in patients who had demonstrated initial therapeutic resistance to the previous treatment suggested that as the 'calcium efficiency' of the signal is improved, so do clinical results. This waveform has now been used on 91 'virgin' ununited fractures, and the results are given in Table 4. Twenty of the most refactory long bone non-unions, selected from these patients show that 80% were healed in an average time of 4.7 months, all had lesions in excess of 2 years, with multiple operations, and all were atrophic, or aligotrophic, including an unhealed femur of 38 years standing with a history of 7 previous operations. It is evident from these data that the new 15 Hz, 200 us single pulse waveform is at least as efficient in this clinical setting as other waveforms used, and appears to be slightly better in terms of case difficulty and time to union.

It has been noted previously that vascular penetration of calcified fibrocartilage is an important step in the repair process. Based on these observations, and on the ability of the single pulse signal to arrest disuse osteoporosis in experimental animals (29), application of PEMF to osteonecrosis (vascular necrosis) of the hip

Table 4 Treatment of Adult Non-Union with a Single Pulse Waveform

Site	Patients	Healed	%
Tibia	43	35	81
Femur	20	16	80
Radius/Ulna	8	5	62
Humerus	8	6	75
Others	12	8	66
TOTALS	91	70	77 (mean)

was planned. As a preliminary step, rabbit femoral heads were surgically devascularized and treated with single pulse signals. Uptake of technetium polyphosphate, 14 days postoperatively was 75% of normal in PEMF-treated hips, as opposed to 25% in untreated controls (30). Subsequently, blood vessels, < 0.03 mm in diameter, were found to be increased in number, in treated femoral heads, and osteogenesis was greater. These results appeared to support the use pf PEMF in trials in humans with osteonecrosis of the hip (31). This problem, once established in young adults leads inexorably to destruction of the femoral head contour and to extensive osteoarthritis of the hip joint. Pain and disability have forced early joint replacement, which in 50% of young adults lasts only 5 years, unless physical activities are greatly restricted. In view of the demontrated safety of this class of PEMF in the laboratory and clinic, the potential benefit/risk ratio appeared to justify a clinical study involving 77 hips treated with PEMF judged against 35 under standard conservative treatment (Table 5). A large majority of the patients had advanced disease, as evidenced by X-ray. These patients, in the past, have failed, routinely to respond to conservative treatment using crutches, bedrest and drugs. Occasional stabilization of their symptoms occurs, but the large majority end with a total hip replacement. This contrasts with PEMF treatment (Table 6) which presents multicenter data, showing an increasing number of hip replacements, as a function of time for the conservatively managed group as opposed to the PEMF treated group that lost only a few advanced cases to surgery early-on in the study.

Perhaps one of the more important differences between the two groups concerns modifications in pain and function. There is continuing improvement in symptoms over at least a 2 year interval with PEMF treatment, whilst in the conservatively-treated group, clinical deterioration forces the decision for operative repair. Radiographically, 55% of the conservative group that avoid surgery, deteriorated at least one Ficat category, as opposed to 7% in the PEMF group judged over a much longer time interval. The etiology of the femoral head necrosis did not affect the clinical benefits to a major degree (Table 7). All groups improved under PEMF, including those on chronic corticosteroids for kidney transplants and lupus. This latter group did not gain relief of symptoms as rapidly as

Table 5 Avascular Necrosis of the Hip - Patient Details and Results

Stage	Treatment Conservative	PEMF	
Ficat II	7	8	14 patients
Ficat III	28	69	with % Hips
Results			
Improved	-	63	
Stabilized	5	4	
Total Hip Replacement	30	10	
Average Age of Patients	41	39	

Table 6 % Survival of Hips from Onset of Symptoms

Follow up time (mths)	Treatment Conservative	PEMF
0 - 12	100	100
13 - 24	69	95
25 - 36	55	88
37 - 48	35	88
48	15	88

the others, however, but have continued to improve as a function of time. In the Columbia programme, none of the 9 patients on chronic corticol steroids and PEMF for an average of 37 months has required hip replacement, and none is anticipated in the foreseeable future.

Relief of pain in osteonecrosis cannot be attributed to an analgesic effect since its onset and progress is highly variable over the first 9 months; the benefits persist for years after PEMF treatment has been discontinued. For the patient with an early lesion (Ficat II) PEMF treatment appears to arrest and reverse the process; in this group, the hip replacement rate has been 0%. In Ficat III patients, deterioration is stopped but 10 have required total hips. In the very far advanced (Ficat IV) lesions, success is limited to the younger age group (i.e. 16-40 years). The overall hip replacement rate for all Ficat IV's treated with PEMF is 40% which is less than half that in all categories of hips treated by conservative means.

The striking survival rate for affected hips (Table 6), in addition to the other data given, supports the contention that PEMF plays an important role in the treatment of osteonecrosis. These results, that were beginning to become apparent earlier in the study, suggested that other skeletal lesions, associated with diminished local circulation, might benefit. This possibility, plus the frequent reports of relief of pain that occurrs during PEMF treatment,

Table 7 Etiology of Avascular Necrosis and Improvement in Clinical Grade*

Etiology	Patients	Day 0	12 mths later
Idiopathic	26	12.3 ± 1.7	15.0 ± 2.5
Hex. Steroid	15	13.7 ± 2.0	15.4 ± 1.8
Alcohol	13	12.3 ± 2.5	15.0 ± 2.5
Chronic Steroid	12	12.5 ± 1.8	13.9 ± 3.0
Trauma	11	12.0 ± 2.5	16.0 ± 2.0

* Merle d'Aubigny clinical grading system. Perfect score = 18 made up of three indices, pain, function and mobility.

led to the initiation of studies into rotator cuff tendinitis at the shoulder clinic in Cambridge in late 1982. Partial interruption of the blood supply to the rotor cuff tendons is believed to be important in initiating tendon degeneration (32,33); healing is slow since the vascular supply to the adult tendon is normally poor. Another factor that pointed to the decision was the knowledge that the formation of collagen was promoted in various model systems (34,35). The value of PEMF for the treatment of persistent rotator cuff tendinitis, a common cause of acute shoulder pain and immobility in the adult population, has been tested in a double-blind study (6, 36) in 29 patients whose symptoms were refactory to steroid injection and treatment with other conventional conservative methods. A previous pilot study of 15 patients showed a marked improvement under PEMF treatment within 4 weeks of the start of therapy. A single pulse waveform, with a peak dB/dt of 7.9 T/s and a repetetion rate of 73 ± 2 Hz was used in conjunction with a single ovoid coil, fitted over the outer aspect of the affected shoulder so that the shoulder protruded through the coil.

For the double-blind study patients who had had their tendon lesions, diagnosed on the criteria of Cyriax (37), persisting for at least 3 months, with no more than transient benefit from previous therapy including steroid injection, were randomly split into 2 groups by allocation of either an active coil or dummy during the first 4 weeks period. Neither patient, nor medical assessor and others involved with the patients were aware of the allocation of active or inactive coils. Subsequently, without breaking the code, all patients were issued with a new set of coils for a further 4 weeks of treatment. Thus some patients had 8, and others only 4 weeks of treatment with active coils.

Assessment of patients was made every 2 weeks and was based on 4 clinical variables. The results of subsequent analysis showed that during the first 4 week period thetreated group (15 patients) with

Table 8 Level of Significance of Assessment of Improvement Versus Time. Treated Compared with Control.

Clinical Variable	2 weeks P <	4 weeks P <
1. Pain score	0.05	0.02
2. Painful Arc on Abduction	0.005	0.005
3. Pain on resisted	0.005	0.005
4. Active Range	0.005	0.005

active coils benefitted significant compared with the control group (14 patients) with dummy coils (Table 8). The rate of recovery from time zero to each follow up visit was compared by use of the Wilcoxon rank sum test. A small placebo effect was evident in pain score, and active range assessments. No significant differences were noted between the groups now all on active coils during the second 4 week phase of the study. At the end of the follow up period, 16 weeks plus for all patients, 19 were judged to be symptomless and 5 others were much improved; 3 were still severely disabled, 1 from the treated group and 2 from the control group. The conclusion was reached from these data that PEMF therapy may be useful in the treatment of severe and persistent rotator cuff and possibly other chronic tendon lesions. Further double-blind trials are now in progress, as well as fundamental studies on tendon lesions in the horse, in collaboration with the Equine Research Station in Newmarket.

This presentation has only been able to briefly outline some of the evidence now available in support of the fact that PEMF may indeed have a direct effect on skeletal events, in that pathological processes normally resistant to a variety of conventional surgical and non-surgical methods appear to have responded to use of this non-invasive method of treatment. It is evident that the 'input' from PEMF signals must be small, but this does not mean they must therefore be ineffectual. The clinical responses achieved appear to reflect specific biological activity at the cellular level, and a considerable body of knowledge of these matters is now beginning to accrue. The previous presentation by one of us has indeed emphasized the way PEMF may modify certain metabolic processes of direct relevance to skeletal repair. Just as the proper expansion of drug usage requires increased knowledge of their pharmacological dynamics, so each new usage of specifically configured PEMF needs to be supported by an appropriate deliniation of factors involved via such adequate studies both in the laboratory and theclinic (38). Without such precision, extension of the therapeutic use of PEMF may well sully their present unique clinical record for safety and - dare one say - effectiveness?

REFERENCES

1. Bassett, C.A.L., Caulo, N. and Kort, J. (1981) Clin. Orthop. 154, 136-149.
2. Bassett, C.A.L., Mitchell, S.N. and Gaston, S.R. (1982) J.A.M.A. 274, 623-628.
3. Bigliani, L.U., Rosenwasser, M.P., Caulo, N. and Bassett, C.A.L. J. Bone Joint Surg. 65A, 480-485.(1983)
4. Bassett, C.A.L., Schink, M.M. and Mitchell, S.N. (1983) In Proceedings of the Third International Symposium on Circulation in Bone. Ed. Hungerford, S.D. Williams & Wilkins, Baltimore.
5. Harrison, M.H.M. and Bassett, C.A.L.(1984) J. Pediatric Orthopedics,4, 579-584.
6. Binder, A., Parr, G., Hazleman, B. and Fitton-Jackson, S. (1984) Lancet 695-698.
7. Bassett, C.A.L. (1983) Surg. Rounds,20-32.
8. Bassett, C.A.L. (1984) Orthopäde 13, 64-77.
9. Bassett, C.A.L. (1984) In'Complications of Fracture Management. I. Ed. Gossling, H.R. Lippincott Philadelphia 1-8.
10. Bassett, C.A.L. and Becker, R.O. (1962) Science 137, 1063-1064.
11. Bassett, C.A.L. (1978) In 'Metabolic Surgery'. Eds. Buchwald, H. and Varco, R.L. Grune & Stratton, New York.
12. Bassett, C.A.L., Pawluk, R.J. and Pilla, A.A. (1974) Ann. N.Y. Acad. Sci. 238, 242-262.
13. Bassett, C.A.L., Pawluk, R.J. and Pilla, A.A. (1974) Science 184, 575-577.
14. Bassett, C.A.L., Pilla, A.A. and Pawluk, R.J.(1977) Clin Orthop. 124 128-143
15. Sutcliffe, M.L. and Goldberg, A.A.J. (1982) Clin. Orthop. 166 45-57.
16. Dick, H.M., Craig, E.V. Norvick, P.D. and Rossenwasser, M.T. (1983) Orth. Trans. 7, 384-5.
17. Pilla, A.A. (1974) Ann. N.Y. Acad. Sci 238, 149-168.
18. Pilla, A.A. (1980) In 'Bioelectrochemistry'. Eds. Reyzer, H. and Gutman, F. Plenum Press, New York p.353-
19. Mulier, J.C. and Spaas, F. (1980) Arch. Orthop. Trauma Surg. 97, 293-297.
20. Sedel, L., Christel, P. and Duriez, J. (1981) Rev. Chir. Orthop. 67, 11-23.
21. Heckman, J.D., Ingram, A.J. and Loyd, D.D. (1981) Clin. Orthop. 161, 58-66.
22. Sharrard, W.J., Sutcliffe, M.L., Robson, M.J. and Maceachern, A.G. (1982) J. Bone Joint Surg. 64B, 189-193.
23. Flint, M.H., Gillard, C.C. and Merrilees, M.J. (1980) In 'Fibrous Proteins Scientific, Industrial and Medical Aspects'. Eds. Parry, D.A.D. and Creamer, L.K. Academic Press. London 2, 107-119.
24. Barker, A.T., Dixon, R.A., Sharrard, W.J.W. and Sutcliffe, M.L. (1984) Lancet, May 5th. 994-996.
25. Le Vay, D. (1984) Lancet, June 9, 1295.

26. O'Connor, B.T. (1984) Lancet, July 21, 171-172.
27. Sherrard, W.J.W. (1984) Lancet, July 21, 172.
28. Fitton-Jackson, S. Manuscript in preparation.
29. Cruess, R.L., Kan, K. and Bassett, C.A.L. (1983) Clin. Orthop. 173, 245-250.
30. Rinsky, L.A., Halpern, A., and Schurman, D.B. (1980) Orthop. Trans. 4, 238.
31. Bassett, C.A.L., Schink, M. and Mitchell, S.N. (1981) Trans. Bioelect. Repair and Growth Soc. 1, 38.
32. Olsson, O. (1953) Acta Scand. Suppl. 181.
33. Kessel, L. and Watson, M. (1977) J. Bone Jt.Surg. 58B, 166-172.
34. Fitton-Jackson, S. and Bassett, C.A.L. (1980) In 'Tissue Culture in Medical Research II'. Eds. Richards, R.J. and Rajan, K.T. 21-28.
35. Luben, R.A., Chin, C.D., Chen, M.C.Y., Rosen, D.M. and Adey, W.R. (1982) Proc. Natl. Acad. Sci. U.S.A. 79 4180-4184.
36. Lunt, M.J. and Barker, A.T. (1984) Lancet, June 9, 1295.
37. Cyriax, J. (1971) In 'Textbook of Orthopaedic Medicine. Bailliere Tindall, London.
38. Bassett, C.A.L. (1984) In 'Complications of Fracture Management' Ed. Gossling, R. Lippincott, Philadelphia 9-39.

SUBTLE EFFECTS OF RADIOFREQUENCY ENERGY ABSORPTION AND THEIR PHYSIOLOGICAL IMPLICATIONS

Sol M. Michaelson

Department of Radiation Biology and Biophysics
School of Medicine and Dentistry, University of Rochester
Rochester, New York 14642

INTRODUCTION

We might define subtle effects of RF energy/biologic tissue interactions as those likely to elude perception or understanding. These can be illusory, unsubstantiated, or tenuous, which would be considered of slight importance or significance. On the other hand, several reported subtle effects are unquestionably of considerable interest and, if substantiated, may indicate basic or fundamental biophysical relationships. For pragmatic considerations the question has to be asked and consideration given as to whether such subtle effects have any physiological implications.

Subtle effects can be categorized as:

1) influence of field generated forces
2) membrane interactions - represented by electrosensitive species; nerve cell stimulation; auditory response
3) biopolymer interactions
4) changes in calcium kinetics
5) millimeter wave interactions
6) orientation, navigation by electrosensing organisms and in ambient electromagnetic fields
7) modification of circadian rhythms
8) behavioral response to "weak" fields.

Schwan (1978) has reviewed the established mechanisms responsible for the interaction of microwaves and other electrical fields with biological systems. Foremost is the heat development which results from the absorption of microwaves. The relative contribution to this heat development caused by the various tissue constituents including ions, water, biopolymers, bound water and lipids are noted. Direct field interactions with various biocomponents are also considered. These include membrane interactions, biopolymer interactions, interactions with biological fluids, and field generated forces acting on biological particles and cells. These forces are frequently neglected in discussions and yet pertain to a large number of published observations.

"Nonthermal" (Athermal) Physiologic Reactions

It is universally accepted that high level radiofrequency (RF) radiation is a source of thermal energy that can, under certain circumstances, elicit subtle (covert) or recognizable (overt) biological and physiological responses consonant with body heating or temperature increase.

When a measurable increase in body temperature occurs, it can be interpreted as an indication that the body is under some "stress." Exposure to RF at SARs below those that cause an increase in core temperature may activate heat dissipating thermoregulatory mechanisms of the body to maintain core temperature within the normal range. Under some conditions of RF exposure discrete regions of the body could experience an increase in temperature due to localized RF energy absorption while the core temperature remains near normal. These, however, are soon dissipated into the physiologic milieu.

The heat development per unit volume caused by the absorption of electrical energy is given by

$$H = E^2\kappa$$

where E is the electrical field strength and κ is the electrical conductivity of the absorber. Hence, the conductivity κ simply expresses the conversion of electrical energy into heat per unit volume. The conductivity largely controls the heat development. The dielectric constant of various tissues, together with the conductivity, determines the distribution of the field patterns in complex tissue arrangements.

Table 1. Mechanisms of Field Interactions

Heat Development
Membrane Interactions
Biopolymer Interactions
Biological Fluids
Field Generated Forces

McAfee (1962) has noted -

"In experiments on the physiological response to electromagnetic radiation in the radiowave and microwave region of the spectrum, the possibility that thermal stimulation of afferent peripheral fibers plays a role in the physiological response observed deserves consideration, particularly since overt manifestation of the response may be masked by homeostatic regulation or by the animal's behavior. When hyperthermia is not produced by localized microwave irradiation but physiological changes are nevertheless observed, mechanisms other than thermal need not be evoked in explanation of the physiological effect observed. For example, because the response to thermal stimulation may be maintained for some time..., adrenal medullary secretions may elicit any one or more of many possible secondary effects. Secretion of adrenocorticotropin (ACTH) may occur, resulting in increased elaboration of adrenal cortical hormones. Changes in the composition of the blood which characterize the stress response are expected to follow. When hyperthermia is produced by microwave irradiation of the whole body, the physiological response observed may include the response associated with thermal stimulation of afferent peripheral nerves as well as the physiological response to hyperthermia. Even though it is not intended that average temperatures reach stimulus levels, afferent peripheral fibers may be stimulated selectively as a result of the inhomogeneity of living tissue to microwave absorption and to local variations in the ability of bodily structures to dissipate heat (Schwan and Piersol, 1954), and a local increase in temperature may be produced at some region sensitive to thermal stimulation that has not been included within the field of detection of a temperature-monitoring instrument. The net response will appear quite different from the response to hyperthermia produced by more ordinary means of heating. Further, thermal stimulation of peripheral nerves may not appear early in the experiment but may occur later, after body temperature has been elevated, further confounding interpretation of the response."

Field-Generated Forces

Electric fields can directly interact with matter and create forces that can act on molecules as well as on cellular and larger structures. Most of these interactions are reversible and do not necessarily have demonstrable biological effects. An example is the movement of ions in an ac field, which is inconsequential, provided that the field is weak enough to prevent undue heating from molecular collisions (e.g., below about 1 V/cm, corresponding to 1 mA/cm^2 in a physiological medium) (Schwan, 1978). Another example is the orientation of polar macromolecules. For field strength values of interest here, only a very partial preferential orientation with the field results. Complete orientation and consequent dielectric saturation requires field strengths of thousands of volts per centimeter. Changes of this magnitude do occur in membranes on depolarization. Hence, field-induced orientation and changes in orientation of membrane molecules appear possible. Corresponding tissue current densities would be in milliamperes per square centimeter (Schwan, 1978).

Electric fields can interact just as well with nonpolar cells and organelles in the absence of any net charge. These "ponderomotoric" forces are well known and understood. Any system exposed to an electric field will tend to minimize its electric potential energy by appropriate rearrangement. This statement is equally true for dc and ac fields, because the potential energy is a function of the square of the field strength. Inasmuch as the induced dipole moment of a cell or large particle depends on both the square of field strength and the volume, it is not surprising that the threshold field to overcome thermal agitation is proportional to $R^{-1.5}$, where R is the effective radius of the particle (Schwan, 1978). Experimental evidence confirms the principle; threshold field values for responses of 10 μm cells are about 10 V/cm. But for 10 nm macromolecules, they are about 10 kV/cm and comparable with the fields needed for complete orientation, owing to the existence of a typical dipole moment of about 10 Debyes (Schwan, 1978).

Table 2 summarizes observed manifestations of field-generated forces. The field effects may manifest themselves as orientation of particles in the direction of the field or perpendicular to it. Or "pearl chain" formation, i.e., the alignment of particles in the field direction may occur. This has long been considered a mysterious demonstration of microwave induced biological effects. Deformation or destruction of cells can be achieved with fields. The movement of cells in inhomogeneous electrical fields can be affected.

Zimmermann et al. (1974) have observed the destruction of red cells and ghost formation. Neumann and Rosenheck (1972) studied the effects of fields on chromaffin vesicles. Friend et al. (1974), as well as Goodman et al. (1976) studied the effects of fields on

Table 2. Field Generated Forces*

Orientation
"Pearl Chain" Formation
Deformation
Movement
Destruction

*from Schwan (1978)

fairly large cellular organisms. Orientation effects have been observed by Teixeira-Pinto et al. (1960), by Sher (1963) and Novak and Bentrup (1973). Pohl (1973) developed "dielectrophoresis" as a tool for separating cells in inhomogeneous fields. And Elul (1967), observed cell destruction phenomena and cell shape changes. Additional discussions have been presented by Schwan (1977) and Bawin and Adey (1977). Some of these field-generated force effects can be very dramatic especially near the tip of small electrodes. Of a similar nature is the movement of magnetotactic bacteria in magnetic fields of fairly low intensity (Blakemore, 1975). Apparently, these bacteria are equipped with magnetic properties and are therefore significantly oriented by the magnetic field to move in the field direction.

Field forces due to the induced dipole moment of the field have been listed as evidence of nonthermal action of electric fields on biologic systems. However, the effects require fairly large field strengths frequently above those that give rise to heating or stimulation of excitable tissues. The field forces also depend on the electric properties of the particular particle and its environment (Schwan, 1978).

In general, available evidence and present understanding indicate that significant effects with field-evoked forces require field strength values above 1 V/cm in the medium, unless cellular dimensions are well above 100 μm (Schwan, 1978).

Membrane Interactions - Nerve Cell Stimulation

There are reports that indicate electrophysiological properties of single cells, especially the firing rates of neurons in isolated preparations, may be affected by RF radiation at SARs as low as 1 W/kg.

Table 3 attempts to summarize information relevant to electrical fields and their effects on biological membranes. Membrane destruction can be achieved with low frequency alternating fields of the order of some hundred millivolts across the membrane. The propagation of action potentials along nerves is initiated or interfered with by pulses or low frequency potentials of roughly 10 mV across the membrane. Corresponding current densities and field strength values in tissues and the medium external to the affected cell are of the order of 1 mA/cm^2 and 1 V/m (Schwan, 1971, 1972; NAS, 1977).

Some reports also suggest effects due to ELF fields of the order of V/cm in air on timing responses and calcium efflux (Bawin and Adey, 1976). Corresponding *in situ* fields would be on the order of 0.1 μV/cm as listed in the table and corresponding fields

Table 3. Electrical Field Effects on Membranes*

	ΔV_M	E, in situ
Membrane Destruction	100-300 mV	
Action Potential (Excitation)	10 mV	1 V/cm
Subtle Effects	0.1 - 1 mV	
Extraordinary Sensitivities		
A. Related to Membranes	0.1 μV	0.01 μV/cm
B. Possibly not Related to Membranes	< 1 nV	0.1 μV/cm
Microwave Sensitivities (1 GHz)	1 μV	1 V/m

*from Schwan (1978)

across membranes below 1 nV. It is, however, not yet obvious if the reported effects are caused by membrane processes and hence the reduction of external fields to in situ fields and then membrane potentials is not necessarily sensible. A more detailed discussion of this topic is given by Schwan (1971) and Bawin and Adey (1976) and the detailed report of the National Academy of Sciences-National Research Council on the Biological Effects of Electric and Magnetic Fields (NAS, 1977).

Microwave sensitivities of the order of 1 V/cm in situ have been frequently reported and correspond to external flux values of the order of 1 to 10 mW/cm^2 (Baranski and Czerski, 1976). Some suspect that these sensitivities correspond to direct interactions with the central nervous system. However, it is straightforward to translate in situ field levels to corresponding membrane potentials and these are at levels of the order of 1 μV or less depending on microwave frequency (Schwan, 1971).

In Table 4 available evidence on the threshold of biological excitation phenomena is summarized for various fields. In cardiology extended experience exists with pacemakers and threshold values range about 0.1-10 mA/cm^2 depending on electrode size and other parameters (Roy et al., 1976). In electrohypnosis, electrosleep and electrical anaesthesia total currents applied are about 10 to 100 mA. Corresponding current densities in the brain may be estimated based on the work by Driscoll (1970). For a total current applied to the head of 1 mA, internal brain current densities are of the order of 10 μA/cm^2. Hence 10-100 mA total current corresponds to brain tissue current densities of the order of 0.1-1 mA/cm^2. Very extended work has been carried out on electrical hazards caused by low frequency potentials applied to the human body (Schwan, 1972). The values quoted in the table as thresholds for sensation, "let go" current and fibrillation are all consistent with a current density of about 1 mA/cm^2. Thus membrane potentials in the mV range as discussed

Table 4. Biological Thresholds*

Cardiology	0.03 - 10 mA/cm^2
Electrosleep and Electrical Anaesthesia	10 - 100 mA
Electro-Hazards:	
Sensation	1 mA
"Let Go"	10 mA
Fibrillation	100 mA

*from Schwan (1978)

above are consistent with the experience gained with pacemakers, and the effects on brain tissue and in the electrical hazards field.

Coupling of ELF fields in the spectrum below 100 GHz will be weak in comparison with radio or microwave fields having the same electric gradients in air (Schwan, 1978).

Evidence that membrane interactions do indeed occur has suffered from inappropriate experiments, often with inadequate controls and lack of repeatability in the hands of others. Nevertheless, certain reports should be noted. Some involve "extraordinary" sensitivities, based on special transductive systems. These include such structural devices as ferromagnetic particles in bacteria (Blakemore, 1975; Kalmijn and Blakemore, 1978) and the tissues of bees (Gould, Kirschvink and Deffeyes, 1978). Similarly, specialized sense organs in the skin of sharks and rays allow detection of oscillating low frequency fields, as weak as 10^{-8} V/cm (Kalmijn, 1971). In mammals, no such specialized receptor systems are known to exist, but equivalent sensitivities to environmental electric fields have been suggested by the reports of modification of biological rhythms (Wever, 1968, 1975) and intrinsic estimates of time intervals (Gavalas-Medici and Day-Magdaleno, 1976).

Biopolymer Interactions

The data for water are very well established. The values quoted for biopolymers and bound water are approximate and may readily vary by a factor of two or more with the type of biopolymer and the amount of bound water as well as the distribution of characteristic frequencies observed for bound water. Bound water properties have been discussed by Schwan (1957,1965), Grant (1965,1966), Pennock and Schwan (1969), Harvey and Hoekstra (1972). Biopolymer properties have been summarized by Oncley (1943) and by Takashima and Minakata (1975).

Based on the analysis of tissue dielectric properties by Schwan (1957, 1965) and Johnson and Guy (1972). The specific volume absorption rate as expressed by κ is greatest for electrolytes up to about 0.5 MHz. Water dominates above 4 or 5 MHz. Bound water may be the major absorber between 0.4 and 4 GHz, depending on its detailed properties. The ionic contributions and water clearly dominate over the total frequency range. Biopolymer and protein bound water contribute only insignificantly to the total absorption of energy.

The bound water and biopolymer contributions were estimated from the bound water work carried out by Grant (1965,1966) and Schwan (1957, 1965), Pennock and Schwan (1969) and their associates and by Oncley (1943) and Takashima and Minakata (1975). The analysis is therefore restricted to an upper frequency of about 10 GHz and more dielectric data are needed at higher frequencies. It is possible that additional vibrational and other modes of interaction may occur above 30 or 100 GHz as reviewed by Illinger (1974, 1978)). But the nonresonant behavior of water has been fairly well established up to nearly 100 GHz (Schwan et al., 1976) and appears to preclude the possibility of resonant types of interactions at frequencies below 100 GHz. It also must be pointed out that the above analysis applies to the abundant tissues with high water content. Little is known for other tissues including bone and fatty tissues; the bound water contribution in cerebral tissues may be larger.

Considerable work has been done on the interaction of electrical fields with biopolymers. The behavior of proteins has been investigated in detail at radiofrequencies by Oncley (1943). Low frequency and radiofrequency work on proteins and nucleic acids have been summarized by Takashima and Minakata (1975). All reported interactions are of the nonresonant relaxation type. These interactions are linear and reversible up to field strength values of many volts per centimeter in the suspending medium (Schwan, 1978). The limit of linearity is in the KV/cm range. Hence interactions in the V/cm range are characterized by responses accompanied by changes in potential energy which are very small in comparison with the thermal energy, KT. More simply expressed, the degree of orientation experienced by a polar protein molecule exposed to a field of 1 V/cm is very small. Thermal collisions with other molecules are far more effective than the field in affecting the orientation and virtually prevent any degree of orientation with the field (Schwan, 1978).

Biopolymer interactions are due to either parallel partial rotation of the total molecule, or, at microwave frequencies, also include relaxation effects due to partial rotation of polar groups or of protein bound water (Schwan, 1965; Grant, 1965). Resonant responses are possible at frequencies above 100 GHz where the viscous properties of water do not prevent resonances. But none have been identified. A detailed discussion of these speculative responses has been presented by Illinger (1974, 1978).

Changes in Calcium Kinetics

Increased mobilization of calcium ions has been reported to occur in brain tissue exposed in vitro to RF energy amplitude modulated at ELF frequencies recorded in the electroencephalogram (EEG) of awake animals. The response appears to be based on the intensity of the electric fields within the tissue which can be related to SAR; the lowest effective SAR in in vitro samples is estimated to be 0.0013 W/kg. Calcium ion efflux under these conditions apears to be a nonlinear effect in terms of both amplitude modulation frequency and field intensity; that is, the response occurs at specific frequencies and electric strengths. The physiological significance of this effect has not been established. Details of the studies on electric and magnetic field influence on calcium fluxes will be present in other chapters of this book.

Millimeter-Wave Interactions

In a series of papers Fröhlich (1968, 1975a, 1975b, 1980, 1983) presented his concepts of long range coherence and energy storage in biological systems. This information applied to very high frequencies of oscillation in the range of 10^{11} to 10^{12} Hz (greater than 100 GHz). Fröhlich's theory of long range coherence and energy storage is based on dipole interactions and the recurrence of certain bonds (such as H-bonds) in macromolecules. It is believed that the frequency of these oscillations lies roughly between 100 and 1000 GHz.

Fröhlich has offered a model of long-range coherence and energy storage in central nervous system tissue based on dipole interactions and on the nature of certain chemical bonds in the macromolecules that comprise the membranes of these cells. This model predicts that a supply of (electromagnetic) energy above a certain threshold could lead to a metastable state in the membrane wherein excitation of coherent electric vibrations in the membrane molecules could alter biological processes in the cell. Selective, very-high-frequency electromagnetic radiation in the millimeter band (i.e., greater than 40 GHz) appears to be required to demonstrate these effects (Devyatkov, 1974), which also seem to require a power density in the range of at least 10,000 microwatts/cm^2 (Potapov et al., 1974). Since transmissions of this nature are highly experimental, involving little or no direct human exposure outside the laboratory, the physiologic pertinence of these models is highly questionable. An extensive review of present knowledge regarding postulated specific tissue interactions with radio frequency electromagnetic fields has been published by Adey (1980, 1981) in which he notes that "without modulation at frequencies below 1 kilohertz, and for the [central nervous system] CNS, below 20 Hz, biological sensitivity to these fields drops sharply, and virtually all observed effects of unmodulated fields appear to have a thermal basis. As predicted from physical considerations, it is only in the millimeter microwave and far infra-red regions that direct molecular resonances are now being detected."

For the past decade a number of investigators have been stimulated to look for the frequency-dependent effects predicted by Fröhlich (1968) and suggested by the experiments of Webb and Booth (1969, 1971) in the US and a group of Soviet workers led by Smolyanskaya and Vilenskaya (1974). Fröhlich suggested that critical oscillations exist in macromolecules which determine the activity and function of the organism. It is believed that the frequencies of these oscillations lie roughly between 100 and 1000 GHz. It is reasonable to conjecture that functional activities occurring at the macromolecular level depend on critical, and perhaps matching oscillations being present at the right place and time for a given reaction to occur. Such oscillations might be finely tuned and "metastable," in which case small inputs of energy to one or both halves could cause disruption in the progress of the function or reaction. Fröhlich and others postulate that there is a type of threshold or limit cycle in metabolic excitation of large-amplitude vibrations (Fröhlich, 1980; Adey, 1981).

It must be remembered that at millimeter-wave frequencies, absorption of energy by genetically important material is essentially precluded due to the very shallow depth of penetration. Millimeter waves only have a penetration depth of 0.5 mm in aqueous solutions.

Yeast and E. coli are favorite unicellular organisms for biological study. The fruit fly, Drosophila melanogaster is the most-studied insect, representing a highly differentiated species that is especially well suited as a model for studies concerned with effects of various agents on genetic material. There have been relatively few millimeter-wave investigations using this insect. Because of its size, it is particularly well suited for high-frequency, short-wavelength studies. However, just as with the yeast, there are some apparent discrepancies in the reported results of the work that has been done. For example, Zalyubovskaya (1974) reported decreased fertility (37 to 55 GHz, 10 mW/cm^2, adult insects), while Dardalhon et al. (1979) found an increase (17 GHz, 60 mW/cm^2, adult females only). No evidence has been found for a mutagenic effect of millimeter waves in the fruit fly.

The most recent study of millimeter-wave effects in Drosophila, has been conducted by Nimtz (1983). Pupae were exposed for 120 hours at an intensity of 10 $\mu W/cm^2$ and a frequency of 40 GHz. In this series of six experiments with over 82,900 flies, millimeter waves had no apparent effect.

The biological action of electromagnetic waves in the millimeter range is of considerable interest. One would not anticipate changes in living organisms upon exposure to millimeter waves. Such an assumption is based on the fact that energy in the millimeter band has low quantum energy, whose action does not break even the weakest chemical bonds of atoms in molecules, or lead to irreversible damage of the latter. Biological shifts that arise upon the action of microwave radiation in the millimeter band are dependent on wavelength, spectral flux of power, and time of exposure. For example, when the wavelength (λ) is changed by 0.01 mm, the biological effect (BE) showed up or disappeared with the change in λ. As has been shown in numerous experiments, the dependence of BE on λ is resonant in nature. In comparing millimeter waves with the longer-wavelength energy, we should note that 70 percent of the energy is absorbed in the tissues of experimental animals at a depth of 300 μm (Vilenskaya et al., 1971). For this reason it is impossible to produce deep local hyperthermia in the millimeter wave band, even with high intensity of radiation, with the exception of elevated temperatures in surface layers of skin. In this connection, it has been suggested that millimeter waves cannot cause serious changes in living organisms, and consequently they can be categorized as harmless forms of action. Results found on bacterial systems (Smolyanskaya and Vilenskaya, 1974), yeasts (Devyatkov, 1974), viruses (Kiselev and Zalyubovskaya, 1978), the hematopoietic system and tumorous processes in experimental animals (Potapov et al., 1974; Sevast'yanova, 1979; Sevast'yanova and Potapov, 1976; Sevast'yanova, 1976; Sevast'yanova and Potapov, 1969; Sevast'yanova and Vilenskaya, 1974) have shown that millimeter waves do not produce gross shifts in cells or tissues, and do not stimulate growth of malignant neoplasms.

Gandhi et al. (1979) described a series of studies dealing with the action of millimeter waves on microorganisms and measurement of the absorption spectra of living tissue in the millimeter wave band. An investigation was made of questions of the action of millimeter waves on the rate of growth of E. coli and the infectivity of La Crosse virus. The paper briefly describes the experimental procedure and gives the parameters of irradiation. In this research, no effect on the growth rate of E. coli or infectivity of La Crosse virus was detected. Considerable research by this group has been devoted to measuring the absorption spectra of biological objects in the millimeter wave band. In this research, attempts to reproduce the results of work by Webb and Booth (1969, 1971) and Stamm et al. (1974) were unsuccessful. The authors failed to get frequency-dependent absorption spectra and felt that there is little likelihood of great differences in absorption between living cells and water, although they do not deny the possibility that there may be such a spectrum with the slightest differences in absorption characteristics.

Several investigators have suggested the athermal action of resonant electromagnetic fields in the millimeter wave band on the life activity of microorganisms. A number of suggestions have been made about the possible mechanisms of such effects. However, both theoretical arguments and their experimental verification are still in the initial stage. It would also be desirable to refine experimental procedures and to standardize them.

Motzkin (1979) has noted that many frequency-dependent biological effects have now been described in relation to the action of millimeter radio waves on biological systems. However, the understanding of the physical and molecular principles of interaction of electromagnetic fields with biological systems remains incomplete. It is necessary to define the mutual relation of physical parameters of the electromagnetic field with cellular species, their orientation and functions. This is impeded by the ambiguity and contradictory nature of results reported by different researchers. Motzkin feels that elucidation of mechanisms of chemical and biophysical interaction necessitates use of simpler systems that enable analysis of the effects of action of the electromagnetic field. It is suggested that in the future, work should be done with inanimate synthetic models of the vesicle membrane type; the complexity of such models can be gradually increased to more exact simulation of biological membranes. For an understanding of the interaction of millimeter waves with membranes at low specific power, it is necessary to determine the shifts that may be produced in biological structures by potential, temperature and phase changes.

<u>Orientation, Navigation and Predation by Electrosensing in Marine Vertebrates</u>

In recent years some extraordinary sensitivities have been reported. Electrosensitive species such as rays and sharks detect fields of intensities as low as 0.01 μV/cm. In order to achieve these sensitivities they sample the field over considerable distances with the aid of special organs, the Ampullae of Lorenzini, and operate over a small frequency range extending from dc to only a few Hertz (NAS, 1977; Kalmijn, 1966).

It may be argued that the microwave-induced membrane potential of about 1 μV is comparable and even higher than the perception level across the endepithelium of the Ampullae of Lorenzini. However, it should be realized that the high sensitivity of this endorgan is only achieved over a narrow bandpath range of some Hertz. If microwave sensitivities existed over such narrow bandpath ranges they would hardly be noticeable experimentally (Schwan, 1978).

Behavioral responses to weak electric and magnetic fields have been reported to occur in certain aquatic vertebrates (Kalmijn, 1979; Bennett, 1971) as well as in other species. Sharks and rays, for example, have been described as responding in various ways to low levels of electromagnetic radiation but not to higher levels, suggesting a "power window" effect with respect to detection of electromagnetic radiation.

From observations of electric fish (Gymnarchus) that emit continuous trains of pulses at rates between 300 and 1,000 Hz, Lissman (1958) proposed that these pulses are essential in locating environmental objects, and that weak electric fish have evolved from a preelectric fish lacking specialized electric organs but already sensitive to electric fields. Subsequent research has shown that catfish, sharks and rays, though lacking electric organs, detect tissue potentials arising in surrounding animals, including predators and prey, as well as fields arising within the animal's own body.

Studies of this electrosensitivity in fish have used weak DC and low frequency oscillating electric fields. Some studies have examined sensitivity to the earth's magnetic field in moving subjects; others have tested sensitivity to electric fields induced in moving sea water at field levels expected to occur through interaction of the moving fluid with the earth's magnetic field. Escape reactions occur with local DC fields of 1-10 mV/cm. Escape reactions also occur in blindfolded sharks when a piece of steel wire comes within several centimeters of the body (Dijkgraff and Kalmijn, 1962). Square-wave 5.0 Hz fields elicited the same response and also caused eyelid contraction. Further studies by Kalmijn (1966, 1971) have strikingly demonstrated the extreme sensitivity of sharks and rays to these fields.

Although there is good evidence that electromagnetic sensing in birds and insects offers a basis for biologically relevant activities, including navigation, uncertainties and controversies surrounding important aspects of these studies exemplify comparable problems in evaluation of related studies in mammals.

The earth's magnetic field affects the orientation of the "waggle run" in the dance of honey bees (Lindauer and Martin,

1968, 1972). The gravitational field appears to be a symbol for the sun in this dance, so that the angle of the dance changes at approximately 15 degrees per hour relative to gravity as the bee compensates in time for the changing azimuth throughout the day. The most accurate runs occur at those times of day when the dance angle coincides with the magnetic vector. Errorless runs can be induced by artificial compensation of the magnetic field. Bees orient the sheets of comb within the hive to the geomagnetic field in absence of other cues (Crane, 1974) and build bizarre combs in stronger or unusual fields (Martin and Lindauer, 1973). With fields changing as slowly as 1.5×10^{-6} G/min, bees are slowed in orienting responses to about 10 percent of normal rates and take about 30 min to adjust to a new field. Although a region of transversely oriented magnetic material has been reported on the front of the bees abdomen, appearing in the pupal state and persisting in adults (Gould, Kirschvink and Deffeyes, 1978), this period of adaptation argues against a role in simple reflex responses for either ferromagnetic particles within sensory cells or electric charge displacement by movement through the field. Termites (Becker, 1976) and beetles (Arendse and Vrins, 1976) are also sensitive to steady magnetic fields (Adey, 1983).

Evidence for insect sensitivities to electric fields is less substantial and the findings are somewhat contradictory. In 50 Hz high voltage fields (3-5 kV/m), bees have higher metabolic rates than controls, with higher food intake and activity and reduced life-span of worker bees (Altmann, 1969). Controls and data in these studies are difficult to evaluate, and the picture is further complicated by reported effects in plastic hives that are absent in wooden hives (Warnke, 1976).

There is now substantive evidence that birds may use a variety of cues in the course of a migratory flight, including landmark recognition, sun altitude, azimuth, a precise time sense, style patterns, the wind and weather, as well as the geomagnetic field. Many early studies of bird orientation in natural and artificial magnetic fields have been questioned on theoretical grounds, or for lack of statistical validation, or because of difficulties in reproducibility. A possible role for the geomagnetic field was first suggested by Middendorf in 1855, and since that time, experiments have tested models based on a comparison of field intensities, travel along a magnetic contour, and movement up or down a gradient (Keeton, 1977). Results have been conflicting, partly because of difficulties in designing experiments that are unequivocal in the evaluation of perception and response to signals that humans may not perceive (Adey, 1977). By contrast, there have been statistically significant results from experiments iin which discrimination behavior in the magnetic environment was associated with unrestrained movement (Bookman, 1977). For example, pigeons were successfully trained in a flight tunnel 3 m long to discriminate between a simulated field of 0.5 G and the field reduced to 0.02 G at paired feeding stations at the termination of the tunnel. Discrimination was statistically significant in trials associated with "fluttering," defined as more than 3 sec of jumping, hovering or flight. "No-flutter" trials yielded random responses (Adey, 1983).

The extreme sensitivity to small differential changes in ambient magnetic field levels has been a striking aspect of these avian studies and a consistent finding among many of them. Minor fluctuations of 10^{-4} - 10^{-5} G associated with perturbations in solar particle fluxes are detected by homing pigeons (Keeton, Larkin, and Windsor, 1974). Orienting by the European robin becomes random after a certain change in the magnetic field strength by 0.1-0.2 G but may resume its directed character after adaptation over several days (Adey, 1983).

No such specialized electroreceptors are known to exist in the mammalian peripheral or central nervous systems, although intracellular ferromagnetic bodies have been described in the skull and muscles of pigeons (Presdi and Pettigrew, 1980).

Circadian Rhythms

Altered behavior has been reported with 60 Hz fields at levels below 10 kV/m, and thus apparently not associated with significant hair movement. For example, during the first days of a 30-day field exposure, motor activity in male albino rats was considerably reduced by exposure to a vertical 1 kV/m 60 Hz fields in the period after midnight. Normally, rats exhibit a major diurnal peak in their motor activity after midnight. This reduction was significant in comparison with animals exposed to 50 and 500 V/m fields and to control conditions. This decreased motor activity was not sustained beyond the first days of field exposure at 1 kV/m (Sagan et al., 1982).

In these same test subjects, the relative motor activity during the epoch from 0900 to 1000 hours was consistently but not significantly greater in the 1 kV/m period than in the 500 V/m and control conditions. The stimulus to this heightened activity related to the presence of animal handlers engaged in feeding procedures at this time (Adey, 1983).

These responses were interpreted as indicating a possible modification in circadian patterns of general motor activity associated with the start of exposure to a 1 kV/m field and a possibly enhanced sensitivity to external stimuli associated with the 1 kV/m condition (Sagan et al., 1982).

Auditory-Acoustic Effect

An acoustic effect of pulsed microwaves was long suspected to represent a direct interaction of the microwaves with the central nervous system. However, subsequent work by Foster and Finch (1974) and by Chou et al. (1975), supports a reasonable classical macroscopic and thermoacoustic explanation. Therefore, reported effects on the behavior of test animals exposed to pulsed fields cannot be considered as evidence for a direct nonthermal microwave interaction with the central nervous system (Schwan, 1978).

Behavior

Some types of animal behavior are disrupted at SARs that are approximately 25 to 50 percent of the resting metabolic rate of many species. For example, changes in locomotor behavior in rats occur at an SAR of 1.2 W/kg and alterations in thermoregulatory behavior in squirrel monkeys occur at an SAR of 1 W/kg. Decreases in other operant or learned behavioral responses during exposures have been found at an SAR 2.5 W/kg in the rat and at 5 W/kg in the Rhesus monkey. The reported behavioral alterations appear to be reversible with time (Adey, 1983).

The search for behavioral correlates of interactions between living organisms and environmental ELF fields presents serious experimental difficulties. The behaviors which may be modified by these interactions are generally subtle, and similarly amenable to modification by other weak environmental stimuli. Intercurrent conditions can therefore seriously influence the statistical validity of many reported reactions (Adey, 1983).

Attempts to establish threshold levels for behavioral sensitivity to environmental ELF fields have lead to sharply differing results that appear related to the specific behavioral criterion under test. For this reason, there is much continuing research in which no final conclusions have yet been provided (Adey, 1983).

Conclusion

Several mechanisms responsible for the interaction of radiofrequency and other electrical fields with biological systems have been described. Foremost is the heat development which results from the absorption of microwaves. The relative contribution to this heat development caused by the various tissue constituents including ions, water, biopolymers, bound water and lipids are noted. Direct field interactions with various biocomponents are also considered. These include membrane interactions, biopolymer interactions, interactions with biological fluids, and field generated forces acting on biological particles and cells. These forces are frequently neglected in discussions and yet pertain to a large number of published observations.

Schwan (1978) believes that reported extraordinary sensitivities to electromagnetic fields should be confirmed by independent investigators and that the experimental regime must be carried out with the greatest care in order to avoid artifacts and provide hard statistical evidence. In the meantime, the reported evidence can neither be accepted or refuted with certainty. It should also be pointed out that for external flux levels of the order of 1 to 10 mW/cm^2 subtle temperature elevations become apparent in man and animals which may well be perceptible but which are not therefore necessarily dangerous (Johnson and Guy, 1972). There is some evidence that the development of "hot spots" is more likely in small animals than in man (Kritikos and Schwan, 1975). If this can be definitely established, higher sensitivities can be expected in small animals than in man and extrapolation from animal experiments to man requires much more circumspection than it has in the past.

ACKNOWLEDGEMENT

This paper is based on work performed under contract No. DE-AC02-76EV03490 with the U.S. Department of Energy and PHS Grant No. ES 03239 at the University of Rochester Department of Radiation Biology and Biophysics and has been assigned report No. DOE/EV/03490-2441.

REFERENCES

Adey, W.R., 1977, The sensorium and the modulation of cerebral states: tonic environmental influences on limbic and related systems, Ann. NY Acad. Sci. 290:396-420.

Adey, W.R., 1980, Frequency and power windowing in tissue interactions with weak electromagnetic fields, Proc. IEEE, 68:119.

Adey, W.R., 1981, Tissue interactions with nonionizing electromagnetic fields, Physiol. Rev., 61:435.

Adey, W.R., 1983, Some fundamental aspects of biological effects of extremely low frequency (ELF), in: "Biological Effects and Dosimetry of Nonionizing Radiation Radiofrequency and Microwave Energies," M. Grandolfo, S.M. Michaelson and A. Alessandro, eds., pp. 561-577, Plenum, NY.

Altmann, G., 1969, The physiological effect of electric fields on organisms, Arch. Met. Geoph. Biokl. Ser. B. 17:269.

Arendse, M.C., and Vrins, J.C.M., 1976, Magnetic orientation and its relation to photic orientation in Tenebrio Molitor L. (Coleoptera, Tenebrioniadae) Netherlands J. Zool. 25:407-437.
Baranski, S., and Czerski, P., 1976, "Biological Effects of Microwaves," Dowden and Ross, Inc, Stroudsburg.
Bawin, S.M. and Adey, W.R., 1976, Sensitivity of calcium binding in cerebral tissue to weak environmental electric fields oscillating at low frequency. Proc. Natl. Acad. Sci. USA 73:1999-2003.
Bawin, S.M., and Adey, W.R., 1977, Brain interactions with weak electric and magnetic fields, Neurosci. Res. Prog. Bull., 15:1.
Becker, G., 1976, Reaction of termites to weak alternating magnetic fields, Naturwissenschaften 63:201-202.
Bennett, M.V.L., 1971, Electrolocation in fish, Ann. NY Acad. Sci., 188:242.
Blakemore, R.P., 1975, Magnetotactic bacteria, Science, 190:377.
Bookman, M.A., 1977, The sensitivity of the homing pigeon, Columbia livia, to an earth strength magnetic field, Nature, London 267:340-342.
Chou, C.K., Galambos, R., Guy, A.W., and Lovely, R.H., 1975, Cochlear microphonics generated by microwave pulses, J. Microwave Power 10:361.
Crane, E., 1974, Directions in which bees build combs, Bee World 55:153-155.
Dardalhon, M. Averbek, D., and Berteaud, A.J., 1979, Determination of thermal equivalent of millimeter waves in living cells, J. Microwave Power 14:307.
Devyatkov, N.D., 1974, Influence of electromagnetic radiation in the millimeter wave band on biological objects, Uspekhi Fizicenskikh Nauk 110(3):453.
Dijkgraff, S., and Kalmijn, A.J., 1962, Verhaltensversuche zur Funktion der Lorenzinischen Ampullen, Naturwissenschaften 49:400.
Driscoll, D.A., 1970, An Investigation of a Theoretical Model of the Human Head with Application to Current Flow Calculations and EEG Interpretation, Ph.D. Thesis, University of Vermont.
Elul, R., 1967, Fixed charges in the cell membrane, J. Physiol., 189:351.
Foster, K.R., and Finch, E.D., Microwave hearing: evidence for thermoacoustic auditory stimulation of pulsed microwaves, Science, 185:256.
Friend, A.W., Finch, E.D., and Schwan, H.P., 1974, Low frequency (1 Hz to 10 MHz) electric field-induced changes in the shape and motility of amoebas, Science, 187:357.
Fröhlich, H., 1968, Long range coherence and energy storage in biological systems, Int. J. Quan. Chem. 2:641.
Fröhlich, H., 1975a, Evidence for bose condensation-like excitation of coherent modes in biological systems, Phys. Lett. 51A:21.

Fröhlich, H., 1975b, The extraordinary dielectric properties of biological materials and the action of enzymes, Proc. Nat. Acad. Sci. USA 72:4211.
Fröhlich, H., 1980, The biological effects of microwaves and related questions, in: "Advances in Electronics and Electron Physics," L. and C. Marton, eds., p. 85, Academic Press, NY.
Fröhlich, H., 1983, Coherence in biology, in: Coherent Excitations in Biological Systems," H. Fröhlich, and F. Kremer, eds., Springer-Verlag, Berlin and Heidelberg.
Gandhi, O.P., Hagmann, M.J., Riazi, A., Hill, D.W., and Partlow, L.M., 1979, Millimeter wave and Raman spectra of living cells--some problems and results, in: "Mechanisms of Microwave Biological Effects," University of Maryland.
Gavalas-Medici, R., and Day-Magdaleno, S., 1976, Extremely low frequency, weak electric fields affect schedule-controlled behavior of monkeys, Nature, London 261:256-258.
Goodman, E.M., Greenebaum, and Marron, M.T., 1975, Effects of extremely low frequency electromagnetic fields on growth and differentiation of Physarum polycephalum, Technical Report Phase I (Continuous Wave), University of Wisconsin and Goodman, E.M., Greenebaum, and Marron, M.T., 1976, Effects of extremely low frequency electromagnetic fields on Physarum polycephalum, Technical Report, Office of Naval Research Contract N-00014-76-C-0180, University of Wisconsin.
Gould, J.L., Kirschvink, J.L., and Deffeyes, K.S., 1978, Bees have magnetic resonance, Science, 201:1026.
Grant, E.H., 1965, The structure of water, neighboring proteins, peptides and amino acids as deduced from dielectric measurements, Ann. N.Y. Acad. Sci., 125:418.
Grant, E.H., 1966, Dielectric dispersion in bovine serum albumin, J. Mol. Biol., 19:133.
Harvey, S.C., and Hoekstra, P., 1972, Relaxation spectra of water adsorbed on lysozyme, J. Phys. Chem., 76:2987.
Illinger, R.H., 1974, Interaction between microwave and millimeter wave electromagnetic fields and biological systems: Molecular mechanisms, in: "Biologic Effects and Health Hazards of Microwave Radiation," P. Czerski et al., eds., Polish Medical Publishers, Warsaw.
Illinger, R.H., 1978, Millimeter wave and infrared absorption in biological systems, pp 43-64, in: "The Physical Basis of Electromagnetic Interactions with Biological Systems," L.S. Taylor, and A.Y. Cheung, eds., HEW Publ (FDA) 78-8055.
Johnson, C.C., and Guy, A.W., 1972, Nonionizing electromagnetic wave effects in biological materials and systems, Proc. IEEE, 60:692.
Kalmijn, A.J., 1966, Electroperception in sharks and rays, Nature, London 212:1232-1233.
Kalmijn, A.J., 1971, The electric sense of sharks and rays, J. Exper. Biol., 55:371.

Kalmijn, A.J., 1979, Electromagnetic guidance systems in fishes, in: "Magnetic Field Effects on Biological Systems," T. Tenforde, ed., Plenum Press, NY.
Kalmijn, A.J., and Blakemore, R.P., 1978, The magnetic behavior of mud bacteria, in: "Animal Migration, Navigation and Homing," K. Schmidt-Konig, and W.T. Keeton, eds., Springer, New York.
Keeton, W.T., 1977, Biological sensitivity to magnetic fields in orientational responses. Neurosci. Res. Program Bull. 15:22-27.
Keeton, W.T., Larkin, T.S., and Windsor, D.M., 1974, Normal fluctuation in the earth's magnetic field influence on pigeon orientation. J. Comp. Physiol. 95:95-103.
Kiselev, R.I., and Zalyubovskaya, M.N., 1978, Investigation of inhibiting action of millimeter microwaves on adenovirus, Vorprosy Virusologii, No. 5:617.
Kritikos, A.N., and Schwan, H.P., 1975, The distribution of heating potential inside lossy spheres, IEEE Trans. Biomed. Eng. 22:457.
Lindauer, M., and Martin, H., 1968, Die Schwereorientierung der Bienen unter dem Einfluss des Erdmagnet feldes. Z. Vgl. Physiol. 60:219-243.
Lindauer, M., and Martin, H., 1972, Magnetic effect on dancing bees, in: "Animal Orientation and navigation," Galler, S.R. Schmidt-Konig, K., Jacobs, G.J. and Belleville, R.E., eds. NASA Publ. SP-262. US Govt. Printing Office, Washington, D.C., pp. 559-567.
Lissman, H.W., 1958, On the function and evolution of electric organs in fish, J. Exper. Biol. 35:156-191.
Martin, H., and Lindauer, M., 1973, Orientierung in Erdmagnetfeld, Fortschr. Zool. 21:211-228.
McAfee, R.D., 1962, Physiological effects of thermode and microwave stimulation of peripheral nerves, Am. J. Physiol. 203:374.
Motzkin, S., Birenbaum, L., Rosenthal, S., Rubenstein, C., Davidow, S., Remily, R., and Melnik, R., 1979, International Radio Science Meeting Bioelectromagnetics Symposium, June 18-22, Seattle, Washington, pp. 460.
Neumann, E., and Rosenheck, K., 1972, Permeability changes induced by electric impulses in vesicular membranes, J. Membrane Biol. 10:279.
Nimtz, G., 1983, On the microwave response of the Drosophila melanogaster, in: "Coherent Excitations in Biological Systems," H. Fröhlich, and F. Kremer, eds., Springer-Verlag, Berlin and Heidelberg.
National Academy of Sciences, USA, 1977, "Biologic of Electric and Magnetic Fields Associated with Proposed Project Seafarer," Washington, D.C.
Novak, B., and Bentrup, F.W., 1973, Orientation of Fucus egg polarity by electric ac and dc fields, Biophysik, 9:253.

Oncley, J.L., 1943, The electric moments and relaxation times of proteins as measured from their influence upon the dielectric constants of solutions, in: "Proteins, Amino Acids and Peptides as Ions and Dipolar Ions," E.J. Cohn, and J.T. Edsall, eds., Reinhold, New York.

Pennock, B., and Schwan, H.P., 1969, Further observations on the electrical properties of haemoglobin bound water, J. Phys. Chem., 73:2600.

Pohl, H.A., 1973, Biophysical aspects of dielectrophoresis, J. Biol. Phys., 1:1.

Potapov, S.L., Sevast'yanova, L.A., and Vilenskaya, R.L., 1974, Restorative processes in bone marrow with exposure to microwave radiation, Biologicheskiye Nauki, No. 3.

Presdi, D., and Pettigrew, J.D., 1980, Ferromagnetic coupling to muscle receptors as a basis for geomagnetic field sensitivity in animals, Nature, London 285:99-101.

Roy, O.Z., Scott, J.R., and Park, G.C., 1976, 60 Hz ventricular fibrillation and pump failure thresholds versus electrode area, IEEE Trans. Biomed. Eng. BME-23:45.

Sagan, P.M., Adey, W.R., Sabbot, I.M., and Bystrom, B.G., 1982, Effects of 60 Hz environmental electric fields on laboratory rats, Bioelectromagnetics, in press.

Schwan, H.P., 1957, Electrical properties of tissue and cell suspensions, in: "Advances in Biological and Medical Physics, Vol. V," Lawrence and Tobias, eds., Academic Press, Inc., New York.

Schwan, H.P., 1965, Electrical properties of bound water, Ann. N.Y. Acad. Sci., 125:344.

Schwan, H.P., 1971, Interaction of microwave and radiofrequency radiation with biological systems, IEEE Trans. MTT-19:146.

Schwan, H.P., 1972, Biological Hazards from Exposure to ELF Electric Fields and Potentials, NWL Technical Report TR-2713, U.S. Naval Weapons Lab., Dahlgren, VA.

Schwan, H.P., 1977, Some tissue determinants of interactions with electric fields, in: "Brain Interactions with Weak Electric and Magnetic Fields," W.R. Adey, and S.M. Bawin, eds., Neurosci. Res. Prog. Bull., 15:88.

Schwan, H.P., 1978, Classical theory of microwave interactions with biological systems, p 91-112, in: "The Physical Basis of Electromagnetic Interactions with Biological Systems," L.S. Taylor, and A.Y. Chenug, eds., Proceedings of a Workshop held at the University of Maryland College Park, Maryland, June 15-17, 1977, HEW Publ. (FDA).

Schwan, H.P., and Piersol, G.M., 1954, The absorption of electromagnetic energy in body tissues: A review and critical analysis, Am. J. Phys. Med. 33:371.

Schwan, H.P., Sheppard, R.J., and Grant, E.H., 1976, Complex permittivity of water, J. Chem. Phys. 64.

Sevast'yanova, L.A., 1976, Change in cellular composition of bone marrow under the action of microwaves combined with imiphos, Biologicheskiye Nauki, No. 7.

Sevast'yanova, L.A., 1979, Characteristics of biological action of millimeter radio waves, and possibilities of using them in medicine, Vestnik Akademii Meditsinskikh Nauk, No. 2:65.
Sevast'yanova, L.A. and Potapov, S.L., 1969, Combined action of X-rays and microwave radiation on bone marrow, Biologicheskiye Nauki, No. 6.
Sevast'yanova, L.A., and Potapov, S.L., 1976, Characteristics of action by radio waves in the millimeter band in combination with fluorafur on the hemogenic system, Biologicheskiye Nauki, No. 12.
Sevast'yanova, L.A., and Vilenskaya, R.L., 1974, Response of mouse bone marrow cells to variation in parameters of microwave irradiation in the millimeter band, Biologicheskiye Nauki, No. 6:48.
Sher, L.D., 1963, Mechanical Effects of AC Fields on Particles Dispersed in a Liquid: Biological Implications. Ph.D. Thesis, University of Pennsylvania, Philadelphia.
Smolyanskaya, A.Z., and Vilenskaya, R.L., 1974, Action of electromagnetic radiation in the millimeter band on functional activity of some genetic components of bacterial cells, Uspekhi Fizicheskikh Nauk, 110:571.
Stamm, M.I., Winters, W.D., and Morton, D.L., 1974, Microwave characteristics of human tumor cells, Oncology 29:294.
Takashima, S., and Minakata, A., 1975, Dielectric behavior of biological macromolecules, in: "Digest of Dielectric Literature," Nat. Acad. Sci. - Nat. Res. Council.
Teixeira-Pinto, A.A., Nejelski, L.L., Cutler, J.L., and Heller, J.H., 1960, The behavior of unicellular organisms in an electromagnetic field, Exp. Cell Res. 20:548.
Vilenskaya, R.L., Sevast'yanova, L.A., and Faleyev, A.S., 1971, Investigation of absorption of millimeter waves in the skin of experimental animals, Elektronnaya Tekhnika, Seriya I. Elektrotekhnika SVCh, No. 7:4.
Warnke, V., 1976, Effects of electric charges on honey bees, Bee World 57:50-56.
Webb, B.J., and Booth, A.D., 1969, Absorption of microwaves by microorganisms, Nature, 222:1199.
Webb, B.J., and Booth, A.D., 1971, Microwave absorption by normal and tumor cells, Science, 174:72.
Wever, R., 1968, Einfluss schwacher elektro-magnetischer Felder auf die circadiane Periodik des Menschen, Naturwissenschaften, 55:29.
Wever, R., 1975, The circadian multioscillatory system of man, Internat. J. Chronobiol., 3:19.
Zalyubovskaya, N.P., 1974, Reactions of living organisms to exposure to millimeter band electromagnetic waves, Soviet Physics-Uspekhi, 110:574.
Zimmermann, U., Pilwat, G., and Riemann, R., 1974, Dielectric breakdown of cell membranes, Biophys. J., 14:881.

WHAT DATA BASES FOR ELECTROMAGNETIC FIELDS EXPOSURE STANDARDS ?

Martino Grandolfo*

Physics Laboratory
Istituto Superiore di Sanità
00161 Rome, Italy

INTRODUCTION

Many concerns have been raised, both by the scientific community and the general public, that quite divergent opinions exist in industrialized countries on the nature and degree of hazard from exposure to electromagnetic fields. This divergence of opinion is reflected in widely differing national exposure standards. In response to this, there has been intense activity, both at the international and national level, on the evaluation of biological effects literature and assessment of health hazards of human exposure to non ionizing electromagnetic radiation. The health authorities of an increasing number of countries become involved in the development of recommendations, regulations or technical advice to limit the exposure of the workers and the general public and they are looking to the international organizations for some guidance in this respect. As a matter of fact, the methodology of protection against ionizing radiation which enabled the safe development of its industrial and medical applications all over the world is often quoted as an example, and every one has in mind the invaluable help provided by the international organizations to resolve the complex scientific and administrative problems encountered in the early management of radiation protection. A standard is a general term incorporating both regulations and guidelines and can be defined as a set of specifications or rules to promote the safety of an individual or group of people. Absolute assurances are rarely if ever practicable and specifying maximum permissible exposure limits to different hazards depends on the degree of risk that is acceptable scientifically and socially and cost versus benefit

*Member of the IRPA International Non-Ionizing Radiation Committee.

analyses are necessary, including the economic impact of such controls. With world opinion varying so widely on what exposures to electromagnetic fields are acceptable it is not surprising that very different criteria have been so far applied in different countries.

Among the many factors that go into the development of an exposure standard, the selection of a good scientific biological effects data base plays quite obviously the most important role: different scientific approaches have produced different philosophies of protection guidelines and thus different exposure limits.

In recent years there have been many comprehensive reviews of the scientific basis for setting standards intended for the control of human exposure to non ionizing electromagnetic fields, and here the ground will not be covered again in detail other than to present some general ideas on the debate which has accompanied the recent approval of new national and international safety standards, which in turn proves the need for a new data base as useful input for a definite risk assessment.

At the international level, the International Radiation Protection Association (IRPA) initiated activities concerned with non-ionizing radiation by forming a Working Group on Non-Ionizing Radiation in 1974. This Working Group late became the International Non-Ionizing Radiation Committee (IRPA/INIRC) at the IRPA International Congress in Paris in 1977. The IRPA/INIRC reviews the literature on non-ionizing radiation and makes assessment of the health risks of human exposure to such radiation.

In the meantime, following the recommendations of the United Nations Conference on the Human Environment held in Stockolm in 1972, and in response to a number of World Health Resolutions, and the recommendation of the Governing Council of the United Nations Environment Programme, a programme on the integrated assessment of the health effects of environmental pollution was initiated by World Health Organization (WHO) in 1973. The programme, known as the WHO Environmental Criteria Programme, has been implemented with the support of the United Nations Environment Programme (UNEP) and incorporated into the International Programme on Chemical Safety. The main result of the Environmental Health Criteria Programme is a series of criteria documents.

Two joint WHO/IRPA Task Groups on Environmental Health Criteria for Extremely Low Frequency Fields (WHO/IRPA, in press) and Radiofrequency and Microwave (WHO/IRPA, 1981), respectively, reviewed and revised existing scientific literature. They made an evaluation of the health risks of exposure to electromagnetic fields, considered rationales for the development of human exposure limits, and gave advice and recommendations for further research, so that on the basis of more definite data bases a critical revision of the existing standards can take place on an international basis, possibly reaching unique and certain safety criteria.

In this paper an attempt will be made to present, according to the main results of the joint WHO/IRPA Task Groups, a comparative analysis of existing or recently proposed standards. Because of difference in physical properties and biological effects data bases the electromagnetic spectrum will be divided into two parts: the extremely low frequency (ELF) electromagnetic fields and the radiowave region, encompassing radiofrequency and microwave radiations.

ELF ELECTROMAGNETIC FIELDS

Unfortunately our state of knowledge of interaction mechanisms operating when biological systems are exposed to ELF field is very limited. It is presently impossible to furnish any theory to predict the effect of exposure to these fields. With this lack of information, one must be very careful in attempting to predict effects in man from effects found in animals. The physical differences (size, shape, fur bearing, etc.) between animals and humans will produce marked differences in the field distribution and the induced currents (Kaune and Phillips, 1980) and in different behavioural and homeostatic responses.

With such limited data available on biological effects and interaction mechanisms, the only practical approach left for making a health risk assessment is to evaluate the available data on exposure levels and bioeffects to determine if thresholds for effects occur. It should be noted that in undertaking such an evaluation, one must be very selective with the data. Only reports that provide adequate information on experimental technique and dosimetry should form part of the evaluation. Ideally, only those data that have been reproduced by independent laboratories and have a direct bearing on health risk should be considered.

Michaelson (1979) has clearly defined what approaches he uses in evaluating the scientific literature. Claims of positive effects in an individual report should be accepted into the body of established scientific knowledge only if the report meets the following criteria:

(1) The experimental techniques, methods and conditions should be as completely objective as possible;

(2) All data analyses should be fully and completely objective, no relevant data deleted from consideration and uniform analytical methods used;

(3) The published description of the methods should be given in sufficient detail that a critical reader would be convinced that all reasonable precautions have been taken to meet requirements 1 and 2;

(4) The results should demonstrate an effect of the relevant

variable at a high level of statistical significance using appropriate tests. The effects of interest should ordinarily be shown by a majority of test organisms and the responses found should be consistent;
and
(5) The results should be quantifiable and susceptible to confirmation by independent researchers.

One would like to be able to make the health risk assessments based on well conceived, conducted and analysed epidemiological studies. Unfortunately all such studies on human exposed to ELF fields have suffered from one or more of the following: small numbers (resulting in low statistical accuracy); a lack of adequate dosimetry or of a properly matched, stable control group that would allow an unequivocal interpretation of the data.

With such a poor scientific data base, any real possibility of determining whether or not a true threshold exposure level exists, below which no adverse health effect occurs, cannot be determined. Thus any health risk analysis for the development of standards must inevitably adopt a phenomenological approach (Kossel, 1982; Repacholi, 1983). In this case, a review of the literature is made to determine the lowest exposure levels at which adverse biological effects occur. Although the biological effect may be some phenomenon or alteration to some biochemical parameter which occurs in vivo and may not appear hazardous, one might presume that the effect could be hazardous. This assumption is made only because insufficient information is available on the effect or the underlying interaction mechanism to make a well substantiated health risk assessment. It should be remembered that if the phenomenological approach is adopted in standards development, the biological effects data used should be selected using the criteria noted above.

One area of study that has produced great interest in recent years has been the change in calcium exchange from brain tissue by specific ELF amplitude modulation of radiofrequency carrier waves (Bawin et al., 1978; Blackman et al., 1979). This calcium effect is windowed in frequency with maximal effect at 16 Hz modulation. This effect occurs at several windows at harmonics of 15 Hz up to 105 Hz in fields of less than 100 V/m in air (Blackman et al., 1982). The calcium efflux effect occurs both in in vitro slices of cerebral tissue and in in vivo experiments (Adey, 1980, 1981). Both the biophysical mechanisms and physiological implications of this phenomenon are nearly unknown. At present insufficient information exists to make any health risk assessment. Although the magnitude of the effect (up to about 20% alteration in calcium exchange) is not of particular concern, the fact that both frequency and amplitude "windows" exist for the effect means that present theory is inadequate to provide explanation. Another problem is the fact that very low electric field strengths are needed to produce the effect. Until more data become available, there does not appear to be any particular cause for concern. Certainly one should not consider today regulatory action on the basis of these experiments.

Some of the recent studies on the health status of linemen and switchyard workers have found no differences from the control group (Stopps and Janischewsky, 1979; Knave et al., 1979). However, Knave et al. (1979) noted that the switchyard workers had fewer sons than daughters compared to the control group. This finding is opposite to that described by Roberge (1976). Knave (1981) suggested that factors other than exposure to electric fields were probably responsible since the difference in the number of children was found to be present 10 to 15 years before work in 400 kV stations began. The observation of a difference in the proportion of male and female children in small groups of people would be expected: in fact, even with large populations it is extremely unlikely to be unity (Bonnell, 1982).

In contrast to the finding of reproductive problems in Swedish switchyard workers, Bauchinger et al. (1981) report no effects on offspring of 32 switchyard workers and no increased chromosomal breakage. These German workers were exposed at levels up to 12 kV/m (50 Hz) for up to 20 years and were matched to 22 men not exposed to electric fields.

Nordstrom et al. (1983) performed a retrospective study on reproductive hazards on 542 employees at Swedish power plants. They found a statistically significant decreased frequency of normal pregnancies, almost exclusively due to an increased frequency of congenital malformations, when the father was a high-voltage switchyard worker. Although the differences in pregnancy outcome could not be explained by any of the confounding factors analysed, the total number of children with malformations (26) and the total number of pregnancies in the study was very small.

Recently, an epidemiologic investigation of HV substation workers has been performed in Italy (Baroncelli et al., 1984; Checcucci, in press).

This study concerns 627 railway workers, active in 50 Hz Interconnection and Conversion substations all over Italy. Measurements performed in both kinds of substation have shown that maximum levels of the electric field strength and of the magnetic flux density are of the order of 5 kV/m and 15 μT, respectively.

A collection of data on the possible health effects of high electric fields exposure has been performed by measuring the following end points: metabolic profile, psychologic performance, ECG, hematogram. A general medical examination and an interview on socio-economic aspects have been performed as well. The study design includes a contrast between three groups of exposed workers (subjects who work on routine maintenance and skilled technicians in charge of difficult repairs) and a group of workers whose exposure is estimated at 0 h/week, that can be regarded as reference entity.

In this recent study on the health status of linemen and switchyard workers it has been found no difference between the control group and those groups exposed to 1 h/week, 10 h/week and 20 h/week, at electric field strength of 5 kV/m.

There are, however, some epidemiological studies that have tended to raise some concerns about exposure to electrical wiring. Wertheimer and Leeper (1979) claimed to have found a relationship between childhood cancer and residence in houses in Colorado with high current flow due to external electrical wiring configurations (e.g. transformers or large primary wires coming from power distribution substations, called high current configurations or HCC's). A high incidence of cancer was found for children who had spent their entire lives at the same address near HCC's. Aside from some preliminary measurements (no electric or magnetic field measurements were made) the authors claimed that the effect appeared to be dose related. The same authors (Wertheimer and Leeper, 1982) also reported that adult cancer was associated with high - current electrical wiring configurations (HCC's) near the patients residence. Again the authors claimed that HCC's and cancer may be causally linked or at least that the magnetic fields act as a promoter or co-agent of the process of carcinogenesis since:
(i) a dose-relationship was found,
(ii) no association with age, urbanicity, neighbourhood or socio-economic level was found,
(iii) association was strongest where cancer caused by urban/industrial factors was at least apt to obscure the effect,
(iv) a pattern of latency between the first exposure to HCC's and cancer diagnosis was seen.

There are a few criticisms that could be levelled at the Wertheimer and Leeper studies (Bonnell, 1982):
(a) the data were not collected blind - therefore the potential for a strong observer bias exists;
(b) the cases were ascertained after death and therefore no account was taken of cancer cases still alive;
(c) both birth and death addresses were used - introducing again the potential for strong observer bias.

Sheppard (1983) has also indicated limitations of these studies:
(i) the techniques used for scoring a residence within the range from HCC may introduce errors;
(ii) difficulty in assuring that the population under study is truly comparable with respect to occupational, social and economic factors, and is not contaminated by extraneous factors known to cause cancer such as chemical pollutants.

Although valid criticisms can be directed at the studies of Wertheimer and Leeper, a similar study conducted in Sweden by Tomenius et al (1982) has also reported an association between childhood cancer and exposure to high tension wires (6-200 kV) substations, transformers or electrified railways within 150 m of the patients dwelling. An analysis of the data provided by Tomenius et al. makes it difficult, however, to see how they arrived at their conclusions.

A study in Rhode Island by Fulton et al. (1980) did not find any correlation between exposure to electrical wiring configurations and childhood leukaemia. However, assumptions made in this latter article make any conclusions questionable (Bryan, 1980). Fulton et al. (1980) assumed the strength of the magnetic field generated by electrical wires (and hence exposure) decreases with the square of the distance from the wires - the magnetic field strength falls off the inverse of the distance. Fulton et al. also assumed that the field strengths are numerically additive. This occurs only in a few wiring configurations, for example for closely spaced wires carrying current in the same direction at the same time. If the parallel wires are carrying the same amount of current in opposite directions at the same instant, and if the distance between the wires is small compared to the distance from the wires to the point of interest, the magnetic fields nearly cancel out. This latter case probably occurred more often in Rhode Island. The only accurate way is to make measurements - as noted in Fulton et al. (1980). Wertheimer and Leeper reanalysed the Rhode Island data using their study methodology and found a slight association of childhood cancer with electrical wiring configuration.

A number of recent studies (Milham, 1982; McDowall, 1983) all suggest an association between electrical occupations (exposure to electric and magnetic fields) and cancer, especially acute myelogenous leukaemia (AML). In general the studies were analyses of occupational mortality data, subject to many sources of errors including:
(i) lack of consistency in designating occupational classification, i.e. the same occupation may be placed in different categories,
(ii) mobility between occupations,
and
(iii) occupational groups sharing exposure to electric and magnetic fields were undoubtedly exposed to other physical and chemical agents.

With the source of errors noted, it is not surprising that the studies appeared internally inconsistent. For example McDowall (1983) found that some of the highest incidences of AML in certain occupational groups were potentially exposed to much lower levels of electric or magnetic fields than other occupational groups showing a lower AML incidence.

At this stage it is not possible to make an unequivocal assessment of health risk from exposure to electrical wiring. However, from the evidence presented, a casual relationship between exposure to alternating electric or magnetic fields and cancer induction has not been demonstrated. In view of a lack of any experimental evidence or any theories suggesting a carcinogenic effect from exposure to power frequency electric or magnetic fields, it appears unlikely that these are capable of carcinogenesis.

At this stage it is impossible to know what the observations mean, although the cluster of reports relating to acute myeloid leukaemia is worrisome. There are inconsistencies in the evidence. The exposing agent implicated is vague; is it the magnetic or electrical fields or some other factors ? Or might the harmful substance be chemicals such as polychlorinated biphenyls or metal fumes with which electrical workers come into contact ? There is no agreement about the cancers involved - might there be an increase in all cancers or just in acute myeloid leukaemia ?

While attempting to arrive at general conclusions concerning the health hazards of ELF fields for protection purposes and following an exhaustive review of the scientific literature, Sheppard suggested a number of factors that influence estimates of risk from chronic ELF electromagnetic field exposure.

Taking all these factors into consideration, one is lead to the conclusions that:

a) expressed adverse biological effects in human are not very likely, and that a pathological effect is even less likely;

b) pathophysiological effects in human beings exposed to 50/60 Hz electric fields at any field strengths are unproven and speculative;

c) subtle effects on the nervous system that may alter mental state, disrupt normal body rhythmicity, alter libido, increase the frequency or severity of headaches, or lead to effects on digestion or other functions influenced by the central nervous system, are not demonstrated by the scientific research or theory to date;

d) effects demonstrated in several animal studies suggest it is reasonable to expect that humans exposed to appropriately scaled electric fields may show physiological or behavioural alterations related to those observed in animals, but the pathological potential of the alterations cannot be foreseen;

e) adverse human health effects have not been established by the few studies of humans. These however are limited in scope and often methodologically flawed. They do not form the basis for a conclusion that chronic exposure to electric fields at arbitrary field strength is without effect for the general population.

To date few regulatory exposure standards have been promulgated limiting human exposure to ELF fields. Guidelines have been developed in a number of countries, mostly as an interim measure until sufficient information on adverse biological effects becomes available to make some reasonable assessment of health risks, and of the exposure levels at which hazards occur.

From a recent review of ELF standards, Repacholi (in press 1) found that the greatest interest in regulations or guidelines was in ELF electric fields at power frequencies. With the growth in number and length of HV transmission lines, increasing concern has occurred among the public, regulatory agencies, and scientists about possible human health effects from exposure to the electric fields associated with these lines. While there is no definitive evidence of such effects, mounting public fear and activism over hypothesized health risks has caused delays in the licesing and construction of major power transmission facilities, and encouraged the formation of regulatory policy in some countries.

The primary basis for public concern was a series of studies conducted in the USSR in the 1960s (Asanova and Rakov, 1966). These studies resulted in the occupational safety standard in the USSR which is summarized in Table 1.

Table 1. Electric field exposure limits for workers in installations of 400 kV and higher in the USSR

Electric field strength/(kV/m)	Permitted exposure duration per day (min)
5	Unrestricted
10	180
15	90
20	10
25	5

The Soviet standard applies to workers in substations or on transmission lines operating at 400 kV and above. The duration of the standard was from January 1st, 1977 to January 1st, 1982, after which it was to be reviewed and either changed or reaffirmed. A new standard for electric 50 Hz power frequency, however, is being discussed by the Council of Mutual Economic Assistance (incorporating Bulgaria, Cuba, Czechoslovakia, the German Democratic

Republic, Hungary, Poland, Romania and the USSR).

The basis for the Soviet standard is that studies conducted since 1962 on the effects on workers of high-voltage power systems revealed electric field influences on human beings. It is believed that the reaction of the human body to the direct influence of an electrical field is non-specific; it can develop after comparatively long times (2-5 months); it has a long-term consequence, pronounced cumulative effects, and strong dependency on individual physiological pecularities of the body (Lyskov et al., 1975).

In the United Kingdom, the National Radiological Protection Board (NRPB, 1982) has issued a proposal for ELF fields. In this consultative document it states: "The Board accepts that exposures to power frequency fields of less than 10 kV/m are safe, although the field may be perceptible at lower values, and that exposures to fields up to 30 kV/m are unlikely to be harmful". The NRPB admits that there is very little information that can be used as a rational basis for limiting exposure and that at 50 Hz, perceptible but harmless effects depend to a large extent on environmental factors and individual sensitivities. However, steps should be taken to prevent such effects from occurring with any degree of regularity. This will generally be achieved if the root mean square field strengths are kept below 10 kV/m. Prolonged exposure to field greater than 20 kV/m that induces currents in excess of 0.5 mA in the body is also undesirable, according to the NRPB.

In Japan, any electric power equipment is subject to the regulation "Technical practices of electrical equipment", an ordinance of the Ministry of International Trade and Industry. The ordinance (Repacholi, in press 2) includes such technical specifications as:

(a) minimum height of electrical conductors;
(b) necessary clearance between a transmission line and buildings; and
(c) the electrical field strength on the ground surface under the line.

In summary, the ordinance states that the unperturbed electrical field strength 1 m above the ground surface must not exceed 3 kV/m rms. In addition, the line must be built so as not to incur any risk for persons. However, in lightly-populated areas such as rice fields, farms, and forests, this limitation is not applied when the line is constructed so as not to incur any risk for any person.

A description of the technical basis for this 3 kV/m standard is provided by the Japan IERE Council (1976). The standard is based on the electrostatic induction sensed by a person who has his cheek or finger in contact with the metallic part of the grip of an umbrella.

In Poland (1980), as part of the electromagnetic radiation standard for frequencies from 0.1 to 300 000 MHz, is a limit on electric field strength at the single frequency of 50 Hz. The standard, effective from 31 January 1980, establishes two "safety" zones. For exposure to 50 Hz electric fields the zones are:

- 1st zone (electric fields above 10 kV/m) - prohibited to everyone except workers in electrical substations and personnel working on power lines;

- 2nd zone (above 1 kV/m to 10 kV/m) - agriculture and recreational activities are allowed, but not the construction of housing, hospitals, schools or kindergartens, except where building and farms existed before the regulations were established.

The standard provides details of administrative controls, approval procedures and electromagnetic field measuring authorities. However, no rationale for the values in the standard appears to have been published.

In the USA, there are a number of different standards with regard to control of electric fields at the edge of rights-of-way for high-voltage transmission lines. General population and media pressure have prompted public hearings and extensive debate over health effects from these lines. The US Department of Energy, Bonneville Power Administration (BPA) has a criterion on electric field exposure levels that results in a low probability of human perception or annoyance from field effects (Lee et al., 1982). BPA allows a maximum of 9 kV/m on the right-of-way, when measured 1 m above the ground (Lee et al., 1982). It would seem that the rationale for setting the 9 kV/m level is so that induced body currents in human beings under the lines will not exceed the current permitted by the National Electric Safety Code (5 mA, rms).

All 50 states in the USA have some legislation for regulating the safety of the general population in the proximity of transmission lines. Of thesem 25 states have enacted legislation requiring the preparation of environmental impact statements for proposed overhead transmission lines with respect to electrical effects. A comprehensive study was completed by the state of New York (1979), in which testimony indicated potential impacts from audible noise and from electrostatic shocks that people can receive when they touch a large vehicle parked under the lines. The testimony failed to demonstrate biological hazards from the field, although further research is necessary to understand better the effects of the fields on biological systems.

Many of the state regulatory agencies have carried out similar studies (Shah, 1979; Montana, 1983) reaching the same basic conclusions as New York (1979). Guidelines from each state on the maximum electric field permitted at the edge of the right-of-way (RoW) differ, but are within the range of 1-3 kV/m (Table 2). Guidelines for maximum field strength within the RoW varied from 7 to 11 kV/m. Most states comply with the National Electric Safety Code (NESC), which restricts currents in the human body to not more than 5 mA (rms).

Table 2 Recommended electric field levels for high voltage transmission lines in the USA[a]

	Max E field/(kV/m)			
State	In RoW	Edge RoW	Comment	Reference
Minnesota	8	-	Resolution	Shah (1979)
Montana	-	1	Resolution	Montana (1983)
New Jersey	-	3	Resolution	New Jersey (1981)
New York	-	1[b]	Temp.Resolut. New EHV Lines	New York (1979)
	7	-	Public roads	Shah (1979)
	11	-	Private roads	Shah (1979)
North Dakota	9	-	Resolution	Shah (1979)
Oregon	9	-	State law	Shah (1979)

[a] From: Repacholi (in press 2).
[b] 1kV/m for flat terrain-use; 1.6kV/m as criterion (Sheppard, 1983).
Note: Most states have adopted NESC (5mA, rms).

From the scientific evidence presently available, its analysis, and the apparent need to develop standards limiting exposure to ELF electric and magnetic fields, the joint WHO/IRPA Task Group on extremely low frequency fields (WHO/IRPA, in press) formulated the following 10 recommendations for new research directions and experiments:

1. In order to relate biological findings from *in vitro* and *in vivo* studies on experimental animals to human beings, it is recommended that dosimetry studies should be undertaken to internal current density distributions in the whole body of animals and human beings.

2. From studies on man and animals, observed sensitivities are consistent with two proposed models, one on the basis of stimulation of peripheral sensory receptors in strong local electric fields at the body surface, and the other on the basis of current densities induced in the extracellular fluid. It is recommended that models be devised that correlate exposure and biological effects in terms of physical factors, such as surface electric field, tissue current density, spark discharges, and waveform.

3. The continuation of basic research on electric and magnetic field interaction mechanisms is strongly recommended. Investigations should be conducted on the possible synergism or antagonism of field influences with physical and chemical agents, since such data are not available.

4. In some studies, restriction of ELF effects to certain "windows" in frequency and amplitude has been reported. It is recommended that the window concept be further investigated to determine applicability of data obtained with various frequencies and waveforms, and to relate the findings to potential health detriment in human beings.

5. Studies have been performed on workers with long-term exposure to electric and magnetic fields, but no adverse health effects have been identified. However, these studies were not designed to evaluate effects on reproductive functions, or long-term carcinogenic risks. In two of the studies, electric field exposure was carefully evaluated, and it was found that average exposures in the occupationally-exposed groups were remarkably low.

A suggestion of increased cancer incidence has been reported by a number of investigators who have examined occupational and general population groups exposed to electric and magnetic fields. The studies performed have serious deficiencies in epidemiological design and do not adequately characterize levels and duration of exposure.

The limited knowledge of the potential human health risk associated with exposure to electric and magnetic fields makes it imperative that well designed epidemiological studies should continue to be undertaken to provide a firmer basis for risk assessment.

6. Occupational exposure to strong electric fields is generally intermittent and of short duration, exposed populations have been identified, and there are some limited data based on practical experience. At field strengths where spark discharges are prevalent,prolonged exposure may impair performance. Such exposures should be avoided, where possible.

7. Linemen working on energized extra- or ultra-high-voltage conductors experience extreme electric field conditions, and the use of appropriate protective clothing or devices is desirable.

8. Whilst it would be prudent in the present state of scientific knowledge not to make unqualified statements about the safety of intermittent exposure to electric fields, there is no need to

limit access to regions where the field strength is below about 10 kV/m. Even at this field strength, some individuals may experience uncomfortable secondary physical phenomena such as spark discharge, shocks, or stimulation of the tactile sense.

9. It is not possible from present knowledge to make a definitive statement about the safety or hazard associated with long-term exposure to sinusoidal electric fields in the range of 1-10 kV/m. In the absence of specific evidence of particular risk or disease syndromes associated with such exposure, and in view of experimental findings on the biological effects of exposure, it is recommended that efforts be made to limit exposure, particularly for members of the general population, to levels as low as can be reasonably achieved.

10. In principle, electric and magnetic field interference with implanted cardiac pacemakers can lead to reversion to a fixed rate, but cessation of stimulation is possible. Direct interference has not been reported in fields below 2.5 kV/m. Although body currents produced by contact with a vehicle in a weaker field may cause interference, the risk of pacemaker reversion is believed to be slight.

Finally, it is recommended that pacemaker designers and manufacturers of other similar electronic equipment ensure that their devices are resistant to failures from electric or magnetic field-induced currents.

RADIOFREQUENCY AND MICROWAVES

The two nations that have made by far the greatest contributions to the study of possible effects of exposure to radiofrequency and microwave energies are the USA and the USSR. The well-known wide disparity between the exposure limits of the USA and of the USSR has, therefore, not unreasonably occasioned distrust and sometimes alarm among people who are, or who could be exposed to radiowaves. It is apparent that very different criteria were used and that the safety factors incorporated into the USSR standard are very much greater.

Baranski and Czerski (1976) suggest 3 possible approaches to determine "safe" limits. Summarized briefly these are:

1. no demonstrable effects;
2. discernible effects but no change in functional efficiency;
3. exposure results in some stress but only within the limits of normal physiological compensation.

In general the USSR subscribes to the principle that all exposure standards should be set at levels such that there is no change in the state of health or well-being as a result of such exposure. In the microwave region (300 MHz to 300 GHz) the maximum permissible level for public exposure is 50 mW/m^2.

100 mW/m^2 is allowed for occupational exposures up to 8 hours, 1 W/m^2 for up to 2 hours, and 10 W/m^2 for 15 to 20 minutes provided that the 100 mW/m^2 limit is not exceeded at any other time during the day. Soviet scientists have stated that these limits were developed on the basis of clinical symptoms developed by microwave workers and of the results of experiments on laboratory animals. They claim that workers exposed for extended periods to levels of between a few 100 mW/m^2 and a few W/m^2 develop functional disturbances and an asthenic condition. The symptoms of the asthenic syndrome are non-specific and include weakness, depression, impairment of memory and an inability to make decisions. The USSR exposure limits have been set a factor 10 below the threshold levels causing any clinical symptoms. In animal experiments the threshold for functional disturbance, as shown by changes in behaviour, was found to be 10 W/m^2 for a 1 hour exposure at 3 GHz. Taking this to be equivalent to 1 W/m^2 for a 10 hour working day and applying a safety factor of 10 to allow for individual variations in susceptibility, results in the Soviet occupational protection standard of 100 mW/m^2.

The limit of 100 W/m^2 adopted for long time in the USA was first proposed by Schwan in 1953 after his pioneering work and is based mainly on physiological considerations, in particular of thermal load and heat balance. The maximum rate of working that can be sustained over a period of hours requires the body to dissipate about 750 W. Total absorption of 100 W/m^2 over the cross-sectional body area of 0.7 m^2 would result in an additional heat load of 70 W which is small compared with that caused by even the lightest manual activity, and is in fact less than the resting metabolic rate. The 100 W/m^2 level is also a factor of at least 10 below the exposures that had been found to be needed to cause injury to the testes or the eye - the two organs widely regarded as being most sensitive to thermal injury. The acknowledged assumption in this standard is that heating of body tissues is the most if not the only significant consequence of absorbing radiofrequency energy. The US standard offers the advice that lower levels should be considered if there are other heat stresses, while higher levels might be appropriate under conditions of intense cold.

Subsequent to the publication of the exposure standards of the USSR and the USA most countries that felt the need for radiofrequency exposure standards elected to work to one or other of these standards. Some other countries have adopted (Poland, Sweden, West Germany, Canada, etc.) or become involved (The European Communities, Italy, etc.) in the development of standards that, in general, incorporate maximum exposure limits that are intermediate between those of the US and Soviet standards. These newer standards also differ in that the limits vary according to the frequency of the radiation. New criteria or standards for limiting the exposure of workers and the general population to radiofrequency and microwave radiations have recently been proposed by WHO, IRPA, the American Conference of Governmental Industrial Hygienists (ACGIH), the American National Standards Institute (ANSI), and the National Radiological Protection Board (NRPB), in the UK.

WHO suggests that exposures to power densities in the range 1-10 W/m^2 are acceptable for occupational exposure throughout a complete working day and that higher exposures might be reasonable for some frequency ranges and occasional exposure. For the general public it is suggested that lower, but unspecified, exposure levels are appropriate and that the levels should be as low as reasonably achievable, according to the ALARA principle.

The IRPA recommendations (IRPA,1984) are more specific and have the purpose of providing guidance on limits of exposure to EM radiation and fields in the frequency range from 100 kHz to 300 GHz. In the IRPA guidelines, the basic limits of exposure formulated for the frequency region of 10 MHz and above are expressed by the quantity specific absorption rate, SAR, i.e. the power absorbed per unit mass. In the frequency region below 10 MHz, basic limits are expressed in terms of the effective electric field strength and the effective magnetic field strength; in the near field or in multipath fields, both the electric and magnetic field strengths must be measured.

The limits of occupational exposure given in Table 3 for the frequencies between 10 MHz and 300 GHz are the IRPA working limits derived from the SAR value of 0.4 W/kg when averaged over any 6 minutes and over the whole body. The same table shows the maximum levels for electric and magnetic field strengths at frequencies below 10 MHz.

Table 3. IRPA occupational exposure limits to radiofrequency and microwaves.

Frequency Range MHz	Unperturbed RMS Electric Field Strength V/m	Unperturbed RMS Magnetic Field Strength A/m	Equivalent Plane Wave Power Density W/m^2	mW/cm^2
0.1-1	194	0.51	*100	*10
>1-10	$194/f^{\frac{1}{2}}$	$0.51/f^{\frac{1}{2}}$	*100/f	*10/f
>10-400	61	0.16	10	1
>400-2000	$3\ f^{\frac{1}{2}}$	$0.008\ f^{\frac{1}{2}}$	f/40	f/400
>2000-300 000	137	0.36	50	5

f = frequency in MHz

* These values are provided for information only and are not to be considered for determining compliance.

In Table 4 the general public exposure limits recommended by IRPA are summarized.

Radiofrequency radiation exposure of the general public at frequencies at or above 10 MHz should not exceed a SAR of 0.08 W/kg when averaged over any 6 minutes and over the whole body. For exposure at frequencies below 10 MHz, the levels of unperturbed root mean square (RMS) electric and magnetic field strength should not exceed the values given in Table 4.

Table 4. IRPA general public exposure limits to radiofrequency and microwaves

Frequency Range	Unperturbed RMS Electric Field Strength	Unperturbed RMS Magnetic Field Strength	Equivalent Plane Wave Power Density	
MHz	V/m	A/m	W/m^2	mW/cm^2
0.1 - 1	87	0.23	* 20	* 2
>1 - 10	$87/f^{1/2}$	$0.23/f^{1/2}$	* 20/f	* 2/f
>10 - 400	27.5	0.073	2	0.2
>400 - 2000	$1.375\ f^{1/2}$	$0.0037\ f^{1/2}$	f/200	f/2000
>2000 - 300 000	61	0.16	10	1

f = frequency in MHz.

* These values are provided for information only and are not to be considered for determining compliance.

The IRPA occupational exposure limits in the frequency ranges above 10 MHz may be exceeded for specific applications provided the SAR remains below 0.4 W/kg when averaged over the whole body, and the spatial average SAR remains below 4 W/kg when averaged over any one gram of tissue. The limits for frequencies at or below 10 MHz may be exceeded (up to 615 V/m or 1.6 A/m) provided workers take the necessary precautions to prevent potentially severe RF burns.

Calculations of the electrical currents induced in the human body making contact with ungrounded metallic objects, suggest that radiofrequency burns and shocks can result from exposure to fields of approximately 615 V/m (1000 W/m^2) in the frequency band 0.3 - 3.0 MHz. In a 3 MHz field, humans touching an ungrounded metal object such as a car will (by induced currents) perceive an E-field of unperturbed strength above 80 V/m and experience a let go current of the object (such as a compact car) above 200 V/m. For larger ungrounded metal objects the let go field strength decreases. The unperturbed E-field strength causing a let go situation remains approximately constant for a given object, for frequencies between 100 kHz and 10 MHz. It was predominantly on the basis of this information that IRPA exposure limits in the frequency ranges below 10 MHz were developed.

Although occupational exposures to 200 V/m at frequencies below 10 MHz can result in minor RF shocks, the IRPA Committee felt that workers could be trained to avoid these. Under special circumstances higher exposures (up to 615 V/m) could be tolerated by workers provided appropriate precautions were taken to avoid potentially severe RF burns. As a general rule however, it was felt an unreasonable strain on workers to have to work continuously in such environments.

For general public exposure below 10 MHz, the IRPA Committee felt that exposure limits should not be so high that RF shocks could occur, since it would be totally unreasonable to require this group to take precautions to avoid such shocks.

In view of the lack of a complete understanding of the biophysics of living systems, and the controversies concerning the non thermal electromagnetic field interactions, no predictive theory based on non-thermal considerations is possible. Empirical data must serve for establishing exposure limits. According to Schwan (1982), non-thermal effects may play a significant role at frequencies below 0.1 MHz. No adequate data on biological effects at these frequencies exist, except for the extremely low frequency range, which has been considered above . Therefore it seemed to be prudent to limit the lower frequency end of the guidelines to 0.1 MHz, and to leave the question of exposure limits below this frequency open. If compelling practical reasons (wide application at a particular frequency) exist, a limit for a narrow range or a particular device (emission standard) can be set on the basis of empirical studies.

Non-thermal effects are expected to be frequency-dependent (Illinger, 1982), and may exhibit complex dose effect relationships, including intensity windows (WHO/IRPA, 1981) both at low and high (above 30 000 MHz) frequencies (Grundler et al.,1982; Schwan,1982a). Threshold data on adverse health effects applicable to humans over the whole frequency range and to all possible modulations do not exist. Most of the present biological data exist for frequencies in the range from 900 MHz to 10 GHz. Thus, where data are lacking, assumptions on possible adverse health effects must be made. These are based on our present knowledge of biophysics of electromagnetic radiation absorption and on analytical or experimental models, as well as on limited epidemiological data. Extrapolation of limits in the range 10 MHz to 300 GHz is based on the frequency dependence of energy absorption in human beings. This may account for thermal effects but it does not allow one to predict non-thermal effects.

The ACGIH recommendations, which apply only to occupational exposures, are given in Table 5 (ACGIH, 1981).

Table 5. ACGIH limit values for radiofrequency and microwave radiation

Frequency range	Power density W/m^2	Electric field strength V/m	Magnetic field strength A/m
0.1 - 3 MHz	1000	614	1.64
3 - 30 MHz	$9000/f^2$	1842/f	4.89/f
30- 100 MHz	10	61.4	0.164
100 MHz -1GHz	f/10	$6.14\ f^{1/2}$	$0.0163\ f^{1/2}$
1 - 300 GHz	100	194	0.514

f = frequency in MHz

Like the IRPA values they apply to exposures averaged over 6 minutes. Emissions from devices with output powers of less than 5 W are exempt. Measurements are restricted to distances greater than 5 cm from any point of the emitting source. By the way, to stimulate exchange of scientific information useful in the establishment of Threshold Limit Values and Biological Indices, and to review procedures for the establishment of legal exposure limits, ACGIH is organizing its first International Symposium on Occupational Exposure Limits in Copenhagen, on 16-19 April, 1985, at the WHO Europe Headquarters.

The American National Standards Institute (ANSI) has recently completed a review of literature (ANSI, 1982) and screened only those reports that produced positive findings, were reproducible, and supplied adequate dosimetric information. Evaluations of these selected reports in terms of whether or not the biological effects constituted a hazard to health, were made. Behaviour in experimental animals was found to be the most sensitive indicator of an adverse health effect (e.g. convulsion activity, work stoppage, work decrement, decreased endurance, perception of the exposing field and aversion behaviour) . On the basis of their review, ANSI concluded that acute (less than 1 hour) exposure to electromagnetic energy that is deposited in the whole body at an average SAR of less than 4 W/kg does not produce an adverse health effect in experimental animals. However, because prolonged exposure (days and weeks) may cause damage, a ten-fold reduction in the permissible SAR (i.e. 0.4 W/kg) was invoked.

The ANSI proposals, shown in Table 6, differ from those of ACGIH in that they apply to both occupational and general population exposures and cover the frequency range 300 kHz - 100 GHz. The first difference is due to the ANSI proposals considering the absorption of radiation in children as well in adults. Exposures are to be averaged over 6 minutes, and levels in excess of those in Table 6 are allowed if it can be demonstrated that the mean specific energy-absorption rate is less than 0.4 W/kg and the peak rate in any part of the body does not exceed 8 W/kg. Emissions from devices radiating less than 7 W and at frequencies below 1 GHz are exempt. Above 1 GHz devices will be treated on a case by case basis. Exposures slightly in excess of the proposed limits are not expected to be harmful but are regarded as undesirable. In uncontrolled situations it is suggested that exposure should be as low as reasonably achievable.

Table 6. ANSI proposals for limiting population and workers exposures to radiofrequency and microwave radiations

Frequency range	Power density W/m^2	Electric field strength V/m	Magnetic field strength A/m
0.3 - 3 MHz	1000	632	1.58
3 - 30 MHz	9000/f	1897/f	4.74/f
30 - 300 MHz	10	63.2	0.158
300 MHz - 1.5 GHz	f/30	$3.65\ f^{1/2}$	$9.12\ 10^{-3}\ f^{1/2}$
1.5 - 100 GHz	50	141.4	0.353

f = frequency in MHz

Guy and his collaborators presented two months ago, in mid-July 1984, at the annual meeting of the Bioelectromagnetics Society (BEMS) in Atlanta the preliminary results of the first long-term study of microwave exposure ever carried out in the USA. The study revealed that exposures to low levels of pulsed 2.45 GHz radiation caused a statistically significant increase in malignant tumors among rats. The new cancer data come from a five-year study headed by Guy and sponsored by the U.S. Air Force. If confirmed, the new study would undermine the 1982 ANSI radiofrequency and microwave radiation exposure standard, because this standard asserts that the level of exposure used in the Guy study is safe. The findings are far from definitive, but a number of experts found the preliminary and yet unpublished data highly suggestive of a link between microwave radiation and cancer, possibly through a stress reaction. These new results lend support to an earlier Polish study which found that microwaves can promote cancer at higher exposure levels. Szmigielski showed, for example, that 2.45 GHz radiation can enhance the carcinogenic action of benzopyrene. The rats in Guy's experiment were exposed to an SAR of 0.4 W/kg, precisely the SAR level specified by ANSI as safe. "Whole body SARs below 4 W/kg were not by consensus associated with effects that demonstrably constitute a hazard", the ANSI committee wrote in 1982. A safety margin of ten was then factored in. The frequency dependent exposure guidelines, 1 mW/cm^2 in the 30-300 MHz band, stem from the 0.4 W/kg level. More than one of the representatives from the federal agencies who attended the Atlanta meeting noted that the new cancer results, if confirmed, would undermine the adequacy of the ANSI standard (MWN, 1984).

The principal basis for the NRPB's proposals, in the UK, is that exposures at levels likely to cause harmful effects should always be prevented. For occupational exposures, levels at which perceptible but harmless effects occur are permissible but should be avoided where reasonably practicable. Levels of exposure of the general population in situations where individuals may be exposed for a substantial fraction of their daily life,i.e. in their own homes and gardens, should be such as to avoid perceptible effects even if harmless.

For microwaves and radiofrequency radiations the Board accepts that exposures resulting in a mean specific energy-absorption rate in the whole body of less than 0.4 W/kg may be regarded as acceptable and providing an adequate margin of safety.

The NRPB limits set out in Tables 7 and 8 are based on calculations of the energy absorption rates in human beings represented as prolate spheroids and the possibility of hot spots at frequencies below 3 MHz. Table 7 is applicable to the adult populations.

Table 7. NRPB proposed limits for the continuous exposure of adult populations

Frequency range	Power density W/m^2	RMS electric field strength V/m	RMS magnetic field strength A/m
3 kHz - 3.0 MHz	-	600	-
3.0 - 30 MHz	$9000/f^2$	1800/f	5.0/f
30 - 100 MHz	10	60	0.16
100 MHz - 1 GHz	f/10	$6.0\ f^{1/2}$	$0.016\ f^{1/2}$
1 - 300 GHz	100	200	0.50

f = frequency in MHz

Table 8 shows the limits proposed for exposure of general population, including children.

Both tables apply to whole day exposures; for shorter exposures, some relaxation is possible.

When exposure is to a mixture of frequencies or a band of frequencies, it should be ensured that the summation

$$\sum P(f)/L(f) < 1$$

is observed. In this expression P(f) is the measured power density and L(f) is the power density limit for the frequency f.

Table 8. NRPB proposed limits for the continuous exposure of general population including children.

Frequency range	Power density W/m^2	RMS electric field strength V/m	RMS magnetic field strength A/m
3 kHz - 3.0 MHz	-	600	-
3.0 - 30 MHz	9000/f	1800/f	5.0/f
30 - 300 MHz	10	60	0.16
300 MHz - 1.5 GHz	f/30	$3.5\ f^{1/2}$	$9.4\ 10^{-3}\ f^{1/2}$
1.5 - 300 GHz	50	140	0.36

f = frequency in MHz

Finally, it is interesting to note that the United State Environmental Protection Agency (USEPA) has delayed the release of its proposed radiofrequency/microwave (RF/MW) radiation exposure guidance, which was scheduled for June, 1984. In the first official statement to indicate rule making may be abandoned, EPA announced in early June 1984 that the agency is considering a number of options for the guidance, including putting it on ice. Meanwhile, the Office of Research and Development's report on RF/MW bioeffects has received final approval from EPA's Scientific Advisory Board. The lengthy document, prepared under the direction of Dr.Elder, will serve as the basis for the guidance.

When one compares all current radiofrequency/microwave standards internationally, a paradox contained in all these standards activities is quite apparent. Presumably, each of these national groups had available to it the same library of reference data, perhaps supplemented to a small degree by some preparatory investigational and research results. And yet the permissible limits for occupational exposures differ by factors of 20 to 100, and those for the general public by factors of 20. Averaging times vary from one minute to a full day. It is obvious, therefore, that several vital factors are at work to create these extreme differences (Michaelson, 1983).

Among the factors that contribute to the extreme differences among national standards intended for the control of human exposure to RF/MW radiation, five are of major importance: (1) the particular physical and biological effects data selected as the basis for standards development, (2) the interpretation of these data into terms significant to humans and their environment, (3) the different purposes to be served by the standards, (4) the compromises made between levels of risk and degrees of conservatism, and (5) the influence of preceding standards activities in each particular nation and in neighboring areas having allied socio-political outlooks.

Governments and industry alike are being compelled to adopt RF/MW safety standards, in spite of the fact that the scientific data base is neither adequate nor complete, nor can much of it be referred usefully to the human situation. The quality and comprehensiveness of the scientific data base determine the confidence with which standards can be derived and defended. To make its data useful to standards-setters, the scientific community needs to agree on descriptive methods of accounting for the wavelength and frequency dependent properties of exposed subjects that determine the patterns of internal RF/MW energy deposition. To make results from related studies intercomparable, common techniques and terminology for internal dosimetry need further development, agreement, and use.

The most unfortunate consequence of this situation is that people have learned to use health effects and safety arguments to try to win in situations where health is not a matter of real concern. This is partly because of the inadequacy and incompleteness of the scientific data base, and largely because much of the data on positive biological effects are incapable of useful interpretation in terms of real human values. Each antagonist can find experts who will interpret the data selectively so as to support his side of the argument.

The traditional response by health agencies to situations like that described here is to adopt or to urge others to adopt restrictive standards, as strict as the supportive data and the competing pressure groups allow.

From the scientific point of view, it can be stated, according to WHO and IRPA (WHO/IRPA, 1981) that basic biophysical mechanisms of interaction of MW and RF with living systems still need clarification and further studies.

Work on both theoretical and experimental dosimetry, the calculation and measurement of fields and of energy deposited within simulated or actual biological systems, should be continued and refined.

Results of animal studies are difficult to extrapolate to man, and these studies alone do not constitute a satisfactory basis for the establishment of health protection criteria. They should, therefore, be supplemented by appropriate epidemiological studies on man. Much useful information may be derived by reevaluation of old animal experimental data using recent dosimetric insights as suggested by Michaelson and Lu (1979).

The existing data on power, amplitude, and frequency "windows" seem to warrant continued investigations.

There is little published information on dose-effect relationships; reports tend to be limited to whether effects are observed at one particular level of exposure rather than over a range. More dose-related information, even covering small subject area, would be valuable. Investigations on the genetic effects and effects on development of MW and RF radiation should have priority. Attention should be given to investigating the different sensitivities to MW/RF exposure of subgroups within the general population.

There is a continuing need for the development of RF/MW measuring instruments that: (a) give direct readings of electric or magnetic field strength, or power density and (b) are sensitive and can be used over a wide frequency range. The problem of the design of personal dosimeters also remains to be solved.

Computation techniques or methods that predict the distribution of fields close to deliberate high-power emitters (in the near field) are needed.

Systematic investigation of the effects of RF/MW exposure at all levels of biological organization are to be encouraged. This includes effects at the molecular level on subcellular components; cells, viruses, and bacteria; organs and tissues; and whole animals. Particular attention should be paid to: (a) long-term,low-level exposures and possible delayed effects; (b) the possibility of differences in sensitivity of various body organs and systems, where specific effects in various animal species are being considered; and (c) the influence of RF/MW exposure on the course of various diseases, including any possible increase in sensitivity to RF/MW that may result because of the disease state.

CONCLUSIONS

In conclusion, different scientific approaches have produced different philosophies of protection guidelines and thus different exposure limits. It is however apparent that, in the light of the continuous advancement of scientific results (Rubino, 1984), the differences are decreasing and the revisions of existing standards or the setting of new ones reflect at least the tendency to merge to a common area.

It is important, however, to maintain an adequate research effort so that on the basis of more definite data bases a critical revision of the existing standards can take place on an international basis, possibly reaching unique and certain safety criteria.

The international cooperation in the development of compatible standards should be encouraged, because the lack of international agreement on the protection standards to be used for non-ionizing radiation constitutes a major drawback for the development of safety regulations in those countries where they do not yet exist. Therefore, it must be hoped that the efforts outlined above to achieve international cooperation in the field of non-ionizing radiation together with progress in our knowledge on the biological effects will allow protection against non ionizing electromagnetic fields to develop in the same climate of international agreement as that which has been reached for ionizing radiation (Duchêne et al., 1984).

ACKNOWLEDGEMENT

I would like to thank Mrs. Basilia Scolaro for her skilled assistance in typing this manuscript.

REFERENCES

ACGIH, 1981, Threshold limit values for physical agents adopted by ACGIH for 1981, American Conference of Governmental Industrial Hygienists, Cincinnati.

ANSI, 1982, American National Standards Institute, Safety levels with respect to Human Exposure to radiofrequency Electromagnetic Fields, 300 kHz to 100 GHz, The Institute of Electrical and Electronic Engineers, ANSI Committee C.95.1, New York.

Adey,W.R., 1980, Frequency and power windowing in tissue interactions with weak electromagnetic fields, Proc. IEEE.,68:119-125.

Adey, W.R., 1981, Tissue interactions with nonionizing electromagnetic fields, Physiological Review, 61: 435-514 .

Asanova, T.P., and Rakov, A.I.,1966, The state of health of persons working in electric field of outdoor 400 and 500 kV switchyards (in Russian), Gig.Tr.Prof. Zabol., 10: 50-52.

Baroncelli, P., Battisti, S., Checcucci, A., Comba, P., Grandolfo, M., Serio, A., and Vecchia, P., 1984, An epidemiologic investigation of HV Substation workers, XXI Int. Congress on Occupational Health, Dublin.

Baranski,S., and Czerski,P., 1976," Biological effects of microwaves", Dowden, Hutchinson & Ross Editors, Stroudsburg.

Bauchinger,M., Hauf,R.,Schmid,E., and Dresp,J., 1981, Analysis of structural chromosome changes and SCE after occupational long-term exposure to electric and magnetic fields from 380 kV systems, Rad. and Env. Biophys., 19: 235-238.

Bawin, S.M., Sheppard,A.,and Adey, W.R., 1978, Possible mechanisms of weak electromagnetic field coupling in brain tissue, Bioelectrochemistry and Bioenergetics, 5: 67-76.

Blackman, C.F.,Elder,J.A., Weil,C.M., Benane,S.G., Eichinger, D.C., and House, D.E.,1979, Induction of calcium-ion efflux from brain tissue by radio-frequency radiation: effects of modulation frequency and field strength, Radio Science, 14: 93-98.

Blackman,C.F.,Benane,S.G.,House,D.E.,Joines,W.T.,and Spiegel,R.J., 1982, Further developments in elf-induced changes in calcium-ion efflux from brain tissues. (Abstract F-4). 4th Annual Meeting, Bioelectromagnetics Society, Los Angeles, June-July, 1982.

Bonnell,J.A.,1982, Effects of electric fields near power-transmission plants, J.Royal Soc.Med., 75: 933-941.

Bryan,F.A., 1980,Letter to the Editor Re: Electrical wiring configurations and childhood leukaemia in Rhode Island, Am. J. Epidemiol., 112: 419-420.

Checcucci,A., in press, An epidemiological Investigation of HV Substation Workers: Study Design and Preliminary Results, In: Biological Effects and Dosimetry of Static and ELF Electromagnetic Fields, Edited by Grandolfo, M.,Michaelson, S.M., and Rindi,A., Plenum Press, New York

Duchêne,A., and Komarov, E.,1984, International Programmes and management of non-ionizing radiation protection, IRPA International Congress, Berlin.

Fulton, J.P., Cobb,S., Preble,L., Leone,L., and Forman,E.,1980, Electrical wiring configurations and childhood leukaemia in Rhode Island, Am. J.Epidemiol., 11(3): 292-296.

Grundler,W., Keilmann, F., Putterlik,V., and Strube,D., 1982, Resonant-like dependence of yeast growth rate on microwave frequencies, Brit.J.Cancer., 45 (supp. V): 206-208.

Illinger K.H., ed., 1982, Biological Effects of Nonionizing Radiation, ACS Symposium Series N° 157, American Chemical Society, Washington D.C.

IRPA, 1984, Interim Guidelines on limits of Exposure to radiofrequency electromagnetic fields in the frequency range from 100 kHz to 300 GHz, Health Physics, 44: 975-984.

Japan IERE council, 1976, Electrostatic induction caused by extra-high voltage overhead trasmission lines, Japan, IERE Council.

Kaune,W.T., and Phillips,R.D., 1980, Comparison of the coupling of grounded humans, swine and rats to vertical 60 Hz electric fields, Bioelectromagnetics, 1: 117-130.

Knave,B., Gamberale,F.,Bergstrom,S., Birke,E., Iregren,A., Kolmodin-Hedman,B., and Wennberg,A.,1979, Long-term exposure to electric fields: A cross-sectional epidemiologic investigation of occupationally exposed workers in high-voltage substations, Scan.J. work environ. and health 5: 115-125.

Knave,B., 1981, In: Transmission lines and the environment (Paper 233 09), Conférence Internationale des Grand Réseaux Electriques a Haute Tension (CIGRE), Stockholm.

Kossel,F., 1982, Regulations and enforcements procedures, In: Non ionizing radiation protection, M.J. Suess, ed., WHO

Regional Publications, European Series No.10,World Health Organization, Regional Office for Europe, Copenhagen.

Lee,J.M., Brunke,J.H., Lee,G.E., Reiner,G.L. and Shon,F.L., 1982, Electrical and biological effects of transmission lines:a review, US Department of Energy, Bonneville Power Administration, Portland, Oregon..

Lyskov,Y.I., Emma,Yu.S. and Stolyarov, M.D., 1975, The factors of electrical field that have an influence on a human. In: Three Russian Papers on EHV/UHV Transmission Line and Substation Design, Uhl, Hall and Rich, Division of Charles T. Main, Inc., Boston, Mass.

McDowall,M.E., 1983, Leukaemia mortality in electric workers in England and Wales. Lancet No.8318, Jan.29, 246.

Michaelson,S.M.,1979, Analysis of studies related to biological effects and health implications of exposure to power frequencies. The Environment Professional 1: 217-232.

Michaelson,S.M., 1983, Microwave/Radiofrequency Protection Guide and Standard, in: "Biological Effects and Dosimetry of Nonionizing Radiation: Radiofrequency and Microwave Energies", Grandolfo,M., Michaelson, S.M., and Rindi,A., eds., Plenum Press, New York.

Michaelson,S.M., and Lu,S.T., 1979, Metabolic and physical scaling in microwave radiofrequency bioeffects studies, BEMS Symposium, Seattle.

Milham,S., 1982, Mortality from leukaemia in workers exposed to electrical and magnetic fields, New England J.Med., 307(4): 249.

Montana ,1983, Preferred and alternative routes for the Bonneville Power Administration 500 kV line from Garrison-West, Helena, Montana, Bonneville Power Administration.

MWN, 1984, Microwave News, Microwaves Promote Cancer, Louis Slesin, ed., New York.

New Jersey (1981), Minutes of the Commission on Radiation Protection, Department of Environmental Health, New Jersey.

New York, 1979, Potential environmental effects of 765 kV transmission lines. Views before the New York State Public Service Commission, Cases 26529 and 26559, Washington DC, US Department of Energy, Environmental Control Technology Division.

Nordstrom,S.,Birke,E., and Gustavsson,L., 1983, Reproductive hazards among workers at high voltage substations, Bioelectromagnetics, 4: 91-101.

NRPB, 1982, Health protection of workers and members of the general public against the dangers of extra low frequency, radiofrequency, and microwave radiations: a consultative proposal, Chilton, Didcot, England, National Radiological Protection Board.

POLAND, 1980, Specific requirements for basic safety to protect against electromagnetic non-ionizing radiation harmful to people and their environment. Gazette of Current Legislation No. 25, Polish People's Republic, Warsaw, 17 Nov., pp.277-278.

Repacholi,M.H., 1983, Development of standards-assessment of health hazards and other factors, in:" Biological effects and dosimetry of nonionizing radiation: Radiofrequency and Microwave Energies", Grandolfo,M., Michaelson, S.M., and Rindi,A., Plenum Press, New York.

Repacholi,M.H., in press 1, Standards on Static and ELF Electric and Magnetic Fields and their Scientific Basis, in:" Biological Effects and Dosimetry of Static and ELF Electromagnetic Fields", Grandolfo, M., Michaelson,S.M., and Rindi,A., eds., Plenum Press, New York.

Repacholi, M.H., in press 2, Health risk assessment of static and ELF electric and magnetic fields, in:"Biological Effects and Dosimetry of Static and ELF Electromagnetic Fields", Grandolfo, M., Michaelson, S.M., and Rindi, A., eds., New York, Plenum Press, New York.

Roberge,P.F., 1976, Study on the state of health of electrical maintenance workers on Hydro-Quebec 735 kV power transmission system, Montreal, Canada.

Rubino,N., 1984, Microwave and Radiofrequency EM Fields Safety Standards, URSI Open Symposium, Florence.

Schwan,H.P., 1982, Microwave and RF hazard standard considerations, J.Microwave Power 17: 1-9.

Schwan,H.P., 1982a, Nonthermal cellular effects of electromagnetic fields: AC-field induced ponderomotoric forces, Brit.J.Cancer 45 (supp.V): 220-224.

Shah,K.R., 1979, Review of state/federal environmental regulations pertaining to the electrical effects of overhead transmission lines: US Department of Energy (Publication ACP/EV-1802).

Sheppard,A., 1983, Biological effects of high voltage AC transmission lines. Report to Montana Dept. of Natural Resources and Conservation. Obtainable from NTIS, 5285 Port Royal Road, Springfield, Va., Report No. PB83 207241.

Stopps,G.J., and Janischewsky,W., 1979, Epidemiological study of workers maintaining HV equipment and transmission lines in Ontario, Canadian Elec. Assn.Research Report,Vancouver, British Columbia, Canada.

Tomenius,L., Hellstrom,L., and Enander,B., 1982, Electrical constructions and 50 Hz magnetic field at the dwellings of tumour cases (0-18 years of age) in the county of Stockholm. In Proc. Int.Symp. on Occup. Health and Safety in Mining and Tunneling, Prague, June 21-25.

Wertheimer,N., and Leeper,E., 1979, Electrical wiring configurations and childhood cancer, Am. J.Epidemiol., 109 (3): 273-284.

Wertheimer,N., and Leeper,E., 1982, Adult cancer related to electrical wires near the home, Int.J.Epidemiol., 11 (4): 345-355.

WHO/IRPA, 1981, Radiofrequency and Microwaves, Environmental Health Criteria 16, WHO,Geneva.

WHO/IRPA, in press, Extremely Low Frequency (ELF) Fields, Environmental Health Criteria 35, WHO, Geneva.

Participants of the international meeting devoted to "Interactions Between Electromagnetic Fields and Cells". (September 17 - 28, 1984).

CONTRIBUTORS

T. ASAKURA
Dept. of Bioengineering
Univ. of Penn. & Children's Hosp.
33rd St. and Walnut St.
PHILADELPHIA, PENN 19104
U.S.A.

C.A. BASSETT
College of Physicians & Surgeons
Columbia Univ.
Black Building, 630 W. 168th Str.
NEW YORK, N.Y. 10032
U.S.A.

F. BELTRAME
Biophys. & Electronic Engr. Dept.
Univ. of Genoa
Via all'Opera Pia 11A
16145 GENOVA
ITALY

B. BIANCO
Biophys. & Electronic Engr. Dept.
Univ. of Genoa
Via all'Opera Pia 11A
16145 GENOVA
ITALY

C.F. BLACKMAN
Health Effects Research Lab. MD-74
U.S. Environmental Protec. Ag.
Environmental Research Center
RESEARCH TRIANGLE PARK, NC 27711
U.S.A.

G. BONANNO
Pharmacology Inst.
Univ. of Genoa
Via Capo Santa Chiara
16146 GENOVA
ITALY

F. CARATOZZOLO
Biophys. & Electronic Engr. Dept.
Univ.of Genoa
Via all'Opera Pia 11A
16145 GENOVA
ITALY

A. CASALEGGIO
Inst. for Electronic Circuits
C.N.R.
Via all'Opera Pia 11
16145 GENOVA
ITALY

G. CASTELLARO
Biophys. & Electronic Engr. Dept.
Univ. of Genoa
Via all'Opera Pia 11A
16145 GENOVA
ITALY

S. CHANG
Dept. of Bioengineering
Univ. of Penn. & Children's Hosp.
33rd St. and Walnut St.
PHILADELPHIA, PENN 19104
U.S.A.

A. CHIABRERA
Biophys. & Electronic Engr. Dept.
Univ. of Genoa
Via all'Opera Pia 11A
16145 GENOVA
ITALY

C. GABRIEL
Dept. of Physics
King's College London
Strand
LONDON WC2 2LS
U.K.

G. GIANNETTI
Biophys. & Electronic Engr. Dept.
Univ. of Genoa
Via all'Opera Pia 11A
16145 GENOVA
ITALY

M. GRANDOLFO
Physics Laboratory
Istituto Superiore di Sanita'
Viale Regina Elena 299
00161 ROMA
ITALY

E.H. GRANT
Dept. of Physics
King's College London
Strand
LONDON WC2 2LS
U.K.

M. GRATTAROLA
Biophys. & Electronic Engr. Dept.
Univ.of Genoa
Via all'Opera Pia 11A
16145 GENOVA
ITALY

W. GRUNDLER
Gesellshaft fuer Strahlen
und Umweltforshung mbH
Ingolstadter Landstrasse 1
D-8042 NEUHERBERG BAVARIA
F.R. GERMANY

E. COUSSAERT
Dept. Scient. Comp.
Hosp. ERASME, Free Univ. of Brussels
Route de Lennick, 808
B-1070 BRUSSELS
BELGIUM

E. DEL GIUDICE
INFN - Dept. of Physics
Univ. of Milan
Via Celoria 16
20133 MILANO
ITALY

A. DIASPRO
Biophys. & Electronic Engr. Dept.
Univ. of Genoa
Via all'Opera Pia 11A
16145 GENOVA
ITALY

S. DOGLIA
GNSM - Dept. of Physics
Univ. of Milan
Via Celoria 16
20133 MILANO
ITALY

H. M. FISHMAN
Dept. of Phys. & Biophysics
Univ. of Texas Medical Branch
GALVESTON, TEXAS 77550
U.S.A.

S. FITTON-JACKSON
Strangeways Res. Lab.
Worts Causeway
CAMBRIDGE CB1 4RN
U. K.

M. HINSENKAMP
Dept. Orthopedic Surg & Traumat.
Hosp. ERASME, Free Univ. of Brussels
Route de Lennick 808
B-1070 BRUSSELS
BELGIUM

F. KAISER
Inst. of Applied Physics, Theory
Technische Hochschule Darmstadt
D-6100 DARMSTADT
F. R. GERMANY

Ad.J. KALMIJN
Scripps Inst. of Oceanography
Univ. of California
San Diego
LA JOLLA, CA 92093
U.S.A.

J.J. KAUFMAN
Bioelectrochem. Lab., Dept. Orthop.Surg.
Mount Sinai School of Medicine
100 thSt. & Madison Ave
NEW YORK, N.Y. 10029
U.S.A.

A.R. LIBOFF
Dept. of Physics
Oakland Univ.
ROCHESTER, MI 48063
U.S.A.

L. MARCONI
Inst. for Electronic Circuits
C.N.R.
Via all'Opera Pia 11
16145 GENOVA
ITALY

T. P. MARSLAND
Cambridge Univ., Engr. Dept. &
Strangeways Res. Lab.
Worts Causeway
CAMBRIDGE CB1 4RN
U.K.

C. MARTINI
Inst. for Electronic Circuits
C.N.R.
Via all'Opera Pia 11
16145 GENOVA
ITALY

V.E.R. McLEAN
Dept. of Physics
King's College London
Strand
LONDON WC2 2LS
U.K.

S. M. MICHAELSON
Dept. Radiation Biol. and Biophys.
The Univ. of Rochester Med. Centre
601 Elmwood Ave
ROCHESTER, N.Y. 14642
U.S.A.

M. MILANI
Dept. of Physics
Univ. of Milan
Via Celoria 16
20133 MILANO
ITALY

M. MORANDO
Inst. for Electronic Circuits
C.N.R.
Via all'Opera Pia 11
16145 GENOVA
ITALY

G. MORGAVI
Inst. for Electronic Circuits
C.N.R.
Via all'Opera Pia 11
16145 GENOVA
ITALY

C. NICOLINI
Biophys. Chair, Fac. of Medicine
Univ. of Genoa
Viale Benedetto XV
16145 GENOVA
ITALY

N.R.V. NIGHTINGALE
Dept. of Physics
King's College London
Strand
LONDON WC2 2LS
U.K.

M. NOORBERGEN
Dept. of Orthopedic Surg. and Traumat.
Hosp. ERASME, Free Univ. of Brussels
Route de Lennick 808
B-1070 BRUSSELS
BELGIUM

F. OOSAWA
Dept. of Biophysical Engr.
Osaka Univ.
Toyonaka
OSAKA 560
JAPAN

M. PARODI
Biophys. & Electronic Engr. Dept.
Univ. of Genoa
Via all'Opera Pia 11A
16145 GENOVA
ITALY

A. A. PILLA
Bioelectrchem. Lab., Dept. Orthop. Surg.
Mount Sinai School of Medicine
100th St. & Madison Ave.
NEW YORK, N.Y. 10029
U.S.A.

H. A. POHL
Francis Bitter Nat. Magnet Lab.
M.I.T.
Building NW14-1102 B
CAMBRIDGE, MA 02139
U.S.A.

A. RAVEANE
Biophys. & Electronic Engr. Dept.
Univ. of Genoa
Via all'Opera Pia 11A
16145 GENOVA
ITALY

G. REIN
Bioelectrochem. Lab.,Dept. Orthop. Surg.
Mount Sinai School of Medicine
100th St. & Madison Ave.
NEW YORK, N.Y. 10029
U.S.A.

S. RIDELLA
Inst. for Electronic Circuits
C.N.R.
Via all'Opera Pia 11
16145 GENOVA
ITALY

C. ROLANDO
Inst. for Electronic Circuits
C.N.R.
Via all'Opera Pia 11
16145 GENOVA
ITALY

M. ROOZE
Dept. of Human Anat. & Embryol.
Hosp. ERASME, Free Univ. of Brussels
Route de Lennick 808
B-1070 BRUSSELS
BELGIUM

F.A. SAUER
Max-Planck-Inst. for Biophys.
Kennedy Allee 70
D-6000 FRANKFURT AM 70
F.R. GERMANY

R.W. SCHLOEGL
Max-Planck-Inst. for Biophys.
Kennedy Allee 70
D- 6000 FRANKFURT AM 70
F.R. GERMANY

R.E. SCHMUKLER
Bioelectrochem. Lab., Dept. Orthop. Surg
Mount Sinai School of Medicine
100th St. & Madison Ave.
NEW YORK, N.Y. 10029
U.S.A.

H.P. SCHWAN
Dept. of Bioengineering
Univ. of Pennsylvania
33rd St. and Walnut St.
PHILADELPHIA, PENN 19104
U.S.A.

G. SCHWARZ
Dept. of Biophysical Chemistry
Biocenter of the Univ. of Basel
Klingelbergstrasse 70
BASEL 4056
SWITZERLAND

M. A. STUCHLY
Radiation Protection Bureau
Health & Welfare Canada
Tunney's Pasture
OTTAWA, ONTARIO K1A 0L2
CANADA

S. S. STUCHLY
Dept. of Electr. Engr.
Univ. of Ottawa
OTTAWA, ONTARIO K1N 6N5
CANADA

S. TAKASHIMA
Dept. of Bioengineering
Univ. of Penn. & Children's Hosp.
33rd St. and Walnut St.
PHILADELPHIA, PENN 19104
U.S.A.

B. TUERLINCKX
Dept. of Electr. Engr.
Hosp. ERASME, Free Univ. of Brussels
Route de Lennick 808
B-1070 BRUSSELS
BELGIUM

R. VIVIANI
Biophys. & Electronic Engr. Dept.
Univ. of Genoa
Via all'Opera Pia 11A
16145 GENOVA
ITALY

INDEX